SOLAR JOURNEY: THE SIGNIFICANCE OF OUR GALACTIC ENVIRONMENT FOR THE HELIOSPHERE AND EARTH

ASTROPHYSICS AND SPACE SCIENCE LIBRARY

VOLUME 338

SOLAR JOURNEY: THE SIGNIFICANCE OF OUR GALACTIC ENVIRONMENT FOR THE HELIOSPHERE AND EARTH

Edited by

PRISCILLA C. FRISCH
University of Chicago, IL, U.S.A.

Springer

A C.I.P. Catalogue record for this book is available from the Library of Congress.

ISBN 978-94-017-8481-8
ISBN 978-1-4020-4557-8 (eBook)

Published by Springer,
P.O. Box 17, 3300 AA Dordrecht, The Netherlands.

www.springer.com

Printed on acid-free paper

Dedicated to the memory of
Professor John Alexander
Simpson, whose
contributions to science, and
dedication to the peaceful
use of that science, were
longstanding and
far-reaching.

Contents

List of Figures

List of Tables

Contributing Authors

Maciej Bzowski is a Research Associate at the Space Research Centre of the Polish Academy of Sciences, in Warsaw. He received his Ph.D. from the Faculty of Physics, Warsaw University, and his interest in the interaction of interstellar material and the heliosphere dates back to his earliest pre-Ph.D. work. He has done research on solar cycle variation of interstellar gas distribution in the heliosphere and more recently on the spatial anisotropy of the solar wind.

Kenneth S. Carslaw is Professor in the Institute for Atmospheric Science, in the School of Earth and Environment, University of Leeds, in Leeds. He received his Ph.D. at the University of East Anglia, UK, and has worked extensively in the fields of atmospheric aerosol processes and the effect of aerosols on clouds.

Hans J. Fahr is Professor of Astrophysics at the Institute for Astrophysics and Extraterrestrial Research of the University of Bonn. He studied Mathematics, Physics and Philosophy and received his Ph.D. in Theoretical Physics at the University of Bonn. He was among the first scientists to study the interaction region between the solar wind and the interstellar medium, the heliosphere. In honour of his research, he recently received the First-Order-Award of Germany.

Horst Fichtner is a Scientist at the Institut für Theoretische Physik, at the Ruhr-Universität Bochum in Bochum. He obtained his Ph.D. from the University of Bonn, and has worked on problems of cosmic ray modulation and the dynamic heliosphere.

Vladimir Florinski is an Assistant Research Scientist at the Institute of Geophysics and Planetary Physics at the University of California, Riverside, California. He obtained his Ph.D. at the University of Arizona, in Tucson. His research interests include modeling the interaction between the solar wind and the interstellar medium, the modulation of galactic cosmic rays by the global heliosphere, and the acceleration of anomalous cosmic rays at the termination shock and beyond.

Priscilla C. Frisch is a Senior Scientist in the Department of Astronomy and Astrophysics at the University of Chicago. She has studied the properties of interstellar material close to the Sun, and the effect of this material on the heliosphere, since receiving her Ph.D. from the University of California in Berkeley.

Jasper Kirkby is an experimental particle physicist at CERN, Geneva. He received his Ph.D. at London University, and has originated several large experiments, including DELCO at SLAC, the BEPC II tau-charm factory in Beijing, FAST at PSI and the CLOUD experiment at CERN, which is applying the techniques of particle physics to understanding cosmic ray interactions with clouds.

Markus Landgraf is a Mission Analyst at the European Space Agency, Darmstadt. He received his Ph.D. at Heidelberg University, and has been studying interstellar dust in the solar system using data from Ulysses, Galileo, and other satellites since that time. At NASA/Johnson Space Center he analysed the signature of the Sun's dust disk in data collected by Pioneer 10 and 11.

Eberhard Möbius is Professor of Physics, in the Department of Physics and the Space Science Center, University of New Hampshire, in Durham. He received his Ph.D. at the Ruhr-Universitt Bochum in Bochum. In addition to his research activities investigating interstellar byproducts and particle acceleration in the heliosphere, he has been extensively involved in educational outreach programs.

Hans-Reinhard Müller is a Research Professor in the Department of Physics and Astronomy, Dartmouth College, in Hanover. He obtained his Ph.D. at Dartmouth College working on early-universe cosmology, and since then has engaged in large scale simulations of the global heliosphere system.

Eugene N. Parker is the S. Chandrasekhar Distinguished Service Professor Emeritus in Physics, Astronomy & Astrophysics and the Enrico Fermi Institute at the University of Chicago. He recieved the National Medal of Science for Physical Sciences in 1989 for his fundamental studies of plasmas, magnetic fields, and energetic particles on all astrophysical scales, as well as his basic theoretical work on stellar winds.

Nikolai V. Pogorelov is a Research Physicist at the Institute of Geophysics and Planetary Physics, University of California, Riverside. He received his Ph.D. and Dr. Sci. degrees at the Institute for Problems in Mechanics of the Russian Academy of Sciences, in Moscow. He has done research on magnetohydronamical models of the heliosphere, astrophysical accretion, supersonic aerodynamics, and mathematical aspects of numerical solution of partial differential equations.

Klaus Scherer is a Scientist at University Bochum, Bochum, and received his Ph.D. from University of Bonn. He has worked on theoretical heliosphere problems, and is a founding editor of the science journal ASTRA.

Nir J. Shaviv is a Senior Lecturer at the Racah Institute of Physics at the Hebrew University in Jerusalem. He received his Ph.D. at the Israel Institute of Technology. In addition to his research on the behavior of extremely luminous atmospheres, he has been investigating possible signatures left by cosmic events on the surface of the Earth.

Jonathan D. Slavin is a Research Astrophysicist at the Harvard-Smithsonian Center for Astrophysics, in Cambridge Massachusetts. He received his Ph.D. from the University of Wisconsin, Madison, and pursues studies of the interstellar medium and galaxy clusters.

Olaf Stawicki is a Postdoctoral Researcher of Physics in the Unit for Space Physics, School of Physics, North-West University, Campus Potchefstroom, Potchefstroom. He received his Ph.D. from the Ruhr-Universitt Bochum, Bochum, and is interested in the transport of charged particles in the heliosphere.

Peter Wurz is Professor of Physics at the Physikalisches Institut, Universität Bern, Bern. He received his Ph.D. from the Technical University of Vienna, and has been engaged in experimental work on neutral particle detection in space.

Ararat Yeghikyan is a Scientist at Byurakan Astrophysical Observatory, Armenia. He received his Ph.D. at the Institute of Astrophysics and Atmospheric Physics, in Tartu Estonia, and has worked on planetary nebula, the modification of the atmosphere by interstellar clouds, and more recently, on astrochemistry.

Gary Zank is Professor of Physics at the University of California at Riverside, and both Director and Systemwide Director of the University of California Institute of Geophysics and Planetary Physics. His Ph.D. is from the University of Natal.

Preface

The distinction between geophysics and astronomy was once clear. Events on Earth constituted the realm of geophysics, while astronomy encompassed objects that are located many light years from the Sun and Earth. Interstellar clouds were "out there", where they could be observed from isolated observatories nestled under the starry skies of the world's deserts. Geology relied on shovels and drill bits to obtain samples of mud and ice that contained clues to the paleoclimate. The space age changed all of this, with the discovery of interstellar gas and dust inside of the heliosphere. Mankind now looks out on the Universe from our vantage point on the spaceship Earth, and the scientific continuum that starts with the mineral and isotope composition on Earth extends back to the creation of the elements and isotopes at the beginning of the Universe. Through geological traces of radioisotopes and interplanetary material with a cosmic origin, the planetary system of the Sun provides a living record of the journey of the Sun through the Milky Way Galaxy. It is this living record, of which we are a part, that makes the discussions of the influence of interstellar matter on the heliosphere and Earth a compelling topic.

The Sun experiences many kinds of Galactic environments on its journey through space. The solar wind bubble, or heliosphere, acts as a buffer between the broad range of interstellar cloud types that are encountered, and the inner portion of the solar system where the Earth is located. The goal of this book is to show how changes in the galactic environment of the Sun affect the heliosphere, solar system, and Earth. It is partly motivated by what may be a purely happenstance coincidence. When I first plotted the solar space trajectory on a map of the Local Bubble,[1] it occurred to me that it may not be a coincidence that our Earth was in the deep vacuum of the Local Bubble during the past $\sim$2.5 million years when the genus *homo* emerged.

Professor John A. Simpson gave me a desk with his group after we moved to Chicago from Berkeley, and I learned about the heliosphere. Convinced that the interstellar hydrogen and helium observed inside of the heliosphere were part of the interstellar cloud seen towards Rasalhague, 14 pc away, I proposed to use the ultraviolet spectrometers on the *Copernicus* satellite to acquire high-resolution data on solar Lyman-alpha photons fluorescing off of interstellar hydrogen inside of the heliosphere[2]. In the world of astronomy, interstellar matter was between distant stars such as Scorpius and Orion, so this may have been the first observational effort to relate interstellar gas inside and outside of the heliosphere.

This book is dedicated to the memory of Professor John Simpson, who helped make the space age a reality. He played an important role in bringing a full-fledged space physics program to fruition at NASA, and was a leader in promoting healthy international scientific collaborations. John was the Principal Investigator for instruments on 10 interplanetary spacecraft and twenty Earth-orbiting satellites. He was an author or coauthor of over 330 scientific papers published between the years 1940–2000. His group made many major scientific contributions, including the discovery of the anomalous cosmic ray component[3]. John also played a vital role in founding and supporting *The Bulletin of the Atomic Scientists*, a periodical dedicated to the peaceful use of nuclear energy. His influence on the world has been profound. This volume is dedicated to John in honor of his scientific and policy contributions.

Notes

1. Frisch, P. and York, D. G. (1986). Interstellar Clouds Near the Sun. In *The Galaxy and the Solar System*, pages 83–100. Eds. R. Smoluchowski, J. Bahcall, and M. Matthews (University of Arizona Press).
2. Adams, T. F. and Frisch, P. C. (1977). High-resolution Observations of the Lyman Alpha Sky Background. *Astrophys. J.*, 212:300–308.
3. Garcia-Munoz, M., Mason, G. M., and Simpson, J. A. (1973). A New Test for Solar Modulation Theory: the 1972 May-July Low-Energy Galactic Cosmic-Ray Proton and Helium Spectra. *Astrophys. J. Lett.*, 182:L81–L84.

PRISCILLA C. FRISCH

Foreword

"An unusual display of the Aurora Borealis was witnessed here on the evening of Oct. 22, 1804...where a luminous cloud was formed, curling and rolling like smoke, and soon after dissipated in quick and repeated coruscations. .... On the 16th of June, 1806, there occurred a remarkable eclipse of the sun, which, at Boston and places farther south, was total. ... This eclipse formed an epoch among farmers, who used to date from it the commencement of those cold seasons, which, with some exceptions, continued with increasing severity, for 10 years."

New England agricultural records during the Little Ice Age, from "Annals of the Town of Warren", by Cyrus Eaton (Hallowell, Masters, Smith & Co.)

Acknowledgments

The editor is appreciative of helpful comments on the articles in this volume provided by James Bjorken, Hans Fahr, Henry Frisch, Tamas Gombosi, Carl Heiles, J. R. Jokipii, Mike Jura, Jasper Kirkby, Timur Linde, Clifford Lopate, Nick Pogorelov, Jonathan Rosner, Gary Thomas, Peter Vandervoort, Adolf Witt, and Donald York.

The editor thanks Andrew Hanson, Phillip Fu and Indiana University, for allowing the use of the cover figure, which shows a heliosphere visualization from the DVD "Solar Journey". The heliosphere model is based on Timur Linde's Ph.D. thesis. The film clip containing this visualization, which was produced with support from NASA Grant NAG5-11999, can be downloaded from *http://cs.indiana.edu/∼soljourn*.

The editor is grateful to Sarah Frisch for her valuable assistance in editing these articles, and to Henry Frisch for his continuous support of this project.

Chapter 1

INTRODUCTION: PALEOHELIOSPHERE VERSUS PALEOLISM

Priscilla C. Frisch
University of Chicago
frisch@oddjob.uchicago.edu

Abstract Speculations that encounters with interstellar clouds modify the terrestrial climate have appeared in the scientific literature for over 85 years. The articles in this volume seek to give substance to these speculations by examining the exact mechanisms that link the pressure and composition of the interstellar medium surrounding the Sun to the physical properties of the inner heliosphere at the Earth.

Keywords: Heliosphere, interstellar clouds, interstellar medium, cosmic rays, magnetosphere, atmosphere, climate, solar wind, paleoclimate

1.1 The Underlying Query

If the solar galactic environment is to have a discernible effect on events on the surface of the Earth, it must be through a subtle and indirect influence on the terrestrial climate. The scientific and philosophical literature of the 18th, 19th and 20th centuries all include discussions of possible cosmic influences on the terrestrial climate, including the effect of cometary impacts on Earth (Halley, 1724), and the diminished solar radiation from sunspots, which Herschel attributed to "holes" in the luminous fluid on the surface of the Sun[1] (Herschel, 1795). The discovery of interstellar material in the 20th century led to speculations that encounters with dense clouds initiated the ice ages (Shapley, 1921), and many papers appeared that explored the implications of such encounters, including the influence of interstellar material (ISM) on the interplanetary medium and planetary atmospheres (e.g. Fahr, 1968, Begelman and

P. C. Frisch (ed.), Solar Journey: The Significance of our Galactic Environment for the Heliosphere and Earth, 1–22.

Rees, 1976, McKay and Thomas, 1978, Thomas, 1978, McCrea, 1975, Talbot and Newman, 1977, Willis, 1978, Butler et al., 1978). The ISM-modulated heliosphere was also believed to affect climate stability and astrospheres (e.g. Frisch, 1993, Frisch, 1997, Zank and Frisch, 1999). Recent advances in our understanding of the solar wind and heliosphere (e.g. Wang and Richardson, 2005, Fahr, 2004) justify a new look at this age-old issue. This book addresses the underlying question:

How does the heliospheric interaction with the interstellar medium affect the heliosphere, interplanetary medium, and Earth?

The heliosphere is the cavity in the interstellar medium created by the dynamic ram pressure of the radially expanding solar wind, a halo of plasma around the Sun and planets, dancing like a candle in the wind and regulating the flux of cosmic rays and interstellar material at the Earth. Neutral interstellar gas and large interstellar dust grains penetrate the heliosphere, but the solar wind acts as a buffer between the Earth and most other interstellar material and low energy galactic cosmic rays (GCR). Together the solar wind and interstellar medium determine the properties of the heliosphere. In the present epoch the densities of the solar wind and interstellar neutrals are approximately equal outside of the Jupiter orbit. Solar activity levels drive the heliosphere from within, and the physical properties of the surrounding interstellar cloud constrain the heliosphere from without, so that the boundary conditions of the heliosphere are set by interstellar material. Figure 1.1 shows the Sun and heliosphere in the setting of the Milky Way Galaxy.

The answer to the question posed above lies in an interdisciplinary study of the coupling between the interstellar medium and the solar wind, and the effects that ISM variations have on the 1 AU environment of the Earth through this coupling. The articles in this book explore different viewpoints, including *gedanken* experiments, as well as data-rich summaries of variations in the solar environment and paleoclimate data on cosmic ray flux variations at Earth.

The book begins with the development of theoretical models of the heliosphere that demonstrate the sensitivity of the heliosphere to the variations in boundary conditions caused by the passage of the Sun through interstellar clouds. A series of *gedanken* experiments then yield the response of planetary magnetospheres to encounters with denser ISM. Variations in the galactic environment of the Sun, caused by the motions of the Sun and clouds through the Galaxy, are shown to occur for both long and short timescales.

The heliosphere acts as a buffer between the Earth and interstellar medium, so that dust and particle populations inside of the heliosphere, which have an interstellar origin, vary as the Sun traverses interstellar clouds. These buffering mechanisms determine the interplanetary medium[2]. The properties of these buffering interactions are evaluated for heliosphere models that have been developed using boundary conditions appropriate for when the Sun traverses different types of interstellar clouds.

The consequences of Sun-cloud encounters are then discussed in terms of the accretion of ISM onto the terrestrial atmosphere for dense cloud encounters, and the possibly extreme variations expected for cosmic ray modulation when interstellar densities vary substantially. Radioisotope records on Earth extending backwards in time for over $\sim$0.5 Myrs, together with paleoclimate data, suggest that cosmic ray fluxes are related to climate. The galactic environment of the Sun must have left an imprint on the geological record through variations in the concentrations of radioactive isotopes.

The selection of topics in this book is based partly on scientific areas that have already been discussed in the literature. The authors who were invited to contribute chapters have previously studied the heliosphere response to variable ISM conditions.

Figure 1.1 shows the heliosphere in our setting of the Milky Way Galaxy. A postscript at the end of this chapter lists basic useful information. I introduce the term "paleoheliosphere" to represent the heliosphere in the past, when the boundary conditions set by the local interstellar material (LISM) may have differed substantially from the boundary conditions for the present-day heliosphere. The "paleolism" is the local ISM that once surrounded the heliosphere.

1.2 Addressing the Query: The Heliosphere and Particle Populations for Different Interstellar Environments

The solar wind drives the heliosphere from the inside, with the properties of the solar wind varying with ecliptic latitude and the phase of the 11-year solar activity cycle. The global heliosphere is the volume of space occupied by the supersonic and subsonic solar wind. Interstellar material forms the boundary conditions of the heliosphere, and the windward side of the heliosphere, or the "upwind direction", is defined by the interstellar velocity vector with respect to the Sun. The leeward side of the heliosphere is the "downwind direction". Figure 1.1 shows a cartoon of the present-day heliosphere, with labels for the major landmarks such as the termination shock, heliopause, and bow shock.

In the present-day heliosphere, the transition from solar wind to interstellar plasma occurs at a contact discontinuity known as the "heliopause", which is formed where the total solar wind and interstellar pressures equilibrate (Holzer, 1989). For a non-zero interstellar cloud velocity in the solar rest frame, the solar wind turns around at the heliopause and flows around the flanks of the heliosphere and into the downwind heliotail. Before reaching the heliopause, the supersonic solar wind slows to subsonic velocities at the "termination shock", where kinetic energy is converted to thermal energy.

The subsonic solar wind region between the termination shock and heliopause is called the inner "heliosheath". The outer heliosheath lies just beyond the heliopause, where the pristine ISM is distorted by the ram pressure of the

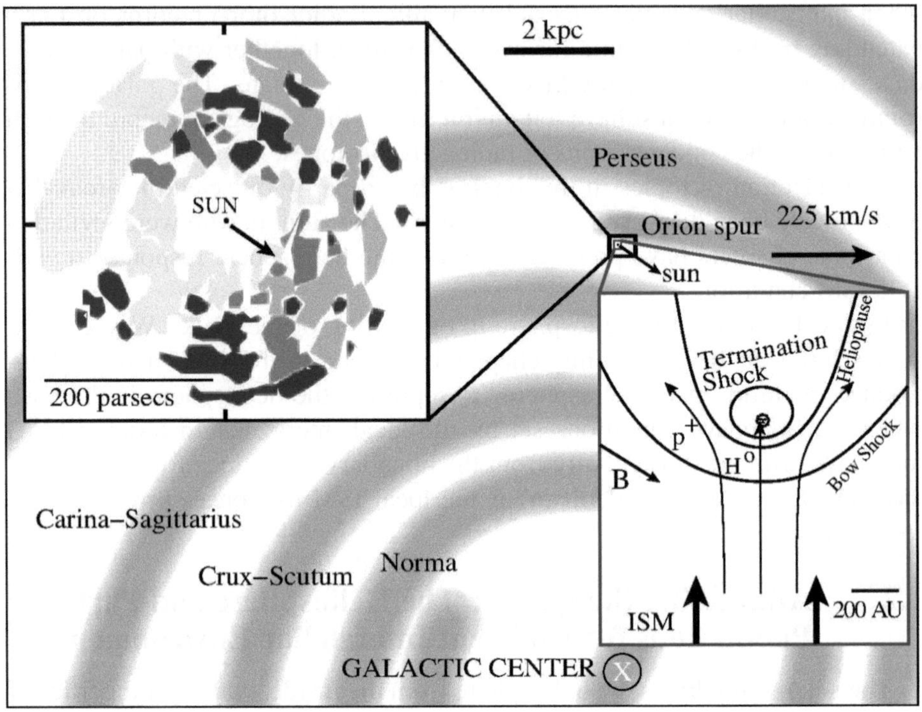

Figure 1.1. The solar location and vector motion are identified for the kiloparsec scale sizes of the Milky Way Galaxy (large image), and for the ~500 parsec scale size of the Local Bubble (medium sized image, inset in upper left hand corner). A schematic drawing of the heliosphere (small image, inset in lower right hand corner) shows the upwind velocity of the interstellar wind ("ISM") as observed in the rest frame of the Sun. Coincidently, this direction, which determines the heliosphere nose, is close to the galactic center direction. The orientation of the plane in the small inset differs from the planes of the large and medium figures, since the ecliptic plane is tilted by 60° with respect to the galactic plane. The Sun is 8 kpc from the center of the Milky Way Galaxy, and the solar neighborhood moves towards the direction $\ell = 90°$ at a velocity of 225 km s^{-1}. The spiral arm positions are drawn from Vallee (2005), except for the Orion spur. The Local Bubble configuration is based on measurements of starlight reddening by interstellar dust (Chapter 6). The lowest level of shading corresponds to color excess values E(B-V) = 0.051 mag, or column densities log N(H) (cm^{-2}) = 20.40 dex. The dotted region shows the widespread ionized gas associated with the Gum Nebula. The heliosphere cartoon shows interstellar protons deflected in the plasma flow in the outer heliosheath regions, compared to the interstellar neutrals that penetrate the heliopause.

heliosphere. A bow shock, where the interstellar gas becomes subsonic, is expected to form ahead of the present-day heliosphere in the observed upwind direction of the ISM flow through the solar system.

Large interstellar dust grains and interstellar atoms that remain neutral inside of the orbit of Earth, such as He, are gravitationally focused in the downwind direction. This "focusing cone" is traversed by the Earth every year in early December, and extends many AU from the Sun in the leeward direction (e.g. Landgraf, 2000, Möbius et al., 2004, Frisch, 2000). The heliotail itself extends $>10^3$ AU from the Sun in the downwind direction, forming a cosmic wake for the solar system.

Of significance when considering the interaction of the heliosphere with an interstellar cloud is that neutral particles enter the heliosphere relatively unimpeded, after which they are ionized and convected outwards with the solar wind. Ions and small charged dust grains are magnetically deflected in the heliosheath around the flanks of the heliosphere (see Figure 1.1).

Space and astronomical data now confirm the basic milestones of the outer heliosphere. Voyager 1 crossed the termination shock at 94 AU on 16 December, 2004 (UT), and observed the signature of the termination shock on low-energy particle populations, the solar wind magnetic field, low-energy electrons and protons, and Langmuir radio emission (Stone et al., 2005, Burlaga et al., 2005, Gurnett and Kurth, 2005, Decker et al., 2005). The present-day termination shock appears to be weak, with a solar wind velocity jump ratio (the ratio of upstream to downstream values) of $\sim$2.6 and a magnetic field compression ratio of $\sim$3. The magnetic wall that is predicted for the heliosphere (Linde, 1998, Ratkiewicz et al., 1998, Chapter 3 by Pogorelov and Zank) appears to have been detected through observations of magnetically aligned dust grains (Frisch, 2005), and the offset between upwind directions of interstellar H° and He° (Lallement et al., 2005). The compressed and heated H° in the hydrogen wall region of the outer heliosheath has now been detected around a number of stars (Wood et al., 2005).

The present-day solar wind is the baseline for evaluating the heliosphere response to ISM variations in the following articles, so a short review of the solar wind is first presented. The remaining part of §1.2 introduces the topics in the following articles in terms of the underlying query of the book.

1.2.1 The Present Day Solar Wind

The solar wind originates in the million degree solar corona that expands radially outwards, with a density $\sim 1/R_S^2$ where R_S is the distance to the Sun, and contains both features that corotate with the Sun, and transient structures (e.g. Gosling, 1996). The properties of the solar wind vary with the phase of the solar magnetic activity cycle and with ecliptic latitude. The best historical indicator of solar magnetic activity levels is the number of sunspots, first detected by Galileo in 1610, which are magnetic storms in the convective zone of the Sun. Sunspot numbers indicate that the magnetic activity levels fluctuate with a $\sim$11 year cycle, or the "solar cycle", and solar maximum/minimum corresponds to the maximum/minimum of sunspot numbers. The magnetic polarity of the Sun varies with a $\sim$22 year cycle. During solar maximum, a low-speed wind, with velocity $\sim$300–600 km s^{-1} and density $\sim$6–10 particles cm^{-3} at 1 AU, extends over most of the solar disk. Open magnetic field lines[3] are limited to solar pole regions. A neutral current sheet $\sim$0.4 AU thick forms between the solar wind containing negative magnetic polarity fields and the solar wind that contains positive magnetic polarity fields. The neutral current sheet reaches its largest inclination ($\geq$70°) during solar maximum. During the conditions of solar minimum, a high speed wind with velocity $\sim$600–800 km s^{-1} and density $\sim$5 cm^{-3} is accelerated in the open magnetic flux lines in coronal holes. During mininum, the high speed wind and open field lines extend from the polar regions down to latitudes of $\leq$40° (Smith et al., 2003, Richardson et al., 1995). The higher solar wind momentum flux associated with solar minimum conditions produces an upwind termination shock that is $\sim$5–40 AU more distant in the upwind direction than during solar maximum conditions (e.g. Scherer and Fahr, 2003, Zank and Müller, 2003, Whang, 2004).

During solar minimum conditions, the magnetic field is dominated by the dipole and hexapole moments, with a small contribution from a quadrupole moment. The alignment and strength of the multipoles depend on the phase of the solar cycle (Bravo et al., 1998). The solar dipole moment is strongest during solar minimum, when it is generally aligned with the solar rotation axis. Sunspots migrate from high to low heliographic latitudes. The magnetic poles follow the coronal holes to the solar equator as solar activity increases. During the solar maximum period, the galactic cosmic rays undergo their maximum modulation, the dipole component of the magnetic field is minimized, and the polarity of the solar magnetic field reverses (Lockwood and Webber, 2005, Figure 1.2). Over historic times, the cosmic ray modulation by the heliosphere correlates better with the open magnetic flux line coverage than with sunspot numbers (McCracken et al., 2004).

Variable cosmic ray modulation produced by a variable heliosphere may be a primary factor in both solar and ISM forcing of the terrestrial climate. The heliosphere modulation of cosmic rays is well established. John Simpson, to whom this book is dedicated, initiated a program 5 solar cycles ago in 1951 to monitor cosmic ray fluxes on Earth using high-altitude neutron detectors (Simpson, 2001). The results show a pronounced anticorrelation between cosmic ray flux levels and solar sunspot numbers, which trace the 11-year Schwabe magnetic activity cycle, and which also show that the polarity of the solar magnetic field affects cosmic ray modulation (see Figure 1.2). The articles in this book show convincingly that the ISM also modulates the heliosphere, and the effect of the solar wind on the heliosphere must be differentiated from the influence of interstellar matter.

Variations in solar activity levels are also seen over $\sim$100–200 year timescales, such as the absence of sunspots during the Maunder Minimum in the 17th century. Modern climate records show that the Maunder Minimum corresponded to extremely cold weather, and radioisotope records show that the flux of cosmic rays was unusually high at this time (see Kirkby and Carslaw, Chapter 12). Similar effects will occur from the modulation of galactic cosmic rays by the passage of the Sun through an interstellar cloud.

These temporal and latitudinal variations in the solar wind momentum flux produce an asymmetric heliosphere, which varies with time. Any possible historical signature of the ISM on the heliosphere must first be distinguished from variations driven by the solar wind itself.

1.2.2 Present Day Heliosphere and Sensitivity to ISM

The ISM forms the boundary conditions of the heliosphere, so that encounters with interstellar clouds will affect the global heliosphere, the interplanetary medium, and the inner heliosphere region where the Earth is located. Today an interstellar wind passes through the solar system at –26.3 km s^{-1} (Witte, 2004). An entering parcel of ISM takes about 20 years to reach the inner heliosphere, so that ISM near the Earth is constantly replenished with new inflowing material. This warm gas is low density and partially ionized, with temperature $T\sim$ 6,300 K, and densities of neutral and ionized matter of $n(\mathrm{H^o})\sim0.2\ \mathrm{cm^{-3}}$, and $n(\mathrm{H^+})\sim0.1\ \mathrm{cm^{-3}}$.

An elementary perspective of the response of the heliosphere to interstellar pressures is given by an analytical expression for the heliopause distance based on the locus of positions where the solar wind ram pressure, P_{SW}, and the total interstellar pressure equilibrate (Holzer, 1989). The solar wind density ρ falls off as $\sim 1/R^2$, where R is the distance to the Sun, while the velocity v is relatively constant. At 1 AU the solar wind ram pressure is $P_{\mathrm{SW,1AU}}\sim\rho\ v^2$ so the heliosphere distance, R_{HP}, is given by:

$$P_{\mathrm{SW,1AU}}/R^2_{\mathrm{HP}} \sim P_{\mathrm{B}} + P_{\mathrm{Ions,thermal}} + P_{\mathrm{Ions,ram}} + P_{\mathrm{Dust}} + P_{\mathrm{CR}}$$

The interstellar pressure terms include the magnetic pressure P_{B}, the thermal, $P_{\mathrm{Ions,thermal}}$, and the ram, $P_{\mathrm{Ions,ram}}$, pressures of the charged gas, and the pressures of dust grains, P_{Dust}, and cosmic rays, P_{CR}, which are excluded by heliosphere magnetic fields and plasma. Some interstellar neutrals convert to ions through charge exchange with compressed interstellar proton gas in heliosheath regions, adding to the confining pressure. An important response characteristic is that, for many clouds, the encounter will be ram-pressure dominated, where $P_{\rm ram} \sim mv^2$ for interstellar cloud mass density m and relative Sun-cloud velocity v, so that variations in the cloud velocity perturb the heliosphere even if the thermal pressures remain constant.

The multifluid, magnetohydrodynamic (MHD), hydrodynamic and hybrid approaches used in the following chapters provide much more substantial models for the heliosphere, and include the coupling between neutrals and plasma, and field-particle interactions. These sophisticated models predict variations in the global heliosphere in the face of changing interstellar boundary conditions, and for a range of different cloud types. Although impossible to model a solar encounter with every type of interstellar cloud, the following articles include discussions of many of the extremes of the interstellar parameter space, including low density gas with a range of velocities, very tenuous plasma, high velocity clouds, dense ISM, and magnetized material for a range of field orientations and strengths. The discussions in these chapters extrapolate from our best theoretical understanding of the heliosphere boundary conditions today to values that differ, in some cases dramatically, from the boundary conditions that prevailed at the beginning of the third millennium in the Gregorian calendar.

The Sun has been, and will be, subjected to many different physical environments over its lifetime. Theoretical heliosphere models yield the properties of the solar wind-ISM interaction for these different environments, which in turn determine the nature and properties of interstellar populations inside of the heliosphere for a range of galactic environments. These models form the foundation for understanding the significance of our galactic environment for the Earth.

The interstellar parameter space is explored by Zank et al. (Chapter 2), where 28 sets of boundary conditions are evaluated with computationally efficient multifluid models. Moebius et al. (Chapter 8), Fahr et al. (Chapter 9), Florinski and Zank (Chapter 10), and Yeghikyan and Fahr (Chapter 11) also develop heliosphere models for a range of interstellar conditions. Together these models evaluate the heliosphere response to interstellar density, temperature, and velocity variations of factors of $\sim 10^9$, $\sim 10^5$, and $\sim 10^2$, respectively.

The interstellar magnetic field introduces an asymmetric pressure on the heliosphere, affecting the heliosphere current sheet and cosmic ray modulation. Pogorelov and Zank (Chapter 3) use MHD models to probe the heliosphere response to the interstellar magnetic field, including charge exchange between the neutrals and solar wind. The resulting asymmetry provides a test of the magnetic field direction, and shows strong differences between cases where the interstellar flow is parallel, instead of perpendicular, to the interstellar magnetic field direction. Since the random component of the interstellar magnetic field is stronger, on the average, than the ordered component, particularly in spiral arm regions where active star formation occurs, a range of interstellar magnetic field strengths and orientations are expected over the solar lifetime (Shaviv, Chapter 5, and Frisch and Slavin, Chapter 6).

1.2.3 Planetary Magnetospheres

The Earth's magnetosphere acts as a buffer between the solar wind and atmosphere, and as such is an ingredient in understanding the effect of our galactic environment on the Earth. The decreasing solar wind density in the outer heliosphere results in an interplanetary medium around outer planets that is more sensitive to ISM variations than for inner planets, with implications for the magnetospheres of Jupiter, Neptune, and Uranus. Most topics in this book are already considered in the scientific literature, but questions about magnetosphere variations from an ISM-modulated heliosphere have received scant attention. In a quintessential *gedanken* experiment, Parker explores the interaction between magnetospheres and the solar wind for variations in the interstellar density, and for inner versus outer planets (Chapter 4).

1.2.4 Short and Long Term Variations in the Galactic Environment

There is every reason to expect that the galactic environment of the Sun varies over geological timescales. The Sun moves through space at a velocity of 13–20 km s^{-1}, and interstellar clouds have velocities ranging up to hundreds of km s^{-1}. The Arecibo Millennium survey showed that $\sim$25% of the mass contained in interstellar H°, including both warm and cold ISM, is in clouds traveling with velocities $\geq$10 km s^{-1} through the local standard of rest (Heiles and Troland, 2003). Thus Sun-cloud encounters with relative velocities exceeding 25 km s^{-1} are quite likely, and for a typical cloud length of $\sim$1 pc the cloud transit time would be $\sim$40,000 years. The many types of ISM traversed by the Sun during the past several million years have affected the heliosphere, the inner solar system, and the flux of anomalous and galactic cosmic rays at Earth (Frisch and York, 1986, Frisch, 1997, Frisch, 1998).

For the past $\sim$3 Myrs the Sun has been in a nearly empty region of space, the "Local Bubble", with very low densities of $<10^{-26}$ gr cm^{-3}. Within the past 44,000–150,000 years the Sun entered a flow of tenuous, partly neutral ISM, nick-named the "Local Fluff", with density $\sim$60 times higher (Chapter 6). This transition was accompanied by the appearance of interstellar dust and neutrals in the heliosphere, along with the pickup ion and anomalous cosmic ray populations. Galactic cosmic ray modulation was affected, providing a possible link between our galactic environment and climate. Intriguingly, the averaged cosmic ray flux at Earth, as traced by ^{10}Be records, was lower in the past $\sim$135 kyrs than for earlier times (Chapter 12). Was the decrease in the galactic cosmic ray flux $\sim$135 kyrs ago caused by an increase in modulation as the Sun entered the Local Fluff?

The galactic environment of the Sun also varies quite dramatically over long time scales, as discussed by Shaviv (Chapter 5). Over its 4.5 billion year lifetime, the Sun traverses spiral arm and interarm regions, with atomic densities varying from less than $10^{-26.1}$ g cm^{-3} to over $10^{-20.1}$ g cm^{-3}, and temperatures ranging over 7 orders of magnitude, 10–10^7 K. The Sun is now in low density space between the Perseus and Sagittarius spiral arms, and on the inner edge of what is known as the Orion spur on the Local Arm. The Local Arm is not shown in Figure 1.1, as is consistent with the usual Galaxy depictions. The Local Arm does not appear to be a grand design spiral shock (Bochkarev, 1984). The Sun has a systematic motion of 13–20 km s^{-1} with respect to the nearest stars, corresponding to $\sim$3–4 AU per year. The Local Interstellar Cloud (LIC) now surrounding the Sun traverses the heliosphere at $\sim$5.5 AU per year. The Sun oscillates vertically through the galactic plane once every $\sim$34 Myrs, and orbits the center of the Milky Way Galaxy once per $\sim$220 Myrs.

Shaviv evaluates variations in the galactic environment of the Sun over long timescales. This bold discussion compares various geologic records of cosmic ray flux variations, based on radioisotope data that sample timescales of $\sim 10^8$ years, with models of the Milky Way Galaxy spiral arm pattern to reconstruct the timing of the Sun's passage through spiral arms. The chapter concludes that star formation in spiral arms leaves a signature on the radioisotope records of the solar system.

Frisch and Slavin (Chapter 6) reconstruct short-term variations of the galactic environment of the Sun using observations of interstellar matter towards nearby stars and inside of the solar system. Radiative transfer models of the LIC show that ionization varies across this low density cloud, so that the heliosphere boundary conditions vary from radiative transfer considerations alone as the Sun traverses the LIC. Cloud transitions are predicted for the past $\sim$3 Myrs, including the departure of the Sun from the Local Bubble interior 44,000–150,000 years ago.

1.2.5 Interstellar Dust

The particle populations formed by the interactions between the solar wind and interstellar dust, gas, and cosmic rays are emissaries between the cosmos and inner heliosphere, varying as the Sun moves through clouds.

About $\sim$1% of the mass of the cloud surrounding the Sun is contained in interstellar dust grains. The largest of these charged grains, mass $>10^{-13}$ g, have large magnetic Larmor radii of $>$500 AU at the heliopause for an interstellar field of $\sim$1.5 μG, and flow into the solar system. The Earth passes through the gravitational focusing cone formed by these grains early each December. The smallest charged grains, mass $<10^{-14.5}$ g and radii <0.01 μm, have Larmor radii of $\sim$20 AU, depending on the magnetic field strength and radiation field, and are deflected around the heliosheath (Frisch et al., 1999). Interstellar dust grains are measured in the inner heliosphere within $\sim$5 AU of the Sun, and over the solar poles, by satellites such as Ulysses, Galileo and Cassini. Landgraf (Chapter 7) reviews the properties of the interaction between interstellar dust and the solar wind, and speculates on the changes that might be expected from an encounter with a dense interstellar cloud.

Should it some day be possible to compare the ratio of large to small interstellar dust grains on the surfaces of the inner versus outer planets, it would become possible to disentangle cloud encounters from solar activity effects.

At the very large end of the dust population mass spectrum we find interstellar micrometeorites, with masses $\sim 3 \times 10^{-7}$ g, open orbits, and inflow velocities greater than the 42 km s^{-1} escape velocity from the solar system at 1 AU. These interstellar objects, detected by radar as they impact the atmosphere, evidently originate in circumstellar disks such as that around β Pictoris, and in the interior of the Local Bubble (Baggaley, 2000, Meisel et al., 2002). These objects do not collisionally couple to the interstellar gas (Gruen and Landgraf, 2000), and should not vary with the type of ISM surrounding the Sun.

1.2.6 Particle Populations in the Inner and Outer Heliosphere

Presently, low energy interstellar neutrals, high energy galactic cosmic rays, and interstellar dust all enter the heliosphere. The characteristics of each of these populations and their secondary products are modified as the Sun transits the ISM, or the cloud ionization changes. The first ionization potential (FIP) of H° is 13.6 eV. Neutral interstellar atoms with FIP $<$ 13.6 eV are ionized in nearly all interstellar clouds because the main source of interstellar opacity is H°. Interstellar ions are deflected around the heliosheath, so the result is that only interstellar atoms with FIP $>$ 13.6 eV enter the heliosphere where they are then destroyed, primarily by charge exchange with solar wind ions.

The density of interstellar neutrals in the inner heliosphere depends on the density and ionization of the surrounding cloud, the ionization (or "filtration") of those neutrals by the heliosheath, and the subsequent interactions with the solar wind inside of the heliosphere. Secondary products produced by solar wind interactions with interstellar neutrals inside of the heliosphere include pickup ions[4], energetic neutral atoms, the gravitational focusing cone formed by helium (also seen in dust), and the anomalous cosmic ray population with energies $<$1 GeV. Interstellar neutrals inside of the heliosphere, and the heliosphere itself, form a coupled system that together respond to variations in the heliosphere boundary conditions.

Moebius et al. (Chapter 8) model the heliosphere for several different conditions, and then probe the response of the inner heliosphere to the density of interstellar neutrals flowing into this ISM-modified heliosphere. At 1 AU, the neutral densities, particle populations derived from interstellar neutrals, and characteristics of the helium focusing cone all respond to variations in the interstellar boundary conditions. For some cases, increased neutral fluxes fall on the atmosphere of Earth (also see Yeghikyan and Fahr, Chapter 11).

The velocity structure of the ISM appears to vary on subparsec scale lengths (Frisch and Slavin, Chapter 6), and these variations may in some cases result in significant modifications of the inner heliosphere, particularly the gravitational focusing cone, when all other interstellar parameters such as thermal pressure are invariant (Zank et al., Chapter 2, Moebius et al., Chapter 8).

The most readily available diagnostics of the paleoheliosphere are radioisotopes, formed by cosmic ray spallation on the atmosphere, interplanetary and interstellar dust, and meteorites. Thus, the evaluation of cosmic ray modulation for various types of interstellar cloud boundary conditions is a key part of understanding the paleoclimate records that might trace the solar journey through the Milky Way Galaxy. Fahr et al. (Chapter 9) and Florinski and Zank (Chapter 10) use our understanding of galactic cosmic ray modulation in the modern-day heliosphere as a basis for making detailed calculations of the response of the paleoheliosphere, or the heliosphere as it once was, to the paleolism, or the local interstellar medium that once surrounded the Sun. The predictions of these calculations are quite intriguing. Both the termination shock compression ratio and the solar wind turbulence spectrum may vary dramatically with different environments, as mass-loading by pickup ions and the heliosphere properties vary. The problem of galactic cosmic ray modulation in an ISM-forced heliosphere is extremely important to understanding the paleoheliosphere signature in the terrestrial isotope record.

Today, galactic cosmic rays (GCR) with energies $\geq$0.25 GeV penetrate the solar system, and anomalous cosmic rays (energies $<$1 GeV) are formed from accelerated pickup ions. The cosmic ray flux at Earth is sampled by geological radioisotope records, as reviewed Kirkby and Carslaw (Chapter 12, also see Florinski and Zank, Chapter 10). Astronomical data indicates that the Sun has emerged from a region of space with virtually no neutral ISM within the past $\sim$0.4–1.5 10^5 years, and entered the Local Fluff (Chapter 6). The GCR modulation discontinuity that accompanied this transition may be in the geologic record, which show lower cosmic ray fluxes at Earth, on the average, for the past 135 kyrs years than the 135 kyrs before that (Christl et al., 2004).

1.2.7 Atmosphere Accretion from Dense Cloud Encounters

Harlow Shapley (1921) suggested that an encounter between the Sun and giant dust clouds in Orion may have perturbed the terrestrial climate and caused ice ages. The discovery of interstellar H° and He° inside the heliosphere was soon followed by studies of the ISM influence on the atmosphere for dense cloud conditions (Fahr, 1968, Begelman and Rees, 1976, McKay and Thomas, 1978, Thomas, 1978, McCrea, 1975, Talbot and Newman, 1977, Willis, 1978, Butler et al., 1978). Yeghikyan and Fahr (Chapter 11), evaluate the density of ISM at the Earth based on models describing the heliosphere inside of an dense cloud, and the interactions between the solar wind and ISM for these dense cloud conditions (also see Chapter 9, by Fahr et al.). These models then yield the concentration of interstellar hydrogen at the Earth, and the flow of water downward towards the Earth's surface, as a function of the dense cloud density. Significant atmosphere modifications are predicted in some cases. Enhanced neutral populations at 1 AU for a somewhat lower interstellar cloud density regime are discussed in Chapter 8, by Moebius et al.

1.2.8 Possible Effects of Cosmic Rays

Both solar activity cycles (Figure 1.2) and ISM variations modulate the cosmic ray flux in the heliosphere, and Kirkby and Carslaw (Chapter 12) compare galactic cosmic ray records with paleoclimate archives. They examine sources of climate forcing such as solar irradiance and cosmic ray fluxes, and conclude that arguments in favor of cosmic ray climate forcing are strong although the mechanism is uncertain. This relation between cosmic ray flux levels and the climate is shown by radioisotope records and climate archives, such as ice cores, stalagmites, and ice-rafted debris, and for modern times, by historical records. Paleoclimate archives include terrestrial records of cosmic ray spallation in the atmosphere, as traced by radioisotopes with short half-lives ($\tau_{1/2}$), e.g. ^{14}C ($\tau_{1/2} = 5,730$ yrs) and ^{10}Be ($\tau_{1/2} = 1.6$ Myrs). Possible mechanisms linking the cosmic ray flux at 1 AU and the climate include cloud nucleation by cosmic rays, and the global electrical circuit (see Chapter 12 and Roble and Hays, 1979). The discussion in Chapter 12 provides persuasive evidence linking the surface temperature to cosmic ray fluxes at Earth. The anti-correlation between sunspot number and cosmic ray fluxes in Figure 1.2 shows the heliosphere role in cosmic ray modulation; this mechanism must have also been a prominent mechanism for relating the ISM-modulated heliosphere with the climate. Fortunately this hypothesis is also verifiable by comparing paleoclimate data with astronomical data on the timing of cloud transitions.

The radioisotope records also indicate that cosmic ray fluctuations have occurred over longer timescales of many 10^8 years. Shaviv compares the ^{36}Cl

($\tau_{1/2}\sim$0.3 Myrs) and ^{40}K ($\tau_{1/2}\sim$1.3 Gyr) cosmic ray exposure records in iron meteorites (Chapter 5), but in this case to obtain cosmic ray flux increases due to the Sun's location in spiral arms where active star formation occurs.

A number of studies, none convincing, have invoked the geological ^{10}Be record, as a proxy for cosmic ray fluxes at Earth, to infer historical encounters with interstellar clouds. As a way of dating the Loop I supernova remnant, it was suggested that the relative constancy of ^{10}Be in sea sediments precluded a strong nearby X-ray source within the past $\sim$2 Myrs (Frisch, 1981). Sonett (1992) suggested that peaks in ^{10}Be layers 35,000 and 65,000 years ago resulted from a compressed heliosphere caused by the passage of a high-velocity interstellar shock. This extreme heliosphere compression expected for a rapidly moving cloud is supported by heliosphere models (Chapter 2). Structure in the ^{10}Be peaks has also been related to spatial structure in the local ISM (Frisch, 1997), and solar wind turbulence caused by mass-loading of interstellar neutrals may supply the required mechanism. Global geomagnetic excursions such as the events $\sim$32 kyr and $\sim$40 kyr ago also affect the ^{10}Be record, and can not be ignored (Christl et al., 2004). Indeed, Figure 1.2 shows the sensitivity of galactic cosmic ray fluxes on Earth to geomagnetic latitude.

1.3 Closing Comments

This brief summary of the scientific question motivating this book does not relay the full significance of the galactic environment of the Sun to the heliosphere and Earth; the following chapters provide deeper insights into this question.

Historical and paleoclimate data show a correspondence between high cosmic ray flux levels and cool temperatures on Earth (Parker, 1996). The disappearance of sunspots for extended periods of time, such as the Maunder Minimum in the years 1645 to 1715, shows up in terrestrial radioisotope records such as ^{10}Be in ice cores (Chapter 12). The solar magnetic activity cycle was present during this period, and cosmic ray modulation by the heliosphere was still evident (McCracken et al., 2004). The ^{10}Be record now extends to $\sim 10^5$ years before present, raising the hope that encounters between the Sun and interstellar clouds can be separated from solar activity effects, and from the global signature of geomagnetic pole wandering.

Sunspots have long been controversial as an influence on the terrestrial climate. Sir William Herschel carefully observed them, and postulated that diminished solar radiation at Earth during sunspot maximum affected the terrestrial climate (1801). Prof. Langley (1876) measured the radiative heat from sunspot umbral and penumbral regions, and concluded the $< 0.1\%$ solar radiation decrease associated with sunspots was inadequate to affect the climate. Climate records show that the Maunder Minimum and other periods of low solar activity levels have been exceptionally cold, which implicates high cosmic ray fluxes with cold climate conditions. Solar activity levels have returned to historic highs in the past few decades (Caballero-Lopez et al., 2004), and the historic correlations indicate these high levels also yield warm climate conditions. Unfortunately, these scientific conclusions also impact the politically loaded issue of global warming.

The possibility that the cosmos has affected the terrestrial climate is a longtime source of speculation, with many of the first discussions focused on explaining the "Universal Deluge". In 1694 Edmond Halley presented his thoughts to the Royal Society as to whether the "casual Shock of a Comet, or other transient Body" might instantly alter the axis orientation or diurnal rotation of the Earth, thus disturbing sea levels, or whether the impact of a comet could explain the presence of "vast Quantities of Earth and high Cliffs upon Beds of Shells, which once were the Bottom of the Sea" (Halley, 1724). Halley's speculation has resurfaced in the hypothesis that the impact of a comet led to the extinction of dinosaurs 65 Myrs ago at the Cretaceous-Tertiary boundary (Alvarez, 1982). The common sense disclaimer that accompanied Halley's discussion is timeless: *"... the Almighty generally making us of Natural Means to bring about his Will, I thought it not amiss to give this Honourable Society an Account of some Thoughts that occurr'd to me on this Subject; wherein, if I err, I shall find myself in very good Company."*

The articles in this volume show firmly that the interaction between the heliosphere and ISM depends on the detailed boundary conditions set for the heliosphere by each type of interstellar cloud encountered by the Sun, and that the galactic environment of the Sun changes over both geologically short time scales of $<10^5$ years, and long time scales of $>10^7$ years. This interaction, in turn, affects the flux of gas, dust, and energetic particles in the inner heliosphere.

The discussions in this book also apply to the study of astrospheres around cool stars, which are expected to have similar properties as the heliosphere. Is the historical astrosphere of a star a factor in climate stability for planetary systems? I think so (Frisch and York, 1986). If so, then the sample of $\sim$100 detected extrasolar planetary systems can be narrowed to those that are the most likely to harbor technological civilizations by evaluating the astrosphere characteristics suitable to the space trajectory of each star (Frisch, 1993). Astrospheres have now been detected towards $\sim$60% of the observed cool stars within 10 pc (Wood et al., 2005), and extensive efforts to detect Earth-sized exoplanets are underway. Perhaps some day these questions will be answered.

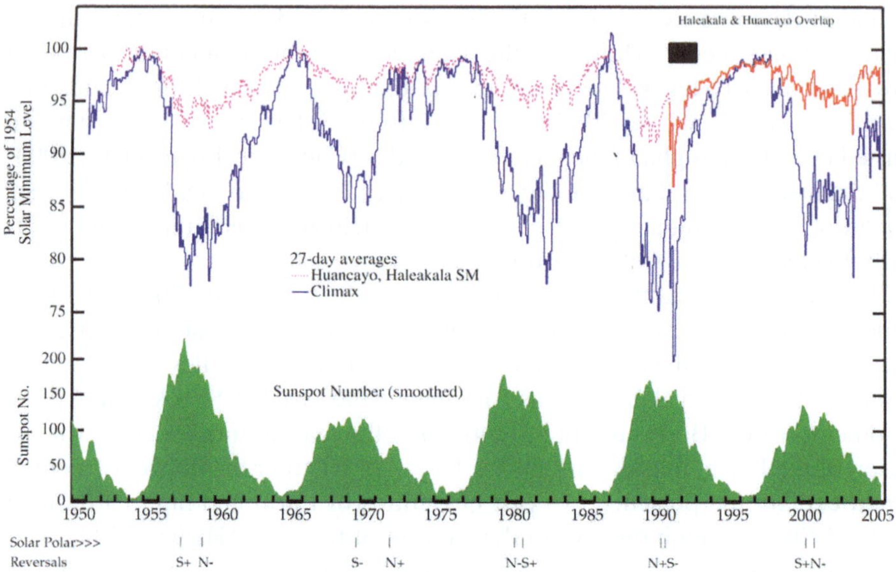

Figure 1.2. Galactic cosmic ray fluxes on Earth versus solar activity levels for sunspot cycles 18–23. Depicted are 27-day averages of the Climax (blue), and Huancayo/Haleakala (pink/red) neutron monitor rates as a percentage of their respective 1954 solar minimum levels. The running averages of the monthly mean sunspot number (green) are a proxy for the level of turbulence in the heliosphere as a function of solar activity. There is a clear anti-correlation between the neutron monitor rates and the sunspot number. The flat-topped versus peaked-top neutron monitor rates seen at successive 11-year solar minimum periods are a function of the polarity of the heliospheric magnetic field, noted at the bottom. The Climax data show solar cycle modulation for >3 GeV GCRs, while the Huancayo/Haleakala data show solar cycle modulation for >13 GeV GCRs (for additional detail please see Lopate, 2005). The geomagnetic latitudes of Climax and Huancayo/Haleakala are 48° and $\sim 0.5^\circ/20^\circ$. The poles are given as N and S, for north and south poles. The terms N+/S- indicate the times when the polarities of the north/south poles became positive/negative, while N-/S+ indicates they became negative/positive instead. The N/S poles do not appear to switch polarities simultaneously. The author thanks Dr. Clifford Lopate for providing this figure, and for maintaining a valuable data stream from an experiment begun by John Simpson in 1950.

Acknowledgments: The author thanks Dr. Clifford Lopate, of the University of New Hampshire, for providing Figure 1.2, and Dr. Lopate thanks NSF Grant 03-39527 for supporting the research displayed in this figure. The author thanks NASA for supporting her research, including grants NAG5-13107 and NNG05GD36G.

Postscript: Definitions

The nine planets of the solar system (including Pluto as a planet) extend out to 39 AU, compared to the distance of the solar wind termination shock in the

Table 1.1. Commonly Used Terms and Acronyms.

Object	Description
Interstellar:	
Interstellar Material, ISM	Atoms in the space between stars
Local Fluff or CLIC	ISM within ~30 pc, density $10^{-24.3}$ g cm^{-3}
	CLIC = Cluster of Local Interstellar Clouds
Local Interstellar Cloud, LIC	The cloud feeding ISM into the solar system
Local Bubble, LB	Nearby ISM with density $<10^{-26.1}$ g cm^{-3}
Heliosphere:	
Solar Wind, SW	Solar plasma expanding to form heliosphere
	Density ~5 ions cm^{-3}, velocity ~450 km s^{-1}
	at Earth
Neutral Current Sheet	Thin neutral region separating SW
	with opposite magnetic polarities
Heliosphere, HS	Region of space containing the solar wind
Termination Shock, TS	Shock where solar wind becomes subsonic
	TS at ~94 AU on 16 December, 2004
Heliosheath	Subsonic solar wind, outside TS
Heliosphere Bow Shock	Shock where LIC becomes subsonic
Focusing Cone	Gravitationally focused ISM dust
	and helium gas downwind of the Sun
Interstellar Products in the Heliosphere:	
Pickup Ions, PUI	Ions from SW-ISM charge exchange
Energetic Neutral Atoms	ENAs, Energetic atoms formed by
	charge exchange with ions
Cosmic Rays:	
Anomalous, ACR	Accelerated pickup ions, energy <1 GeV
Galactic, GCR	From supernova, energy >1 GeV at Earth

upwind direction of 94 AU. The Earth is 8.3 light minutes from the Sun, versus the ~0.5 light day distance to the upwind termination shock of the solar wind. The ecliptic and galactic planes are tilted with respect to each other by ~60°, and the north ecliptic pole points towards the galactic coordinates $\ell = 96.4°$ and $b = +29.8°$. This tilt allows the separation of large scale ecliptic and large scale galactic phenomena by geometric considerations.

Acronyms are used throughout this book, and some of these are listed in Table 1. For those new to this subject, an astronomical unit, AU, is the distance between the Earth and Sun. A parsec, pc, is 206,000 AU, 3.3 light years (ly), or 3.1×10^{18} cm. For comparison, the nearest star, α Cen, is 1.3 pc from the Sun.

Notes

1. In this same paper Herschel commented that "Whatever fanciful poets might say, in making the sun the abode of blessed spirits, or angry moralists devise, in pointing it out as a fit place for the punishment of the wicked, it does not appear that they had any other foundation for their assertions than mere opinion and vague surmise; but now I think myself authorized, *upon astronomical principles,* to propose the sun as an inhabitable world, and am persuaded that the foregoing observations, with the conclusions I have drawn from them, are fully sufficient to answer every objection that may be made against it." These comments show that valuable data are not always interpreted correctly.
2. The buffering processes convert interstellar neutrals into low energy ions, which are convected outwards with the solar wind and accelerated to low cosmic ray energies that have an anomalous composition, including abundant elements with FIP > 13.6 eV. The high energy galactic cosmic ray population incident on the heliosphere is also modulated.
3. Open magnetic field lines are formed in coronal holes that reconnect in the outer heliosphere and contain low density and very high speed, $\sim$700 km s^{-1}, solar wind.
4. The pickup ions are interstellar neutrals formed by charge exchange with the solar wind. Energetic neutral atoms are formed by energetic ions that capture an electron from a low energy neutral by charge exchange. The gravitational focusing cone contains heavy elements (mainly He) that are predominantly ionized inside of 1 AU and therefore gravitationally focused downwind of the Sun (Chapter 8). Large interstellar dust grains are also gravitational focused (Chapter 7). The anomalous cosmic ray population is formed from pickup ions accelerated to low cosmic ray energies, $<$1 GeV, in the solar wind and at the termination shock, and then subjected to the same modulation and propagation processes as galactic cosmic rays (Jokipii, 2004).

References

Alvarez, L. W. (1982). Experimental Evidence that an Asteroid Impact Led to the Extinction of many Species 65 Million Years Ago. *NASA STI/Recon Technical Report N*, 83:33813.

Baggaley, W. J. (2000). Advanced Meteor Orbit Radar Observations of Interstellar Meteoroids. *J. Geophys. Res.*, 105:10353–10362.

Begelman, M. C. and Rees, M. J. (1976). Can Cosmic Clouds Cause Climatic Catastrophes. *Nature*, 261:298.

Bochkarev, N. G. (1984). Large-Scale Bubble Structure of the Interstellar Medium and Properties of the Local Spiral Arm. In *Local Interstellar Medium*, Eds. Y. Kondo, F. Bruhweiler, and B. Savage

Bravo, S., Stewart, G. A., and Blanco-Cano, X. (1998). The Varying Multipolar Structure of the Sun's Magnetic Field and the Evolution of the Solar Magnetosphere Through the Solar Cycle. *Sol. Phys.*, 179:223–235.

Burlaga, L. F., Ness, N. F., Acuña, M. H., Lepping, R. P., Connerney, J. E. P., Stone, E. C., and McDonald, F. B. (2005). Crossing the Termination Shock into the Heliosheath: Magnetic Fields. *Science*, 309:2027–2029.

Butler, D. M., Newman, M. J., and Talbot, R. J. (1978). Interstellar Cloud Material - Contribution to Planetary Atmospheres. *Science*, 201:522–525.

Caballero-Lopez, R. A., Moraal, H., McCracken, K. G., and McDonald, F. B. (2004). The Heliospheric Magnetic Field from 850 to 2000 AD Inferred from ^{10}Be Records. *J. Geophys. Res. (Space Physics)*, 109:12102.

Christl, M., Mangini, A., Holzkämper, S., and Spötl, C. (2004). Evidence for a Link between the Flux of Galactic Cosmic Rays and Earth's Climate during the past 200,000 years. *J. Atmos. Terres. Phys.*, 66:313–322.

Decker, R. B., Krimigis, S. M., Roelof, E. C., Hill, M. E., Armstrong, T. P., Gloeckler, G., Hamilton, D. C., and Lanzerotti, L. J. (2005). Voyager 1 in the Foreshock, Termination Shock, and Heliosheath. *Science*, 309:2020–2024.

Fahr, H. J. (1968). On the Influence of Neutral Interstellar Matter on the Upper Atmosphere. *Astrophys. & Space Sci.*, 2:474.

Fahr, H.-J. (2004). Global Structure of the Heliosphere and Interaction with the Local Interstellar Medium: Three Decades of Growing Knowledge. *Adv. Space Res.*, 34:3–13.

Frisch, P. and York, D. G. (1986). Interstellar clouds near the Sun. In *The Galaxy and the Solar System*, pages 83–100. Eds. R. Smoluchowski, J. Bahcall, and M. Matthews (University of Arizona Press).

Frisch, P. C. (1981). The Nearby Interstellar Medium. *Nature*, 293:377–379.

Frisch, P. C. (1993). G-star Astropauses - A Test for Interstellar Pressure. *Astrophys. J.*, 407:198–206.

Frisch, P. C. (1997). Journey of the Sun. *http://xxx.lanl.gov/*, astroph/9705231.

Frisch, P. C. (1998). Interstellar Matter and the Boundary Conditions of the Heliosphere. *Space Sci. Rev.*, 86:107–126.

Frisch, P. C. (2000). The Galactic Environment of the Sun. *American Scientist*, 88:52–59.

Frisch, P. C. (2005). Tentative Identification of Interstellar Dust in the Magnetic Wall of the Heliosphere. *Astrophys. J. Let.*, 632:L143–L146.

Frisch, P. C., Dorschner, J. M., Geiss, J., Greenberg, J. M., Grün, E., Landgraf, M., Hoppe, P., Jones, A. P., Krätschmer, W., Linde, T. J., Morfill, G. E., Reach, W., Slavin, J. D., Svestka, J., Witt, A. N., and Zank, G. P. (1999). Dust in the Local Interstellar Wind. *Astrophys. J.*, 525:492–516.

Gosling, J. T. (1996). Corotating and Transient Solar Wind Flows in Three Dimensions. *Ann. Rev. Astron. & Astrophys.*, 34:35–74.

Gruen, E. and Landgraf, M. (2000). Collisional Consequences of Big Interstellar Grains. *J. Geophys. Res.*, 105:10291–10298.

Gurnett, D. A. and Kurth, W. S. (2005). Electron Plasma Oscillations Upstream of the Solar Wind Termination Shock. *Science*, 309:2025–2027.

Halley, E. (1724). Some Considerations about the Cause of the Universal Deluge, Laid before the Royal Society, on the 12th of December 1694. By Dr. Edmond Halley, R. S. S. *Philosophical Transactions Series I*, 33:118–123.

Heiles, C. and Troland, T. H. (2003). The Millennium Arecibo 21 Centimeter Absorption-Line Survey. I. Techniques and Gaussian Fits. *Astrophys. J. Supl.*, 145:329–354.

Herschel, W. (1795). On the Nature and Construction of the Sun and Fixed Stars. By William Herschel, LL.D. F. R. S. *Philosophical Transactions Series I*, 85:46–72.

Herschel, W. (1801). Observations Tending to Investigate the Nature of the Sun, in Order to Find the Causes or Symptoms of Its Variable Emission of Light and Heat; With Remarks on the Use That May Possibly Be Drawn from Solar Observations. *Philosophical Transactions Series I*, 91:265–318.

Holzer, T. E. (1989). Interaction between the Solar Wind and the Interstellar Medium. *Ann. Rev. Astron. & Astrophys.*, 27:199–234.

Jokipii, J. R. (2004). Transport of Cosmic Rays in the Heliosphere. In *AIP Conf. Proc. 719: Physics of the Outer Heliosphere*, pages 249–259.

Lallement, R., Quémerais, E., Bertaux, J. L., Ferron, S., Koutroumpa, D., and Pellinen, R. (2005). Deflection of the Interstellar Neutral Hydrogen Flow Across the Heliospheric Interface. *Science*, 307:1447–1449.

Landgraf, M. (2000). Modeling the Motion and Distribution of Interstellar Dust inside the Heliosphere. *J. Geophys. Res.*, 105:10303–10316.

Langley, S. P. (1876). Measurement of the Direct effect of Sun-Spots on Terrestrial Climates. *Mon. Not. Roy. Astron. Soc.*, 37:5.

Linde, T. J. (1998). *A Three-Dimensional Adaptive Multifluid MHD Model of the Heliosphere*. PhD thesis, Univ. of Michigan, Ann Arbor. *http://hpcc.engin.umich.edu/CFD/publications*.

Lockwood, J. A. and Webber, W. R. (2005). Intensities of Galactic Cosmic Rays of 1.5 GeV Rigidity versus Heliospheric Current Sheet Tilt. *J. Geophys. Res. (Space Physics)*, 110:4102.

Lopate, C. (2005). Fifty Years of Ground Level Solar Particle Event Observations, In *Solar Eruptions and Energetic Particles. J. Geophys. Res.*

Möbius, E., Bzowski, M., Chalov, S., Fahr, H.-J., Gloeckler, G., Izmodenov, V., Kallenbach, R., Lallement, R., McMullin, D., Noda, H., Oka, M., Pauluhn, A., Raymond, J., Ruciński, D., Skoug, R., Terasawa, T., Thompson, W., Vallerga, J., von Steiger, R., and Witte, M. (2004). Synopsis of the Interstellar He Parameters from Combined Neutral Gas, Pickup Ion and UV Scattering Observations. *Astron. & Astrophys.*, 426:897–907.

McCracken, K. G., McDonald, F. B., Beer, J., Raisbeck, G., and Yiou, F. (2004). A Phenomenological Study of the Long-Term Cosmic Ray Modulation, 850-1958 AD. *J. Geophys. Res. (Space Physics)*, 109:12103.

McCrea, W. H. (1975). Ice Ages and the Galaxy. *Nature*, 255:607–609.

McKay, C. P. and Thomas, G. E. (1978). Consequences of a Past Encounter of the Earth with an Interstellar Cloud. *Geophys. Res. Lett.*, 5:215–218.

Meisel, D. D., Janches, D., and Mathews, J. D. (2002). Extrasolar Micrometeors Radiating from the Vicinity of the Local Interstellar Bubble. *Astrophys. J.*, 567:323–341.

Parker, E. N. (1996). Solar Variability and Terrestrial Climate. In *The Sun and Beyond*, pages 117.

Ratkiewicz, R., Barnes, A., Molvik, G. A., Spreiter, J. R., Stahara, S. S., Vinokur, M., and Venkateswaran, S. (1998). Effect of Varying Strength and Orientation of Local Interstellar Magnetic Field on Configuration of Exterior Heliosphere. *Astron. & Astrophys.*, 335:363–369.

Richardson, J. D., Paularena, K. I., Lazarus, A. J., and Belcher, J. W. (1995). Radial Evolution of the Solar Wind from IMP 8 to Voyager 2. *Geophys. Res. Lett.*, 22:325–328.

Roble, R. G. and Hays, P. B. (1979). A Quasi-Static Model of Global Atmospheric Electricity. II - Electrical Coupling between the Upper and Lower Atmosphere. *J. Geophys. Res.*, 84:7247–7256.

Scherer, K. and Fahr, H. J. (2003). Breathing of Heliospheric Structures Triggered by the Solar-cycle Activity. *Annales Geophysicae*, 21:1303–1313.

Shapley, H. (1921). Note on a Possible Factor in Changes of Geological Climate. *J. Geology*, 29.

Simpson, J. A. (2001). *The Cosmic Radiation*, pages 117–152. Century of Space Science, Volume I (Kluwer Academic Publishers).

Smith, E. J., Marsden, R. G., Balogh, A., Gloeckler, G., Geiss, J., McComas, D. J., McKibben, R. B., MacDowall, R. J., Lanzerotti, L. J., Krupp, N., Krueger, H., and Landgraf, M. (2003). The Sun and Heliosphere at Solar Maximum. *Science*, 302:1165–1169.

Sonett, C. P. (1992). A Supernova Shock Ensemble Model using Vostok ^{10}Be Radioactivity. *Radiocarbon*, 34:239–245.

Stone, E. C., Cummings, A. C., McDonald, F. B., Heikkila, B. C., Lal, N., and Webber, W. R. (2005). Voyager 1 Explores the Termination Shock Region and the Heliosheath Beyond. *Science*, 309:2017–2020.

Talbot, R. J. and Newman, M. J. (1977). Encounters Between Stars and Dense Interstellar Clouds. *Astrophys. J. Supl.*, 34:295–308.

Thomas, G. E. (1978). The Interstellar Wind and Its Influence on the Interplanetary Environment. *Annual Review of Earth and Planetary Sciences*, 6:173–204.

Vallée, J. P. (2005). The Spiral Arms and Interarm Separation of the Milky Way: An Updated Statistical Study. *Astron. J.*, 130:569–575.

Wang, C. and Richardson, J. D. (2005). Dynamic Processes in the Outer Heliosphere: Voyager Observations and Models. *Adv. Space Res.*, 35:2102–2105.

Whang, Y. C. (2004). Solar Cycle Variation of the Termination Shock. In *AIP Conf. Proc. 719: Physics of the Outer Heliosphere*, pages 22–27.

Willis, D. M. (1978). Atmospheric Water Vapour of Extraterrestrial Origin - Possible Role in Sun-Weather Relationships. *J. Atmos. Terres. Phys.*, 40: 513–528.

Witte, M. (2004). Kinetic Parameters of Interstellar Neutral Helium. Review of Results obtained During One Solar Cycle with the Ulysses/GAS-Instrument. *Astron. & Astrophys.*, 426:835–844.

Wood, B. E., Redfield, S., Linsky, J. L., Müller, H.-R., and Zank, G. P. (2005). Stellar Lyα Emission Lines in the Hubble Space Telescope Archive. *Astrophys. J. Supl.*, 159:118–140.

Zank, G. P. and Frisch, P. C. (1999). Consequences of a Change in the Galactic environment of the Sun. *Astrophys. J.*, 518:965–973.

Zank, G. P. and Müller, H.-R. (2003). The Dynamical Heliosphere. *J. Geophys. Res.(Space Physics)*, 108:7–1.

Chapter 2

HELIOSPHERIC VARIATION IN RESPONSE TO CHANGING INTERSTELLAR ENVIRONMENTS

Gary P. Zank
Institute of Geophysics and Planetary Physics, University of California, Riverside, CA 92521, USA
zank@ucr.edu

Hans-R. Müller
Department of Physics and Astronomy, Dartmouth College, Hanover, NH 03755, USA
hans.mueller@dartmouth.edu

Vladimir Florinski
Institute of Geophysics and Planetary Physics, University of California, Riverside, CA 92521, USA
vflorins@ucr.edu

Priscilla C. Frisch
Department of Astronomy and Astrophysics, University of Chicago, Chicago, IL 60637, USA
frisch@oddjob.uchicago.edu

Abstract While unchanging on human timescales, the interstellar environment has been and will be very different from current conditions. Historically, the heliosphere and solar system have been in regions of the galaxy that were much hotter, or contained higher concentrations of neutral hydrogen, or with interstellar clouds with higher or lower speeds, than is the case today. In this chapter, we describe

P. C. Frisch (ed.), Solar Journey: The Significance of our Galactic Environment for the Heliosphere and Earth, 23–51.

the response of the heliosphere to different interstellar environments, taking into account the basic physics of the coupled neutral hydrogen and plasma self-consistently.

Keywords: Interstellar medium, heliosphere, structure, partially ionized plasma

2.1 Introduction

At the crossroads of space physics and astrophysics, one of the most exciting areas of research today explores the interaction of the solar wind with the interstellar medium. Motivated by the possibility that the aging spacecraft Voyager 1, 2 (and Pioneer 10, 11, although now no longer returning data) might encounter the heliospheric boundaries in the not too distant future, interest in the far outer reaches of our solar system and the local interstellar medium (LISM) has been rekindled. Observations made by the Ulysses spacecraft, as well as old and new spacecraft that make up the flotilla at 1 astronomical unit (AU) (IMP 8, ACE and Wind, etc.), and new kinds of *in situ* and remote observations such as energetic neutral atoms and Lyman-α absorption measurements made by the Hubble Space Telescope, are all providing new and unexpected insights into the solar wind and the local interstellar medium, as well as into the stellar winds of neighboring stars and their local interstellar environment.

The heliosphere is the region of space filled by the expanding solar corona; a vast region extending perhaps 150–180 AU in the direction of the Sun's motion through the interstellar medium and several thousand AU in the opposite direction. Our growing understanding of the physical processes governing the outer heliosphere has led to major discoveries about both our own solar system (the discovery of a "hydrogen wall" bounding our heliosphere, for example), the interstellar medium (the composition of the LISM), and the astrospheres of other stars (such as the discovery of winds blown from stars like our Sun). Since the Sun orbits our galactic center, the solar system has experienced many different interstellar environments, and it is entirely possible that this may have, and may have had, an important impact on the Earth's environment and, ultimately, the habitability of the Earth. Such questions, which have hitherto been largely the preserve of science fiction, have assumed increasing importance with the discovery of extrasolar planets, and inevitably one wonders if some may harbor life of some form or another. The permissibility of this question recognizes the fundamental role the galaxy plays in shaping our own heliosphere, and acknowledges that our heliosphere may serve as a template for understanding stellar and extrasolar planetary systems (Frisch, 1993).

At a more prosaic level, the mutual interaction of large-scale and micro-scale plasma processes is a fundamental ingredient of space physics. In the context of the outer heliosphere, spacecraft such as Voyager, Ulysses and ACE have provided new and important insights into the coupling between the solar wind and the LISM, mediated by interstellar neutrals flowing into the heliosphere. Large-scale, sometimes termed meso- or macro-scale, phenomena contribute to the gross morphology of the heliosphere, in either a time-dependent or a steady-state sense. The solar wind, interplanetary magnetic field (IMF), interplanetary shocks, streams and stream-interaction regions, and heliospheric boundary structures are all examples of macro-/meso-scale structures, whereas turbulence, waves, particle scattering, magnetic field line wandering, dissipation, etc. are micro-scale processes. The mutual feedback between these disparate scales serves to determine almost all aspects of heliospheric physics, and this is particularly true in the context of the interaction of the galaxy with our heliosphere. The complex coupling between the solar wind and the interstellar medium may be the quintessential example of multi-scale, regional, and disparate populations, as neutral atoms, solar wind plasma, pickup ions, and anomalous and galactic cosmic rays, exchange energy, charge, and momentum across complex boundaries. Thus, the study of the solar wind interaction with the LISM is of fundamental interest and importance to space physics, and the ideas developed in this field are being applied to other areas of astrophysics. At a practical level, this interest has been strongly fostered by the Voyager Extended Mission—the last remaining operational spacecraft in the very distant heliosphere, and our only current opportunity to explore the boundaries of the heliosphere *in situ*.

Although considerable progress has been made in the development of theoretical models and much data has been returned by the outer heliospheric spacecraft (Voyager 1, 2; Pioneer 10, 11; Ulysses), many of the key local interstellar parameters are not precisely known.

The interstellar medium (ISM) is characterized by change. Over the course of millions of years, stars stir up the interstellar medium via their radiation, stellar winds and occasional death as supernovae. Interstellar clouds form and are dissipated and the interstellar medium undergoes considerable and constant restructuring. During star formation cavities are created within molecular clouds. If the star is sufficiently massive, its explosive death as a supernova may drive shock waves into the gas surrounding the stellar cavity, creating regions of compressed gas in the form of supershells. The supershell gas may break out of the ambient molecular cloud and drift into low-density regions of the interstellar medium. The gas within these clouds is typically partially ionized, the result of stellar radiation, charged and neutral particle collisions, and radiation from supernova shock heated gas (Spitzer, 1978; Frisch, 1995). As the Sun moves through this dynamically evolving medium, the heliosphere experiences ISM density variations of over 9 orders of magnitude, temperature variations of 6 orders of magnitude, fractional ionizations that vary from $\sim 10^{-4}$ to 1, and relative Sun-ISM velocities that vary by 2–3 orders of

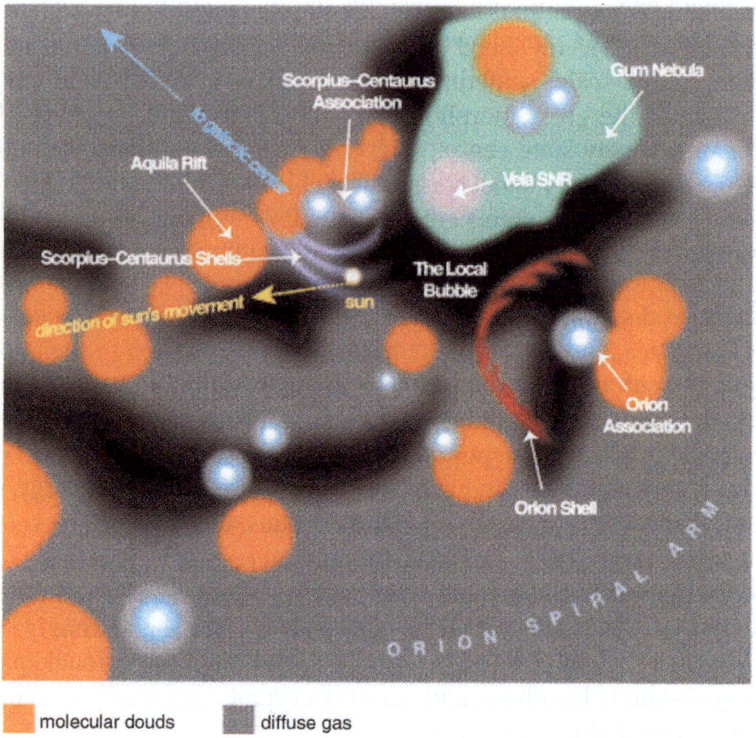

Figure 2.1. A sketch of the galactic environment within 1,500 light-years of the Sun (Frisch, 2000b). The Sun has, over the past several million years, traversed the Local Bubble (black) and it is now embedded in a shell of warm, partly ionized material (violet), the Local Interstellar Cloud, flowing from the Scorpius-Centaurus star-forming region. Cold, dense molecular clouds, such as the Aquila Rift, are depicted in orange. The Vela supernova remnant (SNR, pink) is expected to cool and expand, forming shells of material similar to that surrounding the Sun. The Gum Nebula (green) is primarily ionized hydrogen. Young stellar associations are shown as blue circles with white central dots. The motion of the Sun through the LSR is shown. Fragments of known superbubble shells around the Orion and Scorpius-Centaurus Associations are shown. *Figure copyright, American Scientist*

magnitude. In this chapter we consider the heliosphere response to a few of these variations.

The Sun itself moves through this constantly changing interstellar environment. Relative to nearby stars that define the "Local Standard of Rest" (LSR), the Sun moves at a speed of $\sim$13–20 km/s, and its path is inclined $\sim 25^{\circ}$ to the plane of the galaxy (Chapter 6). The direction of the Sun's path is toward a region in the constellation Hercules near its border with Lyra. The Sun oscillates through the plane of the galaxy with an amplitude of $\sim$230 light-years, crossing the plane every ~33 million years. The Sun, and its galactic neighborhood, orbit the galactic center once every ~250 million years.

The Sun is presently in a tenuous, warm, partially ionized interstellar cloud. The local interstellar cloud, often abbreviated as the LIC, is a mixture of ionized gas, primarily protons and electrons, neutral hydrogen, heavier atoms such as helium, oxygen, neon, nitrogen, carbon, and dust. Approximately 1% of the cloud's mass comprises dust. The elemental composition of the LIC is solar-like, being ~90% hydrogen, ~9.99% helium, with the remaining ~0.01% composed of heavier elements. A relatively crude but typical estimate of the local cloud density and temperature is $n_{\rm H} \sim 0.24$ cm^{-3} for the neutral hydrogen atom number density, $n({\rm e}^-) \sim 0.1$ cm^{-3} for the electron number density, and $T \sim 6,500$ K for the plasma and neutral atom temperature (Lallement, 1996; Slavin and Frisch, 2002; Chapter 6). Of the interstellar particle species, the three principals are the thermal electron-ion plasma, the neutral hydrogen component, and galactic cosmic rays. A measure of their relative importance is the energy density of each species, being approximately 0.38, 0.61, and 0.55 eV cm^{-3} respectively (Frisch, 1995; Ip and Axford, 1985; Florinski et al., 2003b). An estimate for the energy density of the embedded local interstellar field is 0.2 eV cm^{-3}.

The origin of the local interstellar cloud remains unclear. The Sun lies near the edge of both the LIC and Local Bubble, an enormous region of hot, very low-density plasma, ($n({\rm H}^\circ) < 0.0005$ cm^{-3} for neutral H, $n({\rm H}^+) \sim 0.001$–0.005 cm^{-3} for the ionized gas), hot ($T \sim 10^6$ K) gas (Snowden et al., 1998; Lallement, 2004; Chapter 6). The Local Bubble is located within Gould's Belt, a ring of young stars and star-forming regions. The Belt forms a great circle extending from the constellation Orion to Scorpius, and inclined about 20° relative to the galactic plane, illustrated in Figure 2.1.

The Sun has been traveling through the very low-density interior of Gould's Belt for several million years, and has only recently emerged from the Local Bubble into the local ISM (perhaps within the last 1000–250,000 years, see Chapter 6). The LIC itself may well be part of an outflow of material from the Scorpius-Centaurus (Sco-Cen) Association, which is bearing down on the Sun from the galactic center hemisphere (Frisch, 2000a). In the rest frame of the Sun, the upstream direction of the LIC is towards galactic coordinates $l = 3.3^\circ$, $b = +15.9^\circ$ and the relative Sun-LIC velocity is 26.3 km/s (Witte, 2004). In the local standard of rest, the LIC upstream direction is $l = 346^\circ$, $b = -1^\circ$, and the motions of the Sun through space and the LIC are approximately perpendicular.

The Sun may encounter other cloudlets in this flow, with a range of temperatures and ionization, and possibly even denser ISM. The next cloud to surround the Sun is likely to be either the "G" or the "Aql-Oph" cloudlet, both 5 pc from the Sun and with measurable differences in temperature and composition than the LIC (Frisch et al., 2002; Frisch, 2003). Since the nearest stars show about one interstellar absorption component per 1.4–1.6 pc, the observed relative Sun-cloud velocities of $0 - 32$ km/s suggest variations in the galactic environment of the Sun on timescales of $<50,000$ years. By contrast, Talbot and Newman (1977) suggest that higher density ($>10^3$ cm^{-3}) giant molecular clouds will be encountered very infrequently, perhaps only a dozen during the lifetime of the Sun. Thus, the heliosphere may be expected to encounter relatively diffuse interstellar clouds (<10 cm^{-3}) more frequently.

In this chapter, we address heliospheric variations due to encounters with a range of clouds such as expected for the immediate past and future solar location. Different interstellar environments may produce noticeable changes in the interplanetary environment at 1 AU, as indicated by the amount of neutral H, anomalous (ACR), and galactic cosmic rays (GCR) at 1 AU (see the companion Chapter 10 by Florinski and Zank). By using a multi-fluid approach, which is computationally less intensive than corresponding kinetic modeling, we investigate numerically the possible LISM environments (density, temperature, and velocity) the Sun may encounter and may have encountered. Because of the very large parameter space, we consider only a few representative cases here and refer the reader to further studies by Müller et al. (2005).

2.2 Basic physics of the multi-fluid model

Fully three-dimensional kinetic or multifluid models of the interaction of the heliosphere and ISM are computationally challenging and expensive, and it is not currently feasible to study the full parameter space of possible ISM-heliosphere interactions on this basis. Instead, we consider a reduced problem and use 2D multifluid models. Although 2D, the axisymmetric models capture much of the basic ISM-heliosphere physics, from the deceleration of the supersonic solar wind by interstellar neutrals to the formation of the hydrogen wall. Thus, the 2D multifluid models capture the basic morphology of more complex and detailed models and are therefore ideal for large parametric studies of the kind described here. We consider first the basic properties of 2D models, and later discuss the behavior of these models for different ISM boundary conditions.

In general, the heliospheric-LISM plasma environment is composed of three thermodynamically distinct regions: (i) the supersonic solar wind, with a relatively low temperature, large radial speeds, and low densities; (ii) the shock-heated subsonic solar wind with much higher temperatures and densities, and lower flow speeds, and finally (iii) the LISM, where the plasma flow speed and temperature is low but the density is higher than in regions (i) and (ii). As discussed in detail by Zank et al. (1996), each of the thermodynamically distinct regions is a source of a distinct population of neutral atoms produced by charge exchange with the ambient plasma and neutrals entering the region. These three distinct neutral H populations include the "splash" component produced in the fast or supersonic solar wind i.e., fast neutrals, very hot neutrals produced in the inner heliosheath, and the decelerated heated atoms in the hydrogen wall region. Observations of these three populations allow us to test heliospheric models. The self-consistent inclusion of neutral hydrogen in models of the solar wind-LISM interaction is absolutely fundamental to understanding the large-scale structure of the heliosphere.

Two basic classes of models have been developed to describe neutral H in and around the heliosphere: multi-fluid models of varying degrees of sophistication (Pauls et al., 1995; Liewer et al., 1996; Williams et al., 1997; Wang and Belcher, 1998; Pauls and Zank, 1997; Fahr et al., 2000; Florinski et al., 2003b) that treat the neutral atoms as a multi-fluid, and kinetic models that solve the neutral atom kinetic equation, either by a Monte-Carlo technique (Baranov and Malama, 1993, 1995; Izmodenov et al., 1999; Heerikhuisen et al. 2005) or by a particle-mesh method (Lipatov et al., 1998; Müller and Zank, 1999; Müller et al., 2000). The long charge exchange mean free path for neutral hydrogen (up to $\sim$1000 AU in the heliosphere) may mean that the fluid description for neutrals within the heliosphere is not completely justifiable. Multi-fluid and Boltzmann models differ in the detailed predictions that each admits for the neutral atom distribution, and this can lead to 10–15% differences in predicted neutral H densities and temperatures within the heliosphere. Nevertheless, the basic morphological predictions of both models remain the same (Baranov and Malama, 1993, 1995; Pauls et al., 1995; Zank et al., 1996; Müller et al., 2000). These models have been tested using Lyman-α absorption measurements along various sight-lines towards neighboring stars to explain observations of decelerated interstellar neutral H (Gayley et al., 1997; Wood et al., 2000).

The kinetic codes, when coupled self-consistently to the background solar wind and LISM plasma, are computationally intensive, and so both kinetic approaches compromise in the scope of the problems they attack. The Monte-Carlo approach has the potential to yield good neutral atom statistics since it is assumed that the solar wind is inherently steady state. Thus, very long integration times can be used to build up particle statistics. The drawback, however, is that it makes it difficult to undertake a very large parameter study of the kind that Müller et al. (2005) have recently completed. Accordingly, since the multi-fluid models provide a computationally feasible approach to investigating problems in the solar wind, we adopt this approach here.

As discussed by Zank et al. (1996), each of the three thermodynamically distinguishable regions of the heliosphere listed above acts as a source of neutral H atoms whose distribution reflects the character of the plasma distribution in the region. Each of the three neutral atom components is represented by a distinct Maxwellian distribution function appropriate to the characteristics of the source plasma distribution. This allows us to simplify the production and loss terms for each neutral component, and the complete neutral distribution function, the sum of the three components, is a highly non-Maxwellian distribution function. A comparison of the kinetic distribution and the multi-fluid distribution function is given in Zank et al. (2001), and the multi-fluid neutral H distribution is found to reflect the overall broadening of the neutral H distribution function reflected in the kinetic simulation of Müller et al. (2000) and to reproduce the fast or splash component (originating from the supersonic solar wind) rather well. Subject to these assumptions, one obtains immediately a hydrodynamic description for each neutral component (Zank et al., 1996)

$$\frac{\partial \rho_i}{\partial t} + \nabla \cdot (\rho_i \mathbf{u}_i) = Q_{\rho i} \quad (2.1)$$

$$\frac{\partial}{\partial t}(\rho_i \mathbf{u}_i) + \nabla \cdot [\rho_i \mathbf{u}_i \mathbf{u}_i + p_i \mathbf{I}] = \mathbf{Q}_{mi} \quad (2.2)$$

$$\frac{\partial}{\partial t}\left(\frac{1}{2}\rho_i u_i^2 + \frac{p_i}{\gamma - 1}\right) + \nabla \cdot \left[\frac{1}{2}\rho_i u_i^2 \mathbf{u}_i + \frac{\gamma}{\gamma - 1}\mathbf{u}_i p_i\right] = Q_{ei} \quad (2.3)$$

The source terms Q are listed in Pauls et al. (1995) and Zank et al. (1996). The subscript i above refers to the neutral component of interest ($i = 1, 2, 3$ corresponding to each of the regions listed above), ρ_i, $\mathbf{u}_i$, and p_i denote the neutral component i density, velocity, and isotropic pressure respectively, $\mathbf{I}$ the unit tensor, and γ ($= 5/3$) the adiabatic index.

The heliospheric and LISM plasma is described similarly by the hydrodynamic equations

$$\frac{\partial \rho}{\partial t} + \nabla \cdot (\rho \mathbf{u}) = Q_{\rho p} \quad (2.4)$$

$$\frac{\partial}{\partial t}(\rho \mathbf{u}) + \nabla \cdot [\rho \mathbf{u}\mathbf{u} + p\mathbf{I}] = \mathbf{Q}_{mp} \quad (2.5)$$

$$\frac{\partial}{\partial t}\left(\frac{1}{2}\rho u^2 + \frac{p}{\gamma - 1}\right) + \nabla \cdot \left[\frac{1}{2}\rho u^2 \mathbf{u} + \frac{\gamma}{\gamma - 1}\mathbf{u} p\right] = Q_{ep} \quad (2.6)$$

where $Q_{(\rho,m,e)p}$ denote the source terms for plasma density, momentum, and energy. They, too, are listed in Pauls et al. (1995) and Zank et al. (1996). The remaining symbols enjoy their usual meanings. The proton and electron temperatures are assumed equal in the multi-fluid models.

The coupled multi-fluid system of equations (2.1) – (2.6) is solved numerically in the (r, θ) plane, which is aligned so that $\theta = 0°$ is in the direction of solar system motion relative to the LISM. Details regarding the boundary conditions and grid can be found in Zank et al. (1996). The $r \times \theta$ computational domain is [1 AU, 1000 AU] × [0°, 180°].

Multiple models for the LISM will be considered in the following sections, but they fall into two basic categories. These correspond to interstellar conditions that give rise to either a 2-shock model (i.e., supersonic motion of the heliosphere through the LISM) or a 1-shock model (subsonic relative motion). In the two shock model, one of the shocks corresponds to a bow shock (BS) associated with the supersonic relative motion of the heliosphere and the interstellar medium, and the other shock is the termination shock (TS) at which the supersonic solar wind flow is decelerated. The 1-shock model obviously contains only a termination shock. In general, as discussed by Zank et al. (1996), galactic cosmic rays, dust, the local interstellar magnetic field, etc. all contribute to the total LISM pressure against which the solar wind expands. More precisely, one can define a parameter α that describes the effective "temperature" (e.g., Holzer, 1989; Suess, 1990) for the LISM by incorporating the added contribution of cosmic rays, dust, etc. If $p_{\rm tot}$ denotes the total pressure, then α is defined implicitly by

$$\begin{aligned} p_{\rm tot} &= k_B\left(n_p T_p + n_e T_e\right) + p_{\rm CR} + p_{\rm mag} + p_{\rm dust} + p_{\rm turb} + \cdots \\ &\simeq \alpha n_p k_B T_p \end{aligned} \tag{2.7}$$

where the subscripts refer to protons, electrons, cosmic rays, magnetic fields, dust, and turbulence, and p is the pressure associated with these quantities; T is temperature, and k_B is the Boltzmann's constant. For a plasma comprising non-relativistic protons and electrons, $\alpha = 2$ if dust pressure is negligible. For a subsonic LISM where cosmic rays, for example, may contribute to the total pressure, $\alpha > 2$. In the solar wind, $\alpha = 2$ always. In the simulations described here, we simply assign a total interstellar pressure to the system of equations and do not try to identify individual contributions from various components of the ISM. However, the comments are a little simplistic in the manner in which the magnetic field pressure and cosmic ray pressure are included. Depending on energy, cosmic rays do not necessarily couple strongly to the background plasma and often do not experience significant compression at the shocks (Florinski et al., 2003b). Also, MHD models show that even with a large $\mathbf{B}$, a bow shock can form (Florinski et al., 2004). Nonetheless, we proceed on the basis of the simpler representation (2.7).

By way of illustration, and to describe the canonical heliosphere of today, we consider a 2-shock model for which the incident interstellar plasma flows supersonically onto the heliospheric obstacle (see reviews by Holzer, 1989;

Table 2.1. Simulation parameters used to determine the canonical (i.e., at the current time) 2-shock steady state heliosphere.

	LISM H^+	LISM H°	Solar wind (1 AU)
n (cm^{-3})	0.10	0.14	5
u (km/s)	−26	−26	400
T (K)	8000	8000	10^5

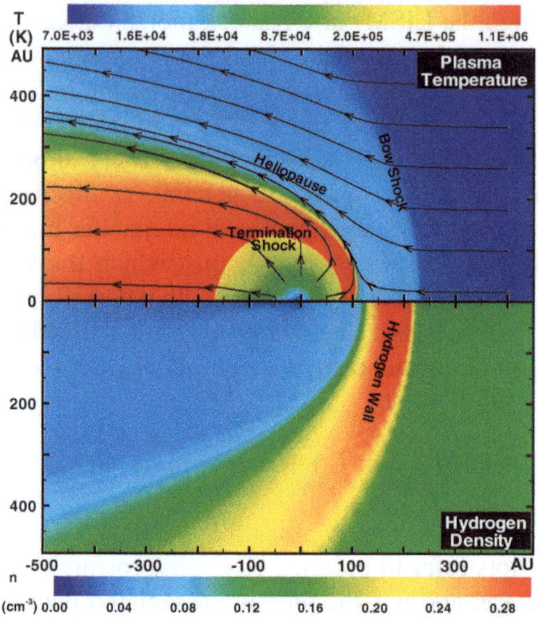

Figure 2.2. The 2D steady-state, 2-shock heliosphere showing, top plot, the temperature distribution in color of the solar wind and interstellar plasma and, bottom plot, the density distribution of neutral hydrogen. The plasma boundaries, termination shock, heliopause, and bow shock are labeled, and the wall of neutral hydrogen is also identified. The solid lines of the top plot show the streamlines of the plasma. The plasma temperature is plotted logarithmically and the neutral density linearly. The distances along the x and y axes are measured in astronomical units (AU).

Suess, 1990; and Zank, 1999 for a theoretical discussion of the physics involved in the interaction). For this case, we adopt $\alpha = 2$. The simulation parameters for the steady-state background state are given in Table 2.1. A color plot of the steady-state 2-shock heliosphere is given in Figure 2.2. The top figure is a 2D plot of the plasma temperature showing the boundaries (labeled): the termination shock, the heliopause (HP), and the bow shock. The extent of the heliosphere in the nose direction (the "stagnation axis") is ~80 AU to the termination shock, ~110 AU to the heliopause, and ~230 AU to the bow shock. The heliopause is a tangential discontinuity (at least in the present hydrodynamic formulation) that separates interplanetary and interstellar material. The solid lines depict the streamlines of both the interstellar plasma and the solar wind plasma. It should be noted that since we use a one-fluid description for the plasma, i.e., we do not model interstellar pickup ions and solar wind plasma separately, the temperature in the outer region of the supersonic solar wind appears hotter in the figure than would be the case if the thermal solar wind plasma were considered separately.

2.3 Possible interstellar environments

In traversing the galaxy, the heliosphere will, and has, encountered many different interstellar environments. For the immediate future and past, we are most interested in the inhomogeneous character of the LISM within a distance of ~30 pc. The properties of the ISM reported for sightlines to stars within 30 pc include clouds with column densities log $N(\mathrm{H}^{\circ})\sim 17.5-19.0$ cm^{-2}, thermal temperatures of $T \leq 15,000$ K, and relative Sun-cloud velocities of -45 to 35 km s^{-1} (Chapter 6); Redfield and Linsky, 2004; Frisch et al., 2002). The ISM within 30 pc is referred to as the Cluster of Local Interstellar Clouds (CLIC), or as the Local Fluff.

In the discussions below, we highlight boundary conditions for the heliosphere that correspond to five types of interstellar clouds that the Sun is likely to have encountered, or will encounter, over time scales of ± 3 Myrs (Table 2.2). The first of these, the hot plasma of the Local Bubble, is inferred to fill the interior of the region surrounding the Sun where very low densities of ISM are seen, this based on the spectra of soft X-ray emission for energies less than 0.5 keV. This very low density ISM will have surrounded the Sun before it entered the present cloud system. The second cloud type, the Local Interstellar Cloud, LIC, is the tenuous partially ionized cloud presently surrounding the solar system, and which has low column densities $N(\mathrm{H}^{\circ}) < 10^{18}$ cm^{-2} so that opacity effects are highly significant for H ionization. A rapidly moving high speed cloud, ~125 km s^{-1}, corresponds to what might be expected for a radiative

Table 2.2. Summary of interstellar environments discussed here. $n_{\rm tot} = n(\mathrm{H}^+) + n(\mathrm{H}^{\circ})$ is the total hydrogen density, B is the magnetic field magnitude, $\eta = n(\mathrm{H}^+)/n_{\rm tot}$ is the ionization ratio, and u_∞ is the cloud speed relative to the Sun.

Model	$n_{\rm tot}$, cm^{-3}	η	T, K	u_∞, km/s	B, μG
Local Bubble (LB)	0.005	1.0	1.2×10^6	12.5	0.0
Local Interstellar Cloud (LIC)	0.21	0.33	7000	25	0.0
High Speed Cloud (HS)	0.24	0.42	8000	127	0.0
Diffuse Cloud (DC)	10	0.0	200	25	0.0
Strong magnetic field (MF)	0.3	0.33	7000	25	4.3

shock moving through space. Similar features are associated with superbubble shells in the Orion-Eridanus region (Welty et al., 1999). A superbubble shell shock propagates at a velocity $V_s \sim n_o^{-0.2}$, where $n_o^{-0.2}$ is the preshock density (Frisch, 1995), and will thus travel relatively unimpeded through the Local Bubble interior. As a consequence, the heliosphere is more likely to have been buffeted by rapidly moving shocks when it was immersed in the Local Bubble interior, than during the present time. We have also included a diffuse cloud with $n(H^o) \sim 10$ cm^{-3}, which is a typical space density for diffuse clouds observed in the UV (Welty et al., 1999). In addition, if the G-cloud is homogeneous, Ca^+ and/or H^o data for the stars α Cen, α Oph, and 36 Oph indicates such a cloud may be within $\sim$1 pc of the Sun in the upstream direction (Frisch, 2003, Chapter 6). The final ISM type we highlight is a cloud with a moderately strong interstellar magnetic field, $B \sim 4.3$ μG. A field of this strength would satisfy the requirements for equipartition of energy at the time the Sun was embedded in the Local Bubble plasma, and is also consistent with estimates of the uniform component of the local interstellar magnetic field, which is strengthened in interarm regions such as around the Sun (Chapter 6).

A much more subtle environmental variation follows from the low opacity encountered by 13.6 eV photons, which ionize H, in the surrounding cloud system where low $N(H^o)$ values prevail. As a result, ionization levels vary locally with depth into the cloud (this is discussed more fully in Chapter 6), and the boundary conditions of the heliosphere can vary with time even though the Sun remains in the same cloud. Radiative transfer models predict equilibrium conditions for low column density clouds, $N(H^o) < 10^{18}$ cm^{-2}, and show that ionization levels and densities vary by the amounts $n(H^o) = 0.16$–0.28 cm^{-3}, and $n(H^+)/(n(H^o)+n(H^+)) = 0.19$ to 0.35, or perhaps even more depending on the exact cloud parameters. The result is that the boundary conditions of the heliosphere will vary simply from the travel of the Sun through low column density ISM, based on opacity considerations alone. For instance, in the downwind direction of the LIC, we expect ionization variations of 10–20% between the Sun and cloud surface.

The discussions in Müller et al. (2005) characterize the recent, and future, significant variations in the boundary conditions of the heliosphere, that are expected from the relative motions of the Sun and nearby ISM. The prominent characteristics of the ISM within 30 pc, are that the CLIC is inhomogeneous, it flows past the Sun with a bulk relative velocity of $\sim -28.1 \pm 4.6$ km s^{-1}, and the Sun is located in the downwind portion of the CLIC. The inhomogeneity of the CLIC is well established by variations in the velocity and temperatures found for ISM within 30 pc (Redfield and Linsky, 2004; Frisch et al., 2002). The flow vector, although slightly biased by the distribution of sample stars on the sky, rather clearly indicates that the CLIC flows through the LSR. After removal of solar apex motion, the bulk velocity of the CLIC transforms into the LSR velocity $\sim -19.4 \pm 4.6$ km s^{-1}from the upwind direction $l \sim 331.4°$, $b \sim -4.9°$ (this value assumes the standard solar apex motion, Chapter 6). The 1σ 4.6 km s^{-1} dispersion of cloudlet velocities around the bulk flow velocity corresponds to macroturbulence, which differs from the nonthermal microturbulence that parameterizes the broadening of individual absorption lines. For typical CLIC temperatures of $T \sim 10^3$–10^4 K, the isothermal sound speed is $V_s \sim$ 3–9 km s^{-1}. For most of these clouds, a collision with the heliosphere would be supersonic, even though the microturbulence within the cloud is subsonic. The location of the Sun close to the downwind edge of the CLIC is shown by the low column densities seen towards downwind stars, N(H°)$\sim$ $10^{17.3}$–$10^{18.3}$ cm^{-2} (α CMa, α Aur, $d < 13$ pc), versus higher column densities in the opposite upwind direction, N(H°)$> 10^{18.6}$ cm^{-2} (e.g. HD 149499B, 37 pc).

Müller et al. (2005) have estimated the timing of the Sun's entry into, and exit from, the LIC, using H° (and H° proxy) data for nearby stars, combined with values for n(H°) obtained from LIC radiative transfer models (also see Chapter 6). These dates are somewhat sensitive to the morphology of the LIC, as well as the solar apex motion for at least one LIC model. The best estimates for the date when the Sun entered the LIC are less than ~40,000 years ago, or sometime within the past ~46,000 years allowing for the many uncertainties in the problem. One LIC morphology even suggests an entry date as late as ~1,000 years ago. The "exit" from the LIC is set more firmly by observations towards 36 Oph and α Cen, and is within the next ~4,000 years.

Within ~350 pc of the Sun, the ISM exhibits a divergent range of properties. Cloud densities range between 10^{-4} and 10^{+5} cm^{-3} and temperatures between 10 and 10^6 K. The ionization n(e$^-$)/n(H°) varies between 10^{-4} and ~1, depending on whether electrons are supplied by trace elements such as C, or H, or cosmic ray ionization. In general, comparisons between the column densities of CLIC components and the low column density cold (~50 K) gas observed in H° 21-cm absorption, show that cold cloudlets may go undetected in the CLIC, under some conditions. Such features, also seen in UV data toward 23 Ori for instance (Welty et al., 1999), have n(H°)~10–15 cm^{-3} and cloud lengths of ~0.02 pc, so the heliosphere would pass through such a cloud in 1,000–2,000 years. On the other hand, if the CLIC could be viewed from a vantage point close to Merope, for instance, it would resemble the warm neutral material that dominates the ISM in our galactic neighborhood, with N(H°) $\sim 10^{19}$ cm^{-2}, $T \sim 10^4$ K, and a 21 cm emission profile showing a full-width-half-maximum of ~20 km s^{-1}.

Although we do not include magnetic fields in our modeling, with the exception of the strong magnetic field case (see the companion Chapters 3 and 6), the interstellar magnetic field data (reviewed by Crutcher et al., 2003) show that the heliosphere may experience field strengths that vary by over three orders of magnitude as the Sun passes through dense clouds. In the solar neighborhood, B is probably $\leq 3 - 4\,\mu$G, although larger values may be possible (Cox and Helenius, 2003; Florinski et al., 2004). The ratio of the uniform to random magnetic field components increases in interarm regions, evidently because star formation disrupts the magnetic field. The closest example of this is the famous Loop I supernova remnant, which was originally discovered as a radio source emitting intense synchrotron emission, and where the interstellar magnetic field is distorted into a giant ($>90^{\circ}$) diameter loop in the northern hemisphere. In fact, parts of this loop appear to have expanded into the low density region surrounding the Sun. For denser clouds, $>10^3$ cm^{-3}, magnetic field strengths correlate with gas density, but this correlation breaks down at lower densities (such as the current solar environment).

2.4 Possible heliospheric configurations

The discussion in the previous sections of the ISM in the solar neighborhood, and close to the Sun, provides the basis for selecting boundary conditions for modeling. Here we consider in some detail the distinctly different heliospheric morphologies that result from strong variation in the ISM boundary conditions.

Müller et al. (2005) have modeled 28 interstellar parameter sets, for clouds with densities varying from 0.005–15 cm^{-3}, ionizations η ranging from almost zero up to 100%, and relative Sun-cloud velocities of up to $\sim$ 130 km s^{-1}. Most of the assumed cloud types are warm and low density clouds, but the possible velocities vary by an order of magnitude. The upstream distances from the Sun to the TS range from 8 to 290 AU, and those to the HP range from 12 to 410 AU. The bow shock associated with a cool, slow, tenuous LISM is as far as 900 AU away from the Sun. Because the overall system is pressure balanced, the locations (heliocentric distances) of TS, HP, and BS are correlated with each other. This correlation, which appears to be a general property of the supersonic models, is an important result of our parametric study of the interaction between the heliosphere and interstellar clouds with different physical properties, providing as it does very simple estimates for the distances to the various boundaries. The correlations found with the parameter study so far involve $r_{\rm TS}$, the distance of the upwind TS, $r_{\rm HP}$ (heliopause distance), upwind bow shock $r_{\rm BS}$, and the downwind distance $r_{\rm TSd}$ of the TS. The correlations are

$$r_{\rm HP} = (1.42 \pm 0.02)\, r_{\rm TS}\ \ (R^2 = 0.99) \quad (4.1)$$
$$r_{\rm BS} = (2.0 \pm 0.08)\, r_{\rm HP}\ \ (R^2 = 0.95) \quad (4.2)$$
$$r_{\rm TSd} = (2.14 \pm 0.10)\, r_{\rm TS}\ \ (R^2 = 0.98) \quad (4.3)$$

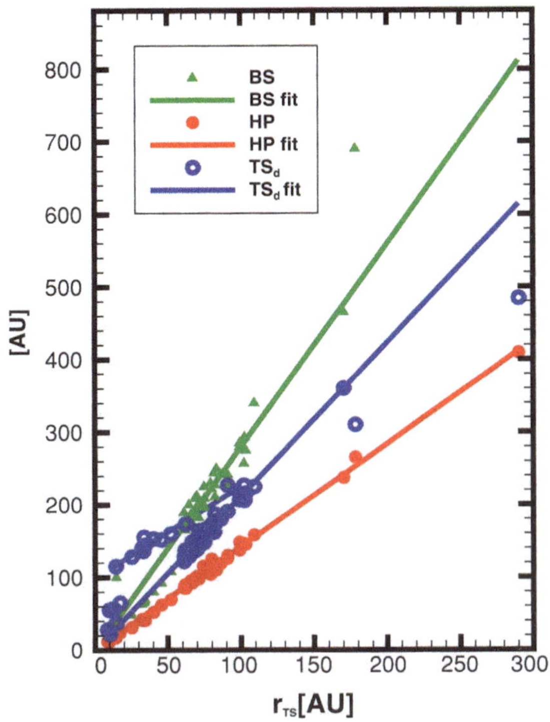

Figure 2.3. Correlation between the upwind termination shock location r_{TS} and other distances, including the upwind heliopause HP (red circles), the upwind bow shock (green triangles), and the tailward TS distance (open blue circles), which are all model results. The empirical linear fits are shown as well, as discussed in the text. Note the aspect ratio of 2:1 (Muller et al., 2005).

obtained with a linear regression analysis after ascribing uncertainties to the location of the boundaries due to grid resolution and heliopause stability (see Fig. 2.3). For high speed cases with a rocket shaped heliosphere, the upwind-downwind asymmetry of the TS changes to $r_{\rm TSd} = (1.21 \pm 0.17)\ r_{\rm TS} + (99.6 \pm 8.7)$AU with $R^2 = 0.96$, where R is the linear correlation coefficient.

It is effectively the upstream solar wind ram pressure that balances the total interstellar pressure (2.7). Hence we define a simple 1D pressure balance, assuming constant solar wind velocity $v_{\rm SW}$ and an r^{-2} dependence of the density for the supersonic solar wind region, as the radial distance determined by

$$r_{\rm pb}/r_1 = \sqrt{p_1/p_{\rm tot}} \tag{4.4}$$

where $p_1 = \rho_1 u_{\rm SW}^2$ is the solar wind ram pressure, and ρ_1 the solar wind density, both taken at $r_1 = 1$ AU. Using the Rankine-Hugoniot shock transition conditions and treating the heliosheath and interstellar flows as incompressible, the TS is calculated to be at

$$\begin{aligned} r_{TS2} &= \sqrt{\frac{2}{\gamma+1}}\, r_{\rm pb} \left[\frac{(\gamma+1)^2}{4\gamma}\left(1-\frac{u_\infty^2}{u_{\rm SW}^2}\right)\right]^{\frac{\gamma/2}{\gamma-1}} \\ &= \frac{16}{5}\frac{1}{\sqrt{3\sqrt{15}}}\, r_{\rm pb} \left(1-\frac{u_\infty^2}{u_{\rm SW}^2}\right)^{5/4} \end{aligned} \tag{4.5}$$

(Suess and Nerney, 1990; Zank, 1999), where u_∞ is the relative velocity between the LISM and heliosphere (Table 2.2). The sum of all plasma pressure contributions should enter into $p_{\rm tot}$ of equation (4.4). If neutral H were tightly coupled to the plasma (i.e. if the mean free paths were very short compared to typical heliospheric length scales), then the neutral H pressure contributions can justifiably be included in $p_{\rm tot}$ as well. However, the neutral-plasma coupling is neither zero nor very tight, such that the solar wind/LISM pressure balance for the 28 heliospheres modeled by Müller et al. (2005) obeys the empirical correlation

$$r_{\rm TS} = (1.352 \pm 0.043)\; r_{\rm TS2} \tag{4.6}$$

(linear correlation coefficient $R^2 = 0.93$). This relation is well suited to predict $r_{\rm TS}$, with the caveat that the neutral-plasma interaction drives the neutrals out of equilibrium, and pressure balance can only be described by an empirical factor of 1.35.

The density $n_{\rm TS}$(H) of neutral hydrogen that crosses the termination shock at the upwind stagnation axis varies over a considerable range in the 28 models, from 0.01 to 0.35 cm^{-3}, with four larger values between 1 and 2 cm^{-3} for the high density models. The filtration ratio f is the neutral density at the TS divided by the interstellar neutral density n(H$^{\rm o}$), well upstream of the BS (to avoid contamination by secondary neutrals), and these relative values vary from 0.1 to 1.0, with a few higher values up to 2.5. The neutral hydrogen at the TS is comprised of original interstellar material, and of slower secondary neutrals, created upwind of the HP, which form the hydrogen wall. After crossing the HP, these neutrals are depleted in the heliosheath by charge exchange that replaces them with neutrals moving mostly in outward directions. For high velocity models, this loss process is inefficient so that filtration ratios larger than one occur. The peak densities in the hydrogen wall range from 1.1 n(H$^{\rm o}$) to massive hydrogen walls of 7 n(H$^{\rm o}$), corresponding to 0.09–105 cm^{-3} in absolute units. The filtration ratios, the absolute TS neutral densities, and the

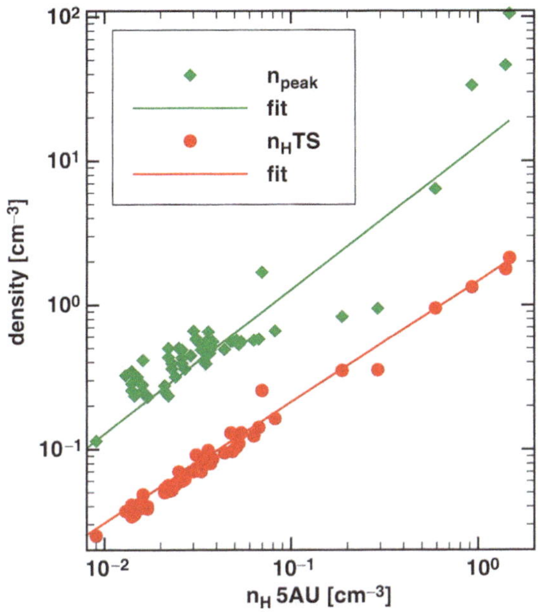

Figure 2.4. Correlations between the neutral H density at 5 AU upwind, and at the TS (circles) and the peak density inside the hydrogen wall, max(n_H(wall)) (diamonds). The solid lines are linear fits (Muller et al., 2006).

peak wall density, correlate well with the neutral density, $n_{5\mathrm{AU}}(\mathrm{H})$, at 5 AU on the upwind stagnation axis. We choose 5 AU as a fixed reference distance with the expectation that photoionization is not yet important at this distance. The correlations are,

$$n_{\mathrm{TS}}(\mathrm{H}) = (1.47 \pm 0.02)\; n_{5\mathrm{AU}}(\mathrm{H})^{0.84\pm0.02} \tag{4.7}$$

$$f = (1.88 \pm 0.04)\; \left[n_{5\mathrm{AU}}(\mathrm{H})/n(\mathrm{H}^0)\right)^{0.83\pm0.02}$$

$$n_{\mathrm{peak}}(\mathrm{H}) = (12.78 \pm 0.66)\; n_{5\mathrm{AU}}(\mathrm{H}) \tag{4.8}$$

(see Fig. 2.4). There is a general trend of the neutral results with the interstellar velocity. The fits are $f = u_\infty/(74\ \mathrm{km\ s^{-1}})$ and $n_{5\mathrm{AU}}(\mathrm{H})/n(\mathrm{H}^0) = u_\infty/(160\ \mathrm{km\ s^{-1}})$, but they are rather poor (uncertainty of a factor of 3).

We do not discuss the very detailed analysis presented in Müller et al. (2005) and present only a subset of the most interesting results. Besides the canonical model already discussed above that reflects current LISM conditions, we consider a hot bubble ISM model, a high-velocity ISM model, and a dense ISM model. In the companion Chapter 10 by Florinski and Zank, a model ISM composed entirely of neutral H is presented and we do not repeat the analysis here.

Example 1: The Hot Local Bubble (LB). In this case, the ISM is hot and nearly completely ionized, making the heliospheric physics considerably simpler since the coupling of neutral H and protons is absent. We adopt ISM parameters of $n(\mathrm{H}^+) = 0.005\ \mathrm{cm}^{-3}$, $n(\mathrm{H}^o) = 0.0$, $u_\infty = 13.4\ \mathrm{km\ s}^{-1}$, and $\log T(\mathrm{K}) = 6.1$. The sound speed in the hot bubble is 190 km s^{-1}, implying that the Sun moves subsonically with respect to the ISM (Mach 0.07). Consequently, no bow shock will form and the ISM will flow adiabatically about the heliospheric obstacle, and the termination shock is highly spherical at a distance of 90 AU from the Sun. This distance is comparable to that of the contemporary heliosphere. The distance to the nose of the heliopause is 300 AU, which makes this sheath very large in comparison to the contemporary heliosphere above. The temperature in the sheath reaches values as high as 2.2×10^6 K. Figure 2.5 shows the plasma temperature profile (top) along the stagnation axis, and the plasma density (bottom), as dashed lines. In the termination shock transition, the density jumps by a factor of 3.8, and the wind speed decreases to 100 km s^{-1}.

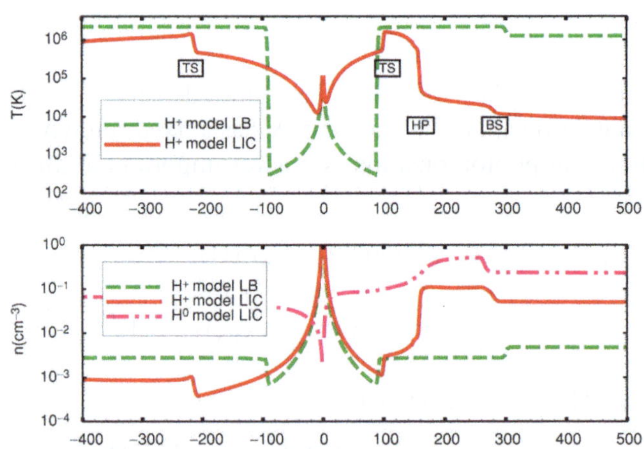

Figure 2.5. One-dimensional profiles along the stagnation axis, with the Sun at center and the LISM coming from right. Top: plasma temperature for the Local Bubble case (dashed), and the model corresponding to contemporary conditions (solid). The heliospheric boundaries are marked in the plot. The bottom panel contains the corresponding densities (hot bubble, dashed; contemporary model, solid; contemporary model neutral H; dash-double dot line) (Müller et al., 2005).

Unlike the contemporary model (which is also shown in Fig. 2.5), the absence of interstellar atoms in the hot ISM implies that no pickup ions and no anomalous cosmic rays are present. Consequently, the outer heliosphere is no longer dominated (by mass) by neutral hydrogen and the physics of the outer heliosphere is completely different from that of today. The thermal component of the solar wind is no longer dominated by the pickup ion pressure, there is no generation of turbulence by the pickup process and the subsequent heating through dissipation of the solar wind, implying quite different cosmic ray modulation characteristics, and there is no deceleration of the solar wind.

The plasma-only model with subsonic interstellar boundary conditions, lends itself to comparison to analytical models of the heliosphere. The analytic models typically assume incompressibility of the flow beyond the TS (Suess and Nerney, 1990; Zank, 1999). For flow around a rigid sphere (the heliopause, in this case), the interstellar velocity on the stagnation axis behaves as $u_\infty(1 - r_{\mathrm{HP}}^3/r^3)$, where r_{HP} is the distance to the HP nose. This analytical description is completely consistent with the numerical simulation. Suess and Nerney (1990) calculate a pressure-balanced TS model and compute the downstream flow streamlines assuming incompressible potential flow. Their TS is located at a distance of 81.8 AU, in reasonable agreement with the simulation TS distance of 90 AU that does not impose the assumptions of incompressibility made by the analytic models.

Example 2: The high-velocity ISM (HS). Müller et al. (2005) consider several models of the heliosphere embedded in a high-velocity ISM environment. The high velocity obviously creates a large ram pressure impinging on the heliospheric obstacle, thereby decreasing the size of the heliosphere in the upwind direction and creating a highly anisotropic structure. Figure 2.6 displays 2D maps of the hydrogen density and plasma temperature of a model for which $u_\infty = 127$ km s^{-1}, showing the heliospheric boundaries and features. However, choosing the same modeling strategy and grid resolution as for the low speed cases leads to numerical errors for the high speed cases, which emphasizes the challenges in modeling the heliosphere for extreme cases of the interstellar parameter space. The grid resolution has been increased in Figure 2.6 to overcome these limitations.

The structure of the global heliosphere now more closely resembles a 2-shock heliospheric structure in the absence of neutral hydrogen (Pauls et al., 1995) in that it has a pronounced bullet- or rocket-shape. As discussed in Baranov and Malama (1993) and Pauls et al. (1995), the decelerated subsonic plasma in the nose region of the heliosheath is accelerated to supersonic speeds in the nozzle-shaped region between the TS and the HP, very much like blunt-body flow. In matching the subsonic heliotail plasma and the supersonic heliosheath plasma, both a shock to decelerate the flow and a tangential

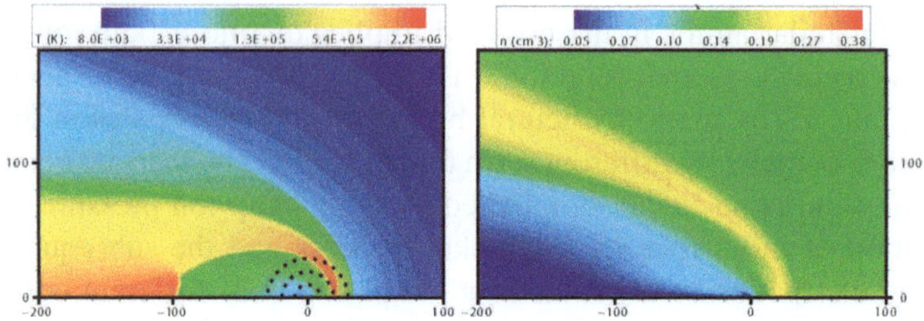

Figure 2.6. The plasma temperature (left) and neutral H density (right) for the high speed case HS, with the Sun at center and the LISM wind impinging from the right. The transition from dark blue to light blue in the plasma temperature is the interstellar bow shock; the dark shades are the hot heliosheath and heliotail. Note the triple point at about (-110, 10), where the heliosheath shock, termination shock, and the tangential discontinuity meet. The orbits of Saturn, Uranus, and Neptune are sketched as dotted lines. In the neutral density (right panel), the hydrogen wall and the depletion of neutrals downwind are clearly visible (Müller et al., 2005).

discontinuity to adjust the density must be inserted, thus creating the characteristic triple point where the heliosheath shock, termination shock, and the tangential discontinuity meet.

For the high-speed case, the TS is highly asymmetric, with a nose distance of 18 AU and a tail distance of 97 AU. The upwind TS shock compression ratio is 3, and the HP and BS are located at 22 AU and 33 AU respectively in the upwind direction. The bow shock is quite strong, with a post shock plasma speed of 33 km s^{-1}, a temperature of 2×10^5 K, and a compression ratio of 3.7.

Because of the large neutral velocity in the post-bow shock region, the neutral mean free path (mfp) for charge exchange is initially $\sim$30 AU, larger than the outer heliosheath, and decreases only gradually as the effective neutral velocity reduces to 50 km s^{-1}. Consequently, the hydrogen wall between the BS and the HP is not very broad. As the TS is so close to the hydrogen wall, neither neutral flow divergence nor charge exchange reduce the neutral density significantly, and the filtration factor is of order unity. These filtration factors of order unity seem possible only when a high interstellar velocity combines with a modest or low density so that the peak hydrogen wall occurs close to the HP, leaving no room for a depletion of neutral H between peak and HP. In the similar sized dense heliosphere (example 3 below), the charge exchange mfps are shorter, the peak hydrogen wall is attained farther away from the HP, and charge exchange upwind close to the HP spreads the H flow and leads to a density decrease already before the H flow crosses the HP (Müller et al., 2005).

Example 3: A diffuse cloud (DC). The possibility of a very dense ISM has been discussed by Zank and Frisch (1999), Florinski et al. (2003a), and most recently by Müller et al. (2001, 2005), Florinski and Zank (Chapter 10), and by Scherer et al. (2002) for a less dense, but strongly ionized cloud. Florinski et al. (2003a) considered a cold ($T = 200$ K) cloud in approximate pressure equilibrium with the surrounding warm low density medium. The hydrogen ionization ratio in the cloud is therefore low, and taken to be zero at the external boundary. Solar photoionization causes the cloud to become partially ionized locally by Ly-α radiation, which is included in the computer model. The model described in Chapter 10 modifies the previous model of Florinski et al. (2003a) by the inclusion of the three neutral hydrogen populations instead of one (Zank et al., 1996; Florinski et al., 2004). The multifluid results for the plasma and neutral distribution are somewhat different from those in Florinski et al. (2003a) because the single fluid approach overestimates neutral atom filtration by the hydrogen wall. In the new model more neutral hydrogen atoms penetrate into the inner heliosphere causing stronger solar wind deceleration by charge transfer. The solar wind is slowed down to 250 km/s by the time it reaches the TS, located at 12 AU (i.e., just past the orbit of Saturn) in the upstream direction. The solar wind is strongly heated by the pickup ions, and the TS is weak with a compression ratio of only 1.6. The HP is located at 24 AU upstream and the thickness of the heliosheath is about equal to the radial extent of the supersonic solar wind region.

Example 4: Strongly magnetized ISM (MF). Both the magnitude and direction of the magnetic field remain to date the poorest known property of the LISM. The uniform component of the mean galactic field in the solar neighborhood was measured to be $|\mathbf{B}| = 1.6\,\mu$G, based on observations of Faraday rotation measures of pulsars within 3 kpc from the Sun (Rand and Kulkarni, 1989). The direction of this field is approximately azimuthal in the galactic frame of reference. The field was also found to undergo random oscillations with an amplitude of $\sim 5.0\,\mu$G on scales of the order of 55 pc, which is comparable to the size of, and may relate to, supernova bubbles. It is not clear if the LIC field is aligned with the mean galactic field. Interestingly, recent Voyager measurements of 2–3 kHz radio emissions exhibit a source distribution with strong clustering in the galactic plane in the direction of the nose of the heliopause (Kurth and Gurnett, 2003). It is believed that these radio emissions are produced by Langmuir waves generated in the dense outer heliosheath following a passage of a large heliospheric disturbance (such as a shock associated with a global merged interaction region or GMIR). The proposed mechanism of wave generation involving pickup-ion-driven electron beams requires strong magnetic fields, such as would be expected from draping of the LISM field around the nose portion of the heliopause (Cairns and Zank, 2002, Mitchell et al., 2004), and the distribution of sources may be interpreted as indicating the direction of the field outside the heliopause. This direction is also consistent with earlier measurements of starlight polarization by interstellar grains (Tinbergen, 1982), which led Frisch (1990) to conclude that the magnetic field in the LIC is in the galactic plane and is inclined by 60° relative to the ecliptic. More recently, it has been shown that this polarization is also consistent with the expected location of the magnetic wall of the heliosphere, formed by interstellar field lines draped over the heliosphere (Frisch, 2005, also see Chapter 3 by Pogorelov and Zank).

A new model of the formation of the LIC (Cox and Helenius, 2003) represents an attempt to explain the apparent pressure imbalance between the LIC and the surrounding hot HII region. In their model, the complex of clouds surrounding the Solar System formed from a part of the shell of the Local Bubble cavity that broke away from the bubble wall, while retaining the strong magnetic field that was compressed by the passage of the supernova blast wave that formed the bubble. This model predicts a rather large magnetic field (5–6 μG) that is aligned with the direction of the interstellar flow. In the absence of such alignment, the excessive magnetic pressure would compress the heliosphere too much and move the termination shock within a distance where it would be already detected by the Voyagers. Conversely, flow-aligned field does not exert extra pressure on the stagnation point of the heliopause, and the termination shock is expected to be located at 90–100 AU, which is in agreement with the accepted value. In any case, it appears entirely possible that the Sun has encountered strongly magnetized interstellar environments during its journey through the Galaxy. Here we only consider axially-symmetric (2D) configurations, while other possibilities for the interstellar field direction are discussed in Chapter 3 of this book.

Parker (1961) was the first to develop a model of the heliospheric interface produced by a strong magnetic field in vacuum. In the absence of MHD effects and interstellar flow, the shape of the heliopause is determined completely by the pressure balance at two points on the interface and consists of a spherical shell with two side channels extending to infinity along the direction of the interstellar magnetic field. Recently, Florinski et al. (2004) used a 2D multifluid numerical model to study the solar wind interaction with a moving partially ionized interstellar cloud with $B = 4.3\,\mu$G (almost three times the "canonical" value) and aligned with interstellar flow velocity vector. The interstellar plasma beta $\beta = 4\pi\gamma p/B^2$ (the ratio of thermal to magnetic pressures) was 0.47, and the interstellar flow was supersonic, but sub-Alfvénic. Their most important results are summarized in Figure 2.7 (left panel). Here the TS is located at 97 AU, the HP at $\sim$140 AU, and the BS at 260 AU in the apex direction. The front portion of the HP displays an oscillatory behavior as a

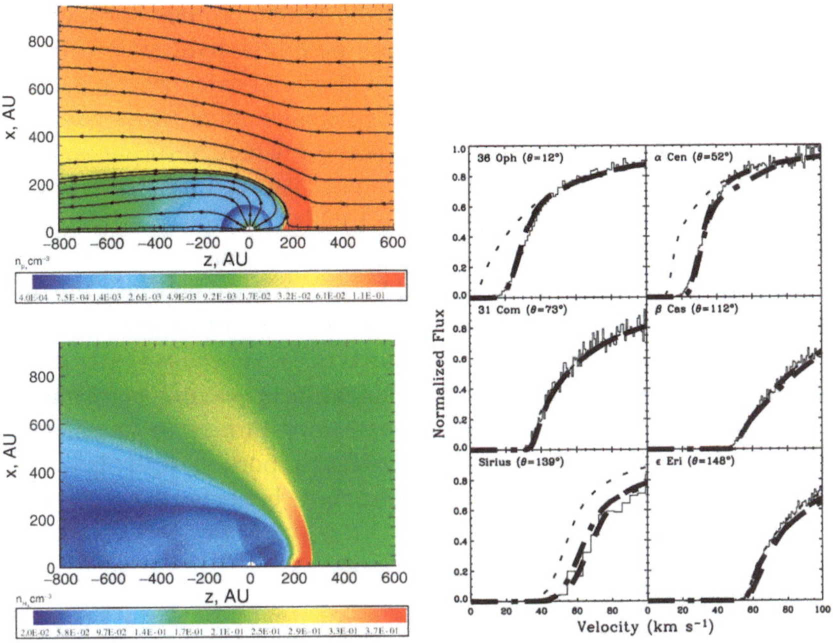

Figure 2.7. Left: Color plots of the plasma density and velocity lines (top panel) and neutral H density (bottom panel) in the presence of a strong interstellar magnetic field that is aligned with the interstellar flow velocity. Right: Plots showing the red side of the H I Lyα line observed for six stars that sample different angles (θ) relative to the upwind direction of the interstellar flow into the heliosphere. The solid lines show Hubble Space Telescope data and the dotted lines show interstellar absorption only (Wood et al., 2000). The dashed lines show the model results for $B = 4.3\,\mu$G and the dash-dotted lines show the results calculated for the reference case with $B = 0$ (Florinski et al., 2004).

consequence of a Rayleigh–Taylor instability driven by charge exchange between the interstellar neutrals and the plasma flow in the heliosheath (Florinski et al., 2005). As shown in Florinski et al. (2004), the bow shock in this case is a slow magnetosonic shock, which means that the magnetic field and flow direction rotate towards the symmetry axis on crossing the shock from the LISM side. The plasma subsequently turns away again from the symmetry axis as it flows around the heliopause. This is quite different from the "standard" model featuring a fast magnetosonic shock that directly deflects the wind away from the axis. In other respects, the general structure of the interface is very similar to the unmagnetized case, and it appears that even the strong interstellar field does not have a particularly significant effect on the heliosphere.

Figure 2.7 (right panel) shows Lyα saturated Voight profiles, for six stars in the solar neighborhood, produced by absorption by the hydrogen wall. Both observed (Hubble Space Telescope data) and model-calculated profiles are shown. The strong field case shows very good agreement between the data and calculations along all lines of sight, especially where absorption is dominated by the hydrogen wall (36 Oph and α Cen). This result indicates that the presence of a strong magnetized field in the LIC is a real possibility. Calculations performed without a magnetic field tend to overpredict absorption toward α Cen and underpredict absorption toward Sirius. Overall, however, the difference in the amount of absorption calculated for $B = 4.3\ \mu$G and $B = 0$ is generally small, and it appears that Lyα absorption is not very sensitive to changes in $\mathbf{B}$ within the parameter range explored. Including an even stronger magnetic field (5–6 μG) may result in a heliosheath extending further upstream, as a result of magnetic tension pulling on the apex-facing portion of the heliopause, with a dominating absorption contribution from the heliosheath neutrals.

2.5 Conclusions

The response of the heliosphere to quite different Galactic environments have been simulated with the use of computationally efficient multifluid models. Although the results discussed here are brief, and do not repeat the complete discussions of the original sources, several general conclusions are apparent.

1 The global morphology of the heliosphere is very strongly determined by the ionization state of the local interstellar medium, being significantly reduced in size by the presence of neutral hydrogen compared to an equivalent heliosphere embedded in a fully ionized ISM. The reduction in size is a consequence of the supersonic solar wind losing momentum and energy to the pickup ion population created from charge exchange with interstellar neutral gas. A very dense partially ionized cloud can reduce the size of the heliosphere so dramatically that the outer planets can find themselves orbiting in the ISM rather than the supersonic solar wind. Correspondingly, the interstellar hydrogen density can be high at the Earth, with possible implications for atmospheric chemistry.

2 Like the ionization state of the LISM, the relative velocity of the Sun-LISM system can play an equally dramatic role in shaping the global morphology of the heliosphere. High velocity ISM models can dramatically reduce the upstream extent of the heliosphere (i.e., distance to the various boundaries) and the overall morphology can be greatly elongated. The overall structure reverts, for sufficiently high velocity models, to one that is bullet- or rocket-shaped, rather like the fully ionized

ISM models, even in the presence of neutral hydrogen. In these models, the outer planets are exposed to the raw ISM and the environment of the inner planets, and possibly of Earth, may be significantly perturbed.

3 A strongly magnetized LISM does not necessarily change the size of the heliosphere, provided the magnetic field lies in the direction of the LISM flow. The bow shock in this case is a slow magnetosonic shock, and the magnetic field and flow direction rotate towards the symmetry axis on crossing the shock from the LISM side. Following that, the plasma turns away again from the symmetry axis as it flows around the heliopause. This is quite different from the "standard" model with a fast magnetosonic shock that directly deflects the wind away from the axis. In other respects, the general structure of the interface is very similar to the unmagnetized case. Even with a very strong magnetic interstellar field, the Ly-α absorption profiles are consistent with observations along a variety of sightlines.

4 Simple empirical scaling laws for the locations of the various boundaries have been established. Similar correlations can be derived for the filtration of interstellar hydrogen. These simple empirical scaling laws can be applied to supersonic heliosphere-ISM interactions for a very wide range of possible ISM states, allowing rough estimates for the location of boundaries and the efficiency of neutral gas filtration as it enters the heliosphere.

Acknowledgements: This research was supported, in part, by NSF grant ATM-0296114 and NASA grants NAG5-11621, NAG5-12879, NAG5-12903, NAG5-13611, NAG 5-13107, and NAG5-11005. Numerical computations were performed on the IGPP/UCR "Lupin" cluster.

References

Baranov, V. B., and Malama, Y. G. (1993). Model of the solar wind interaction with the local interstellar medium: Numerical solution of self-consistent problem, *J. Geophys. Res.*, 98:15,157–15,163.

Baranov, V. B., and Malama, Y. G. (1995). Effect of local interstellar medium hydrogen fractional ionization on the distant solar wind and interface region *J. Geophys. Res.*, 100:14,755–14,762.

Cairns, I. H., and Zank, G. P. (2002). Turn-on of 2–3 kHz radiation beyond the heliopause, *Geophys. Res. Lett.*, 29:1143.

Cox, D. P., and Helenius, L. (2003). Flux-tube dynamics and a model for the origin of the local fluff, *Astrophys. J.*, 583:205–228.

Crutcher, R., Heiles, C., and Troland, T. (2003). Observations of interstellar magnetic fields. In Falgarone, E., and Passot, T., editors, *Turbulence and*

Magnetic Fields in Astrophysics, Lecture Notes in Physics, Vol. 614, pages 155–181.

Fahr, H.-J., Kausch, T., and Scherer, H. (2000). A 5-fluid hydrodynamic approach to model the solar system–interstellar medium interaction, *Astron. Astrophys.*, 357:268–282.

Florinski, V., Pogorelov, N. V., Zank, G. P., Wood, B. E., and Cox, D. P. (2004). On the possibility of a strong magnetic field in the local interstellar medium, *Astrophys. J.*, 604:700–706.

Florinski, V., Zank, G. P., and Axford, W. I. (2003a). The Solar System in a dense interstellar cloud: Implications for cosmic-ray fluxes at Earth and ^{10}Be records, *Geophys. Res. Lett.*, 30:2206.

Florinski, V., Zank, G. P., and Pogorelov, N. V. (2003b). Galactic cosmic ray transport in the global heliosphere, *J. Geophys. Res.*, 108:1228.

Florinski, V., Zank, G. P., and Pogorelov, N. V. (2005). Heliopause stability in the presence of neutral atoms: Rayleigh–Taylor dispersion analysis and axisymmetric MHD simulations, *J. Geophys. Res.*, 110:A07104.

Frisch, P. C. (1990). Characteristics of the local interstellar medium. In Grzedzielski, S. and Page, D. E., editors, *Physics of the Outer Heliosphere*, 19–22.

Frisch, P. C. (1993). G-star astropauses - A test for interstellar pressure. *Astrophys. J.*, 407:198–206.

Frisch, P. C. (1995). Characteristics of nearby interstellar matter, *Space Sci. Rev.*, 72:499–592.

Frisch, P. C. (2000a). The galactic environment of the Sun, *J. Geophys. Res.*, 105:10,279–10,289.

Frisch, P. C. (2000b). The galactic environment of the Sun, *American Scientist*, 88:52–59.

Frisch, P. C., Grodnicki, L., and Welty, D. E. (2002). The velocity distribution of the nearest interstellar gas, *Astrophys. J.*, 574:834–846.

Frisch, P. C. (2003). Local interstellar matter: The Apex cloud, *Astrophys. J.*, 593:868–873.

Frisch, P. C. (2005). Tentative identification of interstellar dust in the magnetic wall of the heliopause, *Astrophys. J.*, 632:L143–L146.

Gayley, K. G., Zank, G. P., Pauls, H. L., Frisch, P. C., and Welty, D. E. (1997). One-versus two-shock heliosphere: constraining models with Goddard High Resolution Spectrograph Ly α spectra towards α Centauri, *Astrophys. J.*, 487:259–270.

Heerikhuisen, J., Florinski, V., Zank, G. P., and Müller, H.-R. (2005). Interaction of the solar wind with interstellar neutral gas: A comparison between self-consistent Monte-Carlo and multi-fluid approaches. In *Solar Wind 11*, B. Fleck and T. H. Zurbuchen (eds.), ESA SP-592:339.

Holzer, T. E. (1989). Interaction between the solar wind and the interstellar medium, *Annu. Rev. Astron. Astrophys.*, 27:199–234.

Ip, W.-H., and Axford, W. I. (1985). Estimates of galactic cosmic ray spectra at low energies, *Astrophys. J.*, 149:7–10.

Izmodenov, V., Geiss, J., Lallement, R., Gloeckler, G., Baranov, V. B., and Malama, Y. G. (1999). Filtration of interstellar hydrogen in the two-shock heliospheric interface: Inferences on the local interstellar cloud electron density, *J. Geophys. Res.*, 104:4731–4741.

Kurth, W. S., and Gurnett, D. A. (2003). On the source location of low-frequency heliospheric radio emissions, *J. Geophys. Res.*, 108:8027.

Lallement, R. (1996). Relations between ISM inside and outside the heliosphere, *Space Sci. Rev.*, 78:361–374.

Lallement, R. (2004). The heliospheric soft X-ray emission pattern during the ROSAT survey: Inferences on Local Bubble hot gas, *Astron. Astrophys.*, 418:143-150.

Liewer, P. C., Karmesin, S. R., and Brackbill, J. U. (1996). Hydrodynamic instability of the heliopause driven by plasma-neutral charge-exchange interactions, *J. Geophys. Res.*, 101:17,119–17,127.

Lipatov, A. S., Zank, G. P., and Pauls, H. L. (1998). The interaction of neutral interstellar H with the heliosphere: A 2.5-D particle-mesh Boltzmann simulation, *J. Geophys. Res.*, 103:20,631–20,642.

Mitchell, J. J., Cairns, I. H., and Robinson, P. A. (2004). Theory for 2–3 kHz radiation from the outer heliosphere, *J. Geophys. Res.*, 109:A06108.

Müller, H.-R., and Zank, G. P. (1999). Self-consistent hybrid simulations of the interaction of the heliosphere with the local interstellar medium. In Habbal, S. R., Esser, R., Hollweg, J. V., and Isenberg, P. A., *Solar Wind Nine*, AIP Conf. Proc. Vol. 471, pages 819–822.

Müller, H.-R., Zank, G. P., and Lipatov, A. S. (2000). Self-consistent hybrid simulations of the interaction of the heliosphere with the local interstellar medium, *J. Geophys. Res.*, 105:27,419–27,438.

Müller, H.-R., Zank, G. P., and Frisch, P. C. (2001). Effect of different possible interstellar environments on the heliosphere: a numerical study. In Scherer, K., Fichtner, H., Fahr, H. J., and Marsch, E., editors, *The Outer Heliosphere: the Next Frontiers*, COSPAR Coll. Ser. Vol. 11, pages 329–332.

Müller, H.-R., Frisch, P. C., Florinski, V., and Zank, G. P. (2005). Heliospheric response to different possible interstellar environments, *Astrophys. J.*, submitted.

Parker, E. N. (1961). The stellar wind regions, *Astrophys. J.*, 134:20–27.

Pauls, H. L., and Zank, G. P. (1997). Interaction of a nonuniform solar wind with the local interstellar medium 2. A two-fluid model, *J. Geophys. Res.*, 102:19,779–19,787.

Pauls, H. L., Zank, G. P., and Williams, L. L. (1995). Interaction of the solar wind with the local interstellar medium, *J. Geophys. Res.*, 100: 21, 595–21,604.

Rand, R. J., and Kulkarni, S. R. (1989). The local galactic magnetic field, *Astrophys. J.*, 343:760–772.

Redfield, S., and Linsky, J. L. (2004). The structure of the local interstellar medium. III. Temperature and turbulence, *Astrophys. J.*, 613:1004–1022.

Scherer, K., Fichtner, H., and Stawicki, O. (2002). Shielded by the wind: the influence of the interstellar medium on the environment of Earth, *J. Atmos. Sol. Terr. Phys.*, 64:795–804.

Slavin, J. D., and Frisch, P. C. (2002). The ionization of nearby interstellar gas, *Astrophys. J.*, 565:364–379.

Snowden, S. L., Egger, R., Finkbeiner, D. P., Freyberg, M. J., and Plucinsky, P. P. (1998). Progress on establishing the spatial distribution of material responsible for the 1/4 keV soft X-ray diffuse background local and halo components, *Astrophys. J.*, 493:715–729.

Spitzer, L. (1978). *Physical Processes in the Interstellar Medium.* Wiley and Sons, New York.

Suess, S. T. (1990). The heliopause, *Rev. Geophys.*, 28:97–115.

Suess, S. T., and Nerney, S. (1990). Flow downstream of the heliospheric terminal shock 1. Irrotational flow, *J. Geophys. Res.*, 95:5403–6412.

Tinbergen, J. (1982). Interstellar polarization in the immediate solar heighbourhood, *Astron. Astrophys.* 105:53–64.

Talbot, R. J., and Newman, M. J. (1977). Encounters between stars and dense interstellar clouds, *Astrophys. J.*, 34:295–308.

Wang, C., and Belcher, J. W. (1998). Numerical investigation of hydrodynamic instabilities of the heliopause, *J. Geophys. Res.*, 103:247–256.

Welty, D. E., Hobbs, L. M., Lauroesch, J. T., Morton, D. C., Spitzer, L., and York, D. G. (1999). The diffuse interstellar clouds toward 23 Orionis, *Astrophys. J. Supp. Ser.*, 124:465–501.

Williams, L. L., Hall, D. T., Pauls, H. L., and Zank, G. P. (1997). The heliospheric hydrogen distribution: a multifluid model, *Astrophys. J.*, 476:366–384.

Witte, M. (2004). Kinetic parameters of interstellar neutral helium. Review of results obtained during one solar cycle with the Ulysses/GAS-instrument, *Astron. Astrophys.*, 426:835-844.

Wood, B. E., Müller, H-R., and Zank, G. P. (2000), Hydrogen Ly α absorption predictions by Boltzmann models of the heliosphere, *Astrophys. J.*, 542:493–503.

Zank, G. P. (1999). Interaction of the solar wind with the local interstellar medium: A theoretical perspective, *Space Sci. Rev.*, 89:413–688.

Zank, G. P., and Frisch, P. C. (1999). Consequences of a change in the galactic environment of the Sun, *Astrophys. J.*, 518:965–973.

Zank, G. P., Müller, H.-R., and Wood, B. E. (2001). The interaction of the solar wind and stellar winds with the partially ionized interstellar medium, *Phys. Plasmas*, 8:2385–2393.

Zank, G. P., Pauls, H. L., Williams, L. L., and Hall, D. T. (1996). Interaction of the solar wind with the local interstellar medium: A multifluid approach, *J. Geophys. Res.*, 101:21,639–21,655.

Chapter 3

THE INFLUENCE OF THE INTERSTELLAR MAGNETIC FIELD ON THE HELIOSPHERIC INTERFACE

Nikolai V. Pogorelov
Institute of Geophysics and Planetary Physics, University of California, Riverside, CA 92521
nikolaip@ucr.edu

Gary P. Zank
Institute of Geophysics and Planetary Physics, University of California, Riverside, CA 92521
zank@ucr.edu

Abstract The properties of the local interstellar cloud (LIC) are rather poorly known, with the exception of the interstellar plasma velocity and temperature. The poorest known quantities are the strength and direction of the interstellar magnetic field (ISMF). We review available observational data and perform three-dimensional, time-dependent parametric modelling of the SW interaction with the local interstellar medium (LISM) for a number of typical strengths and directions of the ISMF. The effect of coupling interstellar and interplanetary magnetic fields on the SW–LISM interaction pattern is investigated for both superfast and subfast magnetosonic interaction regimes. In addition, we consider the case of an ISMF vector perpendicular to the LISM velocity vector and inclined at an angle 60° to the ecliptic plane, as suggested in recent publications relating LIC properties to the radio emission observed by *Voyager 1*. Since the existence of nonevolutionary MHD shocks remains controversial, some attention is paid to the case where the ISMF strength ahead of the bow shock falls into the region of nonevolutionary parallel shocks. It is shown that a complicated, multi-shocked solution that exists for the axially symmetric models acquires a regular evolutionary structure in a genuinely 3D formulation of the problem. We present both the results of purely MHD calculations and those obtained by taking into account charge

P. C. Frisch (ed.), Solar Journey: The Significance of our Galactic Environment for the Heliosphere and Earth, 53–86.

exchange between neutral and ionized hydrogen. It is shown that, besides the usual modifications to heliospheric structure, charge exchange may be a source of heliospheric current sheet instabilities.

Keywords: Interstellar medium, interstellar magnetic field, solar wind, interplanetary magnetic field, heliosphere

Introduction

Although energetic particle data from *Voyager* (Cummings, Stone, & Steenberg, 2002, McDonald et al., 2003, Krimigis et al., 2003) and *Ulysses* (Smith et al., 2003) are beginning to reveal more about the properties of the inner heliosheath region of the solar wind (SW) and its interaction with the surrounding matter, the properties of the local interstellar medium (LISM) still remain insufficiently constrained for the development of a reliable mathematical model of the heliospheric interface. It is accepted that there is no exact definition of the LISM, neither in terms of column densities nor in terms of distances, since it, in fact, consists of many asymmetrically distributed components at different densities and pressures (Ferlet, 1999). The Sun is located in the galactic disk with an average thickness of about 300 pc, inside the so-called Local Bubble (LB), which is a large volume of a very tenuous gas at temperature $T \sim 10^6$ K (Snowden et al., 1998). The characteristic size of this volume is about 200 pc (Lallement, 2001). A few condensations and low-density warm clouds (diffuse clouds) are embedded in the LB. Several such clouds are located in the vicinity of the Sun and are sometimes called the Local Fluff. The typical size of these clouds is 1 pc and our Sun is located in one of them (Frisch, 2000, Lallement, 2001), the Local Interstellar Cloud (LIC). The physical properties of the LIC determine the confinement of the solar wind. Since the Sun moves through space with a velocity of about 13–20 pc per million years in the Local Standard of Rest, it could certainly encounter different clouds. Moreover, observations of dense interstellar structures (Frail et al., 1994) show that environment variations are possible on the time scales of years. Several numerical models have attempted to investigate the possible influence of the variability of the galactic environment of the Sun on the heliospheric interface (see, e.g., Zank & Frisch, 1999, Pogorelov, 2000, Scherer, Fichtner, & Stavicki, 2002, Florinski, Zank, & Axford, 2003). A number of new results are presented in this volume. The outer heliosphere may be identified crudely as a part of the solar wind cavity that extends from $\sim$10 AU outwards, to regions where interstellar material from the LIC is dominant.

Measurements of the kinetic parameters of the interstellar He component have been successfully performed (Witte, 2004) by the *Ulysses* GAS instrument, which gives a neutral helium number density $n_{\rm He} \approx 0.015 \pm 0.003\,{\rm cm}^{-3}$, temperature 6300 ± 340 K, and velocity 26.3 ± 0.4 km s^{-1}. This agrees very well (Möbius et al., 2004) with the data obtained from *Extreme Ultraviolet Explorer (EUVE)* (Vallerga et al., 2004), re-examination of the *Prognoz* data (Lallement et al., 2004), and pick-up ion observations (Gloeckler et al., 2004). Spectroscopic *EUVE* measurements (Dupuis et al., 1995, Wolff, Koestner, & Lallement, 1999) show that the average ratio of neutral hydrogen and helium column densities is about 14.7. Number densities of H^+ and He^+ are determined by the corresponding ionization rates, which are not known with much precision in the LISM. Vallerga (1996) suggests that if we assume that the abundance ratio between hydrogen and helium as chemical elements is 10, the number densities ranges are $0.15 \leq n_{\mathrm{H}} \leq 0.34\,\mathrm{cm}^{-3}$ and $0.004 \leq n_{\mathrm{H^+}} \leq 0.14\,\mathrm{cm}^{-3}$. The densities $n_{\mathrm{H}} = 0.22\,\mathrm{cm}^{-3}$ and $n_{\mathrm{H^+}} = 0.1\,\mathrm{cm}^{-3}$ given by Frisch (2000) correspond to helium and hydrogen ionization ratios of 53% and 31%, respectively. Due to uncertainties in the ionization rates, the LISM parameters, used in particular global models of the outer heliosphere, should be adjusted to account for data observed in the inner heliosheath.

The poorest known LISM parameter is the magnetic field. Frisch (2003a,b) has suggested that the vector of the interstellar magnetic field (ISMF) $\mathbf{B}_\infty$ is nearly perpendicular to the LISM velocity vector $\mathbf{V}_\infty$ and belongs to the galactic plane, making an angle of $\sim 60°$ with the solar ecliptic plane. The identification of this orientation seems to result from the fact that the LISM velocity is directed at $l = 3.6°$ and $b = 15.3°$ in the galactic coordinate plane (Flynn et al., 1998, Witte et al., 2004) and the belief that the magnetic field is directed toward the galactic longitude $l \approx 88 \pm 5°$ (Tinbergen, 1982, Rand & Lyne, 1994). A tilt of 60° has been suggested recently as a possible explanation of the 2–3 kHz emission data (Kurth & Gurnett, 2003). However, most observations used to derive the direction of the magnetic field in the LIC involve averaging data over several parsecs and conclusions should therefore be treated cautiously. The difference in the propagation directions of the He and H atoms discovered recently by Lallement et al. (2005) has been interpreted as a possible tool for determining the ISMF direction (Izmodenov, Alexashov, & Myasnikov, 2005), because the LISM stagnation point on the heliopause can be displaced with respect to the solar ecliptic plane for certain ISMF orientations. However, this interpretation ignores the presence of the interplanetary magnetic field. As will be shown later in this chapter, similar displacements are plausible for nearly all angles between $\mathbf{B}_\infty$ and $\mathbf{V}_\infty$, provided that the heliospheric current sheet bends into one of the hemispheres. Remarkably, in such cases the heliopause becomes asymmetric with respect to the ecliptic plane even for the ISMF aligned with the LISM velocity vector. Thus, the actual angle between $\mathbf{V}_\infty$ and $\mathbf{B}_\infty$ still remains a free parameter. For example, the theory of the LB origin suggested by Cox & Helenius (2003), in order to account for the LIC pressure equilibrium with the surrounding hot gas in the

LB, involves strong magnetic fields of 4–5 μG parallel to the LISM velocity. Such values of magnetic field should probably be excluded (Pogorelov, Zank, & Ogino, 2004), were we to assume $\mathbf{V}_\infty \perp \mathbf{B}_\infty$, because *Voyager 1* has already detected the solar wind termination shock at about 94 AU. However, calculations performed for the field-aligned case by Florinski et al. (2004) exhibit very good agreement with the Lyα absorption profiles (Linsky & Wood, 1996), which are now used frequently to validate theoretical models of the heliosphere. In fact, it appears that magnetic fields up to 4.5 μG cannot modify the heliospheric interface much in the quasi-field-aligned case. It should be mentioned in this connection that the way magnetic pressure can compensate the lack of thermal pressure in the LIC is not quite understood, and seems inconsistent with observations (Lallement et al., 2003, 2004).

Here we summarize the influence of the ISMF strength and direction for certain characteristic angles between $\mathbf{B}_\infty$ and $\mathbf{V}_\infty$. This can hopefully elucidate the interpretation of the observational data (Wood et al., 2002, Kurth & Gurnett, 2003, Krimigis et al., 2003) that appeared after the numerical studies of Washimi & Tanaka (1996), Linde et al. (1998), Pogorelov & Matsuda (1998), Ratkiewicz et al. (1998), and Tanaka and Washimi (1999). We consider ISMF intensities in a wide range corresponding to both superfast and subfast (with velocities greater or less than the fast magnetosonic velocity, respectively) magnetized LISM flows. Calculations have been performed both in a purely MHD and 2-fluid approximations. The latter additionally involves the LISM hydrogen atoms that experience charge exchange with plasma particles.

Of additional theoretical interest is an analysis of the SW–LISM interaction for the range of magnetic fields where parallel MHD shocks are nonevolutionary – unstable to decomposition into other discontinuities – and therefore should be excluded in the ideal magnetohydrodynamic description (Landau & Lifshitz, 1984). Solutions of that kind have been reviewed recently by Kulikovskii, Pogorelov, & Semenov (2001). Nonevolutionary shocks tend to appear in those cases where fully three-dimensional problems are modelled in statements that involve either a symmetry axis or a symmetry plane. It is important to recognize (Barmin, Kulikovskii, & Pogorelov, 1996) that physically inadmissible solutions can persist in numerical calculations where numerical dissipation and resistivity are many orders of magnitude higher than those in physical space plasmas.

The chapter is structured as follows. Section 3.1 presents the physical and mathematical statements of the problem. In Section 3.2 we consider the results of our numerical simulations for weak magnetic fields, assuming that the LISM flow is subfast. Comparison is made of the purely MHD SW–LISM interaction with realistic cases involving neutral hydrogen atoms interacting with the plasma component via charge exchange (Pogorelov & Zank, 2005). We also address the interaction pattern with $\mathbf{V}_\infty \parallel \mathbf{B}_\infty$ with the ISMF strength in a so-called switch-on regime. Section 3.3 describes the case of a strong interstellar magnetic field oriented at an angle to the LISM velocity vector and is followed by the section where we discuss the results obtained.

3.1 SW–LISM Interaction Problem

In the introduction, we specified the range of physical parameters reflecting our current knowledge about the LISM filling the LIC. To focus our study on the influence of the strength and direction of the ISMF on the heliospheric interface in the presence of the interplanetary magnetic field (IMF), we choose the following set of the LISM properties: the number density of H^+ ions is $n_{\mathrm{p}\infty} = 0.07\,\mathrm{cm}^{-3}$, velocity $V_{\mathrm{p}\infty} = 25\,\mathrm{km\,s}^{-1}$, and the plasma speed of sound $c_{\mathrm{p}\infty} = 12.5\,\mathrm{km\,s}^{-1}$, thus implying a LISM temperature $T_{\mathrm{p}\infty} \approx 5679\,\mathrm{K}$ and Mach number $M_{\mathrm{p}\infty} = V_{\mathrm{p}\infty}/c_{\mathrm{p}\infty} = 2$. The magnitude and direction of the ISMF vector $\mathbf{B}_\infty$ remain parameters of the problem. It is assumed that neutral and ionized hydrogen have the same velocity and temperature at the outer boundary, which is located at 1200–1500 AU from the Sun. In the presence of interstellar hydrogen atoms, their density $n_{\mathrm{H}\infty}$ will be either $0.1\ \mathrm{cm}^{-3}$ or $0.2\ \mathrm{cm}^{-3}$. The solar wind plasma is assumed spherically symmetric, which takes place in the vicinity of the solar maximum, with the following parameters: $V_{\mathrm{E}} = 450\,\mathrm{km\,s}^{-1}$, $M_{\mathrm{E}} = V_{\mathrm{E}}/c_{\mathrm{E}} = 10$, $n_{\mathrm{E}} = 7\,\mathrm{cm}^{-3}$ (Baranov & Malama, 1993). The subscript "E" corresponds here to Earth's distance from the Sun (1 AU). Although the magnetic field of the Sun is far from dipole at the solar maximum (Balogh, 1996), for simplicity, we suppose that it has the shape of Parker's spiral (Parker, 1961) at the inner boundary (10–30 AU), with spherical coordinate components (for the z-axis aligned with Sun's rotation axis)

$$B_R = \pm B_{\mathrm{E}} \left(\frac{R_{\mathrm{E}}}{R}\right)^2, \tag{1.1}$$

$$B_\theta = 0, \tag{1.2}$$

$$B_\phi = -\beta B_R \left(\frac{R}{R_{\mathrm{E}}}\right) \sin\theta, \tag{1.3}$$

where $\beta = \Omega R_{\mathrm{E}}/V_{\mathrm{E}} \approx 0.9$ (Ω is the Sun's angular velocity) and the radial component of the IMF at 1 AU is approximately 37.5 μG. It is interesting to note in this connection that the *Voyager 1* observations of the IMF strength agree with Parker's model from 1 to 81 AU and from 1978 to 2000 when one considers the solar cycle variations in the source magnetic field and the latitudinal/time variation of the solar wind speed (Burlaga et al., 2002). Since the IMF is discontinuous across the equatorial plane, a heliospheric current sheet (HCS) naturally forms in accordance with the chosen boundary conditions, where the jump in magnetic field occurs within one computational cell adjacent to the inner boundary chosen at $R = 30$ AU and $R = 10$ AU for purely MHD and 2-fluid calculations, respectively.

We assume that the flow of charged particles is governed by the equations of ideal magnetohydrodynamics that express the conservation laws for mass, momentum, total energy, and magnetic flux:

$$\frac{\partial}{\partial t}\begin{bmatrix}\rho\\ \rho\mathbf{v}\\ e\\ \mathbf{B}\end{bmatrix} + \nabla\cdot\begin{bmatrix}\rho\mathbf{v}\\ \rho\mathbf{v}\mathbf{v} + p_0\hat{\mathbf{I}} - \dfrac{\mathbf{B}\mathbf{B}}{4\pi}\\ (e+p_0)\,\mathbf{v} - \dfrac{\mathbf{B}(\mathbf{v}\cdot\mathbf{B})}{4\pi}\\ \mathbf{v}\mathbf{B} - \mathbf{B}\mathbf{v}\end{bmatrix}$$
$$= -\begin{bmatrix}0\\ \dfrac{\mathbf{B}}{4\pi}\\ \dfrac{\mathbf{v}\cdot\mathbf{B}}{4\pi}\\ \mathbf{v}\end{bmatrix}\times\nabla\cdot\mathbf{B} + \begin{bmatrix}0\\ \mathbf{H}^{\mathrm{m}}_{\mathrm{p-H}}\\ H^{\mathrm{e}}_{\mathrm{p-H}}\\ 0\end{bmatrix}. \tag{1.4}$$

Here density ρ, velocity $\mathbf{v}$, thermal and total pressures p and p_0, and the magnetic field strength $\mathbf{B}$ are normalized by ρ_∞, V_∞, $\rho_\infty V_\infty^2$, and $V_\infty\sqrt{\rho_\infty}$, respectively. Time and length are also dimensionless, with 1 AU and 1 AU/V_∞ being the reference length and time, respectively. The quantity e is the total energy density.

System (1.4) is written in the symmetrizable, Galilean invariant form suggested by Godunov (1972) and successfully applied by a number of authors, e.g. Linde et al. (1998) and Pogorelov & Matsuda (1998), for the purpose of eliminating spurious magnetic charge (see Kulikovskii et al., 2001 and references therein). In the presence of charge exchange between neutral and charged particles, source terms $\mathbf{H}^{\mathrm{m}}_{\mathrm{p-H}}$ and $H^{\mathrm{e}}_{\mathrm{p-H}}$ also appear in the momentum and energy equations (photoionization processes are neglected). The formulas for the charge exchange terms are given by Pauls, Zank, & Williams (1995). In order to apply the multi-fluid approximation, we need to solve also the Euler equations for the neutral components of the mixture. For the purposes of this investigation it suffices to use a two-fluid interaction model, where only primary (interstellar) atoms are included. This approach is somewhat inconsistent, since we need to withdraw secondary neutrals that originate in the inner heliosphere. In order to reproduce a number of important physical effects, at least three populations of neutral hydrogen atoms should be introduced: atoms created in the LISM region ahead of the heliopause (population 1), and in the subsonic (population 2) and supersonic (population 3) SW regions (Zank et al., 1996). Axisymmetric calculations employing this model have been performed by Florinski, Zank, & Pogorelov (2003) using the numerical approach developed by Pogorelov (1995) and Pogorelov & Semenov (1997). The same approach was applied by Pogorelov & Matsuda (1998) to investigate the influence of the ISMF direction on the shape of the global heliopause, and we continue using it in this study.

Parker (1961) originally investigated the SW–LISM interaction, assuming that the solar wind expanded into a magnetized vacuum. Similar solutions, with the heliopause open in both the upwind and downwind LISM directions, were obtained by Myasnikov (1997) and, more recently, by Florinski et al. (2004) for the case of the sub-Alfvénic interstellar medium. Baranov & Krasnobaev (1971) applied a thin-layer Newtonian approximation, suggested for the SW–LISM interaction by Baranov, Krasnobaev, & Kulikovskii (1971) (BKK-model), to analyze the influence of an ISMF parallel to the LISM velocity on the heliopause. Since the LISM flow was assumed to be supersonic and super-Alfvénic, the BKK-model involves two shocks, decelerating the LISM and the SW, respectively (the bow and the termination shocks). The two-shock model of the MHD interaction has been substantially developed further by Matsuda & Fujimoto (1993), Baranov & Zaitsev (1995), Washimi & Tanaka (1996), Pogorelov & Semenov (1997), Pogorelov & Matsuda (1998, 2000, 2004), and Tanaka & Washimi (1999) for a fully ionized gas. A subsonic LISM model has also been considered by a number of authors, such as Parker (1961), Neutsch & Fahr (1982), and Zank et al. (1996). In fact, for present day conditions the plasma LISM flow is always supersonic in terms of the thermal speed of sound, and the use of subsonic models was justified by including additional pressure provided by the ISMF and cosmic rays. However, as has recently been shown convincingly by Florinski et al. (2004), supersonic and sub-Alfvénic LISM flow regimes (that occur in the presence of a strong ISMF) are qualitatively different from subsonic and super-Alfvénic ones (that occur for weak or negligible ISMF strengths). In the presence of neutral particles, the former case is characterized by the presence of a slow MHD bow shock, while the latter case can never include a bow shock. Later in this paper we shall consider the case of the subfast SW–LISM interaction for non-zero angle between the interstellar velocity and magnetic field vectors.

The importance of neutral particles has long been recognized (see the review by Zank, 1999a). Wallis (1971, 1975) first identified most of the now well accepted effects of charge exchange: the weakening of the termination shock, the filtration of hydrogen atoms at the heliopause, and changes to the LISM velocity and temperature due the LISM charge exchange with neutralized solar wind protons. The first self-consistent model involving both charged and neutral particles, the latter being modelled kinetically, was developed by Baranov & Malama (1993) for a nonmagnetized plasma in two space dimensions. It was later modified by Aleksashov et al. (2000) to include the ISMF. Since the hybrid model used in the latter two papers was incapable of managing time-dependent problems and is too time-consuming in three dimensions, less sophisticated approaches were developed (Pauls, Zank, & Williams, 1995, Pauls & Zank, 1996, Zank et al., 1996), where a multi-fluid approach was introduced (see also Fahr, Kausch, & Scherer, 2000, Scherer & Fahr, 2003). Müller, Zank, & Lipatov (2000) applied a mesh-particle method to solve the Boltzmann equation that governs the motion of the neutral particles. A multi-fluid approach has recently been applied to modelling the axisymmetric MHD SW–LISM interaction (Florinski, Zank, & Pogorelov, 2003; Florinski et al., 2004). Linde et al. (1998), performing 3D MHD calculations, used a yet further simplified, non–self-consistent approach suggested by Baranov, Ermakov, & Lebedev (1981) to calculate the charge exchange source terms by modelling interstellar neutrals using a mass conservation equation only.

With the *Voyager 1* spacecraft poised to cross the termination shock, perhaps multiple times, we have an opportunity to distinguish ISMF strengths and directions that are inconsistent with observations. The direction of the magnetic field entered into a recent analysis of the 2–3 kHz emission (Kurth & Gurnett, 2003), thought to originate from the outer heliosheath (the shocked region behind the bow shock and ahead of the heliopause). By investigating the magnetic field orientation and strength in the outer heliosheath, for different orientations of the ISMF, the relation between the ISMF and radio emission directions may be clarified.

3.2 Superfast SW–LISM Interaction

3.2.1 ISMF Perpendicular to the LISM Velocity

As noted in Introduction, the situation with $\mathbf{B}_\infty \parallel \mathbf{V}_\infty$ is often considered nonrealistic, although it is intrinsic for the theoretical explanation of the equilibrium between the LIC and the Local Bubble given by Cox & Helenius (2003). In this subsection, we analyze the cases where these vectors are perpendicular. We discuss, in particular, the location of the heliospheric discontinuities for different ISMF strengths. In addition, we investigate the case where $\mathbf{B}_\infty$ is also tilted 60° to the ecliptic plane.

First, let $\mathbf{B}_\infty$ be parallel to the x-axis, which is aligned with Sun's rotation axis, i. e., $\mathbf{B}_\infty \parallel 0x$. The z-axis is chosen parallel to the LISM flow, while the y-axis is perpendicular to both them. Unless otherwise specifically stated, this orientation of the coordinate axes is preserved throughout the paper. Figures 3.1a and 3.1b show, respectively, the density logarithm distributions in the meridional plane together with the LISM streamlines and the front view of the magnetic field lines draping the heliopause for $B_\infty \approx 1.8\,\mu$G. Figures 3.1$c$ and 3.1d show the same quantities for $B_\infty \approx 2.4\,\mu$G. The surface of the heliopause is colored according to the magnitude of the magnetic field $B_{\rm tot} = (B_x^2 + B_y^2 + B_z^2)^{1/2}$ on it. Its distribution clearly explains the ISMF

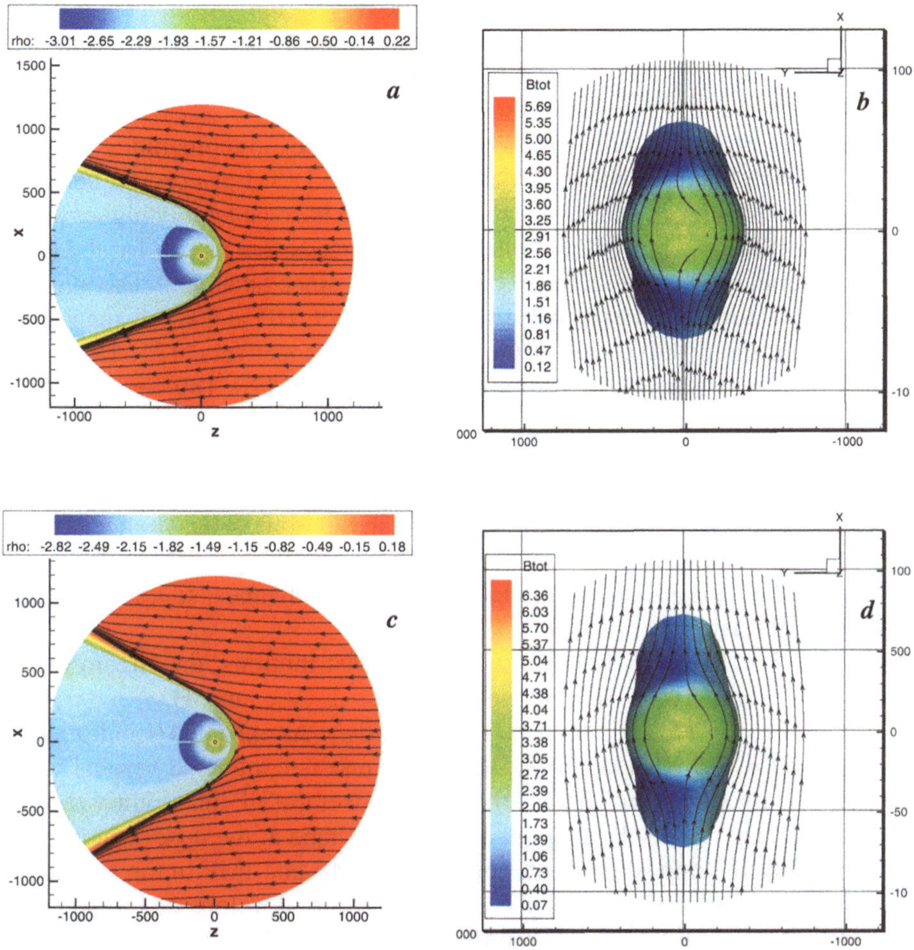

Figure 3.1. Streamlines and the density logarithm distribution in the meridional plane (left) and ISMF lines draping the heliopause (right) for $\mathbf{B}_\infty \perp \mathbf{V}_\infty$, $\mathbf{B}_\infty \parallel 0x$, and $B_\infty \approx 1.8\,\mu$G (*a*, *b*) and $2.4\,\mu$G (*c*, *d*). The x-axis is aligned with the rotation axis of the Sun. [ApJ, 614, 1007. ©2004. The American Astronomical Society.]

effect on the shape and position of the heliopause with respect to the Sun: the HP is pushed closer to the origin along the z-axis and squeezed from the left- and right-hand sides, the effect being greater for larger ISMF strengths. An asymmetry in the solution with respect to the meridional plane in noticeable even ahead of the HP. We do not show the SW streamlines in this figure, since they acquire a considerable component in the y-direction beyond the termination shock. The asymmetry is caused by the interaction of the IMF and ISMF.

We can only observe the global features of the interaction. All magnetic field line reconnection, whenever it happens, is governed by the numerical dissipation and resistivity. It is very difficult to evaluate the importance of this effect, since the ratio between the anomalous and numerical resistivities is unknown. In the case of reconnection, localized increases in temperature can be expected. Once this happens at the heliopause, the LISM gas can be heated. Since the plasma temperature inside the heliopause is considerably higher than that in the LISM, heating will result in a wider dissipative profile for the HP, whose numerical smearing seems to be inevitable, since tangential discontinuities have no stationary dissipative structure. The TS, HP, and BS distances to the Sun along the positive part of the z-axis are, respectively, 125, 187, and 560 AU for $B_\infty \approx 1.5\,\mu$G (this solution is not shown), 120, 180, and 600 AU for $B_\infty \approx 1.8\,\mu$G, and 114, 165, and 670 AU for $B_\infty \approx 2.4\,\mu$G. Clearly the position of the TS depends only slightly on the LISM strength, whereas the distance of the HP to the Sun decreases, and the BS stand-off distance increases substantially. All this is a result of relatively low ISMF strengths that preserve the superfast character of the LISM flow in the perpendicular case. On the other hand, the shape of the HP substantially changes, so affecting the distribution of quantities inside it.

Like Fig. 3.1, Figure 3.2 shows the density distributions, the streamlines, and the ISMF around the HP, but now for $\mathbf{B}_\infty \perp \mathbf{V}_\infty$ and $\mathbf{B}_\infty \perp 0x$ for $B_\infty \approx 1.5\,\mu$G ($a$ and b) and $B_\infty \approx 2.4\,\mu$G ($c$ and d). As expected (Linde et al., 1998, Pogorelov & Matsuda, 1998), the heliopause is contracted vertically in this case. The maximum magnetic field strength over the HP lies slightly above the yz-plane. The HP itself acquires a cusp-like shape at the nose, which can be attributed to the opposite orientation of the ISMF and IMF in the southern hemisphere. For stronger values of B_∞, an instability develops on the bottom of the HP. The cusp-like nose, as reported by Linde et al. (1998), can also be unstable via a Rayleigh–Taylor type instability, which is most likely caused by charge exchange of protons and interstellar neutrals and not included here. This instability was first found by Liewer, Karmesin, & Brackbill (1996) and Zank et al. (1996), and then explicitly calculated by Zank (1999b). Florinski, Zank, & Pogorelov (2005) showed that this instability persists in the presence of the ISMF, its amplitude decreasing with the magnetic field strength. In Fig. 3.3a, we show the side view (with a slight rotation of the picture around the z-axis) of the logarithm density distribution in the ecliptic plane, where one of the spiral IMF lines is also shown together with a set of ISMF lines draped over the HP from its top (here $B_\infty \approx 2.4\,\mu$G). Figure 3.3$b$ shows the distribution of the magnitude of the magnetic field in the meridional plane for the same ISMF strength at a characteristic time. Here we see an increase in the magnetic field pressure caused by ISMF draping over the heliopause and the ridges in the interplanetary magnetic field. The HCS visibly bends into the

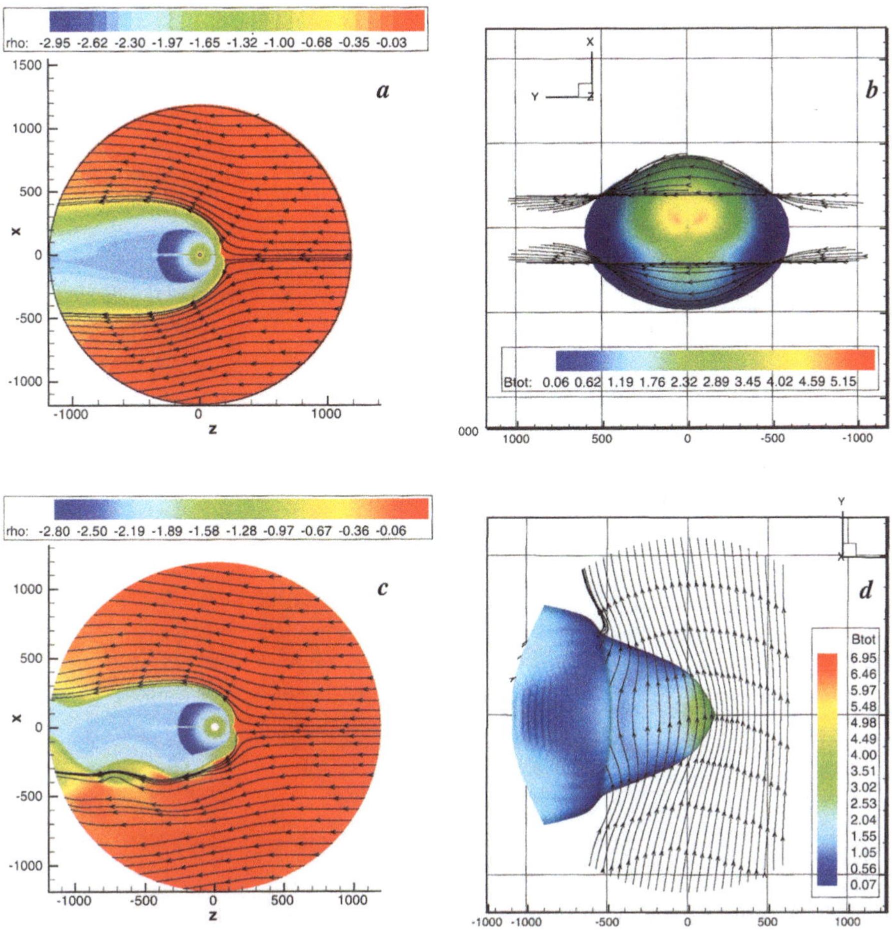

Figure 3.2. Streamlines and the density logarithm distribution in the meridional plane (left) and ISMF lines draping the heliopause (right) for $\mathbf{B}_\infty \perp \mathbf{V}_\infty$, $\mathbf{B}_\infty \perp 0x$, and $B_\infty \approx 1.5\,\mu$G ($a$, b) and $B_\infty \approx 2.4\,\mu$G ($c$, d). [ApJ, 614, 1007. ©2004. The American Astronomical Society.]

southern hemisphere. This occurs because some streamlines begin in the upper hemisphere and penetrate into the lower one. The TS, HP, and BS distances to the Sun along the positive part of the z-axis are equal, respectively, to 120, 161, and 514 AU for $B_\infty \approx 1.5\,\mu$G and 108, 144, and 705 AU for $B_\infty \approx 2.4\,\mu$G.

Consider now the case, where $\mathbf{B}_\infty \perp \mathbf{V}_\infty$ and is tilted 60° to the ecliptic plane. As mentioned earlier, this orientation of the ISMF corresponds to the recent suggestion of Frisch (2003b). It was used by Kurth & Gurnett (2003) to explain the behavior of the 2–3 kHz emission data thought to originate in the outer heliosheath. Figure 3.4 shows the front views of the heliopause for

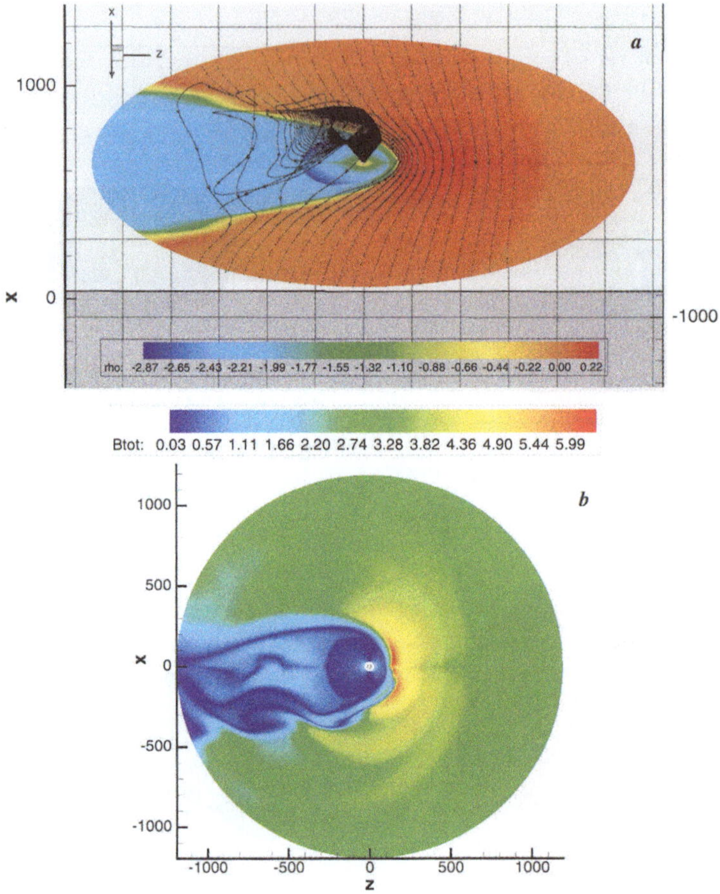

Figure 3.3. Density logarithm distribution in the ecliptic plane, one of the IMF lines, and ISMF lines draping over the heliopause (a, the lines with z = const are spaced at 200 AU) and magnetic field strength distribution in the meridional plane (b) for $\mathbf{B}_\infty \perp \mathbf{V}_\infty$, $\mathbf{B}_\infty \perp 0x$, and $2.4\,\mu$G). [ApJ, 614, 1007. ©2004. The American Astronomical Society.]

$B_\infty \approx 1.5\,\mu$G ($a$ and b) and $B_\infty \approx 2.4\,\mu$G ($c$ and d). In Figs. 3.4a, c we have also added the LISM streamlines that start upwind in the immediate vicinity of the meridional plane xz. Magnetic field line draping over the heliopause is shown in Fig. 3.4b, d. We see that this draping resulted in the HP rotating almost exactly 30° around the z-axis. In the left column of Fig. 3.4, the HP surface is colored according to the strength of the magnetic field on it. We see definite indications of a preferred orientation of the magnetic field on the HP, whose existence was suggested by Kurth & Gurnett (2003). In Fig. 3.5, we show the distribution of the magnetic field magnitude $B_{\rm tot}$ in the meridional plane (a) and in the cross-section of the narrowest flaring of the HP (b), where

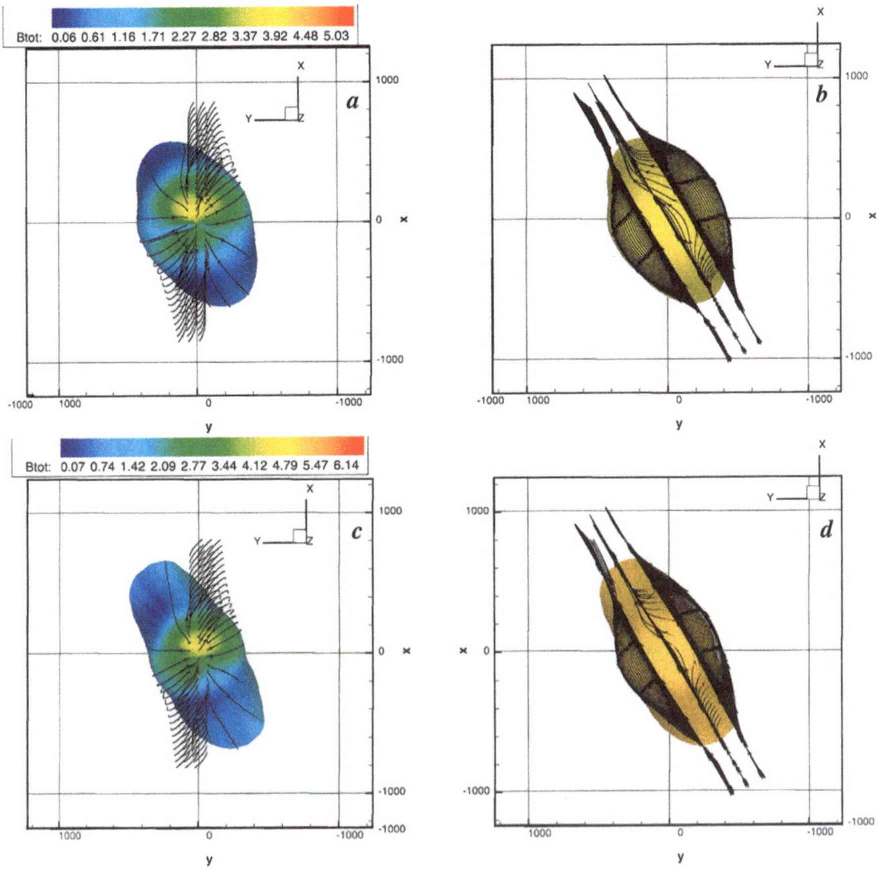

Figure 3.4. Front views of the heliopause for $\mathbf{B}_\infty \perp \mathbf{V}_\infty$ and tilted 60° to the ecliptic plane. The cases with $B_\infty \approx 1.5\,\mu$G (*a*, *b*) and $B_\infty \approx 2.4\,\mu$G (*c*, *d*). The LISM streamlines and ISMF lines are shown on plots *a*, *c* and *b*, *d*, respectively. [ApJ, 614, 1007. ©2004. The American Astronomical Society.]

two of the spiral IMF lines are also shown. It is clearly seen that the HCS rotates and bends down for the chosen orientation of $\mathbf{B}_\infty$. This contrasts with the violently unstable behavior of HCS in the absence of the ISMF reported by Opher et al. (2004). However, physically, there is no doubt that the HCS in the heliosheath should experience instabilities, and they may interact with the hydrodynamically unstable heliopause. Although the presence of the ISMF near the HP may temper these instabilities, this problem still requires a separate detailed investigation. The effect of neutral atoms on the HCS stability will be shown in the subsection dealing with field-aligned LISM flows.

Now we would like to discuss an issue raised by Washimi & Tanaka (2001). In their calculation, in order to model the structure of the HCS, they completely

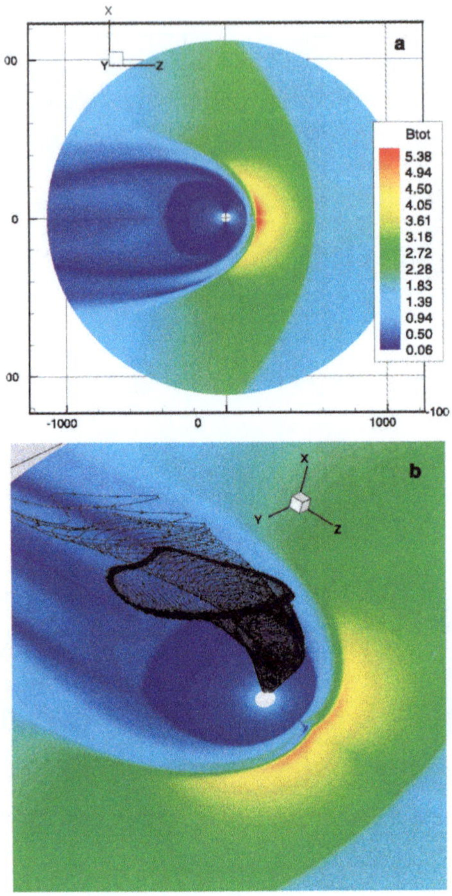

Figure 3.5. Magnetic field magnitude distributions in the meridional plane (a) and in the plane of the narrowest flaring of the heliopause (b) for $\mathbf{B}_\infty \perp \mathbf{V}_\infty$ and tilted 60° to the ecliptic plane. $B_\infty \approx 1.5\,\mu$G. [ApJ, 614, 1007. ©2004. The American Astronomical Society.]

eliminated the IMF inside a thin layer adjacent to both sides of the solar ecliptic plane. Since the width of the HCS is about 10^4 km at 1 AU, a strong refinement of the computational cells is required to resolve this very thin structure. In particular, the HCS was located within $\Delta\theta \approx 1.8°$ in their calculations. For the inner boundary at $R = 50$ AU this results in the minimum width $\sim 7.5 \times 10^7$ km, that is, at least three orders of magnitude greater than the characteristic width of the HCS. The latter, however, increases with distance from the Sun. This increase may be proportional to R, as suggested by Smith (2001), or to $R^{1/5}$ as argued by Opher et al. (2004). In any case, the prescribed value differs very much from the actual width of the HCS (Winterhalter et al., 1994).

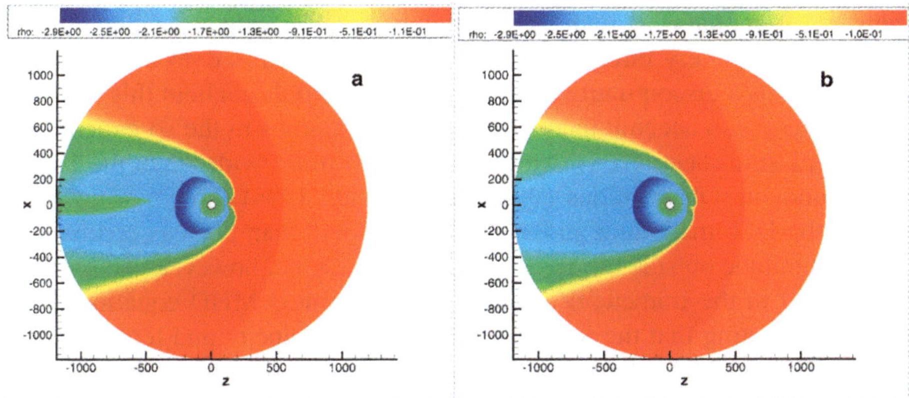

Figure 3.6. Density logarithm distributions in the meridional plane for $\mathbf{B}_\infty \perp \mathbf{V}_\infty$ and tilted 60° to the ecliptic plane with $B_\infty \approx 1.5\,\mu$G for wider (*a*) and narrower (*b*) HCS regions in the statement of Washimi & Tanaka (2001). The orientation of the coordinate axes of that paper is preserved, such that the z-axis is aligned with the solar magnetic axis for this figure.

In Fig. 3.6a, we reproduced a V-shaped structure discovered by Washimi & Tanaka (2001) for the set of parameters corresponding to Figs. 3.4 and 3.5. We assumed a linear distribution of the magnetic field inside a HCS-like structure within $\Delta\theta \approx 0.75°$. In Fig. 3.6 only, for comparison purposes, the z-axis is aligned with the solar magnetic axis, while the x-axis is directed towards the incoming LISM flow. This choice of $\Delta\theta$ results in the minimum width about 1.9×10^7 km. The thickness is twice smaller in the calculation shown in Fig. 3.6b. In both plots, we show the distributions of the density logarithms. We see the presence of the V-shaped structures in both calculations, although they clearly depend on the width of the layer with zero magnetic field near the ecliptic plane. The plasma temperature of the SW usually decreases adiabatically with distance from the Sun in the absence of charge exchange. This behavior does not change in our calculations anywhere except for the region around the ecliptic plane, where it increases with R, being substantially higher than in the regular solar wind region. The latter seems to contradict observations (Smith, 2001). The energy for this increase, of course, is provided by the magnetic field. It is also supplemented by keeping the temperature fixed at the inner boundary, as in the calculations of Washimi & Tanaka (2001), while it should increase since the pressure balance is not preserved at this boundary.

If we look at the typical distribution of the magnetic field in the HCS (Winterhalter et al., 1994), we observe that only the toroidal component of the IMF disappears at the equatorial plane, while the magnitude of $\mathbf{B}$ remains nearly constant. In this case, the HCS should rather be treated as a rotational than as a tangential discontinuity. As stated by Smith (2001), the simplest concept of a current sheet where the field is unidirectional and decreases monotonically to zero and then reappears with the opposite sign (similar to a tangential discontinuity) occurs very rarely, if ever. This means that if we do not want to treat the HCS as a discontinuity (the approach adopted throughout this study), we should take into account the distribution of quantities in the so-called heliospheric plasma sheet (HPS), which is about 30 times wider than the HCS and surrounds it. Observations (Winterhalter et al., 1994) show that the decrease of the IMF magnitude in the HPS can be less than 50% as compared with its magnitude outside the HPS. A study of that kind, however, can hardly be performed in the framework of ideal, dissipationless MHD equations. It is also worth noting that the increase in the plasma velocity and temperature inside the HCS causes extensive charge exchange that smears out the effects discovered by Washimi & Tanaka (2001).

In order to take into account the effect of neutral atoms on the heliospheric interface, we shall adopt a two-fluid model, where the original population of the LISM H atoms is treated as a separate fluid. This is a reasonable approximation whenever we are basically interested in the magnetic field effects. More elaborate approaches that involve a 4-fluid interaction model (Florinski, Zank, Pogorelov, 2003), which has proved to be effective in modelling three-dimensional (Pauls & Zank, 1997) and nonstationary aspects of the SW–LISM interaction (Zank, 1999a, Zank & Müller, 2003), or a kinetic treatment of neutrals similar to Baranov & Malama (1993) and Müller, Zank, & Lipatov (2000), have already been developed. Their results will be reported elsewhere.

The results of our calculations above show that we should expect that the influence of charge exchange is likely to be stronger than the effect of a weak ISMF. In Fig. 3.7, we show the distributions of the magnetic field vector magnitude (*a*) and the neutral hydrogen density (*b*) in the meridional plane for $\mathbf{B}_\infty \perp \mathbf{V}_\infty$ and tilted at 60° to the ecliptic plane for $B_\infty = 1.5\ \mu$G and (1) $n_{\mathrm{H}\infty} = 0.2$ and (2) 0.1 cm^{-3}. An important result of charge exchange is that the velocity of the solar wind decreases as it approaches the termination shock, whereas its temperature increases along the axis aligned with the LISM flow from 1.5×10^4 K at $R = 15$ AU to about 2×10^5 K ahead of the termination shock, in contrast to its adiabatic cooling $\sim 1/R^{4/3}$ in the absence of neutral atoms. This is a manifestation of the fluid plasma model that does not distinguish between pick-up ions (PUIs) and thermal solar wind plasma, and so the temperature increase reflects the hot PUI population in our simulations. One can see a pronounced hydrogen wall, a layer in which hydrogen atom density is enhanced ahead of the heliopause. The "height" of this wall increases with higher $n_{\mathrm{H}\infty}$, while its thickness becomes smaller, which is consistent with the kinetic simulation of Baranov & Malama (1995). Furthermore, a more dense stream of neutral hydrogen produces smaller flaring angle for the heliopause — an effect directly related to the decrease in the

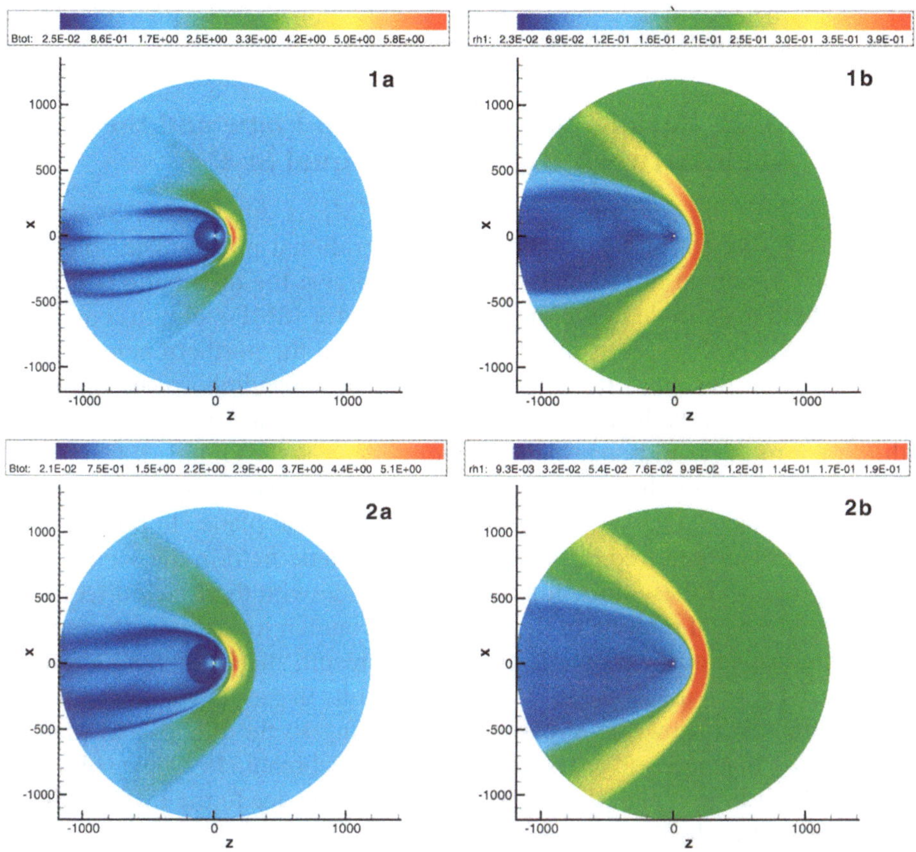

Figure 3.7. Magnetic field magnitude (a) and neutral hydrogen density (b) distributions in the meridional plane for $n_{\mathrm{H}\infty} = 0.2$ (1) and 0.1 (2) cm^{-3}. $\mathbf{B}_\infty \perp \mathbf{V}_\infty$ and tilted 60° to the ecliptic plane. $B_\infty \approx 1.5\,\mu\mathrm{G}$.

ratio of the SW and LISM dynamic pressures due to charge exchange. The behavior of the HCS remains essentially similar to the case without neutrals. The TS, HP, and BS stand-off distances along the z-axis are approximately equal to 80, 123, and 245 AU for $n_{\mathrm{H}\infty} = 0.2\ \mathrm{cm}^{-3}$ and to 100, 146, and 306 AU for $n_{\mathrm{H}\infty} = 0.1\ \mathrm{cm}^{-3}$, respectively. This shows that the former value of the H density is too large to be consistent with the observations from the *Voyager* spacecraft. As in the purely MHD case, there exists an asymmetry in the magnetic field distribution caused by the ISMF tilt with respect to the ecliptic plane. However, it might not be sufficiently sharp to account for the kilohertz emission sources distribution according to Kurth & Gurnett (2003). Additional investigation involving a radio emission model is required to get a decisive answer to that question. It is likely that an acute angle between $\mathbf{B}_\infty$ and $\mathbf{V}_\infty$ will be more favorable to fit observations.

3.2.2 ISMF Parallel to the Meridional Plane and the Angle Between B_∞ and V_∞ Equal to 45°

Consider now the configuration where the ISMF is directed at 45° to the LISM velocity vector and is parallel to the meridional plane. The solution for this case is shown in Fig. 3.8. Plots a and b give the distributions of the magnetic field magnitude in the meridional plane for ISMF strengths equal to 1.5 μG and 2.4 μG, respectively. The IMF spiral in the northern hemisphere is directed from the Sun. It should be noted that depending on the orientation of the IMF spirals the heliopause slightly rotates in an opposite direction with respect to the meridional plane (Pogorelov, Zank, & Ogino, 2004). The maximum strength of the magnetic field on the heliopause also lies in this case on its opposite sides with respect to the meridional plane. In both cases, there exist certain preferred directions of the magnetic field in the vicinity of the heliosphere and these directions do not coincide with the ISMF direction. We observe the usual rotation of the heliopause counterclockwise with respect to its position without a magnetic field. As a result, the bow shock rotates clockwise. The increase in the ISMF strength results in greater rotation angles. These figures also show the streamlines belonging to the meridional plane. There exists a magnetic ridge (Nerney, Suess, & Schmahl, 1993, Washimi & Tanaka, 1996, Linde et al., 1998), which appears to be due to the IMF rollover at the inner side of the heliopause.

It is distinctly seen on both figures that the current sheet leaves the ecliptic plane after crossing the termination shock and bends downwards. The B_y components of the IMF have opposite directions on each side of the ecliptic plane. This feature is similar to that obtained in our previous set of calculations, although in this case there seems to be no substantial rotation of it around the z-axis. Of course, in reality the shape and behavior of the HCS will be much more complicated, since we do not take into account the fact that the Sun's rotation and magnetic axis do not coincide. The change of polarity with the solar cycle also makes the interaction pattern unsteady, as was shown by Tanaka & Washimi (1999). Our analysis, by separating different configurations of the magnetic field that can affect the flow configuration, allows us to realize physical mechanisms contributing to those transient phenomena.

3.2.3 Field-aligned LISM Flows

If the ISMF vector $\mathbf{B}_\infty$ is parallel to the LISM velocity $\mathbf{V}_\infty$, there will be a point on the bow shock where both vectors are parallel to the shock normal,

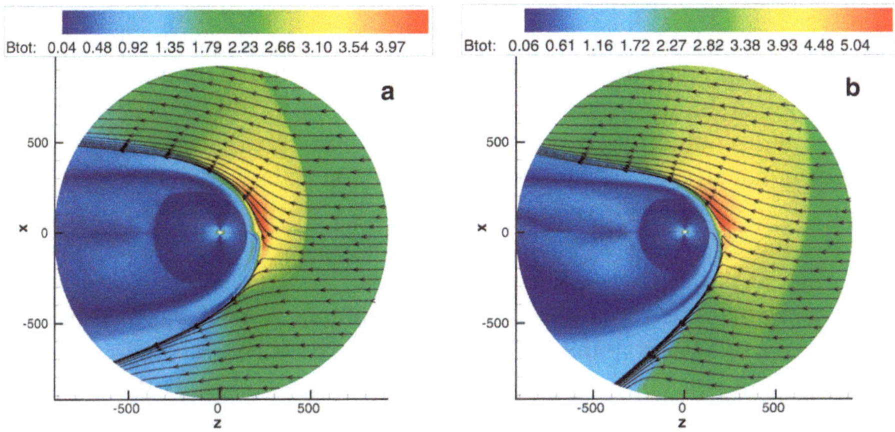

Figure 3.8. Distribution of $B_{\rm tot}$ in the meridional plane for $B_\infty \approx 1.5\,\mu$G ($a$) and $B_\infty \approx 2.4\,\mu$G ($b$) when the angle between $\mathbf{B}_\infty$ and $\mathbf{V}_\infty$ is equal to 45°. Interstellar streamlines are shown. [ApJ, 614, 1007. ©2004. The American Astronomical Society.]

that is, the shock will be parallel in this case. MHD shocks are parallel or perpendicular if the magnetic field ahead of and behind them is parallel or perpendicular to the shock normal, respectively. In singular cases, when the magnetic field is parallel to the shock normal only ahead of (or behind) the shock, the transition is identified as a switch-on (or switch-off) shock. All other shocks are oblique. Possible configurations of the ISMF in the vicinity of the bow shock cannot be understood without some consideration of the issue of shock admissibility. It is well-known that any small-amplitude plane perturbation of the state vector in the hyperbolic system can be represented as a sum of linear simple waves moving with corresponding characteristic velocities λ_i. Each simple wave is characterized by its amplitude. If we consider a shock with velocity W, each wave is either incoming or outgoing, depending on the sign of $\lambda_i - W$. A shock is called *evolutionary* if the problem of its interaction with small perturbations has a unique solution. This occurs if the number of unknowns is equal to the number of Hugoniot relations the shock. The unknowns are represented by amplitudes of outgoing perturbations (incoming perturbations are totally determined by initial conditions) and by a perturbation of the shock velocity. If the number of unknowns is greater than the number of shock relations, the amplitudes cannot be found uniquely and will depend on one or more arbitrary functions of time. Thus, such a discontinuity either cannot exist, or there are physical reasons to impose additional relations on it which are different from those resulting from the corresponding system of conservation laws to be solved. Examples of such "nonclassical" discontinuities were considered by Kulikovskii et al. (2001). If the number of unknowns is less than the number of relations, the solution of the shock interaction with small perturbations does not exist in the linear approximation. Since we expect a well-posed physical problem to have a solution, non-small deviations from the initial state must occur. This usually results in the disintegration of the initial discontinuity into two or more evolutionary discontinuities. It is well-known that all nonevolutionary shocks exhibit trans-Alfvénic transitions, where $v > a_{\mathrm{A}}$ ahead of the shock and $v < a_{\mathrm{A}}$ behind it. Here $v = |\mathbf{v}|$, $B = |\mathbf{B}|$, and $a_{\mathrm{A}} = B/(2\sqrt{\pi\rho})$ is the Alfvén velocity. Vector designations can be omitted in the above formulas, since $\mathbf{B} \parallel \mathbf{v}$. The interaction of small transverse (Alfvén) perturbations with nonevolutionary shocks always has no solutions. It is known from the MHD theory that parallel MHD shocks are not always evolutionary. This occurs (Landau & Lifshitz, 1984) only if

$$1 < A_\infty = V_\infty / a_{\mathrm{A}\infty} < A_* = \sqrt{\frac{(\gamma+1)\,M_\infty^2}{2+(\gamma-1)M_\infty^2}}, \tag{2.1}$$

where γ is a specific heat ratio and the subscript ∞ refers to quantities ahead of the shock.

If we disregard the IMF, the interaction pattern will be axially symmetric. Since in our calculations $M_\infty = 2$, it is easy to check that fast parallel shocks will remain evolutionary only for $B_\infty < 2\,\mu$G. In the range between 2 and 3.03 μG, parallel shocks cannot appear because they are nonevolutionary, while evolutionary (fast) switch-on shocks are inadmissible due to the symmetry restrictions (Pogorelov & Matsuda, 2000, 2004). Indeed, if the flow is continuous in directions tangent to the shock, once the magnetic field switches on in a certain direction at some point, by continuity, it must turn in essentially the same direction in the vicinity of this point, which breaks the axial symmetry of the flow. For this reason, the flow acquires a complicated structure with additional shocks, some of them being nonevolutionary (De Sterck, Low, & Poedts, 1998). If we consider (Pogorelov, Zank, & Ogino, 2004) the SW–LISM interaction in the presence of the interplanetary magnetic field ($\mathbf{B}_\infty$ is parallel to the solar ecliptic plane), the problem becomes genuinely three-dimensional and restrictions on the presence of switch-on shocks are no longer valid. This can be seen by comparison of Fig. 3.9a, where we show density logarithm distributions and magnetic field lines for the axially symmetric SW–LISM interaction in the field-aligned case with $\mathbf{B}_\infty \parallel \mathbf{V}_\infty$ and $B_\infty \approx 2.1\,\mu$G and Figs. 3.9$b$ and c showing the left- and right-hand side views of the heliopause cut by the meridional plane xz. We see that ISMF lines starting in this plane and crossing the bow shock in the vicinity of its forward point gain a considerable out-of-plane component, which was impossible in the axially symmetric case. Due to the magnetic field tension and the presence of the high dynamic pressure flow inside the HCS, the heliopause is elongated along the z-axis. The surface of the heliopause is colored according to the magnitude of the magnetic field on

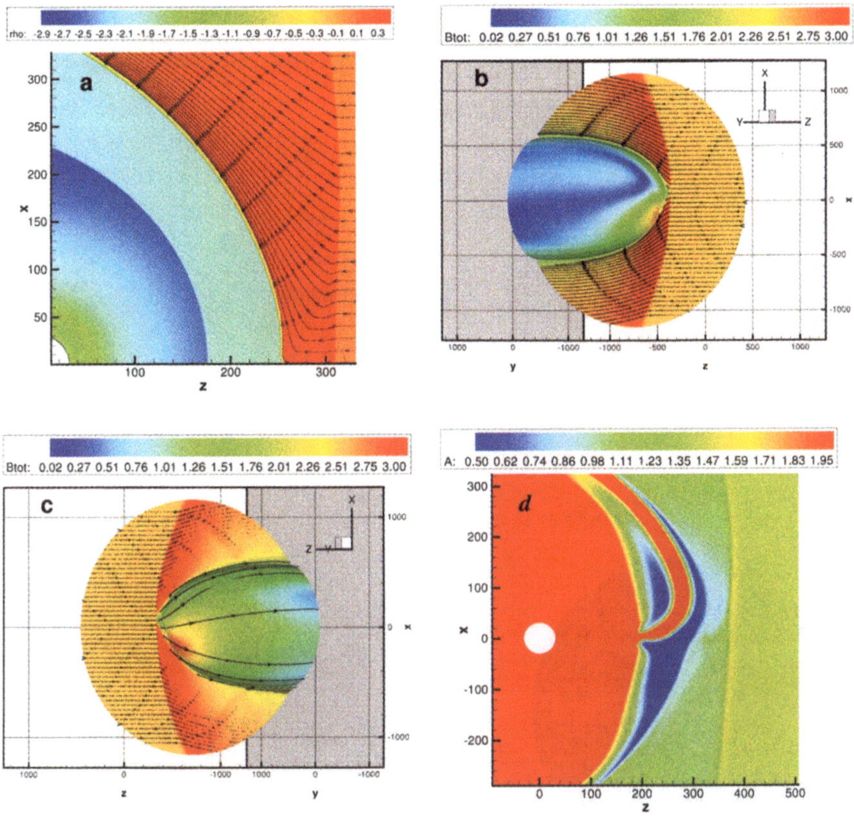

Figure 3.9. Density logarithm distributions and magnetic field lines for $B_\infty \approx 2.1\,\mu G$ in the axially symmetric case (a); left and right-hand side views of the magnetic field lines draping around the heliopause (b and c) and the Alfvén number distribution in the meridional plane (d) in the field-aligned switch-on regime in three dimensions. [ApJ, 614, 1007. ©2004. The American Astronomical Society.]

it. In Fig. 3.9d, we show the distribution of the Alfvén number A in the meridional plane. The range of color variation was restricted between 0.5 and 2.0. For this reason, values of A higher than 2 are shown with the same red color. It is clearly seen that the Alfvén number remains higher than 1 (i. e. shaded green), when the lines cross the bow shock. This implies that the bow shock is a fast mode shock. Clearly, when crossing the termination shock, we move from a region of high-A to a region of low-A flow. However, since the magnetic field substantially decreases inside the HPS, the Alfvén number remains large inside it even beyond the termination shock. Figure 3.9d shows that the HCS experiences substantial bending in this region. Our simulation shows that the HCS is slightly nonstationary, as well as causing a bump on the HP surface. It should be noted that an ideal formulation of the problem does not allow investigation of the properties of the HCS with the desired accuracy, since the HCS is itself

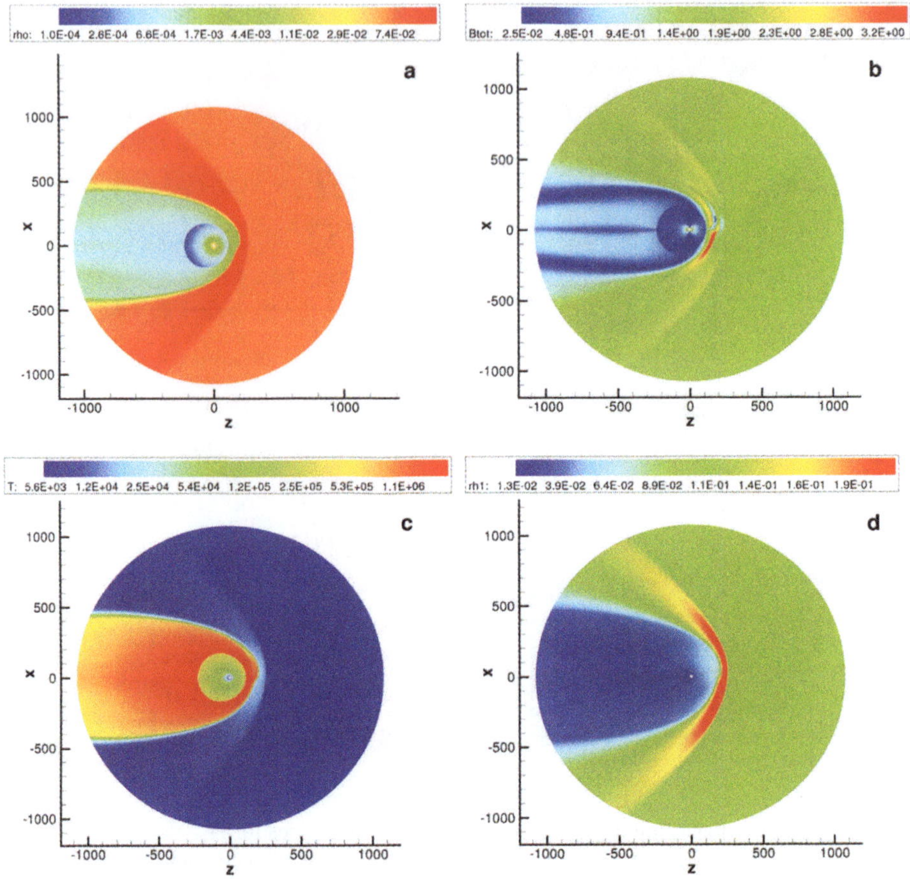

Figure 3.10. The distributions of the plasma density (a), magnetic field strength (b), plasma temperature (c), and neutral hydrogen density (d) in the meridional plane for $n_{\mathrm{H}\infty} = 0.1\ \mathrm{cm}^{-3}$ and $\mathbf{B}_\infty \parallel \mathbf{V}_\infty$. Weak field case, with $B_\infty \approx 1.5\mu$G.

dissipation dominated. Therefore, we can only follow its large-scale behavior, which nevertheless gives us some insight into the distribution of magnetic field lines inside the heliosphere. The latter issue is of considerable importance for describing the transport of galactic cosmic rays throughout the heliosphere (Florinski et al., 2003b). The distribution of magnetic pressure on the surface of the HP correlates with the behavior of the ISMF lines in its vicinity.

The results obtained show that whenever symmetry restrictions do not force magnetic field lines to remain in the meridional plane after crossing the bow shock, the solution with intermediate shocks disappears. To compare our

3D results with the axisymmetric case, we note that the distances of the TS, HP, and BS to the Sun are equal to 186, 359, and 396 AU, respectively. In the axisymmetric case the same distances are 176, 254, and 313 AU. While the TS position was not changed much, the 3D HP was pushed further from the Sun under the influence of the IMF inside the heliopause (Nerney, Suess, & Schmahl, 1993).

It is important to emphasize that mere perturbations of the axially symmetric solution will not necessarily destroy nonevolutionary shocks instantaneously, since numerical dissipation is inevitable when we solve a system of ideal MHD equations by finite-difference or finite-volume methods, and can be many orders of magnitude higher than physical dissipation in space plasmas. For this reason, all variations, which occur inside nonevolutionary shock structures almost instantaneously in nature, can last for a long time in the numerical modelling of the corresponding natural phenomena (Barmin, Kulikovskii, & Pogorelov, 1996). It should also be noted that Alfvén perturbations can simply be absent for certain problems involving either a symmetry axis or a symmetry plane.

Let us consider now the solution of the SW–LISM interaction problem for $\mathbf{B}_\infty \parallel \mathbf{V}_\infty$ and $B_\infty \approx 1.5\,\mu$G modified by the presence of the interstellar neutrals with density 0.1 cm^{-3}. Figure 3.10 shows the distributions of the plasma density (a), magnetic field magnitude (b), plasma temperature (c), and neutral hydrogen density (d). It appears that the presence of neutral particles makes the solution rather unstable in the vicinity of the equatorial plane even for resolutions lower that those in the analysis of Opher et al. (2004). The HCS, which reveals itself as a region of small B beyond the termination shock, penetrates through the magnetic barrier (showed in red color) and lies close to the inner side of the heliopause creating a bulge on it. It appears to be subject to instability and oscillates within the cavity corresponding to the bulge. Such oscillations exist as well in purely MHD calculations, but the HCS very soon bends towards one of the hemispheres and the interaction pattern becomes quasi-stationary. This process, due to stronger instabilities, lasts considerably longer in the presence of neutral atoms. Figure 3.10 shows that, finally, the heliopause for $\mathbf{B}_\infty \parallel \mathbf{V}_\infty$ acquires a typical bulging shape in the hemisphere towards which the HCS bends. The height of the IMF barrier on the inner side of the heliopause is considerably smaller in this hemisphere. This may serve as an indicator of the HCS bending direction. It is worth emphasizing that rotation of the HCS is a non-ideal phenomenon stipulated by reconnection of the IMF and ISMF lines. Once the shape of the heliopause becomes substantially asymmetric with respect to the ecliptic plane through the HCS oscillations, the ISMF begins to support the asymmetry, and indirectly orients the HCS in the corresponding hemisphere. Figures 3.10c and d show that the

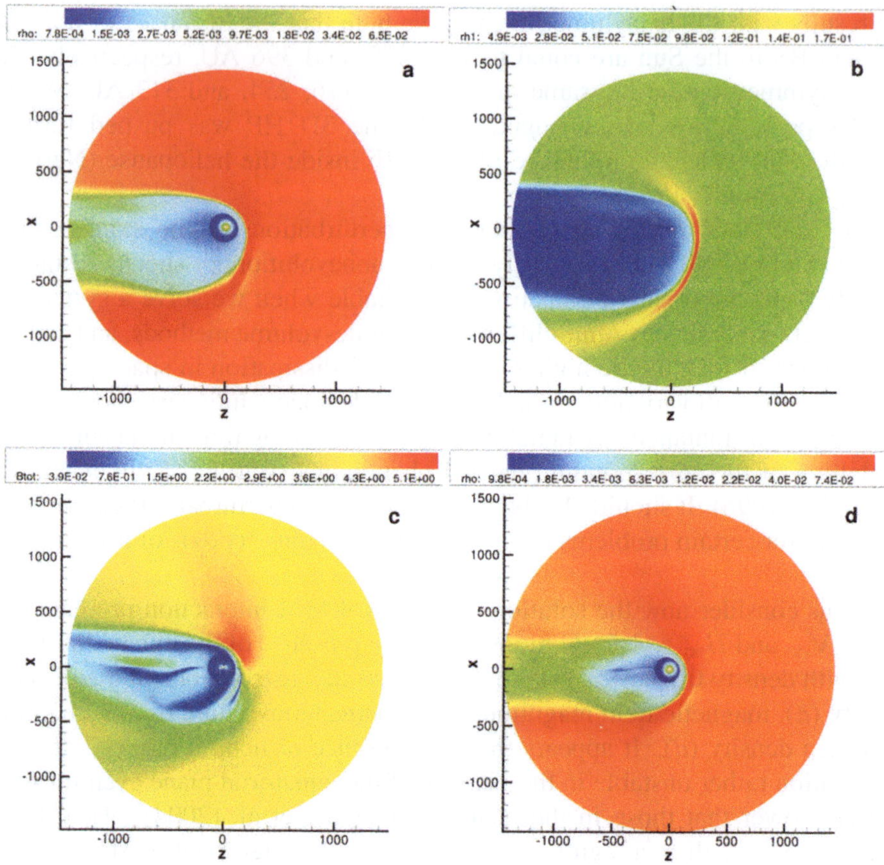

Figure 3.11. The distributions of (a) plasma and (b) neutral hydrogen density distributions, (c) the magnetic field strength for $n_{\rm H\infty} = 0.1\ {\rm cm}^{-3}$, and ($d$) the plasma density for $n_{\rm H\infty} = 0.2\ {\rm cm}^{-3}$ in the meridional plane. Strong field case, with $B_\infty = 4\ \mu$G.

plasma temperature is lower inside the bulge cavity, while the neutral hydrogen density is greater.

3.3 Subfast SW–LISM Interaction

The possibility of stronger ISMF strengths has recently attracted attention (Florinski et al., 2004) in connection with a recent theory for the LIC formation (Cox & Helenius, 2003). Although it is important to note that this scenario remains hypothetical, we would like to determine its possible influence on heliospheric structure on the basis of the MHD theory. By inspecting the linearized shock relations, Kogan (1959) discovered that if the flow is supersonic and sub-Alfvénic, in the field-aligned case, weak MHD shocks (slow) cannot exist, while the shocks themselves are obviously evolutionary. This phenomenon is related to the absence of characteristics in the corresponding elliptic region of the stationary system of MHD equations. This implies that only strong slow MHD shocks of finite spatial extent can exist in the SW–LISM interaction pattern. Another important feature of convex outward, slow MHD shocks is that they decrease the angle between the magnetic field vector and the shock normal direction. For $\mathbf{B}_\infty \parallel \mathbf{V}_\infty$, this is also true for streamlines that focus in the direction of the z-axis, as it was in a superfast switch-on regime with the concave bow shock (see the previous section). Thus, to account for the ISMF draping of the heliopause, one would require the presence of a concave bow shock. This turned out to be untrue (Aleksashov et al., 2000). In reality, there exists a preliminary rotation of the ISMF vector in a rarefaction wave ahead of the bow shock (Florinski et al., 2004). This rotation, as well as the presence of charge exchange, which stops propagation of the heliopause to infinity, as in the solution of Parker (1961), allows us to obtain a steady state solution of the SW–LISM interaction problem with the existence of a slow bow shock.

Here we present results when the angle between $\mathbf{B}_\infty$ and $\mathbf{V}_\infty$ is 15° and $B_\infty = 4\ \mu\text{G}$. The LISM flow, as previously, is parallel to the z-axis. In Fig. 3.11, we show the distributions of (a) the plasma and (b) neutral hydrogen densities, and (c) magnetic field strength for $n_{\mathrm{H}\infty} = 0.1\ \mathrm{cm}^{-3}$ and (d) of the plasma density for $n_{\mathrm{H}\infty} = 0.2\ \mathrm{cm}^{-3}$ in the meridional plane. It is remarkable that the surface of the heliopause is rather compressed from the bottom, as compared with the case of the 45° angle for weaker magnetic fields. For lower $n_{\mathrm{H}\infty}$, the TS, HP, and BS stand-off distances along the z-axis are $\sim$106, 175, and 300 AU, respectively. For higher $n_{\mathrm{H}\infty}$, they are $\sim$86, 140, and 230 AU. It appears that the density of $0.2\ \mathrm{cm}^{-3}$ gives results somewhat inconsistent with the observed location of termination shock at $R = 94$ AU (Stone et al., 2005). It is interesting to observe that the HCS abruptly turns to the southern hemisphere after crossing the termination shock. The hydrogen wall is very asymmetrical and has a limited spatial extent, which should affect Lyα absorption observations along different lines of sight. Worth noting is that the height of the hydrogen wall is smaller than in the case with $\mathbf{V}_\infty \parallel \mathbf{B}_\infty$. This can be explained by the weakness of the bow shock in the subfast regime.

3.4 Discussion

In this chapter, we discussed the solar wind interaction with the local interstellar medium in the presence of both the interplanetary and interstellar magnetic fields in a purely MHD formulation of the problem, where charge exchange of neutral atoms with hydrogen ions was neglected, and in the presence of interstellar neutral atoms. Our study at this stage is parametric, since the strength and direction of the ISMF are not well known. Theoretical predictions of possible ISMF orientations remain controversial, with the angles between $\mathbf{B}_\infty$ and $\mathbf{V}_\infty$ varying from 0 to 90°. The parallel case is the preferred explanation for the LIC and LB being in pressure equilibrium. The perpendicular case is derived from observations towards remote stars, which inevitably gives us only a highly averaged magnetic field direction. In view of the lack of reliable data on the LISM parameters, we explored possible values on the basis of our computational models of the heliosphere. Certain constraints have been established by the comparison of the observed and theoretical Lyα absorption profiles in the direction of nearby stars. However, this comparison is not very sensitive to modifications of the interaction pattern due to the ISMF strength increase, at least, in the field aligned case (Florinski et al., 2004), where B_∞ was as large as $4.3\ \mu$G. The changes induced by the increase in the ISMF strength for its other orientations are more visible. For orthogonal cases, increasing the ISMF results in a very strong contraction of the heliopause on its sides. For the same reason, the HP is pushed closer to the Sun. The bow shock moves farther upstream, since the fast magnetosonic speed approaches the LISM velocity. On the other hand, the increase in B_∞ only slightly affects the position of the termination shock in the LISM upwind direction. Thus, locating the TS will not by itself give us much information about the ISMF.

The 2–3 kHz radio emission data by Kurth & Gurnett (2003) may turn out to be very promising in determining the LISM orientation. Our calculations confirmed the presence of a preferred direction in the magnetic field ahead of the heliopause for $\mathbf{B}_\infty$ perpendicular to $\mathbf{V}_\infty$ and tilted 60° to the ecliptic plane. In this case the heliopause itself rotates by the action of the inclined magnetic field. However, such directions were also shown to exist for other ISMF orientations. We can see them even in the field-aligned case in the switch-on regime ahead of the bow shock. In principle, such directions do not necessarily coincide with the ISMF direction at infinity. Clearly, a more thorough investigation that combines both global modelling and theoretical mechanisms explaining the origin of the mentioned radio emission should be performed.

A particularly important result is the demonstrated bending (and possible rotation) of the HCS after it crosses the termination shock. It is important to note that solutions of ideal MHD problems, were they obtained, might have little physical meaning. For example, if $\mathbf{B}_\infty \perp \mathbf{V}_\infty$ and $\mathbf{B}_\infty \perp 0x$, the solution must be symmetric with respect to the ecliptic plane. In fact, as seen from Fig. 6, the solution becomes asymmetric and the bending of the HCS acts to prevent reconnection in the lower hemisphere. In principle, numerical viscosity and resistivity do not allow us to perform a very detailed investigation of the HCS. Exempli gratia, for certain orientations of the ISMF, solutions exhibit unsteadiness. However, the global behavior of the HCS is quite well resolved and its orientation may be useful in providing some information about the LIC.

Our conclusions about nonevolutionary shocks may be summarized as follows. Strictly speaking, ideal MHD excludes nonevolutionary shocks, but, on the other hand, it is has long been known that some types possess a dissipative structure. Since such structures, for certain types of shocks, are nonunique, they can accumulate perturbations until their type changes to that which does not have a stationary structure. At this stage, the initial nonevolutionary shock splits into an evolutionary shock and a nonevolutionary shock of a different type. Additional Alfvénic perturbations change the structure of the secondary nonevolutionary shock so that it asymptotically approaches an evolutionary (rotational) discontinuity. Thus, by reversing the sign of perturbations we can recover the initial structure. On the other hand, if dissipative coefficients are as small as they are in a great variety of space phenomena, the transition to the evolutionary shock structure occurs almost instantaneously. Of course, nonevolutionary shocks can exist in reality if the development of transverse perturbations is restricted by the physical situation. This takes place more frequently, however, when we reduce the dimension of the physical problem under consideration by introducing either an axis or a plane of symmetry. Solutions obtained in such cases, in reality, will inevitably exhibit a temporally transient character. However, if the problem is genuinely three-dimensional we will, sooner or later, obtain a solution without nonevolutionary shocks.

The presence of neutral atoms is essential for reproducing both quantitative and qualitative features of the heliospheric interface, since it affects plasma parameters in the whole SW–LISM interaction region. Here we included only primary hydrogen atoms that originated in the distant LISM. Numerical results obtained on the basis of a more complete, 4-fluid model will be discussed elsewhere. We do not expect them to alter the magnetic field effects qualitatively.

Finally, we should emphasize that the ISMF, even of the order of the intergalactic value of approximately 1.5 μG, substantially affects the shape of the heliopause by contracting it on the sides that are perpendicular to the magnetic field lines. The heliopause generally loses its symmetry with respect to the plane defined by the LISM velocity vector and the Sun's rotation axis for almost all angles between $\mathbf{B}_\infty$ and $\mathbf{V}_\infty$. The ISMF's effect on the inner heliosphere reveals itself through a bending of the heliospheric current sheet beyond the termination shock, which is the result of the asymmetry of the heliopause. Due to numerical resistivity, the IMF is not shielded from the ISMF at the heliopause, which leads to reconnection in certain regions. Although the effect of neutral hydrogen atoms on the solar wind region is apparently more pronounced than that of the ISMF, magnetic barriers formed on the inner side of the heliopause are of substantial importance in modulating the transport of cosmic rays into the heliosphere. It is worth noting that these magnetic barriers are asymmetric with respect to the ecliptic plane. Thin regions of enhanced magnetic field are also expected to emerge during the propagation of global merged interaction regions from the Sun to the heliopause.

Acknowledgments: This work was supported by NASA grants NAG5-12903, NNG05GD45G, and NAG5-11621 and NSF award ATM-0296114. N. P. was also partially supported by Russian Foundation for Basic Research grant 02-01-00948 and Russian Federation grant 1899.2003.1 for support of leading scientific schools. Calculations were performed on the supercomputers Fujitsu VPP5000, in the framework of the collaborative agreement between the IGPP and Solar-Terrestrial Environment Laboratory of Nagoya University, and IBM Data Star, on the basis of the Academic Associates Program in the San Diego Supercomputer Center. Extensive testing of the code has also been made on the IGPP/UCR "Lupin" cluster. The authors are grateful to Vladimir Florinski and Tatsuki Ogino for fruitful discussions.

References

Aleksashov, D. B., Baranov, V. B., Barsky, E. V., & Myasnikov, A. V. (2000). An axisymmetric magnetohydrodynamic model for the interaction of the solar wind with the local interstellar medium. *Astron. Lett.*, 26:743–749.

Balogh, A. (1996). The heliospheric magnetic field. *Space. Sci. Rev.*, 78:15–28.

Baranov, V. B., & Krasnobaev, K. V. (1971). Model of the interaction of solar wind with the interstellar medium. *Cosmic Research*, 9:568–570.

Baranov, V. B., & Malama, Y. G. (1993). Model of the solar wind interaction with the local interstellar medium - Numerical solution of self-consistent problem. *J. Geophys. Res.*, 98(A9):15157–15163.

Baranov, V. B., & Malama, Y. G. (1995). Effect of local interstellar medium hydrogen fractional ionization on the distant solar wind and interface region. *J. Geophys. Res.*, 100(A8):14755–14761.

Baranov, V. B., & Zaitsev, N. A. (1995). On the problem of the solar wind interaction with magnetized interstellar plasma. *Astron. Astrophys.*, 304:631–637.

Baranov, V. B., Ermakov, M. K., & Lebedev, M. G. (1981). A three-component model of solar wind-interstellar medium interaction: some numerical results. *Soviet Astron. Lett.*, 7:206–209.

Baranov, V. B., Krasnobaev, K. V., & Kulikovskii, A. G. (1971). A model of solar wind–ISM intercation, *Soviet Phys. Dokl.*, 15:791–793.

Barmin, A. A., Kulikovskii, A. G., & Pogorelov, N. V. (1996). Shock-capturing approach and nonevolutionary solutions in magnetohydrodynamics. *J. Comput. Phys.*, 126:77–90.

Burlaga, L. F., Ness, N. F., Wang, Y.-M., & Sheeley, N. R. (2002). Heliospheric magnetic field strength and polarity from 1 to 81 AU during the ascending phase of solar cycle 23. *J. Geophys. Res.*, 107(A11), doi:10.1029/2001JA009217.

Cox, D. P., & Helenius, L. (2003). Flux-tube dynamics and a model for the origin of the Local Fluff. *Astrophys. J.*, 583:205–228.

Cummings, A. C., Stone, E. C., & Steenberg, C. D. (2002). Composition of anomalous cosmic rays and other heliospheric ions. *Astrophys. J.*, 578:194–210.

De Sterck, H., Low, B. C., & Poedts, S. (1998). Complex magnetohydrodynamic bow shock topology in field-aligned low-β flow around a perfectly conducting cylinder. *Phys. Plasmas*, 5(11):4015–4027.

Dupuis, J., Vennes, S., Bowyer, S., Pradhan, A. K., & Thejll, P. (1995). Hot white dwarfs in the local interstellar medium: hydrogen and helium interstellar column densities and stellar effective temperatures from Extreme-Ultraviolet Explorer spectroscopy. *Astrophys. J.*, 455:574–589.

Fahr, H. J., Kausch, T., & Scherer, K. 2000, *Astron. Astrophys.*, 357, 268

Ferlet, R. (1999). The Local Interstellar Medium. *Astron. Astrophys. Rev.*, 9:153–169.

Florinski, V., Zank, G. P., & Axford, W. I. (2003). The Solar System in a dense interstellar cloud: Implications for cosmic-ray fluxes at Earth and ^{10}Be records. *Geophys. Res. Lett.*, 30(23):2206, doi:10.1029/2003GL017566.

Florinski, V., Zank, G. P., & Pogorelov, N. V. (2003). Galactic cosmic rays in the global heliosphere: an axisymmetric model. *J. Geophys. Res.*, 108(A6):1228, doi:10.1029/2002JA00965.

Florinski, V., Zank, G. P., & Pogorelov, N. V. (2005). Heliopause stability in the presence of neutral atoms: Rayleigh–Taylor dispersion analysis and axisymmetric MHD simulations. *J. Geophys. Res.*, 110, A07104, doi:10.1029/2004JA010879.

Florinski, V., Pogorelov, N. V., Zank, G. P., Wood, B. E., & Cox, D. P. (2004). On the possibility of a strong magnetic field in the Local Interstellar Medium. *Astrophys. J.*, 604:700–706.

Flynn, B., Vallerga, J., Dalaudier, F., & Gladstone, G. R. (1998). EUVE measurement of the local interstellar wind and geocorona via resonance scattering of solar He I 584-A line emission. *J. Geophys. Res.*, 103(A4):6483–6494.

Frail, D. A., Weisberg, J. M., Cordes, J. M., & Mathers, C. (1994). Probing the interstellar medium with pulsars on AU scales. *Astrophys. J.*, 436:144–151.

Frisch, P. C. (2000). The galactic environment of the Sun. *J. Geophys. Res.*, 105:10279–10290.

Frisch, P. C. (2003a). Local interstellar matter: The Apex cloud. *Astrophys. J.*, 593:868–873.

Frisch, P. C. (2003b). Boundary conditions of the heliosphere. *J. Geophys. Res.*, 108(A10):8036, doi:10.1029/2003JA009909.

Gloeckler, G., Möbius, E., Geiss, J., Bzovski, M., Chalov, S., et al. (2004). Observations of the helium focusing cone with pickup ions. *Astron. Astrophys.*, 426:845–854.

Godunov, S. K. (1972). Symmetric form of magnetohydrodynamic equations. *Numerical Methods in Mechanics of Continuous Media*, 3(1):26–34 (in Russian).

Izmodenov, V., Alexashov, D., & Myasnikov, A. (2005). Direction of the interstellar H atom inflow in the heliosphere: Role of the interstellar magnetic field. *Astron. Astrophys.*, 437:L35–L38.

Kogan, M. N. (1959). Magnetodynamics of plane and axisymmetric flows of a gas with infinite electrical conductivity. *J. Appl. Math. Mech.*, 23, 92–106.

Krimigis, S. M., Decker, R. B., Hill, M. E., Armstrong, T. P., Gloeckler, G., Hamilton, D. C., Lanzerotti, L. J., & Roelof, E. C. (2003). Voyager 1 exited the solar wind at a distance of $\sim$85 AU from the Sun. *Nature*, 426(6962):45–48.

Kulikovskii, A. G., Pogorelov, N. V., & Semenov, A. Yu. (2001). *Mathematical Aspects of Numerical Solution of Hyperbolic Systems*, Chapman & Hall / CRC, London / Boca Raton.

Kurth, W. S., & Gurnett, D. A. (2003). On the source location of low-frequency heliospheric radio emissions. *J. Geophys. Res.*, 108(A10):8027, doi:10.1029/2003JA009860.

Lallement, R. (2001). Heliopause and asteropauses. *Astrophys. Space. Sci.*, 277:205–217.

Lallement, R., Welsh, B. Y., Vergely, J. L., Crifo, F., & Sfeir, D. (2003). 3D mapping of the dense interstellar gas around the Local Bubble. *Astron. Astrophys.*, 411:447–464.

Lallement, R., Raymond, J. C., Vallerga, J., et al. (2004). Modeling of the interstellar-interplanetary helium 58.5 nm resonance glow: towards a reconciliation with the particle measurements. *Astron. Astrophys.*, 426:875–884.

Lallement, R., Quémerais, E., Bertaux, J. L., Ferron, S., Koutroumpa, D., & Pellinen, R. (2005). Deflection of the interstellar neutral hydrogen flow across the heliospheric interface. *Science*, 307:1447–1449.

Landau, L. D., & Lifshitz, E. M. (1984). *Electrodynamics of Continuous Media*, Pergamon, Oxford.

Liewer, P. C., Karmesin, S. R., & Brackbill, J. U. (1996). Hydrodynamic instability of the heliopause driven by plasma-neutral charge-exchange interactions. *J. Geophys. Res.*, 101(A8):17119–17128.

Linde, T., Gombosi, T. I., Roe, P. L., Powell, K. G., & DeZeeuw, D. L. (1998). Heliosphere in the magnetized local interstellar medium: results of a three-dimensional MHD simulation. *J. Geophys. Res.*, 103(A2):1889–1904.

Linsky, J. L., & Wood, B. E. (1996). The α Centauri line of sight: D/H ratio, physical properties of local interstellar gas, and measurement of heated hydrogen (the "Hydrogen Wall") near the heliopause. *Astrophys. J.*, 463: 254–270

Matsuda, T., & Fujimoto, Y. (1993) MHD interaction between the solar wind and local interstellar medium. In Daiguji, H., editor, *Proc. 5th Int. Symp. on Comput. Fluid Dyn.*, 2:186–193, Japan Soc. Comput. Fluid Dyn., Tokyo.

McDonald, F. B., Stone, E. C., Cummings, A. C., Heikkila, B., Lal, N., & Webber, W. R. (2003). Enhancements of energetic particles near the heliospheric termination shock. *Nature*, 426(6962):48–51.

Möbius, E., Bzovski, M., Chalov, S., Fahr, H.-J., Gloeckler, G., et al. (2004). Synopsis of the interstellar He parameters from combined neutral gas, pickup ion and UV scattering observations and related consequences. *Astron. Astrophys.*, 426:897–907.

Müller, H.-R., Zank, G. P., & Lipatov, A. S. (2000). Self-consistent hybrid simulations of the interaction of the heliosphere with the local interstellar medium. *J. Geophys. Res.*, 105(A12):27419–27438.

Myasnikov, A.V. (1997). *On the Problem of the Solar Wind Interaction with Magnetized Interstellar Plasma*. Preprint No. 585, Institute for Problems in Mechanics, Russian Academy of Sciences, Moscow.

Nerney, S., Suess, S. T., & Schmahl, E. J. (1993). Flow downstream of the heliospheric terminal shock: the magnetic field on the heliopause. *J. Geophys. Res.*, 98(A9):15169–15176.

Neutsch, W., & Fahr, H. J. (1982). The magnetic and fluid environment of an ellipsoidal circumstellar plasma cavity. *Month. Notices Roy. Astron. Soc.*, 202:735–752.

Opher, M., Liewer, P. C., Velli, M., Bettarini, L., Gombosi, T. I., Manchester, W., DeZeeuw, D. L., Toth, G., & Sokolov, I. (2004). Magnetic effects at the sdge of the solar system: MHD instabilities, the de Laval nozzle effect, and an extended jet. *Astrophys. J.*, 611:575–586.

Parker, E. N. (1961). The stellar-wind regions. *Astrophys. J.*, 134:20–27.

Pauls, H. L., & Zank, G. P. (1996). Interaction of a nonuniform solar wind with the local interstellar medium. *J. Geophys. Res.*, 101(A8):17081–17092.

Pauls, H. L., & Zank, G. P. (1997). Interaction of a nonuniform solar wind with the local interstellar medium 2. A two-fluid model. *J. Geophys. Res.*, 102(A9):19779-19788.

Pauls, H. L., Zank, G. P., & Williams, L. L. (1995). Interaction of the solar wind with the local interstellar medium. *J. Geophys. Res.*, 100:21595–21604.

Pogorelov, N. V. (1995). Periodic stellar wind/interstellar medium interaction. *Astron. Astrophys.*, 297:835–840.

Pogorelov, N. V. (2000). Nonstationary phenomena in the solar wind and interstellar medium interaction. *Astrophys. Space. Sci.*, 274:115–122.

Pogorelov, N. V. (2001). MHD modelling of the outer heliosphere: Numerical aspects. In K. Scherer et al., editors, *The Outer Heliosphere: The Next Frontiers. COSPAR Colloquia Series., 11*, pages 33–42, Pergamon, New York.

Pogorelov, N. V., & Matsuda, T. (1998). Influence of the interstellar magnetic field direction on the shape of the global heliopause. *J. Geophys. Res.*, 103(A1):237–245.

Pogorelov, N. V., & Matsuda, T. (2000). Nonevolutionary MHD shocks in the solar wind and interstellar medium interaction. *Astron. Astrophys.*, 354: 697–702.

Pogorelov, N. V., & Matsuda, T. (2004). Comment on "On the interaction of the solar wind with the interstellar medium: Field aligned MHD flow" by R. Ratkiewicz and G. M. Webb. *J. Geophys. Res.*, 109(A2):A02110, doi:10.1029/2003JA009998.

Pogorelov, N. V., & Semenov, A. Yu. (1997). Solar wind interaction with the magnetized interstellar medium. Shock-capturing modeling. *Astron. Astrophys.*, 321:330–337.

Pogorelov, N. V., & Zank, G. P. (2005) Coupling of the interstellar and interplanetary magnetic fields at the heliospheric interface: The effect of neutral hydrogen atoms. *Adv. Space Res.*, doi:10.1016/j.asr.2005.03.124.

Pogorelov, N. V., Zank, G. P., & Ogino, T. (2004). Three-dimensional features of the outer heliosphere due to coupling between the interstellar and interplanetary magnetic fields. I. Magnetohydrodynamic model: interstellar perspective. *Astrophys. J.*, 614:1007–1021.

Rand, R. J., & Lyne, A. G. (1994). New rotation measures of distant pulsars in the inner galaxy and magnetic field reversals. *Month. Notices Roy. Astron. Soc.*, 268:497–505.

Ratkiewicz, R., Barnes, A., Molvik, G. A., Spreiter, J. R., Stahara, S. S., & Vinokur, M. (1998). Effect of varying strength and orientation of local interstellar magnetic field on configuration of exterior heliosphere: 3D MHD simulations. *Astron. Astrophys.*, 335:363–369.

Scherer, K., & Fahr, H. J. (2003) Solar cycle induced variations of the outer heliospheric structures. Geophys. Res. Lett., 30, No. 2, 1045, doi:10.129/2002GL016073.

Scherer, K. H., Fichtner, H., & Stavicki, O. (2002). Shielded by the wind: the influence of the interstellar medium on the environment of Earth. *J. Atmos. Sol. Terr. Phys.*, 64:795–804.

Smith, E. J. (2001). The heliopsheric current sheet. *J. Geophys. Res.*, 106(A8):15819–15831.

Smith, E. J., Marsden, R. G., Balogh, A., Gloeckler, G., Geiss, J., McComas, D. J., McKibben, R. B., MacDowall, R. J., Lanzerotti, L. J., Krupp, N., Krueger, H., & Landgraf, M. (2003). The Sun and heliosphere at solar maximum. *Science*, 302:1165–1169.

Snowden, S., Egger, R., Finkbeiner, D. R., Freyberg, M. J., & Plucinsky, P. P. (1998). Progress on establishing the spatial distribution of material responsible for the 1/4 keV soft X-ray diffuse background local and halo components. *Astrophys. J.*, 493:715–729.

Stone, E. C., Cummings, A. C., McDonald, F. B., Heikkila, B., Lal., N., Webber, W. R. (2005). Voyager 1 explores the termination shock region and the heliosheath beyond. *Science*, in press.

Tanaka, T. & Washimi, H. (1999). Solar cycle dependence of the heliospheric shape deduced from a global MHD simulation of the interaction process between a nonuniform time-dependent solar wind and the local interstellar medium. *J. Geophys. Res.*, 104(A6):12605–12616.

Tinbergen, J. (1982). Interstellar polarization in the immediate solar neighbourhood. *Astron. Astrophys.*, 105:53–64.

Vallerga, J. V. (1996). Observations of the local interstellar medium with the Extreme Ultraviolet Explorer. *Space Sci. Rev.*, 78:277–288.

Vallerga, J. V., Lallement, R., Lemoine, M., Dalaudier, F., & McMullin, D. (2004). EUVE observations of the helium glow: interstellar and solar parameters. *Astron. Astrophys.*, 426:855–865.

Wallis, M. K. (1971). Solar wind – Shock-free deceleration? *Nature Phys. Sci.*, 233:23–24.

Wallis, M. K. (1975). Local interstellar medium. *Nature*, 254:202–203.

Washimi, H. & Tanaka, T. (1996). 3-D magnetic field and current system in the heliosphere. *Space Sci. Rev.*, 78:85–94.

Washimi, H. & Tanaka, T. (2001). A V-shaped gutter on the nose-cone surface of the heliopause caused by MHD processes. *Adv. Space RES.*, 27:509–515.

Winterhalter, D., Smith, E. J., Burton, M. E., Murphy, N., & McComas, D. J. (1994). The heliospheric plasma sheet. *J. Geophys. Res.*, 99(A4): 6667–6680.

Witte, M. (2004). Kinetic parameters of interstellar neutral helium. Review of results obtained during one solar cycle with the ULYSSES/GAS-instrument. *Astron. Astrophys.*, 426:835–844.

Witte, M., Banaszkiewicz, M., Rosenbauer, H., & McMullin, D. (2004). Kinetic parameters of interstellar neutral helium: updated results from the ULYSSES/GAS-instrument. *Adv. Space Res.*, 34:61–65.

Wolff, B., Koestner, D., & Lallement, R. (1999). Evidence for an ionization gradient in the local interstellar medium: EUVE observations of white dwarfs. *Astron. Astrophys.*, 346:969–978.

Wood, B. E., Redfield, S., Linsky, J. L., & Sahu, M. S. (2002). Elemental abundances and ionization states within the local interstellar cloud derived from Hubble Space Telescope and Far Ultraviolet Spectroscopic Explorer observations of the Capella line of sight. *Astrophys. J.*, 581:1168–1179.

Zank, G. P. (1999a). Interaction of the solar wind with the local interstellar medium: a theoretical perspective. *Space Sci. Rev.*, 89:413–688.

Zank, G. P. (1999b). The dynamical heliosphere. In Habbal, S. R. et al., editors, *Solar Wind Nine, American Institute of Physics Conf. Proc., 471*, pages 783–786, AIP, New York.

Zank, G. P., & Frisch, P. C. (1999). Consequences of a change in the galactic environment of the Sun. *Astrophys. J.*, 518:965–973.

Zank, G. P., & Müller, H.-R. (2003). The dynamical heliosphere. *J. Geophys. Res.*, 108(A6):1240, doi:10.1029/2002JA009689

Zank, G. P., Pauls, H. L., Williams, L. L., & Hall, D. T. (1996). Interaction of the solar wind with the local interstellar medium: a multifluid approach. *J. Geophys. Res.*, 101:21639–21655.

Chapter 4

INTERSTELLAR CONDITIONS AND PLANETARY MAGNETOSPHERES

Eugene N. Parker
Depts. of Physics and Astronomy and Astrophysics, and the Enrico Fermi Institute, University of Chicago, Chicago, Illinois

Abstract The size of the heliosphere is determined by the external interstellar wind conditions, with the termination shock in the solar wind presently estimated to lie at about 100AU. Over the 10^{10} year life of the Sun one expects enormous changes in the interstellar wind, from the present 25 km/sec with 0.1 ions/cm^3, 0.24 neutral atoms/cm^3 , and a temperature of the order of 6000 K. In particular, the interstellar wind may occasionally be much denser (10–100 atoms/cm^3), pushing the termination shock in perhaps as close as Saturn. This would expose the magnetospheres of Uranus (at 20 AU) and Neptune (at 30 AU) to the shocked subsonic solar wind gas beyond the termination shock. Present understanding of magnetospheric dynamics (based mostly on detailed studies of the active magnetosphere of Earth) is extrapolated to speculate on the consequences for the activity of the magnetospheres of Uranus and Neptune in such circumstances.

4.1 Introduction

The magnetospheres of the planets may be considered as wind socks, their diameters and downwind extension responding to the strength and Mach number of the supersonic solar wind, and their internal fluttering responding to the turbulence and magnetic activity of the wind. On a larger scale, the heliosphere, with its downstream wake in the local interstellar wind, serves as a wind sock in the galactic environment of the solar system. That galactic environment, or interstellar wind, has varied substantially over the 4.5×10^9 years since the Sun and solar system were formed, and it is expected to change substantially many times again in the remaining 5×10^9 years. The present interstellar wind of some 25 km/sec has an ion density estimated at 0.1 ions/cm^3 and a neutral atom density of 0.24 atoms/cm^3. The kinetic temperature is estimated to be about 6000 K and the speed of sound $(2\gamma kT/M)^{1/2}$ is approximately 12 km/sec. Thus the interstellar wind is supersonic

P. C. Frisch (ed.), Solar Journey: The Significance of our Galactic Environment for the Heliosphere and Earth, 87–98.

at Mach 2. Note then that the wind velocity of 25 km/sec moves the interstellar gas a distance of 1 pc in 4×10^4 years, and in 10^8 years it moves 2.5 kpc. The extreme inhomogeneity of the interstellar gas in the disk of the Galaxy (over scales of 1 pc and more) guarantees future wide variations in the density, temperature, ionization level, and relative speed of the interstellar wind, from densities perhaps as low as 10^{-3} ions/cm^3 at 10^6 K to densities as high as 10–10^2 atoms/cm^3 at temperatures of the order of 10^2 K (Zank and Frisch, 1999; Müller, Frisch, and Zank, 2004). Indeed there are some small regions of interstellar space where molecular hydrogen may greatly exceed densities of 10^3 /cm^3, although the probability of such encounters declines substantially with increasing density. An interstellar wind of, say, 30 atoms/cm^3 at 25 km/sec would push in the heliosphere to about a tenth is present size, with the termination shock in the supersonic solar wind moving inward to 10 AU – at about the orbit of Saturn – from its present position at 10^2 AU (Müller, Frisch, and Zank, 2004). The heliospheric interstellar wind sock would be a miniature copy of its present self.

In particular, the magnetospheres of the outer planets, Uranus and Neptune, would lie outside the termination shock and, in fact, entirely outside the heliosphere, with their magnetospheres exposed directly to the interstellar wind rather than the solar wind. The internal activity (aurorae, magnetospheric convection, and magnetic fluctuations) of the exposed magnetospheres would be significantly altered and the purpose of this article is to speculate on what we might expect.

Zank and Frisch (1999) have studied the dynamics of the heliosphere when subjected to a dense, largely neutral interstellar wind of this character. They discovered the interesting oscillatory state driven by the onset of a Rayleigh - Taylor instability brought on by mass loading of the outer heliosphere by the infalling interstellar neutral atoms. The infalling neutral atoms charge exchange with the supersonic solar wind and are immediately picked up by the transverse magnetic field in the solar wind. Conservation of momentum dictates that this mass burden slows the solar wind, perhaps to subsonic velocities in places, and more neutral atoms have time to tumble into those dense regions. The continuing outflow of wind, whose impact pressure increases inward as $1/r^2$, then drives off the slow dense regions and sets up the supersonic wind to repeat the cycle. The period of this heliospheric oscillation is of the order of a year, and one can imagine that the dynamical state of a planetary magnetosphere exposed in the unstable heliospheric region evolves through a sequence of different states during each oscillation.

Before getting into the variety of effects wrought by long term changes in the interstellar environment, it is not without interest to note the very long term variations in the planetary and heliospheric wind socks themselves, driven by the evolution of the Sun and by the evolution of the dynamos in the respective planetary interiors. Note, then, that during the first 3×10^8 years the Sun evidently produced so massive and magnetic a solar wind as to carry away most of the angular momentum, increasing the rotation period of the Sun from an initial 1–2 days to the present 25 day equatorial period. Nowadays the loss of mass and angular momentum is negligible, indicating that the rate of loss of angular momentum is presently far below what it was when the rotation period first reached 25 days. Hence that subsequent decline in the strength of the wind must be attributed to a decline in the dynamical convection and perhaps in the nonuniformity of the rotation while the Sun has remained at the 25 day overall rotation period. Whatever the cause of the decline, we may conjecture that the magnetic activity of the Sun and the strength of the solar wind will decline further during the remaining $5 x 10^9$ years that the Sun is expected to remain on the main sequence.

It would appear, then, that the planetary magnetospheres were strongly compressed and driven to a high level of activity by the early, massive, vigorous magnetic solar wind. Unfortunately we have no way to estimate the compression because we know neither the strength and turbulent state of the solar wind nor the strength of the individual planetary dipole magnetic fields in that high and far off time.

While the individual planetary wind socks were probably strongly compressed in the initial epoch, the heliospheric wind sock was certainly greatly expanded. The impact pressure NMv^2 of a solar wind with number density N of particles with individual mass M and velocity v declines with radial distance r in the amount r^{-2} as a consequence of the radial decline of N. Thus the radius of the heliosphere varies in proportion to the square root of the ratio of the impact pressure of the solar wind at some fixed distance to the impact pressure of the interstellar wind, noting that the present termination shock in the solar wind is estimated to lie at 100 AU. One can imagine a heliosphere 10 or more times larger in the early years of the Sun, given the same interstellar wind that blows today.

4.2 Future Interstellar Variations

Looking, then, to the future of the heliosphere, we may expect a modest long term decline in the strength of the wind as well as unknown changes in the magnetic fields of the planets. So we ignore these incalculable evolutionary changes and turn attention to the expected, and probably large, future variations in the interstellar wind. In particular, consider how the magnetospheres of Uranus and Neptune might be affected when the termination shock and even the heliopause are pushed inside their orbits.

For this investigation recall that the dipole field of a planet declines with radial distance R in proportion to R^{-3}. Hence the radial distance to the subsolar point on the magnetopause varies in inverse proportion to the sixth root of the impact pressure of the solar wind. It is useful to review the impact pressure of the present day solar wind at different distances from the Sun as a basis for comparison when a planetary magnetosphere finds itself in the shocked solar wind gas beyond the termination shock, or in the shocked interstellar wind gas beyond the heliopause, or even in the supersonic interstellar wind itself. Note, too, that if the interstellar wind were so massive as to penetrate all the way in to Neptune or Uranus, then the wind would probably be composed almost entirely of neutral atoms.

As standard numbers we adopt a typical solar wind density of N = 5 ions/cm^3 at the orbit of Earth with a kinetic ion temperature T = 2×10^5 K, corresponding to an ion thermal velocity of about 60 km/sec. A typical slow wind velocity is v = 400 km/sec. The thermal pressure, assuming a similar temperature for the electrons, is 2NkT = 2.8×10^{-10} dynes/cm^2, to be compared with the impact pressure NMv^2 = 1.6×10^{-8} dynes/cm^2. The density declines outward in proportion to r^{-2}, providing 1.2×10^{-2} ions/cm^3 at 20AU (Uranus) and 0.5×10^{-2} at 30 AU (Neptune), and 5×10^{-4} ions/cm^3 at the termination shock at 100AU. The impact pressure declines in the same proportion, giving 4×10^{-11} dynes/cm^2 at 20 AU, 2×10^{-11} dynes/cm^2 at 30 AU, and 0.2×10^{-11} dynes/cm^2 at 100AU. The density, temperature, and speed of the wind vary substantially, so these numbers are only nominal. The fast tenuous wind may exert twice the impact pressure of the slow wind noted here, and the occasional coronal mass ejection may give even larger transients in the impact pressure.

The magnetic field in the solar wind is typically 6×10^{-5} Gauss at an angle of 30 – 40° to the radial direction at 1 AU. Beyond 1 AU the field is principally azimuthal, declining asymptotically as r^{-1} to about 3×10^{-6} Gauss at 20 AU, 2×10^{-6} Gauss at 30 AU, and 0.6×10^{-6} Gauss at 100 AU. Thus the magnetic pressure $B^2/8\pi$ is 1.5×10^{-10}dynes/cm^{2}at 1 AU, to be compared with the thermal pressure of 2.8×10^{-10} dynes/cm^2 and the impact pressure of 160×10^{-10} dynes/cm^2. Beyond 1AU the impact pressure and the magnetic pressure both decline as r^{-2}, remaining in the ratio of about 50 to 1. The gas density declines as r^{-2} while the temperature declines relatively slowly, but with transient elevation where fast solar wind streams collide with slow streams (Parker, 1963). It is interesting to consider what effects may arise when a planetary magnetosphere is no longer exposed to this standard supersonic solar wind but finds itself beyond the terminal shock.

4.3 Magnetospheric Activity

Consider, then, the present understanding of the activity of the magnetosphere of Earth, whose close observational scrutiny has made it possible to gain some understanding of this complex phenomenon. The principles learned in this provide a means for anticipating the activity of the magnetospheres of Uranus and Neptune when they become exposed to the shocked subsonic solar wind beyond the termination shock. It must be appreciated that the theoretical inference of that future activity of the magnetospheres of Uranus and Neptune involves a double extrapolation from the picture of the active magnetosphere of Earth. First, the activities of the present magnetospheres of Uranus and Neptune, exposed to the same supersonic wind as Earth, are a matter of theoretical inference and extrapolation, for which there are the excellent reviews by Russell (2001, 2004). Second, it requires a further extrapolation to take the inferred present day activity into the hypothetical state of exposure to the subsonic wind when the termination shock is pushed closer to the Sun than one or both of the outer planets. So, in view of the two stage extrapolation, we deal with the activity of the terrestrial magnetosphere only on broad terms, treating the general dynamical effects insofar as they are understood.

Now it must be recognized that the large-scale dynamics of magnetospheric activity is magnetohydrodynamic in character, with the plasma and field moving together as a single elastic fluid with the magnetic stresses pitted against the plasma particle pressure and the plasma momentum (Parker,1996). The dynamics is driven by the rapid relative motion of the solar wind plasma and magnetic field relative to the magnetospheric plasma and field. To begin with the obvious, the magnetosphere of Earth is confined principally by the impact of the solar wind against the sunward boundary (magnetopause). There is a standing shock imme-immediately upstream from the sunward magnetopause, so that the gas in contact with the magnetopause is quite hot (10^6 K), cooling to some degree as it expands and flows around the flanks of the magnetosphere. Particles are accelerated to energies of 100 Kev or more in the standing upstream shock. This impact phenomenon confines the magnetic field of Earth to the familiar comet shape, with a geomagnetic tail stretched out in the anti-solar direction. A magnetosphere subject only to a subsonic shocked wind would seem to be less active, but we will have more to say on this later.

The activity of the magnetosphere of Earth arises from the dynamical instability of the confinement of the magnetic field. First, the solar wind may excite the Kelvin – Helmholtz instability as it streams along the flanks of the magnetosphere, tending to mix solar wind plasma into the magnetosphere and divert an occasional geomagnetic flux bundle out into the solar wind. The rippling instability of the solar wind and the geomagnetic field across the unstable magnetopause provides most of the friction that drives the quiet time magnetospheric convection, at speeds near Earth of the order of 0.1 km/sec. Once caught up in the solar wind the flux bundle is stretched out in the anti-solar direction to become part of the magnetic tail, sketched in Fig. 4.1a. The second effect, contributing even more to the geomagnetic tail, arises from magnetic reconnection between the geomagnetic flux bundles at the magnetopause and any magnetic field in the solar wind that is not precisely parallel to the (northward) geomagnetic field. In particular, the reconnection may proceed rapidly when there is a southward field in the solar wind, sketched in Fig. 4.1b (Dungey, 1961; Russell and Elphic, 1978; Russell and Walker, 1985). The reconnection process is bursty, with individual bundles of field recognized in the space borne magnetometer studies, and referred to as *flux transfer events.* Indeed when the magnetic field in the wind is southward, the bursts of reconnection may proceed so rapidly as to build a substantial accumulation of magnetic flux in the geotail, sketched in Fig. 4.1c. To put it simply, a reconnected field line is tethered to its footpoint in the terrestrial ionosphere while the field line is carried in the solar wind and stretched out along the geotail. This build up of magnetic flux in the northern (sunward directed) lobe of the geotail, and in the southern (anti-solar directed) lobe of the geotail is limited by reconnection of the field across the neutral sheet – the *plasma sheet* – between the northern and southern lobes of the geotail. Plasma pressure may keep the two lobes apart, preventing reconnection at times. But when a southward magnetic field in the solar wind provides a rapid succession of flux transfer events, the magnetic flux builds up in the geotail and sooner or later forces reconnection across the magnetic neutral sheet, restoring the topology of the field lines to their original connection between the northern and southern geomagnetic hemispheres of Earth.

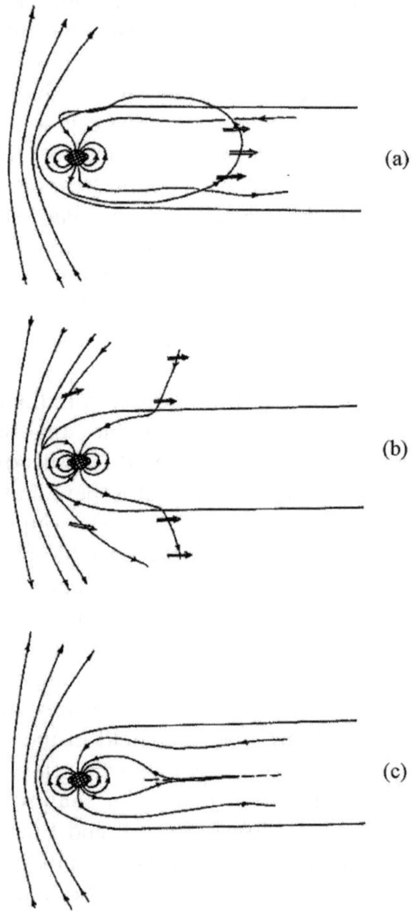

Figure 4.1: (a) Schematic drawing of a flux transfer event without reconnection of the geomagnetic field line with the magnetic field in the solar wind.
(b) With reconnection to a southward magnetic field in the solar wind.
(c) Schematic drawing of the north and south lobes of the geotail with the neutral (current) sheet between.

At the same time the increasing pressure of the magnetic field accumulating on the night side of the magnetosphere drives sunward magnetospheric convection of the field in the inner magnetosphere, at speeds inferred to be of the order of 1 km/sec. This is to be compared with the estimated 0.1 km/sec convection during quiet times, driven by the "friction" of the solar wind streaming along the dynamically unstable magnetopause.

These bursts of reconnection of the outer exposed flux bundles of the geomagnetic field, and the consequent flux transfer into the geotail, create the *substorm* phenomenon, with enhanced magnetospheric convection and enhanced aurora (Zhu, 1993, 1994, 1995). The impact of a blast wave from the Sun, originating as a coronal mass ejection and a maverick magnetic field, provides a transient compression of the magnetosphere, often setting in motion a rapid succession of strong substorms. This enhanced activity heats ambient atmospheric ions to energies of 1 – 10 Kev to such a degree in the vicinity of R = 3 – 5R_E (R_E = radius of Earth = $6.4\text{x}10^8$ cm) as to inflate the geomagnetic field. The outward inflation expands the field throughout R = 1 – 4 R_E, and the horizontal component is observed to decline at the surface of Earth. This sequence of events, starting with the initial compression, represents the classical *magnetic storm.* The initial compressive phase was identified and explained by Chapman and Ferraro (1940) in terms of the impact of intense "solar corpuscular radiation" against the sunward magnetopause. The expansive main phase of the storm was explained more recently in terms of inflation of the geomagnetic field by the pressure of internally trapped particles – ions. The inflation relaxes over periods of many hours through charge exchange between the energetic ions and the background neutral atoms in the outer atmosphere of Earth (Dessler and Parker, 1959, 1968; Dessler, Hanson, and Parker, 1961; Parker, 1962).

Akasofu constructed the empirical form $VB^2\lambda^2\left(\sin\frac{\theta}{2}\right)^4$ (Arnoldy, 1971; Meng, et al., 1973, Akasofu, 1996) for the energy input from the solar wind to the magnetosphere, where V is the wind speed, B is the field strength carried in the wind, and θ is the angle between the geomagnetic field at the sunward magnetopause and the impacting magnetic field in the wind. The scale λ represents the effective impact diameter of the magnetosphere. The dimensions are ergs/sec, and the mathematical form VB^2 represents the magnetic energy flux transported in the solar wind, while the $\left(\sin\frac{\theta}{2}\right)^4$ is a measure of the magnetic reconnection between the magnetosphere and the solar wind field, essentially the fraction of the incident magnetic energy flux that is injected into the magnetosphere by reconnection. This basic empirical relation fits the estimates of the energy input, based on the AE index, etc. of magnetospheric activity quite well. The physical concept of reconnecting the magnetic energy flux is simple and plausible, and one is tempted to extrapolate the relation to the outer planets.

Note, finally, that the curious sheet structure of the aurora appears to be a direct consequence of the flux transfer events from the sunward magnetopause. The haphazard transfer of flux bundles into the geotail produces an untidy field line topology in the tail, with the initial bundles randomly interlaced, sketched in Fig. 4.2.

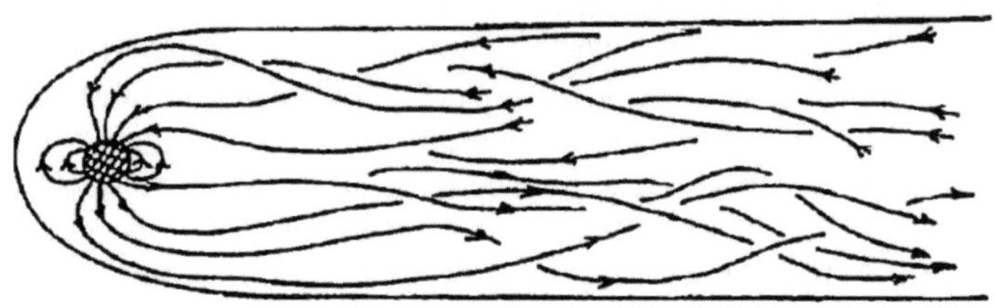

Figure 4.2: Schematic drawing of the interlaced untidy topology of the flux bundles accumulated in the geotail through flux transfer events.

In particular, the strong shear between the region of closed dipole like field lines and the flux stretched out into the geotail is a place of untidy field topology. The relaxation to equilibrium between neighboring flux bundles gives rise to surfaces of tangential discontinuity (current sheets) between confluent bundles (Parker, 1994). These surfaces of discontinuity extend all the way along the field to the ionosphere, where the concentrated electric current sheets give rise to the visible aurora (see illustrations in Akasofu, 2004). The observed squirming of the auroral sheets is presumably a consequence of the dynamical state in the magnetosphere and geotail.

4.4 Magnetic Activity at Uranus and Neptune

The next step is to employ the present understanding of these diverse phenomena to develop some idea of the activity of the magnetospheres of Uranus and Neptune while they are subject to the present supersonic solar wind. The magnetospheres of Uranus and Neptune are large in relation to the diameter of the planets themselves. While the radial distance to the sunward magnetopause of Earth is approximately 10 R_E, for Uranus and Neptune the distance is 18 and 35 times the planetary radius, respectively. There is then the additional complication that the magnetic dipole axis is inclined to the spin axis of the planet by 59° and 47°, respectively, providing a strong diurnal wobble of the magnetospheres.

Turning to Russell (2004), his Table 1 emphasizes that the magnetosonic Mach number of the solar wind increases outward from 5 or 6 at the orbit of Earth to about 10 at Uranus and Neptune, while the ratio of the thermal pressure to the magnetic pressure declines by a factor of the order of 3 as a consequence of the declining temperature of the solar wind plasma (see the numbers quoted in section II of this paper). Russell suggests that the higher Mach number at the outer planets makes reconnection between the magnetosphere and the magnetic field in the solar wind less effective, while presumably enhancing particle acceleration in the bow shock. Magnetic reconnection is the principal driver of magnetospheric convection and the cause of the substorm phenomenon, leading to the conclusion that the magnetospheres of the outer planets are substantially less active than the magnetosphere of Earth, i.e. weaker substorms. Then the impact of the coronal mass ejections is subdued because the blast wave has picked up more mass by the time it encounters the outer planets, and the impact is not as violent compared to the impact of the ambient solar wind. The rotation velocity of the magnetopause is large, of the order of 80 km/sec, but not as large as the solar wind speed of 400 km/sec or more, so it probably adds little to the activity. There are observable reconnection events (Huddleston, et al., 1996), but evidently greatly subdued. Consequently, a weaker aurora is expected. It is not immediately obvious how the rotating and wobbling inclined magnetic axis of the planet may affect the flux transfer phenomenon and the substorm activity. So the magnetospheres of Uranus and Neptune would seem to be relatively quiet under present conditions.

What happens, then, if Uranus and Neptune find themselves in the shocked solar wind gas, when the termination shock is pushed in by a large interstellar pressure? The solar wind would be subsonic, with a speed of perhaps 160 km/sec, and a sound speed of 200 km/sec. The magnetic field would be enhanced by the same factor as the density in passing through the shock, perhaps a factor of three. The confinement of the planetary magnetic field to form the magnetosphere would be largely the consequence of the pressure of the surrounding shocked gas and magnetic pressure, rather than the impact of a supersonic flow on the sunward magnetopause. We expect that magnetic reconnection between the planetary magnetosphere and the magnetic field in the wind is significantly enhanced, compared to the situation in the supersonic solar wind. The reconnected field lines are not whisked around into the tail so rapidly, but that may not affect the final result that their flux is added to the tail. Reconnection across the neutral sheet in the magnetic tail has to keep up with the net flux input to the tail. Unfortunately it is not possible to calculate the rates of the many different effects to determine the net effect. The greater rate of reconnection of the sunward magnetopause suggests enhanced activity of the general nature of the substorm, with the proportions of the different effects, reconnection and magnetospheric convection, unlike the substorms at Earth. The empirical expression $VB^2\lambda^2\left(\sin\frac{\theta}{2}\right)^4$ for the power input to the magnetosphere, that works so well for Earth, diminishes because B is smaller by 1/20 and 1/30 at Uranus and Neptune, respectively, while λ is larger by a factor of about 7 and 15, respectively. It is not at all clear how to handle the sine factor, with the strong diurnal wobble of the magnetic axis and the subsonic relative wind velocity. The sine appears to the fourth power and could be quite influential, so we will say no more.

There is another effect that may over shadow these concerns. That is the fact that an interstellar wind that is sufficient to push the termination shock inside the orbit of Neptune or Uranus is relatively dense, 10–100 atoms/cm^3, suggesting that it is mostly un-ionized gas. Neutral hydrogen blows and falls freely into the heliosphere, and much of it may get all the way to the neighborhood of Jupiter before it becomes ionized, by charge exchange with the solar wind ions and by photo-ionization in the solar UV. At the present time the interstellar neutral density is estimated at 0.24 atoms/cm^3. Upon ionization near Jupiter the atoms are picked up by the solar wind and swept out through the termination shock at 100AU. The very high thermal velocity of the pickup ions in the wind – essentially equal to the wind velocity where they were picked up – favors them for acceleration to an Mev/nucleon or more in the termination shock. From there their high velocity may cause them to diffuse inward, in opposition to the outward sweeping magnetic fields in the wind, to be observed in the inner solar system and referred to as the *anomalous cosmic rays.* The situation becomes rather different with the high density of neutral interstellar atoms that push the termination shock inside the orbits of Uranus and Neptune.

Suppose, then, that the interstellar neutral atoms were 50 times denser than at present forming an interstellar wind of, say, 10 km/sec. The number density of about 10/cm^3 yields a neutral particle flux of 10^7 atoms/cm^3, which is equal to the solar wind particle flux at the orbit of Jupiter. Converting the infalling interstellar atoms to ions through charge exchange with the outflowing solar wind ions introduces a large mass burden on the solar wind. Conservation of momentum reduces the wind velocity to about half while the new pick up ions in the wind have temperatures, i.e. thermal velocities, comparable to the wind speed of 400 km/sec or more. The temperature increase combined with the decrease in the solar wind speed would reduce the supersonic solar wind to subsonic velocity without passing through a terminal shock.

There are many effects associated with this situation. For instance, there would be none of the usual anomalous cosmic rays because there would be no shock acceleration of the pick up ions. The magnetospheres of Uranus and Jupiter would be confined more by the thermal pressure of the hot subsonic gas than by the modest impact pressure. However there would be the additional effect that the magnetospheres and ionospheres of Uranus and Neptune would be subject to infalling 10^7 neutral atoms/cm^2sec, and, in particular, to the outward moving (400–800 km/sec) neutral atoms created in the solar wind by charge exchange with the infalling interstellar neutral atoms. We estimate as many as 10^6 of these fast neutral atoms/cm^2 sec plunging into the ionosphere of Uranus and Neptune. It would be interesting to formulate quantitatively the kinetic problem posed by these diverse particle fluxes into planetary magnetospheres.

Finally, as already noted, Zank and Frisch (1999) found that the heliosphere is unstable when the interstellar wind involves a large fraction of neutral atoms, driving a forced oscillation with a period of the order of a year. So the planetary magnetospheres would be confined and driven in different ways throughout the year-long heliospheric variation. The problem becomes very complicated indeed, and the exploring theoretician cannot look for observational guidance because this oscillatory state does not exist in the present epoch. We can only speculate on the state of the planetary ionospheres and their agitation and substorm activity in the diverse wind conditions that may one day beset them.

References

Akasofu, S. I. 2004, Secrets of the aurora borealis, *Alaska Geographic* **29**, No. 1, 1–112.
Akasofu, S. I. 1996, Search for the "unknown" quantity in the solar wind: A personal account, *J. Geophys. Res.* **101**, 10531–10540.
Arnoldy, R. C. 1971, Signature in the interplanetary medium for substorms, *J. Geophys. Res.* **76**, 5189–5201.
Chapman, S. and V. C. A. Ferraro 1940, The theory of the first phase of a geomagnetic storm, *Terrestrial Magnetism and Atmospheric Electricity* **45**, 245–268.
Dessler, A. J., W. B. Hanson, and E. N. Parker 1961, Formation of the geomagnetic storm main-phase ring current, *J. Geophys. Res.* **66**, 3631–3637.
Dessler, A. J. and E. N. Parker 1959, Hydromagnetic theory of geomagnetic storms, J. Geophys. Res. 64, 2239–2252.
Dessler, A. J. and E. N. Parker 1968, Corrections to paper, Hydromagnetic theory of geomagnetic storms, *J. Geophys. Res.* **73**, 3091.
Dungey, J. W. 1961, Interplanetary magnetic field and the auroral zones, *Phys. Rev. Lett.* **6**, 47–48.
Huddleston, D. E., C. T. Russell, and G. Le, et al., 1996, Magnetopause structure and the role of reconnection at the outer planets, *J. Geophys. Res.* **102**, 24,289–22,4302.
Meng, C. I., B. Tsuratani, K. Kawasaki, and S. I. Akasofu, 1973, Cross correlation analysis of the AE index and the interplanetary magnetic field by cross correlation, J. Geophys. Res. 78, 617–629.
Müller, H. R., P. C. Frisch, and G. P. Zank 2004, Heliospheric response to different possible interstellar environments, *Astrophys. J.* in press.
Parker, E. N. 1962, Dynamics of the geomagnetic storm, *Space Sci. Rev.* **1**, 62–99.
Parker, E. N. 1963, Interplanetary Dynamical Processes, Interscience Division of John Wiley and Sons, New York, p 151.
Parker, E. N. 1994, Spontaneous current sheets in magnetic fields, Oxford University Press, New York.
Parker, E. N. 1996, The alternative paradigm, *J. Geophys. Res.* **101**, 10,587–10,625.
Russell, C. T. 2001, The dynamics of planetary magnetospheres, *Planet. Space Sci.* **49**, 1005–1030.
Russell, C. T. 2004, Outer planet magnetospheres: a tutorial, *Advances in Space Sci.* **33**, 2004–2020.
Russell, C. T. and R. C. Elphic 1978, Initial ISEE magnetometer results: magnetopause observations, *Space Sci. Rev.* **22**, 681–718.
Russell, C. T. and R.J. Walker 1985, Flux transfer events at Mercury, *J. Geophys. Res.* **90**, 11,067–11,074.
Zank, G. P. and P. C. Frisch 1999, Consequences of a change in the galactic environment of the Sun, *Astrophys. J.* **518**, 965–973.
Zhu, X. 1993, Magnetospheric convection pattern and its implications, *J. Geophys. Res.* **98**, 21, 291.

Zhu, X. 1994, The relation between magnetospheric convection and ionospheric heating, J. Geophys. Res. 99, 11,327–11,330.

Zhu, X. 1995, How the magnetosphere is driven into a substorm, *J. Geophys. Res*, **100**, 1847–1856.

Chapter 5

LONG-TERM VARIATIONS IN THE GALACTIC ENVIRONMENT OF THE SUN

Nir J. Shaviv

Racah Institute of Physics, Hebrew University, Jerusalem 91904, Israel

Abstract We review the long-term variations in the galactic environment in the vicinity of the solar system. These include changes in the cosmic ray flux, in the pressure of the different interstellar components and possibly even gravitational tides. On very long time scales, the variations arise from the variable star formation rate of the Milky Way, while on shorter scales, from passages through the galactic spiral arms and vertical oscillations relative to the galactic plane. We also summarize the various records of past variations, in meteorites, in the ocean sea floor and even in various paleoclimatic records.

Keywords: Milky Way, Solar System, Cosmic Ray Flux, Interstellar Medium, Star Formation Rate, Spiral Arms, Meteorites, Climate

5.1 Introduction

What do we mean by long term variations in the galactic environment? On the time scale of million of years, various parameters characterizing our galactic environment can vary considerably. These include the flux of cosmic rays, the thermal and non-thermal pressure components, the density of interstellar dust and its composition. Even galactic tides, which can perturb the Oort cloud, can be considered as part of the nearby galactic environment. Some of these factors will be shown to be relatively important.

On various time scales, variations in the galactic environment can arise from different physical processes. For example, very long time scales are governed by variations in the global star formation rate (SFR) of the Milky Way (MW), perhaps affected by nearby passages of the Large Magellanic Cloud (LMC).

P. C. Frisch (ed.), Solar Journey: The Significance of our Galactic Environment for the Heliosphere and Earth, 99–131.

On somewhat shorter scales, the galactic "geography" and dynamics, and in particular, its spiral structure become important. This is because the environment within spiral arms is notably more "active" than the inter-arm regions. On time scales shorter still, it is the vertical oscillations relative to the galactic plane that become important, as do local inhomogeneities in the intergalactic medium. We shall limit ourselves here to time scales longer than about 10^7 years, and therefore will not discuss the "local" inhomogeneities. These short time scale variations are elaborated in Chapter 6, and their consequences for the heliosphere are discussed in other chapters in this book.

We begin by reviewing the relevant parameters characterizing the galactic environment and their effect on the solar system. We will then continue with a lengthy description of the different processes governing the *variations* in the galactic environment and the various records we have of them.

5.2 Characterizing the Physical Environment

In this section, we briefly review the main environmental parameters influencing the solar system on long time scales. This part is merely intended as a short background. Different processes are elaborated in subsequent chapters. In particular, many of the processes are notably more important on shorter scales, and deserve more elaborate attention.

5.2.1 Solar Wind and Interaction with the ISM

The interstellar medium consists of various components, each supplying a similar contribution to the total pressure. These include the gas thermal pressure, turbulent pressure, magnetic energy density, as well as cosmic rays and ram pressures from the motion of the solar system. The ram pressure is interesting because it acts very asymmetrically, while the cosmic rays, as we will discuss further below, have additional ways of interacting with the solar system. Before detailing the components, it is important to note that the "partial" pressures of the components are comparable (e.g., Chapter 6, or Boulares and Cox, 1990). This is most likely not a coincidence, but instead the result of various equilibrating processes.

The main effect that the ISM pressure has on the solar system is through the interaction with the solar wind. At the simplest level, this can be seen as an equilibrium between the solar wind ram pressure $\dot{M} v_{wind}/4\pi R_{helio}^2$ and the total ISM pressure, $P_{\rm ISM}$, which determines the heliopause. The equilibration of the pressures yields a rough estimate of the location of the heliopause:

$$R_{helio} \sim \sqrt{\frac{v_{wind}\dot{M}}{4\pi P_{\rm ISM}}}. \tag{2.1}$$

A higher ISM pressure therefore implies a smaller heliopause, while a higher mass loss rate implies a larger heliopause.

The largest long-term variations in the total ISM pressure arise from the stratified structure, perpendicular to the galactic plane. Various models yield average profiles for each of the components (e.g., Boulares and Cox, 1990, Ferriere, 1998). The total pressure also decays with height. Over the first 100 pc, it drops by about 20%, or 35% over the first 200 pc. Thus, as the solar system oscillates around the galactic plane (with its current amplitude of about 100 pc, e.g., Bahcall and Bahcall, 1985), this stratification will result with an ISM pressure that varies with time.

More variations arise from the spiral structure. The spiral density waves are generally $\mathcal{O}(10\%)$ perturbations in the overall density, necessarily giving rise to similar variations in the pressure. However, cold components can be more notably affected by the spiral density wave and even form shocks (e.g., Binney and Tremaine, 1987).

Since typical velocities relative to the ISM are $\sim$10 km/s (for example, from the vertical oscillation), the ram pressure can be comparable to the "static" components. It becomes very important if the solar system happens to cross a molecular cloud.

Another interesting effect can take place if the ISM's gas density becomes high enough. The hitherto neglected gravitational field of the Sun can become important. As ISM gas approaches the solar system, it accelerates such that an additional ram pressure component is obtained. Once this effect becomes important, the solar wind cannot overcome the infalling ISM material, since both the solar wind and the ISM's accretion ram pressure scale as r^{-2}. Namely, if the effect begins, the solar wind quenches altogether and Bondi accretion of interstellar material takes place unhindered (Begelman and Rees, 1976). Talbot and Newman (1977) estimated that the solar system may have passed 10-100 times, during its existence, through dense enough clouds for this effect to have taken place. For a 30°K cloud, the critical density is $n_\infty \sim 100\ \mathrm{cm}^{-3}$.

Under normal conditions, however, when the solar wind is intact, it acts to shield the solar system from low energy cosmic rays and small dust grains.

Interestingly, as the solar rotation slowly decreases with time, both its non-thermal activity and mass loss, $\dot{M}$, will also decrease. This is the source of the largest change in the heliopause over time.

By now, it is well established that the non-thermal activity (as measured with X-rays), stellar rotation and winds of late type stars are related (Pallavicini et al., 1981). Thus, by observing nearby stars, it is possible to reconstruct the mass loss history of solar-like stars. Wood et al. (2002) find a mass loss which scales as:

$$\dot{M}(t) \propto t^{-2.0\pm0.5}. \tag{2.2}$$

Thus, earlier in the solar system's history, the solar wind was stronger and the heliopause located further out, such that the various effects of the solar wind were more prominent.

5.2.2 Cosmic Rays

Cosmic Rays are one of the constituents of the interstellar medium, contributing an average pressure comparable to the other ISM components. However, because they "penetrate" the solar wind, they are different from the other components.

In the frame of reference of a random magnetic field, the cosmic rays can be seen to have a random walk. However, since the cosmic rays that succeed in reaching the inner solar system necessarily had net scattering in the upwind heliosphere, these cosmic rays have lost energy, typically of order $\sim$1 GeV (e.g., see Chapters 9, 10 for detailed discussions).

Their loss of energy can be parameterized using the modulation parameter Φ. In the limit of high energies ($\gtrsim$1 GeV), the modulation parameter corresponds to the potential in the force-field approximation (Gleeson and Axford, 1968), given by $\Phi = (R_{helio} - R_E)v_{wind}/3\kappa_0$, where v_{wind} is the solar wind velocity, κ_0 is the cosmic ray diffusion constant, while R_{helio} is the size of the heliosphere, and R_E is the solar-terrestrial distance ($=$ 1 AU). Using Φ, we can relate the differential particle density flux near Earth $J(E_k)$ (as a function of the proton kinetic energy E_k) to that outside the solar system $J_0(E_k)$ (e.g., Boella et al., 1998):

$$J(E_k) = J_0(E_k + \Phi)\left[\frac{E_k(E_k + 2m_0c^2)}{(E_k + \Phi)(E_k + \Phi + 2m_0c^2)}\right]. \tag{2.3}$$

At high energies, the differential particle flux outside the solar system can be approximated with a modified power law $J_0(E_k) = C(E_k + x)^{-2.5}$, where $x = \exp(-2.5 \times 10^{-4} E_k)$ (e.g., Castagnoli and Lal, 1980). Over the solar cycle and long term solar activity variations, the modulation parameter Φ typically varies between 0.2 to 1.2 GeV, with the long-term average being $\Phi = 0.55$ GeV (Ready, 1987).

We can see that varying Φ will have a larger effect on lower energies. Typical energies of interest to us are $\sim$1 GeV, which are the energies recorded in

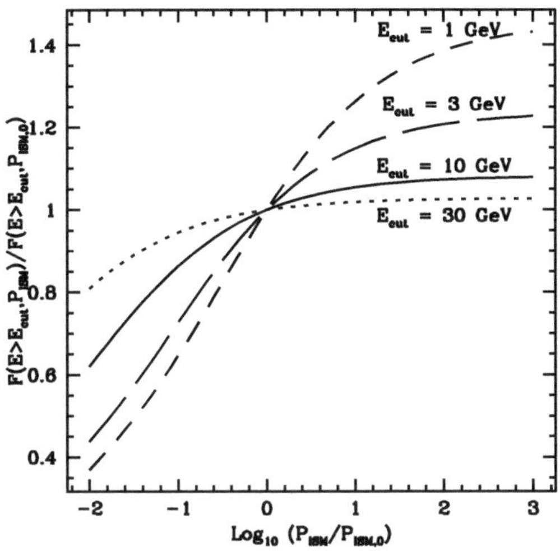

Figure 5.1. The effect that the changed ambient pressure has on the integrated energy flux of cosmic rays reaching Earth, above energies larger than $E_{cut} = 1, 3, 10$ and 30 GeV. At higher energies, the effect is smaller and becomes asymmetric. Moreover, the effect is capped as the pressure is increased—at the value of the flux outside the system.

spallation products within meteorites or the atmosphere, and $\sim$10 GeV, which presumably affect climate through modulation of the tropospheric ionization (see §5.4.3). Since some of the solar wind parameters, such as the solar wind, regularly change by as much as $\mathcal{O}(1)$ over the solar cycle or due to secular changes in solar activity, we expect variations of order $\mathcal{O}(1)$ in the spallation products, while only $\mathcal{O}(10\%)$ for the higher energies that penetrate the troposphere.

The modulation parameter, Φ, also varies through changes in $P_{\rm ISM}$ because of the dependence of Φ on R_{helio}, and R_{helio} on $P_{\rm ISM}$. The ensuing dependence of the integrated energy flux F reaching Earth, as a result of changes in the ISM pressure, is depicted in Fig. 5.1. We see that for high energies, those which are presumably responsible for climate variations, the effect of the varying ISM pressure is relatively small. On the lower energies that affect the rate of spallation, the variations can be large. Interestingly, flux changes following the reduction in the ISM pressure are more important than increases in $P_{\rm ISM}$. This is because irrespective of the increase in $P_{\rm ISM}$, the ensuing increase in J is capped by J_0. On the other hand, arbitrary reductions in $P_{\rm ISM}$ can arbitrarily increase R_{helio} and with it decrease J relative to J_0.

As an example, changing the pressure by a factor of 2 or 0.5 will change the flux by 10% or -10% at energies of order 1 GeV, but only by 2.5% or -3% at energies of order 10 GeV. Thus, the vertical oscillations relative to the galactic plane, with an amplitude of $\sim$100 pc, or $\sim$20% in the pressure, will be a -3% effect at energies larger than 1 GeV or 1% for a 10 GeV cutoff.

In addition to the above variations, which are apparently small and arise from modulations of the heliopause, long time scale variations will also arise from *intrinsic* variations in the cosmic ray flux (CRF) sources. Namely, there will be variations because our solar neighborhood may have more or less supernova taking place in its vicinity. This flux, which reaches the outskirts of solar system, is going to be proportional to the rate of nearby supernovae. This, in turn, will depend on various factors. For example, it will depend on the SFR in our galactic neighborhood, on the overall SFR in the Milky Way, or even on local inhomogeneities in the magnetic field which affect the propagation of the cosmic from their sources. This flux will have only a small energy dependence arising from the weak energy dependence of the diffusion parameters describing the diffusion of cosmic rays in the ISM.

5.3 Variations in the Galactic Environment

We continue with the study of the physical processes governing the long term variations. These are summarized in Fig. 5.2 and include global changes in the SFR (perhaps through gravitational tides induced by LMC perigalactic passages), rotation around the galaxy and the related passages through the galactic spiral arms, as well as vertical oscillations relative to the plane of the Milky Way.

5.3.1 Star Formation Rate

The local and overall SFR in the MW are not constant. These variations will in turn control the rate of supernovae. Moreover, supernova remnants accelerate cosmic rays (at least with energies $\lesssim 10^{15}$ eV), and inject fresh high-Z material into the galaxy. Thus, cosmic rays and galactic nuclear enrichment are proportional to the SFR.

Although there is a lag of several million years between the birth and death of massive stars, this lag is small when compared to the relevant time scales at question. Over the 'short term', i.e., on time scales of 10^8 yr or less, the record of nearby star formation is 'Lagrangian', i.e., the star formation in the vicinity of the moving solar system. This should record passages through galactic spiral arms. On longer time scales, of order 10^9 yr or longer, mixing is efficient enough to homogenize the azimuthal distribution in the Galaxy (Wielen, 1977). In other words, the long-term star formation rate, as portrayed by nearby stars, should record the long term changes in the Milky Way SFR activity. These

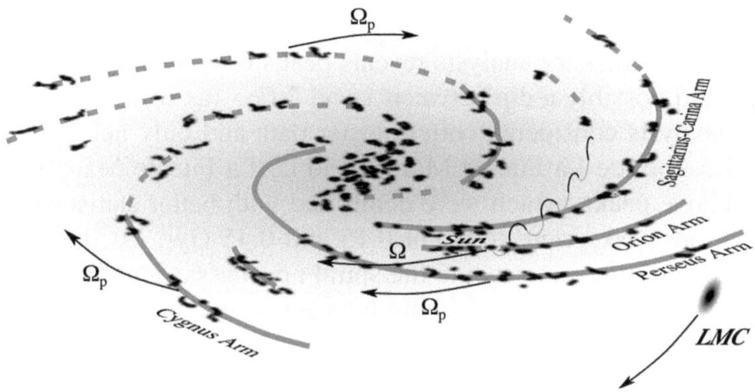

Figure 5.2. A simplified heuristic cartoon of the Milky Way and the causes for long term variations in the galactic environment. On the longest time scales (few 10^9 yr or longer), there are global variations in the MW's SFR (any variations measured using nearby stars is necessarily global through azimuthal diffusion). These may be influenced from nearby passages of the LMC—the last two perigalactic passages do appear to coincide with increased SFR in both the MW and the LMC. On time scales of 10^8 yr, we have passages through the galactic spiral arms. The actual spiral structure is probably more complicated than illustrated. In particular, it is hard to reconstruct the spiral structure within the solar circle because of the azimuthal ambiguity of velocity-longitude maps, which yield a clear structure only outside the solar circle (e.g., Blitz et al., 1983, Dame et al., 2001). Thus, spiral arms marked with a "dashed line" should be considered cautiously. Secondly, there is mounting evidence that the spiral arms observed are actually part of at least two sets (which happen to roughly coincide today), each having a different pattern speed (Shaviv, 2003a, Naoz and Shaviv, 2004). Compounded on that, we are currently located on the Orion "spur", moving at roughly the same speed as the solar system (Naoz and Shaviv, 2004). We may have passed through other such spurs in the past. On the shortest time scale, the solar system performs vertical oscillations, such that every few 10^7 yr, it crosses the galactic plane.

variations may arise, for example, from a merger with a satellite or a nearby passage of one.

Scalo (1987), using the mass distribution of nearby stars, concluded that the SFR had peaks at 0.3 Gyr and 2 Gyr before present (b.p.). Barry (1988), and a more elaborate and recent analysis by Rocha-Pinto et al., 2000, measured the star formation activity of the Milky Way using chromospheric ages of late type dwarfs. They found a dip between 1 and 2 Gyr and a maximum at 2-2.5 Gyr b.p.[1] (see also Fig. 5.3).

---

[1]There are contradicting results by Hernandez et al. (2000), but this analysis assumes that the stars in the solar vicinity have the same metalicity as the Sun, and it also assumes implicitly that the SFR today necessarily approaches zero. Thus, this discrepancy is not a source of concern.

Another approach for the reconstruction of the SFR, is to use the cluster age distribution. A rudimentary analysis reveals peaks of activity around 0.3 and 0.7 Gyr b.p., and possibly a dip between 1 and 2 Gyr (as seen in Fig. 5.3). A more recent analysis considered better cluster data and only nearby clusters, closer than 1.5 kpc (de La Fuente Marcos and de La Fuente Marcos, 2004). Besides the above peaks, which were confirmed with better statistical significance, two more peaks were found at 0.15 and 0.45 Gyr. At this temporal and spatial resolution, we are seeing the spiral arm passages. On longer time scales, cluster data reveals a notable dip between 1 and 2 Gyr (de La Fuente Marcos and de La Fuente Marcos, 2004, Shaviv, 2003a).

5.3.2 Spiral Arm Passages

On time scales shorter than those affecting global star formation in the Milky Way, the largest perturber of the local environment is our passages through the galactic spiral arms.

The period with which spiral arms are traversed depends on the relative angular speed around the galaxy, between the solar system with $\Omega_\odot$ and the spiral arms with Ω_p:

$$\Delta T = \frac{2\pi}{m\,|\Omega_\odot - \Omega_p|}, \tag{3.1}$$

where m is the number of spiral arms.

Our edge-on vantage point is unfortunate in this respect, since it complicates the determination of both the geometry and the dynamics of the spiral arms. This is of course required for the prediction of the spiral arm passages. In fact, a consensus has not been reached regarding the spiral structure of the Milky Way Galaxy.

Claims in the literature for a 2-armed and a 4-armed structure are abundant. There is even a claim for a combined $2 + 4$ armed structure (Amaral and Lepine, 1997). Nevertheless, if one examines the $v - l$ maps of molecular gas, then it is hard to avoid the conclusion that *outside* the solar circle, there are four arms[2]. Within the solar circle, however, things are far from clear. This is because $v - l$ maps become ambiguous for radii smaller than $R_\odot$, such that each arm observed can be "unfolded" into two solutions. Shaviv (2003a), has shown that if the outer 4-arms obey the simple density wave dispersion relation, such that they cannot exist beyond the 4:1 Lindblad resonances, then two sets of arms should necessarily exist. In particular, the fact that these arms are apparent out to $r_{out} \approx 2R_\odot$ necessarily implies that their inner extent, the inner Lindblad radius, should roughly be at $R_\odot$. Thus, the set of

[2]Actually, three are seen, but if a roughly symmetric set is assumed, then a fourth arm should simply be located behind the galactic center.

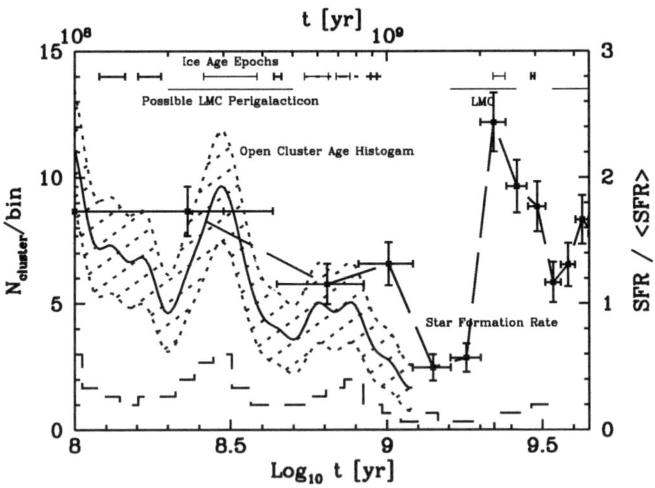

Figure 5.3. The history of the SFR. The squares with error bars are the SFR calculated using chromospheric ages of nearby stars (Rocha-Pinto et al., 2000), which is one of several SFR reconstructions available. These data are corrected for different selection biases and are binned into 0.4 Gyr bins. The line and hatched region describe a 1-2-1 average of the histogram of the ages of nearby open clusters using the Loktin et al. (1994) catalog, and the expected 1-σ error bars. These data are not corrected for selection effects (namely, the upward trend with time is a selection effect, favorably selecting younger clusters more of which did not yet dissolve). Since the clusters in the catalog used are spread to cover two nearby spiral arms, the signal arising from the passage of spiral arms is smeared, such that the graph depicts a more global SFR activity (i.e., in our galactic 'quadrant'). On longer time scales (1.5 Gyr and more), the galactic azimuthal stirring is efficient enough for the data to reflect the SFR in the whole disk. There is a clear minimum in the SFR between 1 and 2 Gyr BP, and there are two prominent peaks around 0.3 and 2.2 Gyr BP. Interestingly, the LMC perigalacticon should have occurred sometime between 0.2 and 0.5 Gyr BP in the last passage, and between 1.6 and 2.6 Gyr BP in the previous passage. This might explain the peaks in activity seen. This is corroborated with evidence of a very high SFR in the LMC about 2 Gyr BP and a dip at 0.7-2 Gyr BP (Gardiner et al., 1994, Lin et al., 1995). Also depicted are the periods during which glaciations were seen on Earth: The late Archean (3 Gyr) and mid-Proterozoic (2.2-2.4 Gyr BP) which correlate with the previous LMC perigalacticon passage (Gardiner et al., 1994, Lin et al., 1995) and the consequent SFR peak in the MW and LMC. The lack of glaciations in the interval 1-2 Gyr b.p. correlates with a clear minimum in activity in the MW (and LMC). Also, the particularly long Carboniferous-Permian glaciation correlates with the SFR peak at 300 Myr BP and the last LMC perigalacticon. The late Neo-Proterozoic ice ages correlate with a less clear SFR peak around 500-900 Myr BP. Since both the astronomical and the geological data over these long time scales have much to be desired, the correlation should be considered as an assuring consistency. By themselves, they are not enough to serve as the basis of firm conclusions.

arms internal to our galactic radius should belong to a set other than the outer four arms.

The dynamics, i.e., the pattern speed of the arms, is even less understood than the geometry. A survey of the literature (Shaviv, 2003a) reveals that about half of the observational determinations of the relative pattern speed $\Omega_\odot - \Omega_p$ cluster around $\Omega_\odot - \Omega_p \approx 9$ to 13 km s^{-1} kpc^{-1}, while the other half are spread between $\Omega_\odot - \Omega_p \approx -4$ and $5\,\text{km s}^{-1}$ kpc^{-1}. In fact, one analysis (Palous et al., 1977) revealed that both $\Omega_\odot - \Omega_p$ = 5 and 11.5 km s^{-1} kpc^{-1} fit the data equally well.

Interestingly, if spiral arms are a density wave (Lin and Shu, 1966), as is commonly believed (e.g., Binney and Tremaine, 1987, ch. 6), then the observations of the 4-armed spiral structure in HI outside the Galactic solar orbit (Blitz et al., 1983) severely constrain the pattern speed to satisfy $\Omega_\odot - \Omega_p \gtrsim 9.1 \pm 2.4$ km s^{-1} kpc^{-1}, since otherwise the four armed density wave would extend beyond the outer 4-to-1 Lindblad resonance (Shaviv, 2003a).

This conclusion provides theoretical justification for the smaller pattern speed. However, it does not explain why numerous different estimates for Ω_p exist. A resolution of this "mess" arises if we consider the possibility that at least two spiral sets exist, each one having a different pattern speed. Indeed, in a stellar cluster birth place analysis, which allows for this possibility, it was found that the Sagittarius-Carina arm appears to be a superposition of two arms (Naoz and Shaviv, 2004). One has a relative pattern speed of $\Omega_\odot - \Omega_{P,Carina,1} = 10.6^{+0.7}_{-0.5sys} \pm 1.6_{stat}$ km s^{-1} kpc^{-1} and appears also in the Perseus arm external to the solar orbit. The second set is nearly co-rotating with the solar system, with $\Omega_\odot - \Omega_{P,Carina,2} = -2.7^{+0.4}_{-0.5sys} \pm 1.3_{stat}$ km s^{-1} kpc^{-1}. The Perseus arm may too be harboring a second set. The Orion "armlet" where the solar system now resides (and which is located in between the Perseus and Sagittarius-Carina arms), appears too to be nearly co-rotating with us, with $\Omega_\odot - \Omega_{p,Orion} = -1.8^{+0.2}_{-0.3sys} \pm 0.7_{stat}$ km s^{-1} kpc^{-1}.

For comparison, the combined average of the seven previous measurements of the 9 to 13 km s^{-1} kpc^{-1} range, which appear to be an established fact for both the Perseus and Sagittarius-Carina arms, gives $\Omega_\odot - \Omega_p = 11.1 \pm 1$ km s^{-1} kpc^{-1}. At reasonable certainly, however, a second set nearly co-rotating with the solar system exists as well.

The relative velocity between the solar system and the first set of spiral arms implies that every ~150 Myr, the environment near the solar system will be that of a spiral arm. Namely, we will witness more frequent nearby supernovae, more cosmic rays, more molecular gas, as well as other activity related to massive stars. We will show below that there is a clear independent record of the passages through the arms of the first set. On the other hand, passages through arms of the second set are too infrequent for them to have been reliably recorded.

5.3.3 Vertical Motion in the Galactic Disk

In addition to the revolutions around the galaxy, the solar system also performs vertical oscillations relative to the disk. Since the potential is not Keplerian, the period of the vertical oscillations is different from the orbital period. Because mass is concentrated towards the galactic plane, the vertical motions depend primarily on the amount of mass in the disk. If one assumes a constant mass density ρ_m, then the frequency will be given by:

$$\Omega_v = \sqrt{4\pi G \rho_m}. \tag{3.2}$$

More than two dozen estimates for ρ_m range from between 0.05 to 0.25 $M_\odot\ pc^{-3}$, such that Ω_v is not known accurately enough. Although it is not clear that averaging of the different results obtained in various methods is legitimate, it yields a half period (i.e., plane crossing intervals) of $P_v/2 = 35 \pm 8$ Myr (Matese et al., 1995), or even 37 ± 4 Myr (Stothers, 1998).

Estimates for the phase of this oscillation are better constrained. This is because we are near the galactic plane, having recently crossed it and now moving upwards, away from it. This implies that we last crossed the plane roughly before $T_{cross} \approx 0 - 5$ Myr, based on the estimates of 1.5 ± 1.5 Myr (Bahcall and Bahcall, 1985), 5 ± 5 Myr (Rampino and Stothers, 1986) and 4 ± 2 Myr (Shoemaker and Wolfe, 1986).

At this periodicity, we should witness several effects. First, the stratified structure implies that the pressure closer to the center of the plane is higher. For the current amplitude, which is $\sim$100 pc, the pressure variations are $\sim 20\%$. This implies a heliopause radius which varies by typically 10%.

Second, the vertical oscillations will also imply variations in the CRF. The smaller source of variations originates from the varying size of the heliopause, thereby modulating the CRF. However, this effect is going to be of order $\sim 3\%$ at the low energies that affect spallation, and negligible at the higher energies that affect tropospheric ionization. The larger variations in the CRF will come from the vertical stratification in the CRF distribution. Typical estimates for the CRF density give a 10% decrease over the 100 pc amplitude of the vertical oscillations (e.g., Boulares and Cox, 1990).

Third, passages through the galactic plane impose galactic tides that can perturb the Oort cloud. This, in turn, can send a larger flux of comets into the inner solar system (Torbett, 1986, Heisler and Tremaine, 1989), which can either hit directly (and leave a record as craters or cause mass extinction,

Rampino and Stothers, 1984), or affect climate through disintegration in the atmosphere (Napier and Clube, 1979, Alvarez et al., 1980) or disintegrate and be accreted as dust (Hoyle and Wickramasinghe, 1978).

5.4 Records of Long Term Variations

In this section, we will concentrate on the various records registering the long period variations in the solar system's environment. These include different records of the CRF, and of accreted interstellar material.

5.4.1 Cosmic Ray Record in Iron Meteorites

Various small objects in the solar system, such as asteroids or cometary nuclei, break apart over time. Once the newly formed surfaces of the debris are exposed to cosmic rays, they begin to accumulate spallation products. Some of the products are stable and simply accumulate with time, while other products are radioactive and reach an equilibrium between the formation rate and their radioactive decay. Some of this debris reaches Earth as meteorites. Since Chondrites (i.e., stony) meteorites generally "crumble" over $\lesssim 10^8$ yr, we have to resort to the rarer Iron meteorites, which crumble over $\lesssim 10^9$ yr, if we wish to study the CRF exposure over longer time scales.

The cosmic ray exposure age is obtained using the ratio between the amount of the accumulating and the unstable nuclei. Basically, the exposure age is a measure of the integrated CRF, as obtained by the accumulating isotope, in units of the CRF "measured" using the unstable nucleus. Thus, the "normalization" flux depends on the average flux over the last decay time of the unstable isotope and not on the average flux over the whole exposure time. If the CRF is assumed constant, then the flux obtained using the radioactive isotope can be assumed to be the average flux over the life of the exposed surface. Only in such a case, can the integrated CRF be translated into a real age.

Already quite some time ago, various groups found that the exposure ages of Iron meteorites based on "short" lived isotopes (e.g., ^{10}Be) are inconsistent with ages obtained using the long lived unstable isotope ^{40}K, with a half life of $\sim$1 Gyr. In essence, the first set of methods normalize the exposure age to the flux over a few million years or less, while in the last method, the exposure age is normalized to the average flux over the lifetime of the meteorites. The inconsistency could be resolved only if one concludes that over the past few Myr, the CRF has been higher by about 30% than the long term average (Hampel and Schaeffer, 1979, Schaeffer et al., 1981, Aylmer et al., 1988, Lavielle et al., 1999).

More information on the CRF can be obtained if one makes further assumptions. Particularly, if one assumes that the parent bodies of Iron meteorites tend to break apart at a constant rate (or at least at a rate which only has slow variations), then one can statistically derive the CRF history. This was done by Shaviv (2003a), using the entire set of ^{40}K dated Iron meteorites. To reduce the probability that the breaking apart is real, i.e., that a single collision event resulted with a parent body breaking apart into many meteorites, each two meteorites with a small exposure age difference (with $\Delta a \leq 5 \times 10^7$ yr), and with the same Iron group classification, were replaced by a single effective meteor with the average exposure age.

If the CRF is variable, then the exposure age of meteorites will be distorted. Long periods during which the CRF was low, such that the exposure clock 'ticked' slowly, will appear to contract into a short period in the exposure age time scale. This implies that the exposure ages of meteorites is expected to cluster around (exposure age) epochs during which the CRF was low, while there will be very few meteors in periods during which the CRF was high.

Over the past 1 Gyr recorded in Iron meteorites, the largest variations are expected to arise from our passages through the galactic spiral arms. Thus, we expect to see cluster of ages every $\sim$150 Myr. The actual exposure ages of meteorites are plotted in Fig. 5.4, where periodic clustering in the ages can be seen. This clustering is in agreement with the expected variations in the cosmic ray flux. Namely, Iron meteorites recorded our passages through the galactic spiral arms.

Interestingly, this record of past cosmic ray flux variations and the determination of the galactic spiral arm pattern speed is different in its nature from the astronomical determinations of the pattern speed. This is because the astronomical determinations assume that the Sun remained in the same galactic orbit it currently occupies. The meteoritic measurement is 'Lagrangian'. It is the measurement relative to a moving particle, our solar system, which could have had small variations in its orbital parameters. In fact, because of the larger solar metalicity than the solar environment, the solar system is more likely to have migrated outwards than inwards. This radial diffusion gives an error and a bias when comparing the effective, i.e., 'Lagrangian' measured $\tilde{\Omega}_p$, to the 'Eularian' measurements of the pattern speed:

$$\tilde{\Omega}_p - \Omega_p = 0.5 \pm 1.5 \text{ km s}^{-1} \text{ kpc}^{-1}. \tag{4.1}$$

Taking this into consideration, the observed meteoritic periodicity, with $P = 147 \pm 6$ Myr, implies that $\Omega_\odot - \Omega_p = 10.2 \pm 1.5_{sys} \pm 0.5_{stat}$, where the systematic error arises from possible diffusion of the solar orbital parameters.

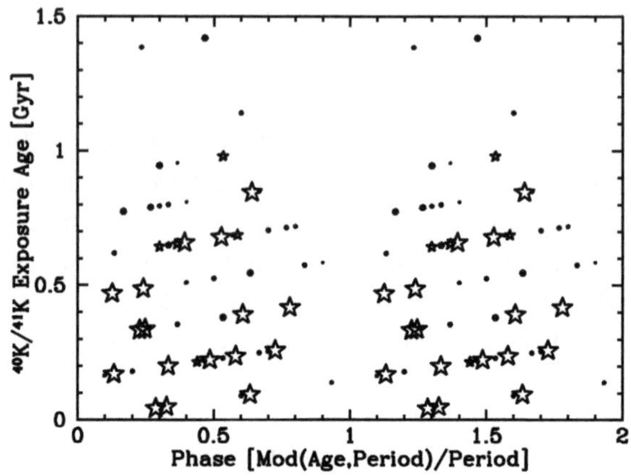

Figure 5.4. The exposure age of Iron meteorites plotted as a function of their phase in a 147 Myr period. The dots are the ^{40}K exposure ages (larger dots have lower uncertainties), while the stars are ^{36}Cl based measurements. The K measurements do not suffer from the long term "distortion" arising from the difference between the short term (10 Ma) CRF average and the long term (1 Gyr) half life of K (Lavielle et al. 1999). However, they are intrinsically less accurate. To use the Cl data, we need to "correct" the exposure ages to take into account this difference. We do so using the result of Lavielle et al. (1999). Since the Cl data is more accurate, we use the Cl measurement when both K and Cl are available for a given meteorite. When less than 50 Ma separates several meteorites of the same Iron group classification, we replace them with their average in order to discount for the possibility that one single parent body split into many meteorites. We plot two periods such that the overall periodicity will be even more pronounced. We see that meteorites avoid having exposure ages with given phases (corresponding to epochs with a high CRF). Using the Rayleigh Analysis, the probability of obtaining a signal with such a large statistical significance as a fluke from random Poisson events, with *any* period between 50 and 500 Ma, is less than 0.5%. The actual periodicity found is 147 ± 6 Myr, consistent with both the astronomical and geological data.

This result is consistent with the astronomically measured pattern speed of the first set of spiral arms.

5.4.2 Accretion of Extraterrestrial Dust

One of the interesting effects of the solar wind is the shielding against the penetration of small ISM grains. Namely, grains smaller than about 10^{-13} gr are blocked from entering the solar system through electromagnetic interaction with the solar wind's magnetic field. The flux of the more massive dust particles, which can penetrate the solar system and be accreted on Earth, is typically

$\sim$1 ton/yr globally (Frisch, 1999, Landgraf et al., 2000). However this should vary depending on the state of the heliosphere.

The interstellar origin of the dust can be identified through its kinematics (high velocity and retrograde orbits). For comparison, the flux of accreted interplanetary dust, is much higher at $\sim$100 kton/yr (Kyte and Wasson, 1986). The latter is also comparable to the long term average injection of volcanic ash into the stratosphere.

Because of the small fraction of ISM dust from the total accretion, such material can be detected only by identifying extra-solar fingerprints. Such a record exists in the form of non-cosmogenically produced radioactive isotopes. For example, ^{244}Pu has a half live $\sim 10^8$ yr. Any natural occurrence must originate from extraterrestrial dust coming from a nearby SN. However, nominal estimates for its flux are 10 times larger than the actual measurement from scraped deposition on the ocean floor. Only an upper limit could be placed, because the detected amount of ^{244}Pu is consistent with the fallout from manmade nuclear experiments (Paul et al., 2001). This is yet an unresolved problem. Ongoing measurements of ^{244}Pu in ocean floor cores will help us understand the history of the interstellar dust around us, and its relation to SNe.

As for the much larger amounts of accreted interplanetary material, the historic flux of fine planetary dust can be reconstructed using the concentration of ^{3}He extracted from ocean floor cores (Farley, 1995). This is possible once the sedimentation rate is accounted for. This reconstruction, over the past 70 Myr is plotted in Fig. 5.9.

From the figure, it is evident that the extraterrestrial dust flux is not constant. In fact, notably higher accretion rates took place recently, (in the past few Myr), at $\sim$35 Myr and $\sim$70 Myr. This is consistent in phase and period with our passages through the galactic plane, though unfortunately not yet statistically significant.

One plausible explanation is that comets are more often injected from the Oort cloud into the inner solar system, as a result of gravitational perturbations. The first suggestion was of perturbations from a gravitational impulse of nearby passage of a molecular cloud (Rampino and Stothers, 1984). Perturbations by the potential of the galactic disk were later found to be more important (Heisler and Tremaine, 1986, Torbett, 1986, Heisler et al., 1987, Matese and Whitman, 1989).

Nevertheless, because the link with plane crossing is not firm, other mechanisms are possible. For example, the Oort cloud could be periodically perturbed by a companion ("Nemesis") each time it reaches perihelion (Davis et al., 1984). In this case, the coincidence with the phase and period of the galactic plane passages should be considered a coincidence. In any case, both type of solutions still lack firm evidence for support, and therefore should not be considered more than hypotheses. Although the popularity of "Nemesis" has greatly waned, it was not ruled out altogether (Muller, 2002).

5.4.3 Climate Record and the Cosmic Ray Flux

Recent empirical evidence suggests that the cosmic ray flux may be affecting the terrestrial climate. If so, it would imply that any long term climate record could be registering the "historic" CRF variations, and with it, the environment of the solar system. We begin with a short summary of this empirical evidence (see Chapter 12 for a lengthy review of the topic), and continue with the geological record of long term climate variations.

Accumulating evidence suggests that solar activity is responsible for at least some climatic variability. These include correlations between solar activity and either direct climatic variables or indirect climate proxies over time scales ranging from days to millennia (e.g., Herschel, 1796, Eddy, 1976, Labitzke and van Loon, 1992, Lassen and Friis-Christensen, 1995, Svensmark and Friis-Christensen, 1997, Soon et al., 1996, Soon et al., 2000, Beer et al., 2000, Hodell et al., 2001, Neff et al., 2001). It is therefore difficult at this point to argue against the existence of any causal link between solar activity and climate on Earth. However, the climatic variability attributable to solar activity is larger than could be expected from the typical 0.1% changes in the solar irradiance observed over the decadal to centennial time scale (Beer et al., 2000, Soon et al., 2000). Thus, an amplifier is required unless the sensitivity to changes in the radiative forcing is uncomfortably high.

The first suggestion for an amplifier of solar activity was suggested by Ney (1959), who pointed out that if climate is sensitive to the amount of tropospheric ionization, it would also be sensitive to solar activity since the solar wind modulates the CRF, and with it, the amount of tropospheric ionization. An indeed, over the solar cycle, the solar wind strength varies considerably, such that the amount of tropospheric ionization changes by typically 5%-10% (see §2). Moreover, several recent analysis (Svensmark, 1998, Svensmark, 2000, Marsh and Svensmark, 2000, as well as Palle Bago and Butler, 2000) have shown that the variations in the amount of low altitude cloud cover (LACC) nicely correlate with the CRF reaching Earth over two decades of solar variations.

One of the points raised as a critique of the CRF→LACC link is that the empirical correlation between CRF variations and the LACC arise when comparing any solar activity index, such that a mere correlation is not sufficient to prove the link. However, as we shall see below, intrinsic variations in the CRF also correlate with climate. Moreover, a recent analysis (Usoskin et al.,

2004) reveals that geographically, the CRF/LACC correlation is as predicted. Namely, the amount of cloud cover change over the solar cycle at different latitude ranges is proportional to the change in tropospheric ionization averaged over the same ranges. Secondly, there is growing theoretical and empirical evidence linking tropospheric ionization to the formation of condensation nuclei. In particular, Yu (2002) has shown how modification of the ionization rate could affect the formation of condensation nuclei (CN), primarily at low altitudes where ions are scarce. Eichkorn et al. (2002) have shown, using airborne observations, that the formation of CN is indeed linked to charge, while Harrison and Aplin (2001) have shown that the formation of CN correlates with natural Poisson variability in the CRF. It has yet to be proved that the formation of CN does affect cloud condensation nuclei (CCN), or more specifically that CN grow by mutual coalescence as opposed to being scavenged by the already large CCN.

Chapter 12 includes a more elaborate discussion of the possible cosmic ray flux link. Here we shall explore the ramifications of this link on long, i.e., geological time scales.

Earth's glacial activity and the Star Formation Rate. Considering the longest time scales, i.e., over Earth's entire history, the largest variations in the cosmic ray flux arise from of the Milky Way's star formation activity. Over these time scales, however, there are other large sources of climate variations that we should consider if we are to look for the signature of the SFR variability.

According to standard solar models, the solar luminosity increased from about 70% of the present solar luminosity at 4.5 Gyr b.p., to its present value. If Earth were a black body, its temperature would have been 25°K lower, enough to have kept large parts of it frozen until about 1-2 Gyr b.p. Besides, however, the past Eon (0 to 1 Gyr), and the Eon between 2 and 3 Gyr b.p., it appears that glaciations were altogether absent from the global surface. This is the crux of the so called faint Sun paradox (Sagan and Mullen, 1972, Pollack, 1991, Sagan and Chyba, 1997).

Various solutions have been presented to explain this paradox. Some utilize various greenhouse gases. In particular, it was suggested that small amounts of NH_3 could have supplied the required Greenhouse gas (GHG) warming (Sagan and Mullen, 1972, Sagan and Chyba, 1997). Although not impossible, it is difficult to prevent NH_3 from irreversibly photolyzing into H_2 and N_2. Another suggestion is that CH_4 itself was the major GHG warmer (Pavlov et al., 2000). This solution requires a long residency time of methane in the atmosphere, and probably dominance of methanogenic bacteria.

Interestingly, CO_2 was also suggested as a possible greenhouse solution (Kuhn and Kasting, 1983, Kasting, 1993). However, it probably cannot resolve by itself the faint Sun paradox because the amounts required to keep Earth warm enough are huge (several bars), and more than some geological limits (e.g., Rye et al., 1995, find $p\mathrm{CO}_2 \leq 10^{-1.4}$ bar between 2.2 and 2.7 Gyr before present).

If cosmic rays do affect climate, then the faint Sun paradox can be notably extenuated, through the following scenario:

Younger Sun rotated faster and was more active
↓
The solar wind was stronger and heliopause further out
↓
CRF reaching earth was lower, as was the tropospheric ionization
↓
CRF/climate effect contributed a warming, compensating the faint Sun.

For nominal numbers obtained over the Phanerozoic (past 550 Million years), this effect translates to a compensation of about half of the faint Sun paradox (Shaviv, 2003b).

On top of the above processes, we are now in a position to consider the effect of a variable SFR. This is done by assuming that the CRF reaching the outskirts of the solar system is not constant, but proportional to the Milky Way's SFR. This is done in Fig. 5.5 while considering the SRF obtained by Rocha-Pinto et al. (2000).

The figure demonstrates that (a) the weakening solar wind and ensuing increased CRF can compensate for about half of the faint Sun paradox. Together with moderate greenhouse warming (e.g., with reasonably higher levels of CO_2), the faint Sun paradox can be resolved. (b) The varying star formation introduces a large effect. It explains why the temperature was on average colder over the last Eon and between 2 and 3 Gyr ago, enough to explain the appearance of ice-epochs, while the lower SFR between 1 and 2 Gyr before present explains the increased temperature and total lack of glacial activity. And (c) the compensation that the reduced solar wind/increased CRF has on the increasing solar luminosity is reaching its end, since almost all the CRF at $\sim$10 GeV is reaching Earth. The temperature will therefore increase in the future, albeit on very long time scales!

Ice-Age Epochs and Spiral arm passages. If the proposed CRF/climate link is real, then the periodic CRF variations arising from our passages through the galactic spiral arms, and measured in Iron meteorites, necessarily imply that cold epochs should arrive every $\sim$145 Myr on average.

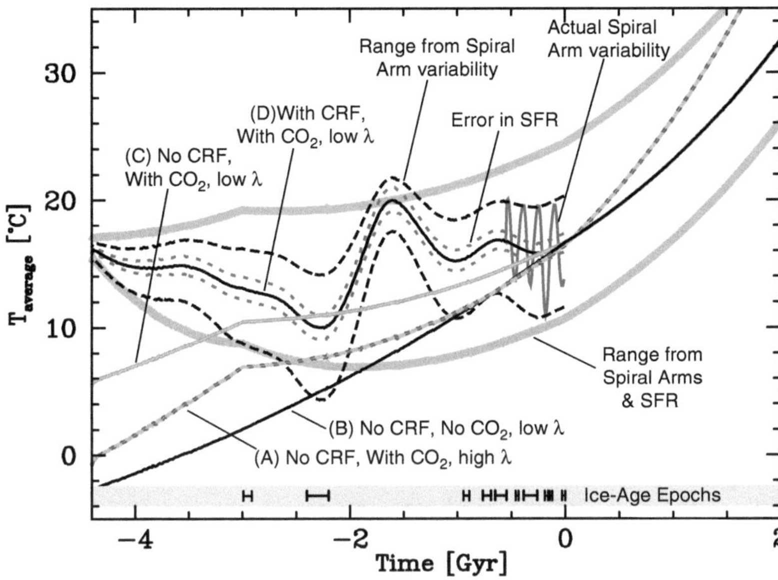

Figure 5.5. Predicted temperature as a function of time before present for various models (as elaborated in Shaviv, 2003b) but using more recent data on climate sensitivity). Model A includes a nominal global temperature sensitivity of $\lambda = 0.54°C/(W/m^2)$, obtained from various paleoclimatic data (Shaviv, 2004), while neglecting a CRF / climate link (λ is the change in the average global temperature associated with a change of 1 W/m^2 in the radiative budget). The model also assumed GHG warming by modest levels of CO_2, not larger than the geological constraints (0.01 bar of CO_2 before 3 Gyr and exponentially decreasing afterward to current levels). Model B includes no CO_2 contribution, but a lower sensitivity of $\lambda = 0.35°C/(W/m^2)$ (close to that of a black body Earth). It is the nominal sensitivity obtained from paleoclimate data while assuming the CRF/climate link is valid (Shaviv, 2004). Model C has both CO_2 and the sensitivity of model B, but still has no CRF contribution. Model D includes a CRF contribution using nominal values for its effect ($\mu \equiv -\Phi_0(dT_{global}/d\Phi_{CR}) = 6.5°$K, Shaviv, 2004). The additional lines give the variations expected or observed from spiral arm passages and SFR variations. Since these cannot be predicted, we only plot the total expected range of variations in the future. Since almost all the 10 GeV CRF (causing tropospheric ionization) currently reaches the inner solar system, this compensation cannot continue, and the temperature will start to significantly increase in the future.

Under typical diffusion parameters, the cosmic ray density contrast between spiral arms and the inter-arm region is expected to be between about 2–6 (Shaviv, 2003a). This is consistent with the meteoritic exposure age data, which gives a contrast larger than 2.5 (Shaviv, 2003a). An upper limit cannot be placed because of the statistical error in the exposure ages of the meteorites.

The large $\mathcal{O}(1)$ variations in flux imply large variations in temperature, of order $\sim 5°$K globally. Thus, each spiral arm crossing the average temperature is expected to decrease sufficiently to either hinder or trigger the appearance of ice-age epochs (IAE) on Earth.

The correlations between the different celestial and terrestrial records is seen in Figs. 5.6-5.8. The statistical significance of the correlations is difficult to assess. This is because various assumptions on the assumed priors necessarily affect the implied significance.

For example, if one *only* compares the occurrence of ice-age epochs as obtained using the sedimentation record, with the CRF signal reconstructed using the meteoritic data, then a minimum χ^2/ν of about 0.5 is obtained (with $\nu = 6$ being the effective number of degrees of freedom). The ice-age epochs lag behind the spiral arm passage by 33 ± 20 Myr, while the prediction is a lag of 28 ± 7 Myr, arising from the skewness of the CRF distribution towards the trailing side of the arm and the lag between an arm passage and the average SN event (Shaviv, 2003a).

To estimate the significance of this correlation, one can try and calculate the probability that a random distribution of IAEs could generate a χ^2 result, that is as small as previously obtained. To do so, glaciation epochs where randomly realized and compared with the occurrence of IAEs (Shaviv, 2002). To mimic the effect that nearby glaciations might appear as one epoch, glaciations that are separated by less than 60 Myr were bunched together (which is roughly the smallest separation between observed glaciations epochs). The fraction of random configurations that surpassed the χ^2 obtained for the best fit, found using the real data, is of order 0.1% for *any* pattern speed. (If glaciations are not bunched, the fraction is about 100 times smaller, while it is about 5 times larger if the criterion for bunching is a separation of 100 Myr or less). The fraction becomes roughly 6×10^{-5} (or a 4-σ fluctuation), to coincidentally fit the actual period seen in the Iron meteorites.

We see that irrespective of the exact assumptions on the priors, we find a statistically significant result. Also important is the redundancy in the correlations. Namely, there are two independent celestial signals and two independent geological signals all correlating with each other, such that the conclusions do not rest on the validity of any single set of data.

A possible record of the vertical oscillations. Various spectral analyses of paleoclimate variability have shown that climate may be exhibiting oscillations at different periods. One of these oscillations has a periodicity around $\sim$32 Myr, and appears to be more pronounced during the Mesozoic and late Paleozoic (e.g., Prokoph and Veizer, 1999, and references therein). It is of particular interest to us here because it may correspond to a signal formed as a

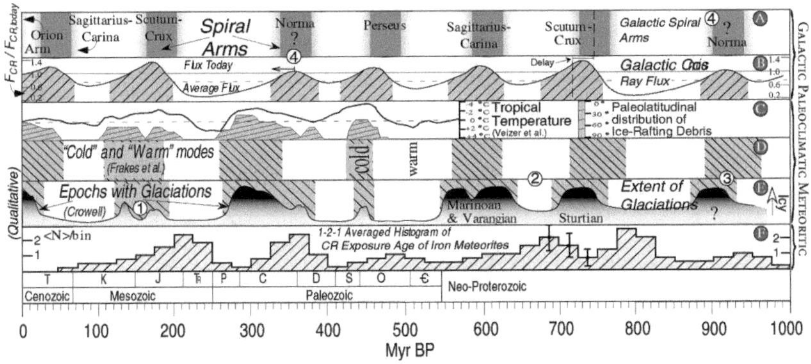

Figure 5.6. Earth's recent history. The top panel (A) describes passages through the Galactic arms assuming a relative pattern speed of $\Omega_p - \Omega_\odot = -11.0$ km s^{-1} kpc^{-1}, which best fits the ice-age epochs (IAEs). Panel (B) describes the Galactic CRF reaching the solar system using a CR diffusion model, in units of the current day CRF. An important feature is that the flux distribution around each spiral arm is lagging behind spiral arm crossings. This can be seen with the hatched regions in the second panel, which qualitatively show when IAEs are predicted to occur if the critical CRF needed to trigger them is the average CRF. Two dashed lines mark the middle of the spiral crossing and to the expected mid-glaciation point. Panels (C) (D) and (E) describe the paleoclimatological record of the past Eon. The solid line in panel (C) depicts the tropical sea surface temperatures relative to today, as inferred from calcite and aragonite shells in the past $\sim$550 Myr. The filled areas describe the paleolatitudinal distribution of ice rafted debris (both from Veizer et al., 2000). Panel (D) and (E) qualitatively describes the epochs during which Earth experienced ice-ages, the top part as described by Frakes et al. (1992), while the bottom one by Crowell (1999). The Phanerozoic part of panel E is directly taken from Crowell (1999). Note that (1) The mid-Mesozoic ice-ages did not have true polar caps but were certainly colder than average. (2) Around 650 Myr, Earth still was relatively cold, but still warmer than nearby epochs. (3) The existence of an IAE at 900 Myr is inconclusive. (4) The Norma arm's location is actually a logarithmic spiral extrapolation from its observations at somewhat smaller Galactic radii (Leitch and Vasisht, 1998, Taylor and Cordes, 1993). However, since it is now clear that the inner set probably has nothing to do with the outer spiral set, a more probable location for the "outer" Norma arm would be symmetrically between its neighboring arms. Note also that the correlations do not have to be absolute since additional factors may affect the climate (e.g., continental structure, atmospheric composition, etc.). Panel (F) is a 1-2-1 smoothed histogram of the exposure ages of Fe/Ni meteors. The meteor exposure ages are predicted to cluster around epochs with a lower CRF flux.

response to the vertical oscillations of the solar system, relative to the galactic plane.

Recently, the δ^{18}O data (Veizer, 1999), with which the Phanerozoic tropical temperature was reconstructed, was expanded such that previous gaps were

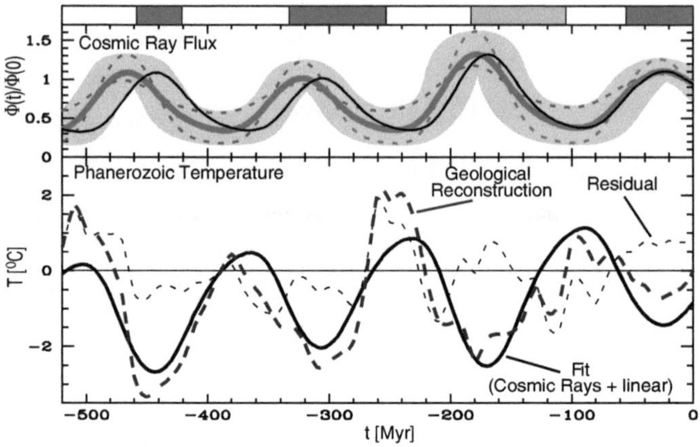

Figure 5.7. The cosmic ray flux (Φ) and tropical temperature anomaly (ΔT) variations over the Phanerozoic (Shaviv and Veizer, 2003). The upper curves describe the reconstructed CRF using iron meteorite exposure age data (Shaviv, 2003a). The heavy gray line depicts the nominal CRF, while the shading delineates the allowed error range. The two dashed curves are additional CRF reconstructions that fit within the acceptable range. The solid black curve describes the nominal CRF reconstruction after its period was fine tuned, within the measurement error, to best fit the low-latitude temperature anomaly. The bottom dashed curve depicts the temperature reconstruction, measured with a 10 Myr bin and smoothed with a 5-bin top hat averaged (Veizer et al., 2000). The solid line is the predicted ΔT based on the nominal CRF model above while also taking into account a secular long-term linear contribution. The light dashed line is the residual. The largest residual is at 250 My b.p., where only a few measurements of δ^{18}O exist due to the dearth of fossils subsequent to the largest extinction event in Earth history. The bars at the top represent cool climate modes (icehouses) and the white bars are the warm modes (greenhouses), as established from sedimentological criteria (Frakes and Francis, 1998, Frakes et al., 1992). The lighter shading for the Jurassic-Cretaceous icehouse reflects the fact that true polar ice caps have not been documented for this time interval.

filled (Prokoph et al., 2004). This allowed analysis of the temperature reconstruction at a higher resolution, with the result that in addition to the large oscillation with a 145 Myr periodicity, the 32 Myr oscillation appears very pronounced over 400 Myr (see Fig. 5.9).

Whether or not this oscillation is indeed linked to plane crossing is yet to be proved. Empirically, however, it is consistent both in phase and period with the expectation. Irrespectively, the stability of the period indicates that it is most likely of celestial origin.

The physical link to the vertical galactic oscillations is even more interesting to address. We expect that CRF variations will result from the vertical stratification, but under nominal parameters for the CRF/climate effect, the expected signal is at least two to three times smaller than actually observed (Prokoph et al., 2004). Another possibility is that the Oort cloud is periodically perturbed, and the injected comets later disintegrate and accreted. However, if the

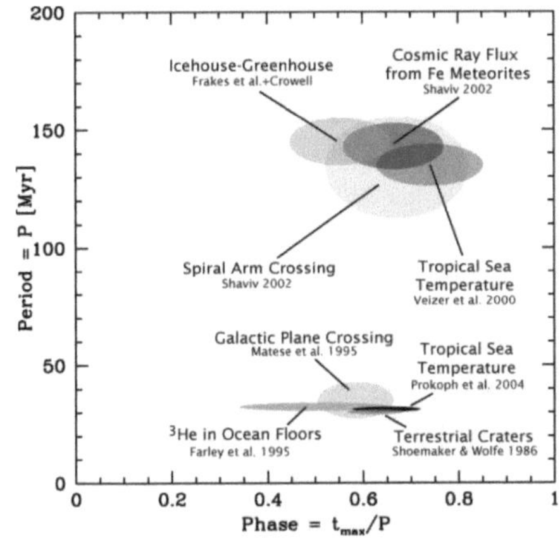

Figure 5.8. The period and phase of various measured signals. The phase is defined as the time (before present) of either observed or predicted maximum warmth, relative to the period. The four signals plotted at the top all have a $\sim$ 145 Myr periodicity. Two signals are extraterrestrial "signals". They have the same periodicity and phase as two independent terrestrial records. Plotted at the bottom are three extraterrestrial signals that are consistent with the observed 32 Myr climatic oscillation. The crater record includes a not very significant periodicity at 31 ± 1 Myr, but it is coincident with other periods.

dust has the same radiative properties as volcanic ash, which is occasionally injected into the stratosphere, the effect is about 10 times smaller than required to explain the observed δ^{18}O oscillations (Prokoph et al., 2004). Interestingly, the ^{3}He based measurements do indicate that the flux of interplanetary dust does vary by a factor of a few, and the three largest peaks do coincide with minimum temperatures of the 32 Myr oscillations (Farley, 1995). If we are underestimating the effect of the dust (e.g., if the particle size is different than volcanic ash, its residence time in stratosphere and its radiative properties will be different), we may be seeing a combined signature of the gravitational tides perturbing the Oort cloud and of a variable CRF.

5.5 Crater Record

We have seen that a periodic modulation of the accretion rate of interplanetary dust could in principle arise from the periodic perturbation to the Oort cloud. The same periodicity may also manifest itself in the record of terrestrial impact craters. Although still highly controversial, there were claims for the existence of such periodicity.

Shoemaker & Shoemaker (1993) have argued that extinct comets are more numerous than previously estimated, and concluded the about 50% of the ter-

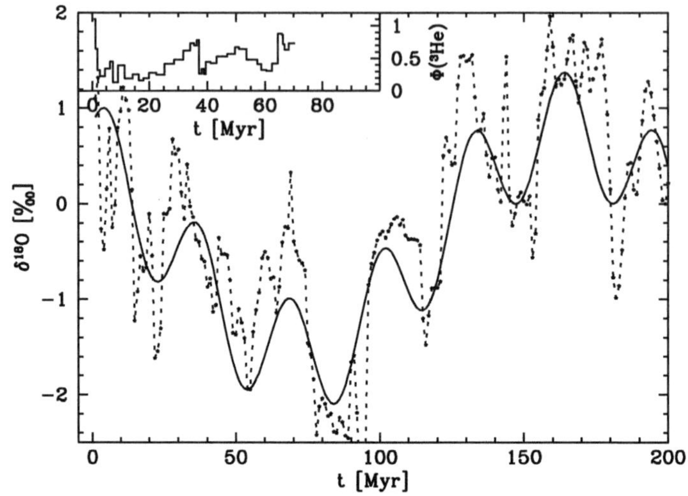

Figure 5.9. Plotted are the tropical ocean δ^{18}O record (which is primarily sensitive to the average temperature—higher δ^{18}O values imply lower temperatures) over the past 200 Myr, and a double oscillation model. The model δ^{18}O (and temperature) is assumed to have a long period oscillation, with a 145 Myr periodicity, and a short period oscillation, with a 32 Myr periodicity (Prokoph et al., 2004). The long periodicity is most likely part of the spiral arm passages. The short periodicity is consistent in both period and phase with the vertical oscillations, perpendicular to the galactic plane. The amplitude of the short period oscillation is, however, somewhat larger than would be expected from just a CRF/climate link. The inset shows the flux of ^{3}He (Farley, 1995, in units of 10^{-12} pcc cm^{-2} kyr^{-1}), which is a proxy of the extraterrestrial dust reaching Earth. The three highest peaks in the dust accretion coincide with minima temperature according to the δ^{18}O reconstruction (but not the fourth peak). These could arise from Oort cloud comets injected into the inner solar system and disintegrating each plane crossing. It could also be a coincidence.

restrial craters $\gtrsim 20$ km in diameter were due to long and short period comets and that the fraction increases with crater diameter. Moreover, Heisler et al. (1987), have argued using a Monte Carlo analysis that the null hypothesis of randomness in the cratering record can be ruled out, at the 90% c.l., if the r.m.s. of the dating error is 4 Myr or less. Thus, a periodic cratering signal could exhibit periodicity if larger accurately dated craters are considered.

Shoemaker & Wolf (1986) considered 25 young ($\lesssim 220$ Myr) accurately dated craters, with a relatively low diameter cutoff of 5 km. They noted a periodicity of 32 ± 1 Myr, with a last peak at 4 ± 2 Myr. They conclude the four most recent peaks in the crater rate (at $\approx 2, 32, 65$ and 99 Myr) appear to be too close to the predicted plane crossing events, presumably ruling out more stochastic mechanisms such as Oort cloud perturbation by the gravitational impulses of molecular clouds, particularly in view of the observed distribution of molecular clouds (Thaddeus, 1986).

In an improved statistical analysis (Yabushita, 1994), it was argued that the 31 Myr periodicity is significant at the 0.2% level for a 5 km diameter cutoff, and 0.4% level for a 10 km cutoff. Yabushita also compared the 31 Myr period with the 26 Myr period apparent in the extinction record (Raup and Sepkoski, 1984) but did not find any conclusive link.

It is interesting to note that the periodicity and phase found by Shoemaker & Wolf (1986) are consistent with the three major peaks in the Farley (1995) dust accretion data, and the $\delta^{18}O$ inferred temperature variations. Thus, craters too may be registering Oort cloud perturbations. Nevertheless, it should be stressed that the existence of the cratering periodicity is not uncontested.

5.6 Summary

Over long time scales, the three most important causes of change in the environment of the solar system, on times scales of 10^7 yr or longer, are the following:

1 Star formation rate: Over time scales of a few 10^8 yr and longer, the SFR in the solar vicinity varies. The SFR can be reconstructed using stellar clusters (typically on time scales up to several 10^8 yr, since stellar clusters disperse over longer time scales), or using the distribution of individually estimated stellar ages. The various reconstructions reveal elevated rates over the past Eon (with increased activity at $\sim$0.3 and $\sim$0.7 Gyr), before 2 to 3 Gyr, and before 4 to 5 Gyr. Estimates for the LMC's orbit give perigalactic passages coincident with the above peaks. This is probably not sufficient to prove a cause, but it is interesting to note that the LMC too appears to have had a paucity of star formation after 2 Gyr b.p. During periods of higher star formation activity, more SN produce more cosmic rays, and other cataclysmic events could happen more frequently.

2 Passage through galactic spiral arms: Every $\sim$145 Myr, the solar system passes through one of the four arms belonging to the outer spiral set. The last passage was of the Saggitarius-Carina arm. Since we are close to the co-rotation radius of a second set (which today happens to coincide with arms of the first set), arms of the second set are traversed very infrequently. Since spiral arms are the main regions of star formation activity in our galaxy, most SNe and subsequent cosmic ray acceleration takes place there. The cosmic ray density is therefore expected to be

higher near the arms. Because the ISM is more diverse inside the arms, more effects can take place while crossing them, including for example, passages through a molecular cloud or in the vicinity of massive stars.

3 Oscillations perpendicular to the galactic plane: Since the galactic potential is far from Keplerian, different perturbations to the circular motion have different periods. In particular, vertical oscillations relative to the galactic plane have a relatively "short" period, estimated at 35 ± 8 Myr (Matese et al., 1995) for example. The relatively large inaccuracy arises because of the uncertainties in the amount of matter, and in particular dark matter, in the Milky Way's disk. The current vertical amplitude is about 100 pc, notably smaller than other objects with the solar system's age. The phase is better known—the last plane crossing took place a few Myr before present.

The above variations in the solar system's environment are more than hypothetical. In fact, very diverse records from different disciplines (other than the astronomical data used to "predict" these variations) registered the above variability. In particular,

1 Iron Meteorites record their exposure to cosmic rays on time scales of $\gtrsim 10^8$ yr, and can therefore be used to reconstruct variations in the CRF. The inconsistency between ^{40}K based exposure ages of individual meteorites, and those which are derived from "short" half-lived radioisotopes, has been long used to show that the CRF over the past few million years has been higher than the average over the past 1 Gyr. Using a statistical analysis of the Iron meteorites exposure ages, it is possible to reconstruct the actual CRF variations and obtain a clear ~145 periodicity in the CRF history. Thus, Iron meteorites recorded our spiral arm passages.

2 ^3He in sea floor sedimentation is sensitive to the flux of extraterrestrial dust. Measurements sampling these data for the past 70 millions years reveal four major peaks. The larger three have a ~33 Myr periodicity and a phase consistent with the vertical oscillations. One plausible explanation is that, with each plane-crossing, the Oort cloud was gravitationally perturbed so that more comets were injected into the inner solar system, where they subsequently disintegrated and accreted as dust. The peaks are unlikely to result from variations in accreted ISM dust, because the ISM dust flux is minute relative to accreted solar system material. Deeper sea floor excavations may prove or disprove this picture, if peaks at ~102 Myr and ~145 Myr could be detected. Direct

reconstruction of the accretion of interstellar dust will soon be available through measurement of ^{244}Pu in sea floor cores, thereby providing a real temporal record of the dusty environment outside the solar system.

3 Paleoclimate Records: A growing body of evidence suggests that climate is sensitive to the amount of tropospheric ionization, governed by the changes in the cosmic ray ionization. Namely, climate proxies could be registering changes in the CRF. The paleoclimate itself can be reconstructed using sedimentation data (e.g., deposits left by glaciers, by evaporating salt lakes, etc.) or using isotopic data (i.e., δ^{18}O in fossils, which is sensitive to the oceanic water temperature and amount of ice-sheets). And indeed, the paleoclimatic data appears to have recorded all the three predicted temperature variations arising from the varying galactic environment. Long term glacial activity on Earth (in the past eon and between 2.2 and 2.7 Gyr before present) correlates with star formation activity. Over the past eon, seven spiral arm passages, as recorded in the meteoritic data, coincide with ice-age epochs on Earth during which glaciations were present. Last, the δ^{18}O temperature reconstructions reveal a remarkable $\sim$32 Myr periodicity over at least 12 cycles. This periodicity is consistent with the vertical oscillations (in particular their phase). If indeed these variations correspond to the vertical oscillations, then an unresolved problem is their amplitude, which is over twice as large as naively expected from the CRF/climate link. One possibility is that we underestimate the vertical stratification of the cosmic ray flux distribution. Another possibility is that the accreted dust can markedly affect climate as well.

References

Alvarez, L. W., Alvarez, W., Asaro, F., and Michel, H. V. (1980). Extraterrestrial cause for the Cretaceous Tertiary extinction. *Science*, 208:1095–1108.

Amaral, L. H. and Lepine, J. R. D. (1997). A self-consistent model of the spiral structure of the galaxy. *Mon. Not. Roy. Astr. Soc.*, 286:885–894.

Aylmer, D., Bonanno, V., Herzog, G. F., Weber, H., Klein, J., and Middleton, R. (1988). ^{26}Al and ^{10}Be production in iron meteorites. *Earth and Plan. Sci. Lett.*, 88:107–118.

Bahcall, J. N. and Bahcall, S. (1985). The sun's motion perpendicular to the galactic plane. *Nature*, 316:706–708.

Barry, D. C. (1988). The chromospheric age dependence of the birthrate, composition, motions, and rotation of late F and G dwarfs within 25 parsecs of the sun. *Astrophys. J.*, 334:436–448.

Beer, J., Mende, W., and Stellmacher, R. (2000). The role of the sun in climate forcing. *Quatenary Science Reviews*, 19:403–415.

Begelman, M. C., Rees, M. J. (1976). Can cosmic clouds cause climatic catastrophes. *Nature*, 261: 298–301.

Binney, J. and Tremaine, S. (1987). *Galactic dynamics*. Princeton, NJ, Princeton University Press, 747 p.

Blitz, L., Fich, M., and Kulkarni, S. (1983). The new Milky Way. *Science*, 220:1233–1240.

Boella, G., Gervasi, M., Potenza, M. A. C., Rancoita, P. G., and Usoskin, I. (1998). Modulated antiproton fluxes for interstellar production models. *Astroparticle Physics*, 9:261–267.

Boulares, A. and Cox, D. P. (1990). Galactic hydrostatic equilibrium with magnetic tension and cosmic-ray diffusion. *Astrophys. J.*, 365:544–558.

Castagnoli, G. C. and Lal, D. (1980). Solar modulation effects in terrestrial production of carbon 14. *Radiocarbon*, 22:133–158.

Crowell, J. C. (1999). *Pre-Mesozoic ice ages: their bearing on understanding the climate system*, volume 192. Memoir Geological Society of America.

Dame, T. M., Hartmann, D., and Thaddeus, P. (2001). The Milky Way in molecular clouds: A new complete CO survey. *Astrophys. J.*, 547:792–813.

Davis, M., Hut, P., and Muller, R. A. (1984). Extinction of species by periodic comet showers. *Nature*, 308:715–717.

de La Fuente Marcos, R. and de La Fuente Marcos, C. (2004). On the recent star formation history of the Milky Way disk. *New Astronomy*, 9:475–502.

Eddy, J. (1976). The Maunder minimum. *Science*, 192:1189–1202.

Eichkorn, S., Wilhelm, S., Aufmhoff, H., Wohlfrom, K. H., and Arnold, F. (2002). Cosmic ray-induced aerosol formation: first observational evidence from aircraft based ion mass spectrometer measurements in the upper troposphere. *Geophys. Res. Lett.*, 29:10.1029/2002GL015044.

Farley, K. A. (1995). Cenozoic variations in the flux of interplanetary dust recorded by ^{3}He in a deep-sea sediment. *Nature*, 376:153–156.

Ferriere, K. (1998). Global model of the interstellar medium in our galaxy with new constraints on the hot gas component. *Astrophys. J.*, 497:759–776.

Frakes, L. A., Francis, E., and Syktus, J. I. (1992). *Climate modes of the Phanerozoic; the history of the Earth's climate over the past 600 million years*. Cambridge: Cambridge University Press, 1992.

Frakes, L. A. and Francis, J. E. (1998). A guide to Phanerozoic cold polar climates from high-latitude ice rafting in the Cretaceous. *Nature*, 333:547–549.

Frisch, P. C. et al. (1999). Dust in the local interstellar wind. *Astrophys. J.*, 525:492–516.

Gardiner, L. T., Sawa, T., and Fujimoto, M. (1994). Numerical simulations of the Magellanic system - part one - orbits of the Magellanic Clouds and the global gas distribution. *Mon. Not. Roy. Astr. Soc.*, 266:567–582.

Gleeson, L. J. and Axford, W. I. (1968). Solar modulation of galactic cosmic rays. *Astrophys. J.*, 154:1011–1026.

Hampel, W. and Schaeffer, O. A. (1979). ^{26}Al in iron meteorites and the constancy of cosmic ray intensity in the past. *Earth Planet. Sci. Lett.*, 42:348–358.

Harrison, R.G. and Aplin, K.L. (2001). Atmospheric condensation nuclei formation and high-energy radiation. *J. Atmos. Solar-Terr. Phys.*, 63:1811–1819.

Heisler, J. and Tremaine, S. (1986). The influence of the galactic tidal field on the Oort comet cloud. *Icarus*, 65:13–26.

Heisler, J. and Tremaine, S. (1989). How dating uncertainties affect the detection of periodicity in extinctions and craters. *Icarus*, 77:213–219.

Heisler, J., Tremaine, S., and Alcock, C. (1987). The frequency and intensity of comet showers from the oort cloud. *Icarus*, 70:269–288.

Hernandez, X., Valls-Gabaud, D., and Gilmore, G. (2000). The recent star formation history of the Hipparcos solar neighbourhood. *Mon. Not. Roy. Astr. Soc.*, 316:605–612.

Herschel, W. (1796). Some remarks on the stability of the light of the sun. *Philosophical Transactions of the Royal Society, London*, page 166.

Hodell, D. A., Brenner, M., Curtis, J. H., and Guilderson, T. (2001). Solar forcing of drought frequency in the Maya lowlands. *Science*, 292:1367–1370.

Hoyle, F. and Wickramasinghe, C. (1978). Comets, ice ages, and ecological catastrophes. *Astrophys. Sp. Sci.*, 53:523–526.

Kasting, J. F. (1993). Earth's early atmosphere. *Science*, 259:920–926.

Kuhn, W. R. and Kasting, J. F. (1983). Effects of increased CO_2 concentrations on surface temperature of the early earth. *Nature*, 301:53–55.

Kyte, F. T. and Wasson, J. T. (1986). Accretion rate of extraterrestrial matter - Iridium deposited 33 to 67 million years ago. *Science*, 232:1225–1229.

Labitzke, K. and van Loon, H. (1992). Association between the 11-year solar cycle and the atmosphere. Part V: Summer. *Journal of Climate*, 5:240–251.

Landgraf, M., Baggaley, W. J., Grün, E., Krüger, H., and Linkert, G. (2000). Aspects of the mass distribution of interstellar dust grains in the solar system from in situ measurements. *J. Geopys. Res.*, 105:10343–10352.

Lassen, K. and Friis-Christensen, E. (1995). Variability of the solar cycle length during the past five centuries and the apparent association with terrestrial climate. *J. Atmos. Terr. Phys.*, 57:835–845.

Lavielle, B., Marti, K., Jeannot, J., Nishiizumi, K., and Caffee, M. (1999). The ^{36}Cl-^{36}Ar-^{40}K-^{41}K records and cosmic ray production rates in iron meteorites. *Earth Planet. Sci. Lett.*, 170:93–104.

Leitch, E. M. and Vasisht, G. (1998). Mass extinctions and the sun's encounters with spiral arms. *New Astronomy*, 3:51–56.

Lin, C. C. and Shu, F. H. 1966. On the Spiral Structure of Disk Galaxies, II. Outline of a Theory of Density Waves. *Proc. Natl. Ac. Sci.*, 55: 229–234.

Lin, D. N. C., Jones, B. F., and Klemola, A. R. (1995). The motion of the Magellanic clouds, origin of the Magellanic Stream, and the mass of the Milky Way. *Astrophys. J.*, 439:652–671.

Loktin, A. V., Matkin, N. V., and Gerasimenko, T. P. (1994). Catalogue of open cluster parameters from UBV-data. *Astronomical and Astrophysical Transactions*, 4:153.

Marsh, N. and Svensmark, H. (2000). Cosmic rays, clouds, and climate. *Space Science Reviews*, 94:215–230.

Matese, J. J. and Whitman, P. G. (1989). The galactic disk tidal field and the nonrandom distribution of observed Oort cloud comets. *Icarus*, 82:389–401.

Matese, J. J., Whitman, P. G., Innanen, K. A., and Valtonen, M. J. (1995). Periodic modulation of the Oort cloud comet flux by the adiabatically changing galactic tide. *Icarus*, 116:255–268.

Muller, R.M̃. (2002). Measurement of the lunar impact record for the past 3.5 b.y. and implications for the Nemesis theory. In "Catastrophic Events and Mass Extinctions: Impacts and Beyond", eds., Koeberl C., and MacLeod K. G., (*Geo. Soc. Am.*), p. 659.

Naoz, S. and Shaviv, N. J. (2004). Open star clusters and the Milky Way's spiral arm dynamics. Submitted to *New Astron.*

Napier, W. M. and Clube, S. V. M. (1979). A theory of terrestrial catastrophism. *Nature*, 282:455–459.

Neff, U., Burns, S. J., Mangnini, A., Mudelsee, M., Fleitmann, D., and Matter, A. (2001). Strong coherence between solar variability and the monsoon in Oman between 9 and 6 kyr ago. *Nature*, 411:290–293.

Ney, E. P. (1959). Cosmic radiation and weather. *Nature*, 183:451.

Pallavicini, R., Golub, L., Rosner, R., Vaiana, G. S., Ayres, T., Linsky, J. L. 1981. Relations among stellar X-ray emission observed from Einstein, stellar rotation and bolometric luminosity. *Astrophys. J.*, 248:279–290.

Palle Bago, E. and Butler, J. (2000). The influence of cosmic terrestrial clouds and global warming. *Astronomy & Geophysics*, 41:18–22.

Palous, J., Ruprecht, J., Dluzhnevskaia, O. B., and Piskunov, T. (1977). Places of formation of 24 open clusters. *Astron. Astrophys.*, 61:27–37.

Paul, M. et al. (2001). Experimental limit to interstellar ^{244}Pu abundance. *Astrophys. J.*, 558:L133–L135.

Pavlov, A. A., Kasting, J. F., Brown, L. L., Rages, K. A., and Freedman, R. (2000). Greenhouse warming by CH_4 in the atmosphere of early earth. *J. Geophys. Res.-Plan.*, 105 (E5):11981–11990.

Pollack, J. B. (1991). Kuiper prize lecture - present and past climates of the terrestrial planets. *Icarus*, 91:173–198.

Prokoph, A., Shaviv, N. J., and Veizer, J. (2004). A 32 Ma climate periodicity and its possible celestial link. Submitted to Earth Plan. Sci. Lett.

Prokoph, A. and Veizer, J. (1999). Trends, cycles and nonstationarities in isotope signals of Phanerozoic seawater. *Chem. Geology*, 161:225–240.

Rampino, M. R. and Stothers, R. B. (1984). Terrestrial mass extinctions, cometary impacts and the sun's motion perpendicular to the galactic plane. *Nature*, 308:709–712.

Rampino, M. R. and Stothers, R. B. (1986). Geologic periodicities and the galaxy. *The Galaxy and the Solar System*, pages 241–259.

Raup, D. M. and Sepkoski, J. J. (1984). Periodicity of extinction in the geologic past. *Proc. Natl. Acad. Sci.*, 81:801–805.

Ready, R. C. (1987). Nuclide production by primary-ray protons. *J. Geophys. Res. Supp.*, 92:E697–E702.

Rocha-Pinto, H. J., Scalo, J., Maciel, W. J., and Flynn, C. (2000). Chemical enrichment and star formation in the Milky Way disk. II. Star formation history. *Astron. Astrophys.*, 358:869–885.

Rye, R., Kuo, P. H., and Holland, H. D. (1995). Atmospheric carbon dioxide concentrations before 2.2 billion years ago. *Nature*, 378:603–605.

Sagan, C. and Chyba, C. (1997). The early faint sun paradox: Organic shielding of UV-labile greenhouse gases. *Science*, 276:1217–1221.

Sagan, C. and Mullen, G. (1972). Earth and mars - evolution of atmospheres and surface temperatures. *Science*, 177:52.

Scalo, J. M. (1987). The initial mass function, starbursts, and the Milky Way. In *Starbursts and Galaxy Evolution*, pages 445–465.

Schaeffer, O. A., Nagel, K., Fechtig, H., and Neukum, G. (1981). Space erosion of meteorites and the secular variation of cosmic rays over 10^9 years. *Planet. Sp. Sci.*, 29:1109–1118.

Shaviv, N. J. (2002). Cosmic ray diffusion from the galactic spiral arms, iron meteorites, and a possible climatic connection. *Physical Review Letters*, 89(5):051102.

Shaviv, N. J. (2003a). The spiral structure of the Milky Way, cosmic rays, and ice age epochs on earth. *New Astronomy*, 8:39–77.

Shaviv, N. J. (2003b). Toward a solution to the early faint sun paradox: A lower cosmic ray flux from a stronger solar wind. *J. Geophys. Res.–Space Physics*, 108:1437.

Shaviv, N. J. (2004). On climate response to changes in the cosmic ray flux and radiative budget. submitted to *J. Geophys. Res.–Atmos.*

Shaviv, N. J. and Veizer, J. (2003). Celestial driver of Phanerozoic climate? *GSA Today*, 13(7):4–10.

Shoemaker, E. M. and Shoemaker, C. S. (1993). The flux of periodic comets near earth. In *Asteroids, Comets, Meteors*. IAU Symposium 160, abstract 270.

Shoemaker, E. M. and Wolfe, R. F. (1986). Mass extinctions, crater ages and comet showers. *The Galaxy and the Solar System*, pages 338–386.

Soon, W. H., Posmentier, E. S., and Baliunas, S. L. (1996). Inference of solar irradiance variability from terrestrial temperature changes, 1880-1993: an astrophysical application of the sun-climate connection. *Astrophys. J.*, 472:891.

Soon, W. H., Posmentier, E. S., and Baliunas, S. L. (2000). Climate hypersensitivity to solar forcing? *Annales Geophysicae*, 18:583–588.

Stothers, R. B. (1998). Galactic disc dark matter, terrestrial impact cratering and the law of large numbers. *Mon. Not. Roy. Astr. Soc.*, 300:1098–1104.

Svensmark, H. (1998). Influence of cosmic rays on earth's climate. *Physical Review Letters*, 81:5027–5030.

Svensmark, H. (2000). Cosmic rays and earth's climate. *Space Science Reviews*, 93:175–185.

Svensmark, H. and Friis-Christensen, E. (1997). Variation of cosmic ray flux and global cloud coverage–a missing link in solar-climate relationships. *J. Atmos. Terr. Phys.*, 59:1225–1232.

Talbot, R. J. and Newman, M. J. (1977). Encounters between stars and dense interstellar clouds. *Astrophys. J. Supp.*, 34:295–308.

Taylor, J. H. and Cordes, J. M. (1993). Pulsar distances and the galactic distribution of free electrons. *Astrophys. J.*, 411:674–684.

Thaddeus, P. (1986). Molecular clouds and periodic events in the geologic past. In "The Galaxy and the Solar System", eds. R. Smoluchowski, J. Bahcall, & M. Matthews (Tucson: Univ. Arizona Press), p. 61–82.

Torbett, M. V. (1986). Injection of Oort cloud comets to the inner solar system by galactic tidal fields. *Mon. Not. Roy. Astr. Soc.*, 223:885–895.

Usoskin, I. G., Marsh, N., Kovaltsov, G. A., Mursula, K., and Gladysheva, O. G. (2004). Latitudinal dependence of low cloud amount on cosmic ray induced ionization. *Geopys. Res. Let.*, 31:6109.

Veizer, J., Godderis, Y., and Francois, L. M. (2000). Evidence for decoupling of atmospheric CO_2 and global climate during the Phanerozoic eon. *Nature*, 408:698.

Veizer, J. et al. (1999). ^{87}Sr/^{86}Sr, ^{13}C and ^{18}O evolution of Phanerozoic seawater. *Chemical Geology*, 161:59–88.

Wielen, R. (1977). The diffusion of stellar orbits derived from the observed age-depen dence of the velocity dispersion. *Astron. Astrophys.*, 60:263–275.

Wood, B. E., Müller, H., Zank, G. P., and Linsky, J. L. (2002). Measured mass-loss rates of solar-like stars as a function of age and activity. *Astrophys. J.*, 574:412–425.

Yabushita, S. (1994). Are periodicities in crater formations and mass extinctions related? *Earth Moon and Planets*, 64:207–216.

Yu, F. (2002). Altitude variations of cosmic ray induced production of aerosols: Implications for global cloudiness and climate. *J. Geophys. Res.*, 248:248.

Chapter 6

SHORT-TERM VARIATIONS IN THE GALACTIC ENVIRONMENT OF THE SUN

Priscilla C. Frisch
University of Chicago
frisch@oddjob.uchicago.edu

Jonathan D. Slavin
Harvard-Smithsonian Center for Astrophysics
jslavin@cfa.harvard.edu

Abstract The galactic environment of the Sun varies over short timescales as the Sun and interstellar clouds travel through space. Small variations in the dynamics, ionization, density, and magnetic field strength in the interstellar medium (ISM) surrounding the Sun can lead to pronounced changes in the properties of the heliosphere. The ISM within ~30 pc consists of a group of cloudlets that flow through the local standard of rest with a bulk velocity of ~17–19 km s^{-1}, and an upwind direction suggesting an origin associated with stellar activity in the Scorpius-Centaurus association. The Sun is situated in the leading edge of this flow, in a partially ionized warm cloud with a density of ~0.3 cm^{-3}. Radiative transfer models of this tenuous ISM show that the fractional ionization of the ISM, and therefore the boundary conditions of the heliosphere, will change from radiative transfer effects alone as the Sun traverses a tenuous interstellar cloud. Ionization equilibrium is achieved for a range of ionization levels, depending on the ISM parameters. Fractional ionization ranges of χ(H) = 0.16–0.27 and χ(He) = 0.19–0.34 are found for tenuous clouds in equilibrium. In addition, both temperature and velocity vary between clouds. Cloud densities derived from these models permit primitive estimates of the cloud morphology, and the timeline for the Sun's passage through interstellar clouds for the past and future ~10^5 years. The most predictable transitions happen when the Sun emerged from the near vacuum of the Local Bubble interior and entered the cluster of local interstellar clouds flowing past the Sun, which appears to have

P. C. Frisch (ed.), Solar Journey: The Significance of our Galactic Environment for the Heliosphere and Earth, 133–193.

occurred sometime during the past 44,000–150,000 years, and again when the Sun entered the local interstellar cloud now surrounding and inside of the solar system, which occurred sometime within the past 45,000 years, possibly a 1000 years ago. Prior to ~150,000 years ago, no interstellar neutrals would have entered the solar system, so the pickup ion and anomalous cosmic ray populations would have been absent. The tenuous ISM within 30 pc is similar to low column density material observed globally. In this chapter, we review the factors important to understanding short-term variations in the galactic environment of the Sun. Most ISM within 40 pc is partially ionized warm material, but an intriguing possibility is that tiny cold structures may be present.

Keywords: Interstellar Matter, Heliosphere, Equilibrium Models

6.1 Overview

In 1954 Spitzer noted that the "Study of the stars is one of [mankind's] oldest intellectual activities. Study of the matter between stars is one of the youngest." Comparatively, the study of interstellar matter (ISM) at the heliosphere is an infant. Still younger is the study of the interaction between the heliosphere and ISM that forms the galactic environment of the Sun. The total pressure of the ISM at the solar system is counterbalanced by the solar wind ram pressure, but since both the Sun and clouds move through space, this balance is perturbed as the Sun passes between clouds with different velocities or physical properties. In this chapter, we focus on variations in the galactic environment of the Sun over timescales of ± 3 Myrs, guided by data and models of local interstellar clouds. In essence, we strive to provide a basis for understanding the "galactic weather" of the solar system over geologically short timescales, and in the process discover recent changes in the Sun's environment that affect particle populations inside of the heliosphere, and perhaps the terrestrial climate. The heliosphere, the bubble containing the solar wind plasma and magnetic field, dances in the wind of interstellar gas drifting past the Sun. This current of tenuous partially ionized low density ISM has a velocity relative to the Sun of $\sim$26 km s^{-1}. It would have taken less than 50,000 years for this gas to drift from the vicinity of the closest star α Cen, near the upwind direction in the local standard of rest (LSR, Tables 6.1), and into the solar system.

Star formation disrupts the ISM. The nuclear ages of massive nearby stars in Orion (at a distance of $\sim$400 pc) and Scorpius-Centaurus (at $\sim$150 pc) are 4-15 Myr, so the solar system has been bombarded by high energy photons and particles many times over the past 10^7 years. Recent nearby supernova events include the formation of the Geminga pulsar $\sim$250,000 years ago, and possibly the release of ζ Oph as a runaway star $\sim$0.5-1 Myr ago. The dynamically evolving interstellar medium inhibits the precise description of the solar Galactic environment on timescales longer than $\sim$1–3 Myrs.

Variations in the galactic environment of the Sun for time scales of $\sim$3 Myrs are short compared to the vertical oscillations of the Sun through the galactic plane, and short compared to the disruption of local interstellar clouds by star formation. The oscillation of the Sun in the Galactic gravitational potential carries the Sun through the galactic plane once every $\sim$34 Myrs, and the Sun will reach a maximum height above the plane of $\sim$78 pc in about 14 Myrs (Bash, 1986, Vandervoort and Sather, 1993, also see Chapter 5 by Shaviv). The motion of the Sun compared to the kinematics of some ensemble of nearby stars is known as the solar apex motion, which defines a hypothetical closed circular orbit around the galactic center known as the Local Standard of Rest (LSR, §6.1.2). Note that the LSR definition is sensitive to the selection of the comparison stars, since the mean motions and dispersion of stellar populations depend on the stellar masses. The transformation of the solar trajectory into the LSR for comparison with spatially defined objects, such as the Local Bubble, introduces uncertainties related to the LSR.

A striking feature in the solar journey through space is the recent emergence of the Sun from a near vacuum in space, the Local Bubble, with density $\rho < 10^{-26}$ g cm^{-3} (Frisch, 1981, Frisch and York, 1986, §6.2). We will show that the Sun has spent most of the past $\sim$3 Myrs in this extremely low density region of the ISM, known as the Local Bubble. The Local Bubble extends to distances of over 200 pc in parts of the third galactic quadrant ($l = 180^\circ \rightarrow 270^\circ$, Frisch and York, 1983), and is part of an extended interarm region between the Sagittarius/Carina and the Perseus spiral arms. The proximity of the Sun to the Local Bubble has a profound effect on the solar environment, and affects the local interstellar radiation field, and dominates the historical solar galactic environment of the Sun (§6.2.2). Neutral ISM was absent from the bubble, and byproducts of the ISM-solar wind interaction, such as dust, pickup ions and anomalous cosmic rays, would have been absent from the heliosphere interior.

The Local Bubble is defined by its geometry today, so the solar space motion is a variable in estimating the departure of the Sun from this bubble. For all reasonable solar apex motions, sometime between 44,000 and 150,000 years ago the Sun encountered material denser than the nearly empty bubble. During the late Quaternary geological period, the Sun found itself in a flow of warm low density ISM, $n \sim 0.3$ cm^{-3}, originating from the direction of the Scorpius-Centaurus Association (§§6.3, 6.6.4). The Sun is presently surrounded by this warm, $T \sim 7{,}000$ K, tenuous gas, which is similar to the dominant form of ISM in the solar neighborhood. The physical characteristics of this cluster of local interstellar cloudlets (CLIC) are discussed in §6.3.

The solar entry into the CLIC would correspond to the appearance of interstellar dust, pickup ions and anomalous cosmic rays in the heliosphere. The dust filtration by the solar wind and in the heliosheath is sensitive both to dust mass and the phase of the solar cycle (see Landgraf chapter), so that these two effects are separable. In principle the solar journey through clouds could be

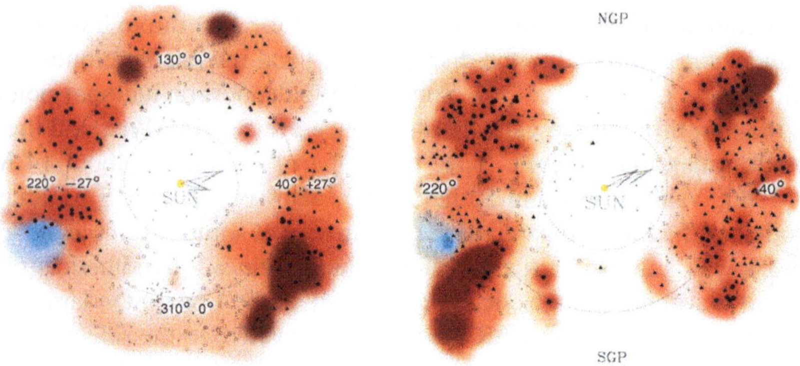

Figure 6.1. The solar position and motion are compared to the Local Bubble for two planes aligned with the solar apex motion (Table 6.1). The left figure gives the Local Bubble ISM distribution in a plane that is tilted by 27° to the Galactic plane with the northern surface normal pointing towards (220°, +63°). Stars with latitudes within about ±15°of that plane are plotted. The right figure shows a meridian slice perpendicular to the galactic plane, and aligned with the ℓ = 40°– 220° axis. Stars with longitudes within ±25° of the meridian slice are plotted. These ISM distributions were constructed from cleaned and averaged photometric and astrometric data for O, B, and A stars in the Hipparcos catalog. For a solar LSR velocity of 13 – 20 pc/Myrs, the Sun has been within the Local Bubble for over 3 Myrs. The symbols show $E(B-V)$ values 0.017–0.051 mag (tiny x's), 0.051–0.088 mag (boxes), 0.088–0.126 mag (dots), and >0.126 mag (triangles) so that shading levels give $N(\mathrm{H}) \sim 10^{20.4} \rightarrow 10^{20.7}$ cm^{-2}, $\sim 10^{20.7} \rightarrow 10^{20.9}$cm^{-2}, and $>10^{20.9}$cm^{-2}, after assuming log N(H°+2H$_2$)/$E(B-V)$=21.76 cm^{-2}mag^{-1}. The solar motion is shown for both Standard and Hipparcos values. The large intensely shaded regions show dust clouds from Dutra and Bica (2002). The CLIC column densities are below the minimum value of $N(\mathrm{H}^{\circ}) \sim 10^{20.40}$ cm^{-2} displayed here.

traced by comparative studies of interstellar dust settling onto the surfaces of inner versus outer moons.

In §6.5.2 we discuss the Sun's entry into, and exit from, the interstellar cloud now surrounding the solar system, and show that it can be determined from a combination of data and theoretical models. The velocity of this cloud, known as the local interstellar cloud (LIC), has been determined by observations of interstellar He° inside of the solar system (Witte, 2004). We test several possibilities for the three-dimensional (3D) shape of the LIC, and conclude that the Sun entered the surrounding cloud sometime within the past 40,000 years, and possibly very recently, in the past ∼1000 years.

The velocity difference between interstellar He° inside of the solar system, and interstellar D°, Fe^+ and other ions towards stars in the upwind direction, including the nearest star α Cen, together indicate the Sun will emerge from the cloud now surrounding the Sun within ~4,000 years. This raises the question of "What is an interstellar cloud?", which is difficult to answer for low column density ISM such as the CLIC, where cloudlet velocities may blend at the nominal UV resolution of 3 km s^{-1}.

Determining the times of the solar encounter with the LIC, and other clouds, requires a value for $n(H^o)$ in the cloud because distance scales are determined by $N(H^o)/n(H^o)$. Theoretical radiative transfer models of cloud ionization provide $n(H^o)$ and $n(H^+)$ for tenuous ISM such as the LIC, and supply important boundary constraints for heliosphere models. CLIC column densities are low, $N \lesssim 10^{19}$ cm^{-2}, and radiative transfer effects are significant (§6.4). Models predict $n(H^o) \sim 0.2$ cm^{-3} and $n(H^+) \sim 0.1$ cm^{-3} at the Sun, with ionization levels rising slowly to the cloud surface so that $\langle n(H^o) \rangle_{LIC} \sim 0.17$ cm^{-3}.

In §6.6.1 we discuss the relative significance of diffuse warm partially ionized gas (WPIM) and traditional Strömgren spheres for the heliosphere. Distinct regions of dense fully ionized ISM are infrequent near the Sun, although regions of ionized gas surround distant massive stars bordering the Local Bubble, and surround β Cen and λ Sco, which helps energize the Loop I superbubble interior. The Sun's path is unlikely to traverse, or have traversed, an H II region around an O–B1 star for ±4 Myrs because no such stars or H II regions are within 80 pc of the Sun. Hot white dwarf stars are more frequent nearby, and may in some cases be surrounded by small Strömgren spheres. In contrast, diffuse ionized gas is widespread and has been observed extensively using weak optical recombination lines from H^+ and other species, synchrotron emission, and by the interaction of pulsar wave packets with ionized interstellar gas. The CLIC and diffuse ionized gas are part of a continuum of ionization levels predicted and observed for low density interstellar gas.

The heliosphere should vary over geologically short timescales, for reasons unrelated to the Sun or solar activity, when encountering the small scale structure of the nearby ISM. These variations in the cosmic environment of the Sun may be traced by terrestrial radioisotope records, or by variable fluxes of interstellar dust grains deposited on planetary moons (see other chapters in this volume). Over 3 Myrs timescales the Sun travels 40-60 pc through the LSR, and the interval between cloud traversals should be less than $\leq 80,000$ years. Differing physical properties of cloudlets in the flow will affect the heliosphere and interplanetary medium. If cloud densities are relatively uniform at $n(H^o) \sim 0.2$ cm^{-3}, then data on ISM towards nearby stars indicate these clouds fill ~33% of space within 10 pc, with mean cloud lengths of ~1.0 pc, and with a downwind hemisphere that is emptier than the upwind hemisphere (§6.3.3).

The ISM within ~2 pc is inhomogeneous. The ratio Fe^+/D^o increases as the viewpoint sweeps from the downwind to upwind direction, signaling either grain destruction in an upwind interstellar shock, or increased cloud ionization. The Fe^+/D^o gradient is consistent with the destruction of CLIC dust grains in a fragment of the expanding superbubble around the Scorpius-Centaurus Association (§6.6.4). There are several possibilities for the next cloud to be encountered, and the most likely candidate is the cloud towards the nearest star α Cen (§6.5.3), where only modest consequences for the heliosphere are expected, since the temperature and velocity of the cloud differ only by $\sim$800 K and $\sim$3 km s^{-1}, respectively.

The properties of the CLIC are similar to low column density ISM observed in the solar vicinity, and observed in the Arecibo Millennium survey of very low N(H^o) clouds (Heiles and Troland, 2004). The warm neutral medium (WNM) in the Arecibo survey appears to be similar to the CLIC, but so far, counterparts of the exotic cold, low column density, neutral clouds, $T \sim 50$ K and N(H^o)$\sim 10^{18}$ cm^{-2}, are not found locally (§6.6). If they are hidden in the upwind ISM, they would be far from ionization equilibrium (having much lower ionization than the surrounding warm, ionized gas), in contrast to the LIC gas (§6.4). Cold ISM towards 23 Ori indicate densities of $>$10 cm^{-3} for the cold neutral medium gas (CNM).

We do not discuss high-velocity, $v \sim 100$ km s^{-1}, low column density radiative shocks. Such disturbances would cross the heliosphere on timescales of $\sim$5–10 years, seriously perturbing it (Sonett et al., 1987, Frisch, 1999, Mueller et al., 2005), and would be evolving as they cool. Low column density high velocity shocked gas, N(H) $\sim 10^{16}$–10^{17} cm^{-2}, $n_o \sim$ 0.5–5 cm^{-3}, could be common in ISM voids. However, with high velocities of $\sim$100 km s^{-1}, such gas will traverse voids such as the Local Bubble in less than 1 Myr. Similar high-velocity gas is associated with the Orion's Cloak superbubble shell around Orion and Eridanus, with $v \sim 100$ km s^{-1} and N(H) $\sim 10^{16}$–10^{17} cm^{-2} (Welty et al., 1999). Rapidly moving low column density shocks similar to these would pass over the heliosphere quickly, in 50–1,500 years, with strong implications for heliosphere structure (Zank et al., this volume). The fairly coherent velocity and density structure and relatively weak turbulence in the CLIC argues against any such disturbance having occurred recently (see §6.3.4. We also do not discuss dense molecular clouds, which are discussed in the Shaviv chapter.

As a way of summarizing short term solar environment, we show the solar position inside of the Local Bubble in Figure 6.1 (see 6.2.1). We will now briefly review relevant fundamental material.

6.1.1 Fundamental Concepts

Basic concepts needed for understanding the following discussions are summarized here. We use the local standard of rest, LSR, based on the Standard solar motion, unless otherwise indicated. Velocities quoted as heliocentric (HC) are with respect to the Sun, while LSR velocities are obtained by correcting for the motion of the Sun with respect to the nearest stars, or the solar apex motion. The basic principles underlying obtaining the properties of interstellar clouds are summarized in §6.1.3.

The ultraviolet and X-ray parts of the spectrum play an essential role in the equilibrium of interstellar clouds. The spectral regions relevant to the following discussions are the ultraviolet (UV), with wavelengths 1200 Å $< \lambda$ $<$ 3000 Å; the far ultraviolet (FUV), 912 Å $< \lambda <$ 1200 Å; the extreme ultraviolet (EUV), 100 Å $< \lambda <$ 912 Å; and the soft X-ray, $\lambda <$ 100 Å. The spectrum of the soft X-ray background (SXRB), photon energies <0.5 keV, has been an important diagnostic tool for the distribution of ISM with column density N(H°) $> 10^{19.5}$ cm^{-2}. The first ionization potential (FIP) of an atom is the energy required to ionize the neutral atom. The FIP of H is 13.6 eV. Elements with FIP <13.6 eV are predominantly ionized in all clouds, since the ISM opacity to photons with energies $E <$ 13.6 eV is very low. The ionization potential, IP, of several ions are also important. For instance IP(Ca^+) $=$ 11.9 eV, and N(Ca^{++})/N(H°) may be large in warm clouds. For $T >$ 4,000 K and n(e) $<$ 0.13 cm^{-3}, Ca^{++}/Ca^+ $>$ 1. Variations in the spectrum of the ionizing radiation field, as it propagates through a cloud caused by the wavelength dependence of the opacity, $\tau(\lambda)$, are referred to as radiative transfer (RT) effects.

Color excess is given by $E(B-V)$, and represents the observed star color, in magnitudes, measured in the UBV system, compared to intrinsic stellar colors (e.g. Cox, 2000). Color excess represents the intrinsic star color combined with reddening by foreground dust. The column density, N(X), is the number of atoms X contained in a column of material with a cross section of 1 cm^{-2} and bounded at the outer limit by an arbitrary point such as a star. The volume density, n(X), is the number of atoms cm^{-3}. The notation $n_{HI,0.2}$ indicates the total H° space density in units of 0.2 cm^{-3}. The fractional ionization of an element X is χ(X) $=$ N(X^+)/(N(X°)+N(X^+)). The filling factor, f, gives the percentage of the space that is filled by ISM denser than the hot gas in the Local Bubble (i.e. warm or cold gas).

Among the various acronyms we use are the following: The interstellar cloud feeding neutral ISM and interstellar dust grains (ISDG) into the solar system is the local interstellar cloud (LIC). A cluster of local interstellar clouds (CLIC) streams past the Sun, forming absorption lines in the spectra of nearby stars. Alternative names for the CLIC are the "Local Fluff" and the very local ISM, and the LIC is a member of the CLIC. The upwind portion of the CLIC has been called the "squall line" (Frisch, 1995). In the global ISM we find cold neutral medium (CNM), warm neutral medium (WNM), and warm ionized medium (WIM). Both observations and our radiative transfer models of tenuous ISM, $N(\mathrm{H}) \lesssim 10^{18}$ cm^{-2}, indicate that warm partially ionized medium (WPIM) is prevalent nearby.

The Scorpius-Centaurus Association (SCA) is a region of star formation that dominates the distribution of the ISM in the upwind direction of the CLIC flow, including the Loop I supernova remnant. The Local Bubble (LB) is a cavity in the interstellar dust and gas around the Sun that may merge in some regions with the bubble formed by stellar evolution in the SCA, Loop I, near the solar location (Frisch, 1995). The Local Bubble is used here to apply to those portions of nearby space, within $\sim$70 pc, with densities $\rho < 10^{-26}$ g cm^{-3}.

The densities of neutral interstellar gas at the solar location are provided by data on pickup ions (PUI) and anomalous cosmic rays (ACR). PUIs are formed from interstellar neutrals inside the heliosphere, which have become ionized and captured by the solar wind (Moebius et al., this volume). ACRs are PUIs that are accelerated in the solar wind and termination shock to energies of $\leq$1 GeV/nucleon (Fahr et al., Florinski et al., this volume).

Element abundances in the CLIC gas are similar to warm gas in the Galactic disk, and the missing atoms are believed to be incorporated in dust grains (Frisch et al., 1999, Welty et al., 1999). We will assume that the solar abundance pattern provides the benchmark abundance standard for the local ISM, although this issue has been debated (Snow, 2000), and although the solar abundance of O and other elements are controversial (e.g. Lodders, 2003). Using solar abundances as a baseline, the depletion δ of an element represents the amount of the element that is missing from the gas. Depletion is defined as $\delta_{\mathrm{X}} = \log \frac{[\mathrm{X}]_{\mathrm{ISM}}}{[\mathrm{X}]_{\mathrm{Solar}}}$ for an element X with abundance [X] (Savage and Sembach, 1996). The depletion and condensation temperature of an element correlate fairly well. This is shown by the three observed interstellar depletion groups $\delta_{\mathrm{Ti},Ca} \sim -3.3$, $\delta_{\mathrm{Fe},Cr,Co,Ni,V} \sim -2.3$, and $\delta_{\mathrm{Mg},Si} \sim -1.3$, which condense at temperatures of $\sim$1,500 K, $\sim$1,330 K, and $\sim$1,300 K respectively during mineral formation in an atmosphere of solar composition and pressure (Ebel, 2000). The interstellar gas-to-dust mass ratio and grain composition can be reconstructed from the atoms missing from the gas phase (Frisch et al., 1999). Depletions vary between warm tenuous and cold dense ISM, an effect originally parameterized by velocity and known as the Routly-Spitzer effect (McRae Routly and Spitzer, 1952), and now attributed to partially ionized warm ISM that has been shocked (§6.4.3). The largest variations are found for refractory elements, and result from ISDG destruction by interstellar shocks (Jones et al., 1994, Slavin et al., 2004). Depletion estimates are sensitive to uncertainties in $N(\mathrm{H})$, which in cold clouds must include $N(\mathrm{H_2})$, and in warm tenuous clouds must include H^+, both of which can be hard to directly measure.

Satellites that have revolutionized our understanding of the galactic environment of the Sun include *Copernicus*, the International Ultraviolet Explorer, IUE, the Extreme Ultraviolet Explorer, EUVE, the shuttle launched Interstellar Medium Absorption Profile Spectrograph, IMAPS, the Far Ultraviolet Explorer, FUSE, and the Hubble Space Telescope, HST (with the Goddard High Resolution Spectrometer, GHRS, and the Space Telescope Imaging Spectrograph, STIS).

Several papers in the literature have provided important insights to the physical properties of the local ISM, and are referred to throughout. Among them are: Frisch (1995, hereafter F95), which reviews the physical characteristics of the Local Bubble and the CLIC; Gry and Jenkins (2001, hereafter GJ) and Hebrard et al. (1999), which present the physical characteristics of the two nearby interstellar clouds observed in the downwind direction, the LIC and blue-shifted cloud; Slavin and Frisch (2002, hereafter SF02) and Frisch and Slavin (2003, 2004, 2005, hereafter FS), which present detailed radiative transfer model calculations of the surrounding ISM; Frisch et al. (2002, hereafter FGW), which probes the kinematics of nearby ISM; and a series of papers by Redfield and Linsky (2000, 2002, 2003, 2004, hereafter RL,), and Wood et al. (2005, hereafter W05), which assemble a wide range of UV data on the CLIC. Recent papers presenting results from FUSE also contribute significantly to our understanding of the tenuous ISM close to the Sun.

6.1.2 The Solar Apex Motion and Local Standard of Rest

Comparisons between the solar position and spatially defined objects, such as the LB, require adoption of a velocity frame of reference. The "Local Standard of Rest" (LSR) represents an instantaneous inertial frame for a corotating group of nearby stars (<500 pc) on a closed circular orbit around the galactic center, and is commonly used for this purpose. The solar motion with respect to the LSR, known as the solar "apex motion", is dynamically defined with respect to a sample of nearby stars. All reasonable values for the LSR indicate that the Sun is traveling away from the ISM void in the third quadrant of the Galaxy, $\ell = 180° \rightarrow 270°$, at a velocity of $\sim$13–20 pc/Myrs.

The Sun is $\sim$8 kpc from the galactic center. The mean orbital motion around the galactic center of nearby stars is $\sim$220 km s^{-1}, and the solar velocity is 225 ± 20 km s^{-1} (e.g., see proper motion studies of extragalactic radio sources beyond the galactic center, Reid et al., 1999). Accurate astrometric data for stars is provided by the $Hipparcos$ catalog of proper motions and positions of $\sim$120,000 nearby stars (Perryman, 1997). For the coordinate system $\hat{x}$, $\hat{y}$, $\hat{z}$, which represent unit vectors towards the galactic center, direction of galactic rotation, and north galactic pole respectively, the speeds $V_{\rm x}$, $V_{\rm y}$, and $V_{\rm z}$ then represent the mean solar velocity in the $\hat{x}$, $\hat{y}$, and $\hat{z}$, directions compared to some arbitrarily selected set of nearby stars. A kinematically unbiased set of stars from the Hipparcos Catalog shows that the mean vertical and radial components of nearby star velocities have no systematic dependence on stellar mass (Dehnen and Binney, 1998). However V_y determined for stars hotter than A0 ($B-V < 0.0$ mag) exceeds the mean LSR by $\sim$6 km s^{-1}, and the velocity dispersion for these relatively young stars is smaller than for evolved low mass stars. Since molecular clouds may share the motions of massive stars, the result is a lack of clarity concerning the best LSR transformation for studying the solar motion with respect to spatially defined objects.

The general practice is to transform radio data into the LSR velocity frame using the "Standard" solar motion, which is based on the weighted mean velocity for different populations of bright nearby stars irrespective of spectral class. The velocity of the Sun with respect to the Standard LSR (LSR_{Std}) is (V_x, V_y, V_z) = (10.4, 14.8, 7.3) km s^{-1} (Mihalas and Binney, 1981), giving a *solar apex motion*, of 19.5 km s^{-1} towards $l = 56°$, $b = +23°$ (or LSR_{Std}, Table 6.1). The solar motion with respect to a kinematically unbiased subsample of the *Hipparcos* catalog (omitting stars with $B - V < 0.0$ mag) is (V_x,V_y,V_z) = (10.0,5.3, 7.2) km s^{-1}, corresponding to a solar speed of $V = 13.4$ km s^{-1} towards the apex direction $\ell = 27.7°$ and $b = 32.4°$ km s^{-1} (or LSR_{Hip}).

Table 6.1. The LSR Velocities of the Sun and Nearby Interstellar Clouds.

	HC $V, \ell\, b$ (km s^{-1}, ° °)	LSR_{Std} $V, \ell\, b$ (km s^{-1}, ° °)	LSR_{Hip} $V, \ell\, b$ (km s^{-1}, ° °)
Sun	...	19.5, 56° 23°	13.4, 27.7° 32.4°
CLIC	–28.1±4.6, 12.4° 11.6°	–19.4, 331.0° –5.1°	–17.0, 357.8° –5.1°
LIC	–26.3±0.4, 3.3° 15.9°	–20.7, 317.8° –0.5°	–15.7, 346.0° 0.1°
GC	–29.1, 5.3° 19.6°	–21.7, 323.6° 6.3°	–17.7, 351.2° 8.5°
Apex	–35.1, 12.7° 14.6°	–24.5, 341.3° 3.4°	–23.3, 5.5° 4.1°

Notes: The first row lists the Standard and Hipparcos values for the solar apex motion (§6.1.2). The LIC, CLIC, GC (G-cloud), and Apex cloud velocity vectors refer to the upwind direction. Columns 2, 3, and 4 give the heliocentric velocity vector, and the LSR velocities using corrections based on the Standard and Hipparcos-derived solar apex motions, respectively.

6.1.3 Finding Cloud Physics from Interstellar Absorption Lines

Sharp optical and UV absorption lines, formed in interstellar clouds between the Sun and nearby stars, are the primary means for determining the physical properties of the CLIC. UV data provide the best look at cloud physics because resonant absorption transitions for the primary ionization state of many abundant elements fall in the UV portion of the spectrum. However, results based on UV data are limited by a spectral resolution, which typically is $\gtrsim 3.0$ km s^{-1}, and are unable to resolve the detailed cloud velocity structure seen in high resolution (typically $\sim$0.3–0.5 km s^{-1}) optical data. Ultimately high resolution UV data, $R \sim$300,000, are required to probe the CLIC velocity structure and to separate the thermal and turbulent contributions to absorption line broadening. The details of interpreting absorption line data can be found in Spitzer (1978); see alternatively the classic application of the "curve of growth" technique to the ISM towards ζ Oph (Morton, 1975).

The classic target objects for ISM studies are hot rapidly rotating O, B and A stars, where sharp interstellar absorption features stand out against broad stellar lines. However O and B stars are relatively infrequent nearby, and interstellar line data can be contaminated by sharp lines formed in circumstellar shells and disks around some A and B stars. Hence care is required to separate ISM and circumstellar features (e.g. Ferlet et al., 1993). White dwarf stars of DA and DO spectral types provide a hot far UV continuum that can, in many cases, be observed in the interval 912–1200 Å (e.g. Lehner et al., 2003). Cool stars are very frequent, and there are $\sim$30 times more G and K stars than A stars. The cool stars have a weak UV flux and active chromospheres. Strong chromospheric emission lines can be used as continuum sources for interstellar absorption lines, although great care is required in the analysis (e.g. Wood et al., 2005). The disadvantage of cool stars is that uncertainties arise from blending of the interstellar absorption and stellar chromosphere emission features. For example, only the Sun has an unattenuated H$^{\circ}$ Lyman-α emission feature, and its strength and shape vary with the phase of solar magnetic activity cycle. Heeding these limitations, cool stars provide an important source of data on local ISM.

Absorption Line Data. Observations of interstellar absorption lines in the interval 912–3000 Å are required to diagnosis the physical properties of the ISM, such as composition, ionization, temperature, density, and depletions (e.g., Jenkins, 1987, York, 1976, Snow, 2000, Savage, 1995). Most elements in diffuse interstellar clouds are in the lowest electronic energy states, and have resonant absorption lines in the UV and FUV. The type of ISM traced by H°, D°, H_2, C^{+}, C^{+*}, C^{+3}, N°, N^{+}, O°, O^{+}, Mg°, Mg^{+}, Si^{+}, Si^{+2}, Ar°, S^{+}, Fe^{+} depends on the element FIP and interaction cross-sections. Neutral gas is traced by D°, H°, N°, O°, and Mg°, although Mg° is also formed by dielectronic recombination in WPIM (§6.3.1). Charge exchange couples the ionization levels of N, O, and H. The CLOUDY code documentation gives sources for charge exchange and other reaction rate constants (§6.4). Ions with FIP<13.6 eV, such as C^{+}, Mg^{+}, Si^{+}, S^{+}, and Fe^{+}, are formed in both neutral and ionized gas. Highly ionized gas is traced by Si^{+2} and C^{+3}. The ratios $N(Mg^{+})/N(Mg^{o})$ and $N(C^{+})/N(C^{+*})$ are valuable ionization diagnostics (§6.3.1).

Local ISM has been studied at resolutions of 2.5–20 km s^{-1} for ~30 years by *Copernicus*, IUE, HST, HUT, EUVE, IMAPS, FUSE, and various rocket and balloon experiments. The best source of UV data in the 1190–3000 Å region are data from the GHRS and STIS on the Hubble Space Telescope, while FUV data are available from FUSE, IMAPS, and *Copernicus*.

Ground-based optical observations have an advantage over space data in two respects — instruments provide very high resolution, ~0.3–1.0 km s^{-1}, and there are fewer constraints imposed on telescope time. Lines from Ca^{+}, Na^{o} and Ti^{+} are useful diagnostics of the nearby ISM. The Ca II K line (λ3933 Å) is usually the strongest interstellar feature for nearby stars because Ca depletion is lower (abundances are higher) in warm gas (e.g. Savage and Sembach, 1996), and the lines are strong. Calcium and titanium are refractory elements with highly variable abundances. Both Ca^{+} and Na^{o} are trace ionization states in the local ISM, while Ti^{+} (FIP = 13.6 eV) and H^{o} have similar distributions. High resolution optical data provide the best diagnostic of the cloud kinematics of ISM near the Sun and in the upwind direction. Optical lines, including Ca II, are generally too weak for observations in the downwind direction.

Interpreting Absorption Data. Interstellar absorption features in a stellar spectrum provide a fundamental diagnostic of ISM physics. Cloud temperatures are determined from these data, but the ~3 km s^{-1} resolution of the best UV data is inadequate to separate out closely spaced velocity components, or fully distinguish turbulent and thermal line broadening.

The wavelength-dependent opacity of an absorption line traces the atomic velocity distribution and the underlying strength of the transition between the energy levels, the Einstein transition probability. The atomic velocity distribution has contributions from the bulk ISM motion, turbulence, and the thermal dispersion of atomic velocity. In principle, thermal and turbulent contributions to the line broadening can be separated, given data for atoms with a range of atomic mass. In practice, data on either D^{o} or H^{o}, as well as heavier elements, are needed. The temperature of the cloud is found from line widths, so understanding the limitations of this method are important.

Absorption line data are analyzed with the assumption that particles, of velocity v and mass $m_{\rm A}$, obey a Maxwellian distribution at a kinetic temperature $T_{\rm k}$, $f(v) \sim T_{\rm K}^{-3/2} \exp(-m_{\rm A} v^2/2kT_{\rm K})$. The kinetic temperature also parameterizes the energy of ion-neutral and neutral-neutral elastic scattering. The velocity dispersion is characterized by the Doppler line broadening parameter, which is $b_{\rm D}= (\frac{2kT_{\rm k}}{m_{\rm A}})^{1/2}$ for a purely thermal distribution of velocities. The line full-width-at-half-maximum is FWHM $\sim$ 1.7 $b_{\rm D}$. Few interstellar absorption lines are Maxwellian shaped, however, because the line opacity is a non-linear function of column density, and because multiple unresolved clouds may contribute to the absorption. In this case, lines are interpreted using a best-fitting set of several velocity components, each representing a separate cloud or "velocity component", at velocity v, that are found to best reproduce the line shape given the instrumental resolution.

Each velocity component is broadened by thermal motions and turbulence. The non-thermal line broadening is incorporated into a general term "turbulence", ξ, so that $b_{\rm D}^2 = (2kT_{\rm K}/m_{\rm a}) + \xi^2$. The turbulence term ξ is supposed to trace the mass-independent component of line broadening. The main technique for separating turbulence from thermal broadening uses $b_{\rm D}$ as a function of atomic mass, although this technique does not distinguish unresolved velocity structure.

For low opacity lines, column density is directly proportional to the equivalent width W, and $N \sim W/\lambda^2$. For high line center opacities, $\tau_{\rm o} \propto N\lambda/b_{\rm D}$, the line strength increases in a non-linear way, $W \sim \lambda b_{\rm D}(\ln \tau_{\rm o})^{1/2}$. Column densities derived from observations of partially saturated lines are highly insensitive to line strength. These important limitations in determining the component properties from absorption line data are well known (Spitzer, 1978).

Resolution Limitations: Unresolved Clouds. The resolutions of UV instruments, $R < 10^5$, are well below those of the best optical spectrometers, $R > 3{\times}10^5$. The result is the loss of information about the velocity structure of the clouds forming UV absorption lines. High resolution, 0.3–0.5 km s^{-1}, optical data on Ca^+, Na^o, and K^o show that the number of adjacent components separated by velocity δv increases exponentially as $\delta v \rightarrow 0$ km s^{-1} (Welty et al., 1994, Welty et al., 1996, Welty and Hobbs, 2001). The result is that $\sim$60% of cold clouds may be missed at the $\sim$3 km s^{-1} resolution of STIS and GHRS. Cold clouds have a median $b_{\rm D}$(Na^o) $\sim$ 0.73 km s^{-1}, and typical temperature $\sim$80 K. Fig. 6.2 shows the distribution of component separations, δv, for Na^o, K^o, and Ca^+. The reduction of the numbers of components with small separations, $\delta v \lesssim 1.5$ km s^{-1}, compared to the best fit exponential distribution, indicate unresolved velocity structure, while the increase of component separations for $\delta v > 6-7$ km s^{-1} indicates that the distribution of cloud velocities is not purely Gaussian (Welty and Hobbs, 2001).

6.2 Solar Journey through Space: The Past 10^4 to 10^6 Years

The Sun has spent most of the Quaternary era, the past 2–3 Myrs, in the Local Bubble, a region of space with very low ISM density, which extends to distances of over 200 pc in parts of the third galactic quadrant ($\ell = 180° \rightarrow 270°$,

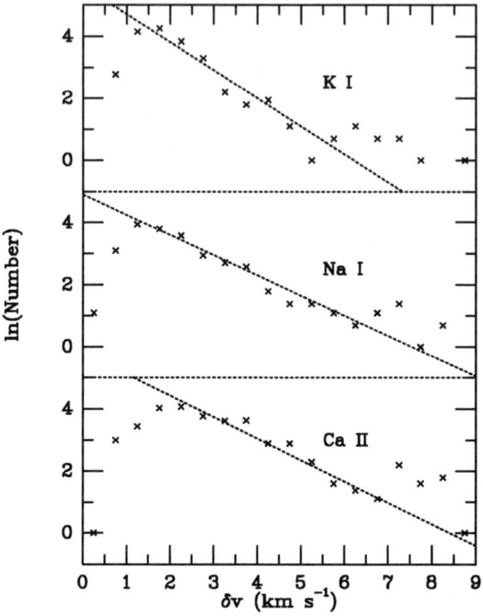

Figure 6.2. The distribution of velocity separations (δv) between adjacent absorption components for Na°, K°, and Ca^+. The dotted line shows the best fit over the range 2 km $s^{-1} \leq \delta v \leq$ 6.5 km s^{-1}, with slopes –0.9, –0.6, and –0.7, for K°, Na°, and Ca^+ respectively. The turnover in the numbers of components for small separations is the result of unresolved velocity structure, while the increase of component separations for $\delta v > 6 - 7$ km s^{-1} shows that the distribution of cloud velocities is not purely Gaussian (Welty and Hobbs, 2001). Figure courtesy of Dan Welty.

Frisch and York, 1986). Sometime within the past ~100,000 years the Sun encountered a substantially denser region, and the galactic environment of the Sun changed dramatically. The Local Bubble appears to be part of an extended interarm region between the Sagittarius/Carina and the Perseus spiral arms (Beck, 2001). For all viable solar apex motions, the Sun is leaving the emptiest part of this bubble and now entering low density ISM approaching us from the direction of the Scorpius-Centaurus Association (SCA, Frisch, 1981, 1995). The solar motion with respect to the LB and to the CLIC are shown in Figs. 6.1 and 6.5. The LB dominates the historical solar galactic environment of the Sun and affects the local interstellar radiation field. The absence of interstellar neutrals in the near void of the LB interior would lead to an absence of pickup ions and anomalous cosmic rays in the heliosphere for periods when the Sun was in the bubble. Reconstructing the solar galactic environment for the past several million years requires knowledge of both solar motion and the LB properties.

6.2.1 Inside the Local Bubble

The LB surrounding the Sun was discovered through the absence of interstellar dust grains, which redden starlight, in the nearest ~70 pc (Eggen, 1963, Fitzgerald, 1968, Lucke, 1978, Vergely et al., 1997). Reddening data is typically sensitive to color excesses $E(B-V)$>0.01 mag, corresponding to $N(\mathrm{H}) = 10^{19.76}$ cm^{-2}, depending on photometric quality. The LB overlaps the interior of Gould's Belt (Frogel and Stothers, 1977, Grenier, 2004). About 1% of the ISM mass is in dust grains, and interstellar gas and dust are well mixed in space.

Data showing starlight reddening by interstellar dust are used here to determine the Local Bubble configuration (LB, Figs. 6.1), Fig. 6.1 displays this configuration for two planes aligned with the solar apex motion (Table 6.1). The right figure shows a slice perpendicular to the galactic plane, where only stars with longitudes within 25° of the slice are plotted. The left figure shows the distribution of dust in a plane that is tilted by 27° to the Galactic plane with the northern surface normal pointing towards $(\ell,b) = (220°, +63°)$ (thus intersecting the galactic plane along $\ell = 130°$ and $\ell = 310°$). Only stars with latitudes within about ±15°of that plane are plotted. The $E(B-V)$ values are smoothed for stars with overlapping distances and within 13° of each other. Fig. 6.1 shows that in no direction is a column density of $N(\mathrm{H}) > 10^{19.7}$ cm^{-2} identified close to the Sun in the anti-apex direction (for the gas-to-dust ratio $N(\mathrm{H})/E(B-V) = 5.8 \times 10^{21}$ atoms cm^{-2} mag^{-1}, Bohlin et al., 1978). The reddening data are from color excess values, $E(B-V)$, calculated for O, B, and A0–A3 stars in the Hipparcos catalog, after cleaning the data to omit variable or otherwise peculiar stars (Hipparcos flag H6 = 0, Perryman, 1997). Both projections show that an ISM column density of $N(\mathrm{H}) > 10^{19.7}$ cm^{-2} is not reached in the nearest ~60–70 pc of the Sun along the solar trajectory, and this conclusion is unaltered by the selection of the solar apex motion.

The Local Bubble cavity is found in optical, UV, and EUV data, with different sensitivities to the ISM. The LB is seen in UV data sensitive to $N(\mathrm{H}) > 10^{17}$ cm^{-2} (Frisch and York, 1983), in the distribution of white dwarf and cool stars observed in the EUV and sensitive to $N(\mathrm{H}) > 10^{18}$ cm^{-2} (Warwick et al., 1993), in Na^{+} absorption line data sensitive to $N(\mathrm{H}) \gtrsim 10^{19}$ cm^{-2} (Sfeir et al., 1999, Vergely et al., 2001), and in dust polarization data sensitive to $N(\mathrm{H}) \gtrsim 10^{19}$ cm^{-2} (Leroy, 1999). The sensitivities and sample densities differ between surveys, but the results are consistent in showing that there is no slowly moving dense ISM in the path of the Sun for the previous 3 Myrs, or ~60 pc. UV and EUV data are more sensitive to low column densities, and detect upwind CLIC gas. Figure 1 in Chapter 2 displays the Local Bubble compared to nearby stellar associations, the Gum Nebula, which is a giant H II region, and the distribution of molecular clouds in the solar neighborhood.

Are there nearby interstellar clouds that are inside of the Local Bubble but not shown in Fig. 6.1? CLIC column densities are too low for detection in $E(B-V)$ data, but the CLIC is seen in EUV, UV, and optical data as local ISM with N(H) $\lesssim 10^{19}$ cm^{-2} (§6.3). Looking towards the third galactic quadrant, the only ISM close to the anti-apex direction, and in the distance interval $5 < d < 80$ pc, are Ca$^+$ components in γ Ori ($\ell = 197°$, $b = -16°$, d = 75 pc) showing N(H°) $\sim 10^{19.50}$ cm^{-2} if Ca$^+$/H° $\sim 10^{-8}$. The LSR motions of these Ca$^+$ components correspond to $0 \rightarrow 13$ km s^{-1}(Standard LSR), so that the Sun has moved 20–33 pc with respect to this ISM over ~1 Myrs. Only if one of the components towards γ Ori is within 35 pc of the Sun will they have surrounded the Sun recently. This argument is consistent with Ca$^+$ limits towards the star β Eri (27 pc), near the anti-apex direction, indicating N(H°) $< 10^{19.30}$ cm^{-2} (Vallerga, 1993, Frisch et al., 1990). There is not enough data to rule out encounters with very low column density clouds near the anti-apex direction, N(H°) $\lesssim 10^{18}$ cm^{-2}, particularly since the Arecibo survey shows that such ISM may have high LSR velocities (§6.6).

Data on CO molecular clouds indicate the Sun is unlikely to have crossed paths with a large dusty cloud over the past 3 Myrs. Nearby CO clouds, $d < 130$ pc, typically have low LSR velocities. Several dust clouds are found at ~120 pc (dark shading, Fig. 6.1) towards the anti-apex direction, $\ell \sim 200°$, $b \sim -36°$. They are 3C105.0 ($V = 8.0$ km s^{-1}), MBM20 ($V = 0.3$ km s^{-1}), and LDN1642 ($V = 1.3$ km s^{-1}, Standard LSR motions). The clouds would have traveled less than 24 pc over 3 Myrs, compared to the distance traveled by the Sun of 40–60 pc, so Sun-cloud separations remained over 40 pc.

6.2.2 Radiation and Plasma of the Local Bubble

The location of the Sun inside of a cavity in the ISM has a profound affect on the immediate solar environment, including the ambient radiation field and the properties of the interstellar plasma at the solar location. Prior to the entry of the Sun into the CLIC, the very low density material of the Local Bubble surrounded the Sun. We now go into some detail about the physical characteristics of that material, and the resultant radiation field, because the Local Bubble radiation field is the dominant factor in the ionization state of the ISM close to the Sun, and because the physical properties of the heliosphere are extremely sensitive to the LIC ionization. The following discussion is based on equilibrium assumptions for the Local Bubble plasma. Alternate models, including models of non-equilibrium cooling, are found in the literature but not discussed here.

While the nature and origin of the Local Bubble is the subject of ongoing debate, at least a few facts are agreed upon. The first is that the Solar System and the low density clouds around it are embedded within a large, ~50–200 pc radius, cavity that has a very low density of both neutral and ionized gas. Even allowing for large thermal pressure variations in the ISM, we expect that this volume is filled with a very low density ionized plasma, which provides some degree of pressure support for the cavity. The view that held sway for many years was that the Local Bubble was filled with hot, relatively high pressure gas with a temperature of $\sim 10^6$ K and pressure of $P/k_B \sim 10^4$ cm^{-3} K. These conclusions were drawn from observations of the diffuse soft X-ray background (SXRB), which was mapped over the entire sky at low spectral and spatial resolution but high sensitivity at energies from $\sim$70 eV to several keV by the Wisconsin group using sounding rockets (e.g. McCammon et al. 1983).

The ROSAT all-sky survey map, which has higher spatial resolution and sensitivity than the Wisconsin group maps, has made it clear that a substantial amount of the SXRB at high latitude comes from beyond the boundaries of the Local Bubble. X-ray shadows made by clouds outside the Local Bubble have provided information about the fraction of the emission coming from within the cavity.

Another complication regarding the source of the soft X-ray emission is that it may be contaminated by X-ray emission from gas much closer to Earth. As discussed by Cox (1998), and further explored by Cravens (2000), Cravens et al. (2001), and Pepino et al. (2004), a potentially substantial contribution to the SXRB may be coming from the heliosphere and geocorona. The emission mechanism involves charge transfer reactions of highly charged ions such as O^{+7} in the solar wind with neutral H or He either in the interstellar wind or geocorona. After the charge transfer, the ion is in a highly excited state and radiatively decays to ground by emitting Lyman or K-shell photons. The heliospheric emission is expected to peak in emissivity at about 1–10 AU because of the combination of the spatial dependence of the ionization of inflowing neutral atoms and the density of the solar wind. Current estimates are still quite uncertain, but put the contribution of the charge transfer photons to the observed SXRB at about 10–50% (Wargelin et al., 2004, Cravens et al., 2001).

It is important to note that even if the heliospheric and geocoronal soft X-ray emission is at the upper end of current estimates, they are unimportant for LIC ionization because the flux in the LIC, which is much further away ($\gtrsim 100\times$), is insignificant. Moreover, the emission observed by *ROSAT* and the Wisconsin sounding rockets ($\gtrsim$70 eV) is substantially harder than the emission that is directly responsible for LIC ionization ($\sim$13.6–54.4 eV). Thus the observed SXRB is relevant to the photoionization of the cloud only insofar as it provides information on the hot gas that is emitting EUV photons that ionize the LIC.

The available information on the temperature of the hot gas in the Local Bubble, which we still believe to be responsible for most of the observed

SXRB, particularly near the Galactic plane, is mostly at low spectral resolution. The all sky maps of *ROSAT* in the energy band near 1/4 keV have been used to infer an effective plasma temperature near 10^6 K (e.g. Kuntz and Snowden, 2000) under the assumption of a plasma in collisional ionization equilibrium (CIE). The two higher spectral resolution data sets that are available (DXS, Sanders et al., 2001 and XQC, McCammon et al., 2002) are for very limited portions of the sky and at very low spatial resolution, and are not consistent with this model. In fact these spectra do not fit any currently known model for the temperature and ionization state of the plasma. One thing that does seem clear at this time is that iron must be significantly depleted in the plasma because the bright Fe line complex near 70 eV is observed to be much fainter than predicted. Perhaps this should not be surprising, since in the LIC $\sim$95% of the Fe is missing from the gas and presumably depleted onto dust grains, while in cold clouds $\sim$99% of the Fe is depleted.

Disregarding the problems with models of the emission spectrum, under the assumption that there is a CIE hot plasma filling the LB, the implied density is roughly 5×10^{-3} cm^{-3} and pressure of more than $P/k_B = 10^4$ cm^{-3}K. This is a high thermal pressure for the ISM, though certainly not unheard of, and may be typical of hot gas in the ISM. If the LIC is only supported by thermal pressure, however, there would seem to be a substantial mismatch between its pressure, $P/k_B \sim 2300$ cm^{-3}K, and that of the LB (§6.3.4). This mismatch may be fixed by magnetic pressure, though the size of the heliopause appears to require a field somewhat below that necessary to make up the difference (Ratkiewicz et al., 1998). That limit can also be avoided for a magnetic field that is nearly aligned with the direction of the interstellar wind (Florinski et al., 2004), since in this case the heliosphere confinement is much less affected by the field. If a substantial portion of the SXRB originally attributed to the hot gas is from charge transfer emission, then models that properly take that into account will necessarily imply lower pressures for the LB, which helps to resolve this discrepancy. Note, though, that the emissivity goes as P^2, so a 50% reduction in emission results in only a factor of $1/\sqrt{2}$ reduction in P.

6.3 Neighborhood ISM: Cluster of Local Interstellar Clouds

The ISM surrounding the Sun is flowing through the LSR with a bulk velocity of -19.4 ± 4.6 km s^{-1} and an upstream direction $\ell \sim 331°$, $b \sim -5°$, towards the Scorpius-Centaurus Association (SCA). This warm, low density gas consists of cloudlets defined by velocity, which we denote the cluster of local interstellar cloudlets, CLIC. The upwind direction of the bulk velocity of the CLIC in the LSR suggests an origin related to stellar activity in the SCA (Frisch, 1981, Frisch and York, 1986, Frisch, 1995, also see §6.6.4). Low column densities are found for ISM within 30 pc, $N(\mathrm{H}^\circ) < 10^{19}$ cm^{-2}. The CLIC temperature, $N(\mathrm{H}^\circ)$, and kinematics resemble global warm neutral medium (WNM) detected by radio H° 21 cm data (§6.6).

If we restrict the star sample to objects within ~10.5 pc, the mean $N(\mathrm{H}^\circ)$ for CLIC cloudlets is $\langle N(\mathrm{H}^\circ)\rangle \sim 6.4 \times 10^{17}$ cm^{-2} based on H° or D° UV data (§6.1.1, and Table 6.2). For $n(\mathrm{H}^\circ)$ similar to the LIC density at the solar location, $n(\mathrm{H}^\circ) = 0.2$ cm^{-3}, the mean cloud length is 1.0 pc, and the mean time for the Sun to cross them is $\lesssim$68,000 years (Table 6.2). The LSR solar velocity is 13–19 km s^{-1} (§6.1.2), and the solar galactic environment over timescales of 3 Myrs will be regulated by ISM now within ~60 pc, provided there are no undiscovered high velocity ($V > 20$ km s^{-1}) clouds with suitable trajectories.

Table 6.2. The Characteristics of ISM within 10 pc of the Sun.

LSR Upwind direction [(a)]	$\ell \sim 331.0°, b \sim -5.1°$
LSR Velocity[(a)]	$V = -19.4 \pm 4.6$ km s^{-1}
No. stars sampling ISM within 10 pc[(b)]	20
Sightline averaged H° space density	$\langle n(\mathrm{H}^\circ)\rangle = 0.07$ cm^{-3}
No. velocity components[(c)]	30
Component averaged, $N(\mathrm{H}^\circ)$	$\langle N(\mathrm{H}^\circ)\rangle = 6.4 \times 10^{17}$ cm^{-2}
Component averaged cloud thickness, L, for $n(\mathrm{H}^\circ) = 0.2$ cm^{-3}	$\langle L\rangle = 1.0$ pc
Component averaged cloud crossing time, T_L	$\langle T_\mathrm{L}\rangle \lesssim 68{,}000$ years
Galactic center hemisphere filling factor, f	0.40
Anti-center hemisphere filling factor, f	0.26
Solar entry into the CLIC	(44,000–140,000)/$n_{HI,0.2}$ yra
Solar entry into the LIC	<40,000/$n_{HI,0.2}$ yra
Solar exit from LIC	next ~3700/$n_{HI,0.2}$ years

Notes: The unit "years ago" is represented as "yra". $n_{HI,0.2}$ is the average H° density in units of 0.2 cm^{-3}. (a) From Table 6.1. (b) The HD numbers of these stars within 10 pc are: 10700, 17925, 20630, 22049, 23249, 26965, 39587, 48915, 61421, 62509, 48915B, 115617, 128620, 128621, 131156, 155886, 165341, 187642, 197481, 201091, 209100. Sources of data for these stars are given in §6.1.1. (c) Based on UV data with resolution ~3 km s^{-1}, generally.

The first evidence of interstellar gas within 15 pc was an anomalously strong Ca^+ line observed towards Rasalhague (α Oph), formed in what is now known as the G-cloud that is widespread in the galactic-center hemisphere (Adams, 1949, Munch and Unsold, 1962). The first spectral data of Lyα emission from interstellar gas inside of the heliosphere, obtained by *Copernicus*, showed the similarity of ISM velocities inside of the heliosphere and towards nearby stars (Adams and Frisch, 1977, McClintock et al., 1978).

The basic properties of ISM forming the galactic environment of the Sun were discovered with *Copernicus*, including the widespread presence of partially ionized gas (§6.6.1), the asymmetrical distribution of local ISM showing higher column densities towards the galactic center hemisphere (Bruhweiler and Kondo, 1982, Frisch and York, 1983), and the discrepancy between the velocity of ISM inside of the solar system and towards the nearest star α Cen (Landsman et al., 1984, Adams and Frisch, 1977). The shocked history of the nearest ISM was revealed by enhanced abundances of Fe and other refractory elements, which indicated the destruction of interstellar dust grains by interstellar shocks (Snow and Meyers, 1979, Frisch, 1979, Frisch, 1981, Crutcher, 1982, York, 1983). The first data showing the shift between the velocities of H^o inside of the heliosphere and towards nearest star α Cen, now interpreted as partially due to the hydrogen wall, were obtained by *Copernicus* and IUE (Landsman et al., 1984), although the hydrogen wall contribution was not recognized as such until the models of H^o deceleration in the heliosheath were constructed (Gayley et al., 1997, Linsky and Wood, 1996).

Optical or UV data are now available for $\sim$100 stars sampling nearby ISM (see Frisch et al., 2002, Redfield and Linsky, 2004b, Wood et al., 2005, and references in these papers). Very high-resolution optical data, $\sim$0.3–0.5 km s^{-1}, are available for $\sim$40 nearby stars, and high resolution UV data, $\sim$3 km s^{-1} for an additional $\sim$65 stars. Component blending between local and distant ISM usually prevent the use of data from distant stars. The H^o Lα line is always saturated, even for low column density sightlines, so we use D^o as a proxy for H^o, with a ratio $D^o/H^o = 1.5 \times 10^{-5}$ that is valid for local ISM (Vidal-Madjar and Ferlet, 2002, Linsky, 2003, Lehner et al., 2003). These data, combined with radiative transfer models of ionization gradients, give the velocity, composition, temperature, and morphology of the CLIC.

6.3.1 Warm Partially Ionized Medium, WPIM

Charged and neutral particles have different interactions with the heliosphere, and the ionization gradient of the cloud affects the heliosphere as it traverses a tenuous cloud. Over 30 years ago *Copernicus* discovered that N^+ is widespread towards nearby stars, $N^+/N^o \sim 1$, indicating that EUV photons capable of ionizing hydrogen, nitrogen, and oxygen penetrate to the interiors of tenuous clouds and showing that the nearest ISM is partially ionized, $\chi(H) > 0.1$ (Rogerson et al., 1973). This discovery contradicted the classic view of fully neutral or fully ionized ISM derived from observations of denser ISM, $\log N(H^o)$ (cm^{-2}) $>$ 20, where electrons originate from low FIP, $<$ 13.6 eV, abundant elements such as C. The ionizations of H, O, and N, with

FIP = 13.6, 13.6, and 14.5 eV respectively, are coupled by charge exchange. The exception is near a cloud edge where high EUV flux causes N^o/H^o to dip because photoionization dominates charge exchange ionization for N^o (§6.4, Fig. 6.8). Partially ionized gas in the CLIC is established by N fractional ionizations of χ(N) $\sim$ 0.27–0.67, derived from FUV and UV data on N^o and N^+ towards stars with $\log N(H^o)(cm^{-2}) < 19$, such as HZ 43, HD 149499B, WD 0549+158, WD 2211-495, and η UMa (Lehner et al., 2003, Frisch et al., 2005).

Among these stars the most highly ionized sightlines are HD 149499B and WD2211-495. Both of these hot stars have detected O^{+5} absorption, which appears to be interstellar, although a stellar origin can not be ruled out entirely (Oegerle et al., 2005). These stars are in the upwind hemisphere of the CLIC (§6.3.2), with HD 149499B at $\ell = 330°$, $b = -7°$, d = 37 pc, and WD2211-495 at $\ell = 346°$, $b = -53°$, d = 53 pc. In contrast, ISM towards WD 1615-154 is primarily neutral with χ(N) $\sim$ 0.05 ($\ell = 359°$, $b = +24°$, d = 55 pc). Several stars sample mainly the LIC, such as WD0549+158 near the downwind direction ($\ell = 192°$, $b = -5.3°$, d = 49 pc), where $N(H^o) \sim 5 \times 10^{17}$ cm^{-2} is comparable to the expected LIC column density.

The H and He ionizations are found from EUV spectra of white dwarf stars, and comparisons of fluxes at the He^o and He^+ ionization edges at 504 Å and 229 Å with atmosphere models. Well observed stars such as HZ 43, GD 153, and WD 0549 + 158 show $N(H^o)/N(He^o) \sim$9.8–15.8, instead of the ratio of 10/1 expected for neutral gas with a $H/He = 10$ cosmic abundance. When lower quality data are included, the variation is larger, $N(H^o)/N(He^o)$ = 9–40 (Dupuis et al., 1995, Kimble et al., 1993, Frisch, 1995, Vallerga, 1996, Wolff et al., 1999). Ratios of $H^o/He^o > 10$ indicate that He is more highly ionized than H, and the hardness of the radiation field implied by this discovery is discussed in §6.4.1.

The ratios $N(Mg^+)/N(Mg^o)$ and $N(C^+)/N(C^{+*})$ serve as ionization diagnostics and give $n(e)$ values that are independent of abundance uncertainties. The Mg^o abundance is enhanced by dielectronic recombination in warm ISM, $T > 6000$ K, and Mg^+ is the dominant ionization state. The electron densities are given by $n(e) = C_{Mg}(T,\Lambda)\ N(Mg^o)/N(Mg^+)$ and $n(e) \sim C_C(T)$ $N(C^{+*})/N(C^+)$. The quantity $C_{Mg}(T,\Lambda)$ is the ratio of the photoionization to recombination rates, and depends on cloud temperature (T) and the FUV radiation field (Λ). $C_C(T)$ is the temperature sensitive ratio of the C^+ fine-structure de-excitation to collisional excitation rates, and is independent of the radiation field (York and Kinahan, 1979). For more details, see York and Kinahan (1979). The observed ratios in the CLIC of $N(C^+)/N(C^{+*})$ = 50–200 and $N(Mg^+)/N(Mg^o)$ = 200–450 are consistent with the predictions of equilibrium radiative transfer models of low column density gas such as the LIC (section §6.4). These equilibrium models predict $n(e)$ = 0.06–0.12 cm^{-3}, with the best value for the CLIC being $\sim$0.1 cm^{-3}. Some caution is required when using observed C^+ column densities, however, since the C^+ 1335 Å line is generally saturated, which may bias $N(C^+)/N(C^{+*})$ towards smaller values.

Neutral Ar, FIP $=$ 15.8 eV, is a valuable ionization diagnostic since it is observed inside the solar system, in the form of the anomalous cosmic ray component seeded by interstellar Ar^o, and also in FUV observations of nearby stars. The ratio log Ar^o/H^o in the ISM, $\sim -5.8 \pm 0.3$ dex, is below values seen towards B-stars by factors of two or more (Jenkins et al., 2000, Lehner et al., 2003). The recombination coefficients of Ar^o and H^o are similar, but the Ar^o photoionization cross section exceeds that of H^o by $\sim$10. Argon depletion is expected to be minimal, so Ar/H towards nearby stars is interpreted as conversion to unobserved Ar^+. The RT models predict $Ar^+/Ar^o = 1.8$–2.9, $Ar^o/Ar = 0.16$–0.30, and log $Ar^o/H^o = -5.98$ to -6.17 for the LIC at the solar location, compared to observed ACR values of log $Ar^o/H^o = -5.97 \pm 0.16$ after filtration effects in heliosheath regions are included (see FS for more discussion and references).

A classic ionization diagnostic uses the strong optical lines of the trace ionization species Na^o and $Ca^+$and the assumption that ionization and recombination rates balance:

$$n(\mathrm{Na}^\circ)\Gamma(\mathrm{Na}^\circ) = n(\mathrm{Na}^+)\alpha(\mathrm{Na}^+)n(\mathrm{e}) \tag{6.1}$$

Implicit is the assumption that $N(\mathrm{Na}^o) = L\, n(\mathrm{Na}^\circ)$ cm^{-2}, or analogously that $n(\mathrm{Na}^o)/n(\mathrm{Na}^+)$ is constant, where L is the cloud length. $\Gamma(\mathrm{Na}^\circ)$ is the total ionization rate for $Na^o \rightarrow Na^+$, and $\alpha(\mathrm{Na}^\circ)$ is the total recombination rate for $Na^+ \rightarrow Na^o$ (e.g. Spitzer, 1978, Pottasch, 1972). In principle Ca^+ and Ca^{++} can be substituted for Na^o, and Na^+, respectively. Na^+ and Ca^{++} are unobservable. Since Na^o (and Ca^+) are trace ionization states in warm gas, some assumption is required for the element abundances, $[\frac{Na}{H}]$ (and $[\frac{Ca}{H}]$):

$$n(\mathrm{Na}^\circ)\Gamma(\mathrm{Na}^\circ) = n(\mathrm{H})[\frac{Na}{H}]\alpha(\mathrm{Na}^+)n(\mathrm{e}) \tag{6.2}$$

The electron densities derived from Na^o and Ca^+ data may be unreliable for low column density warm clouds because of large ionization corrections and highly variable abundances, of $\sim$40 and $\sim$1.6, respectively, for Ca and Na (Welty et al., 1999, §6.1.1). Calcium ionization levels are temperature sensitive, and $Ca^{++}/Ca^+ > 1$ for $T > 4,000$ K and $n(\mathrm{e}) < 0.13$ cm^{-3}. Electron densities of $n(\mathrm{e}) = 0.04$–0.19 cm^{-3} are found for the CLIC within 30 pc from Mg^+/Mg^o and Na^o. (Frisch et al., 1990, Lallement and Ferlet, 1997).

The ratio Fe^+/D^o presents clear evidence for variations in the ISM over spatial scales of 1–3 pc, however, it is unclear whether the variation arises from an ionization gradient or from a gradient in the Fe abundance close to the Sun. Ionization and depletion both affect Fe^+/D^o, since Fe^+ is the dominant ionization state of Fe for both neutral and warm ionized gas (SF02) and D^o traces only neutral gas. Fig. 6.3 shows Fe^+/D^o for CLIC velocity components, plotted as a function of the angle, θ, between the star and the LSR upwind direction of the CLIC (Table 6.1). The ratio $N(Fe^+)/N(D^o)$ for cloudlets within $\sim$4 pc differs by $\sim$54% over spatial scales of 4 pc between upwind stars such as α Cen AB, $\theta \sim 16°$, and downwind stars such as Sirius, $\theta \sim 103°$, although uncertainties overlap. Similar Fe abundance variations are found in thin ($\sim$0.1–0.9 pc) disk clouds towards Orion (Welty et al., 1999). The Fe abundance is highly variable in the ISM, and varies between cold dense and warm tenuous disk clouds by factors of $\sim$4–6 (Savage and Sembach, 1996, Welty et al., 1999). The local Fe^+/D^o trend may be from a combination of ionization and abundance variations. The first explanation for the physical properties of local ISM attributed the CLIC to a superbubble shell around the SCA that is now approaching the Sun from the upwind direction (Frisch, 1981, §6.6.4). Dust grain destruction in the shocked superbubble shell would have restored elements in silicates, which can be destroyed by shocks with speeds $>$50 km s^{-1}, back to the gas phase (Slavin et al., 2004, Jones et al., 1994). This effect is echoed in extended distances to the upwind edge of the CLIC obtained from Ca^+ data (Figure 6.5).

A diffuse H II region appears to be close to the Sun in the LSR upwind direction of the CLIC, based on ISM towards λ Sco, and the high value χ(N) $\sim$0.66 towards HD 149499B (WD 1634-593, Lehner et al., 2003). About 98% of the neutral gas towards λ Sco belongs to the CLIC, and shows a HC velocity of -26.6 km s^{-1}. A diffuse H II region, with a diameter of $\sim$30° and $n(e)$ $\sim$0.1–0.3 cm^{-3}, is also present at ~ -17.6 km s^{-1} (using optical data to correct *Copernicus* velocities to the HC scale, York, 1983). This diffuse H II region may extend in front of both the HD 149499B and λ Sco sightlines. The HD 149499B sightline samples the ISM approaching the solar system from the LSR CLIC upwind direction.

Highly ionized gas is observed in the downwind CLIC. Interstellar Si^{+2} is seen towards ϵ CMa but not towards α CMa (Sirius), $\sim$12° away (Gry and Jenkins, 2001, Hebrard et al., 1999). The limit on Si^{+2} towards Sirius is a factor of 15 below the ϵ CMa LIC component, with $N(Si^{+2}) = 2.3 \pm 0.2 \times 10^{12}$ cm^{-2}. This ion is particularly interesting because it is predicted to have quite a low column density both in the warm LIC gas, because of its high charge transfer rate with H^o, and in an evaporative boundary where it becomes quickly ionized in the interface (see SF02). Gry and Jenkins speculate that Si^{+2} is formed in an outer layer of the LIC behind Sirius (2.7 pc). However this scenario requires that the second local cloud observed towards both stars, which is blue-shifted by ~ -7 km s^{-1} from the LIC (hence the name Blue Cloud, BC), is a clump embedded in an extended LIC. Since Si^{+2} is also detected for the BC towards ϵ CMa but not Sirius (but with large uncertainties), the origin of

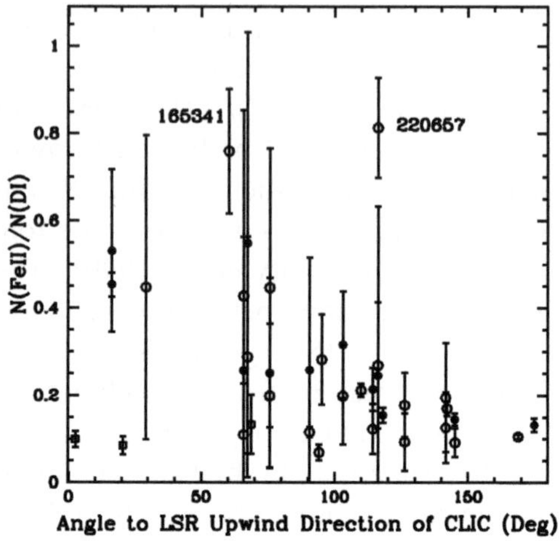

Figure 6.3. The ratio $N(\mathrm{Fe}^+)/N(\mathrm{D}^\circ)$ varies between the upwind and downwind direction of the CLIC, and over spatial scales of $\sim$3 pc. $N(\mathrm{Fe}^+)/N(\mathrm{D}^\circ)$ is plotted against the angle (in degrees) between the background star and the LSR upwind direction of the CLIC ($\ell = 331.4^\circ, b = -4.9^\circ$, Table 6.1). Clouds with observed velocities within ± 1 km s^{-1} of the projected LIC velocity are plotted as filled circles, while open circles represent CLIC components at other velocities. The open squares represent lower resolution data towards α Oph, λ Sco and HD 149499B. The ratio $N(\mathrm{Fe}^+)/N(\mathrm{D}^\circ)$ for the LIC differs between the direction of α Cen AB, at a distance 1.3 pc and angle of 16°, and Sirius, at a distance of 2.7 pc and angle of 103°, although uncertainties overlap. Velocities are drawn from the data compilations in the RL and Frisch et al. (2002) surveys.

the Si^{+2} towards ϵ CMa may instead require nonuniform interface layers such as turbulent mixing layers (see below).

6.3.2 Dynamical Characteristics of Nearby ISM

The heliosphere radius in the upwind direction is approximately proportional to the relative velocity between the Sun and ISM, so that the heliosphere is modified by variations in the ISM velocity. Absorption line data give the centroid of the radial component of a cloud velocity, integrated over a cloud length, but UV data do not resolve all of the velocity structure (§6.1.3). The coherent motion of nearby ISM is found from these velocity centroids. With the exception of the LIC, which is detected inside of the heliosphere, only radial velocities can be measured. Thus the 3D motions of other cloudlets must be inferred from observations towards several stars.

It has been known for some time that the Sun is immersed in ISM flowing away from the Scorpius-Centaurus Association and towards the Sun (Frisch, 1981, Crutcher, 1982). The CLIC kinematical data can be interpreted as a coherent flow (Frisch and York, 1986, Vallerga et al., 1993), or individual clouds can be identified with similar velocities towards adjacent stars (Lallement et al., 1986). Clouds that have been identified are listed in Table 6.3. Among these clouds is the G-cloud, in the galactic center hemisphere, which was first identified in optical data long ago (Adams, 1949, Munch and Unsold, 1962). The Sun is located at the leading edge of the stream of CLIC gas.

Kinematics of nearby ISM show three characteristics. The first is that the ISM within ~30 pc flows past the Sun. The bulk flow velocity, $V_{\rm CLIC}$, can be derived from the Doppler-shifted radial velocities of ~100 interstellar absorption line components towards ~70 nearby stars (Frisch et al., 2002). The resulting $V_{\rm CLIC}$ corresponds to a heliocentric velocity -28.1 ± 4.6 km s^{-1} and upwind direction (ℓ,b) = (12.4°,11.6°) (Frisch et al., 2002). An additional uncertainty of ~1 km s^{-1} may be introduced by a biased sample, since there are more high-resolution optical data for stars in the upwind than in the downwind directions. The CLIC motion in the LSR is then found by subtracting the solar apex motion from the CLIC heliocentric vector. Table 6.1 gives the LSR CLIC velocity for two choices of the solar apex motion. Figure 6.4 (top) displays the velocities of individual absorption components in the rest frame of the CLIC versus the rest frame of the LSR. Obviously the CLIC velocity is more representative of the nearby gas.

The second property is that distinct clouds in the flow contribute to the ±4.6 km s^{-1} dispersion of $V_{\rm CLIC}$ (Tables 6.3 and 6.1). Among the clouds are the Apex Cloud, within 5 pc and centered towards the direction of solar apex motion, and the G-cloud, which is seen towards many stars in the upwind hemisphere. The Apex and G-cloud data indicate that temperature varies by a factor of 2–6 inside of the clouds (Table 6.3, §6.3.4, RL). Evidently, for constant pressure the clouds are either clumpy, or there is a dynamically significant magnetic field. For densities similar to the LIC, n(H°) $\sim$ 0.2 cm^{-3}, the clouds fill only ~33% of nearby space and the flow of warm gas is fragmented, rather than a streaming turbulent homogeneous medium (§6.3.3). The interpretation of CLIC kinematics as a fragmented flow is supported by the fact that the upwind directions for the CLIC, LIC, Apex Cloud, and GC are all within 20° of each other.

A third characteristic is that the flow appears to be decelerating. This is shown in Fig. 6.4, bottom. Stars in the upwind and downwind directions are those with projected CLIC velocities of <0 km s^{-1}and >0 km s^{-1}, respectively. The velocities of clouds closest to the upwind direction are approaching the Sun compared to $V_{\rm CLIC}$, while those closest to the downwind direction lag $V_{\rm CLIC}$, as would be expected from a decelerating flow. This deceleration is seen even for the nearest stars, $d < 6$ pc, indicating the pileup of ISM is close to the Sun, which is consistent with the fact the Sun is in the leading edge of the flow.

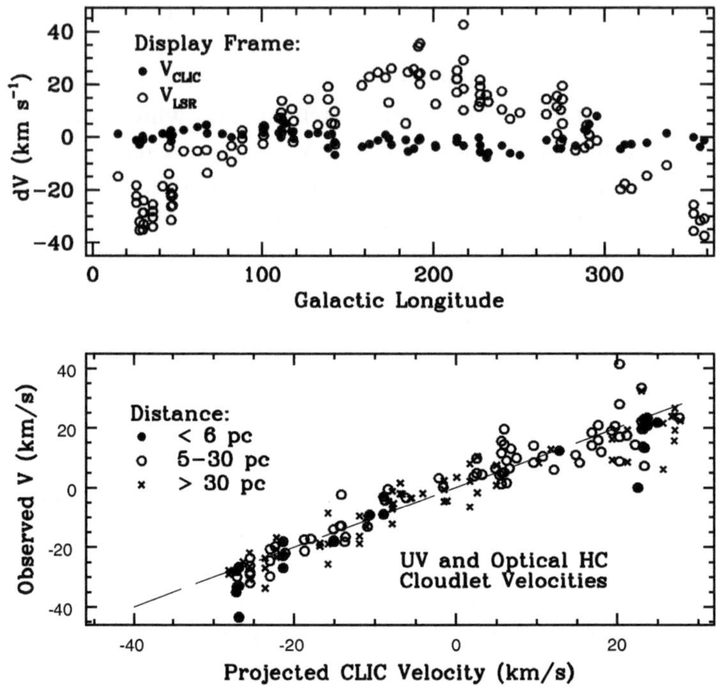

Figure 6.4. Top: The star longitude is plotted against the velocities, dV, of CLIC absorption components after transforming the absorption components into the V_{CLIC} frame (filled circles), and into the LSR (open circles, Table 6.1). Clearly V_{CLIC} provides a better fit to the CLIC data than the LSR velocity frame. LSR_{Std} is used here, and only components within 50 pc of the Sun are plotted. Bottom: The observed heliocentric velocities, V, of absorption components tracing the CLIC are plotted against the expected CLIC velocity projected towards each star (Table 6.1). The components near the upwind (approaching gas, $V < 0$ km s^{-1}) and downwind (receeding gas, $V > 0$ km s^{-1}) directions, which are expected to have the maximum values of projected $|V|$, show indications of a decelerating flow.

Note that because the 3D LIC velocity vector is known, a 3D correction for the solar apex motion is made, and the "true" downwind direction in the LSR differs from the direction measured with reference to the solar system barycenter.

6.3.3 The Distribution of Nearby ISM

The distribution of ISM within 5–40 pc is dominated by the CLIC, while over larger scales it is dominated by the Local Bubble (§6.2.1). The $\geq$15°

Table 6.3. Nearby Interstellar Clouds.

Cloud	Velocity		Location	Temp.
	HC	LSR	D, ℓ b	
	(km s^{-1})	(km s^{-1})	(pc,° °)	(10^3 K)
LIC	–26.3±0.3	–20.7	0., 50°→250° (all b)	6.4±0.3
G	–29.3±4.0	–21.7	1.4, 0°±90° 0°±60°	2.7–8.7
Blue	10±1	...	3, 233°±7° 10°±2°	2.0–5.0
Apex	–35.1±0.6	–24.5	5, 38°±10° 9°±14°	1.7–13.0
Peg/Aqr	–4.5±0.5	...	30, 75°±13° –44°±5°	...

Notes: The LSR velocities are based on the Standard solar apex motion. The cloud distances are upper limits. The Apex cloud is also known as “Panoramix” or the “Aql-Oph cloud”.
References: Table 6.1, Witte (2004), FGW, Frisch (2003), Lallement et al. (1986), Lallement and Bertin (1992),GJ, RL.

diameter cloud observed towards the Hyades stars, ~40–45 pc away, is not included in the CLIC discussion here because $N(\mathrm{H^o})$ data are unavailable, and because the nonlocal ISM may be inside the cluster (Redfield and Linsky, 2001).

The solar apex motion is compared to the distribution of CLIC gas in Fig. 6.5, where the distance to the cloud edge for the CLIC is shown projected onto the galactic plane, and in a vertical plane perpendicular to the galactic plane and aligned along the solar apex motion. The angular width of the vertical display, which extends from $\ell = 40° \pm 25°$ to $\ell = 220° \pm 25°$, includes the directions of both the Hipparcos and Standard solar apex motions (Table 6.1). The downwind direction of V_{CLIC} and direction of solar motion (Table 6.1) are shown by the arrows. The distance of the CLIC edge in the direction of a star is given by $N(\mathrm{H^o})/\langle n(\mathrm{H^o})\rangle$, where $N(\mathrm{H^o})$ is the total interstellar column density towards the star, and a uniform density similar to LIC values is assumed, $\langle n(\mathrm{H^o})\rangle = 0.2$ cm^{-3}. This distribution is based on D^o, H^o, (dots) and Ca$^+$ (crosses) data. The high resolution optical Ca$^+$ data provide excellent velocity resolution, but conversion to $N(\mathrm{H^o})$ is uncertain because of possible variable abundances. The value $N(\mathrm{Ca^+})/N(\mathrm{H^o}) = 10^{-8}$ cm^{-3} is used. The solar apex motion is shown by the arrows pointing right, while the arrows pointing left show the two LSR velocity vectors of the CLIC bulk flow, for the Standard (solid) and Hipparcos (dotted) solar apex motions, respectively.

The CLIC shape is based on data in Hebrard et al. (1999), RL, Dunkin and Crawford (1999), Crawford et al. (1998), FGW, and Frisch and Welty (2005). The $N(\mathrm{H^o})$ column densities are either estimated from $N(\mathrm{D^o})$ ($\mathrm{D^o/H^o} = 1.5 \times 10^{-5}$), or based on the saturated H^o Lα line, or estimated from $N(\mathrm{Ca^+})$ using

$N(\mathrm{Ca}^+)/N(\mathrm{H}^\circ) = 10^{-8}$. The $N(\mathrm{Ca}^+)/N(\mathrm{H}^\circ)$ conversion factor is based on CLIC data towards nearby stars such as α Aql, η UMa, α CMa, and is uncertain because Ca depletion varies strongly. Most Ca is Ca^{++} for warm gas, if $T > 4,000$ K and $n(\mathrm{e}) < 0.13$ cm^{-3}. Radiative transfer models of the LIC predict that $\mathrm{Ca}^{++}/\mathrm{Ca}^+ \sim 40$–50, so small temperature uncertainties may produce large variations in $\mathrm{Ca}^+/\mathrm{H}^\circ$.

With the possible exceptions of α Oph, interstellar column densities for stars within 30 pc are less than $\log N(\mathrm{H}^\circ)(\mathrm{cm}^{-2}) \sim 18.5$ (Frisch et al., 1987, Wood et al., 2000a). Neutral CLIC gas does not fill the sightline to any nearby star if the ISM density is $n(\mathrm{H}^\circ) \sim 0.2$ cm^{-3}. The CLIC extends farther in the galactic center hemisphere than the anti-center hemisphere, as traced by D° and Ca^+ data. The most puzzling sightline is towards α Oph (14 pc), where Ca^+ is anomalously strong and the 21-cm H° emission feature at the same velocity suggests $\log N(\mathrm{H_{tot}})(\mathrm{cm}^{-2}) \sim 19.5$ dex (see 6.5.3). High column densities ($\log N(\mathrm{H}^\circ)(\mathrm{cm}^{-2}) \sim 18.75$) are also seen towards HD 149499B, 37 pc away in the LSR upwind CLIC direction, and towards LQ Hya, where strong ISM attenuation of a stellar H° Lα emission feature causes the poor definition of the line core and wings.

The percentage of a sightline filled with ISM offers insight into the ISM character, and is given by the filling factor, f. Restricting the discussion of the ISM filling factor to the nearest 10 pc, we find that ~67% of space may be devoid of H°. If $n(\mathrm{H}^\circ) \sim 0.2$ cm^{-3}, then $f \sim 0.33$ for ISM within 10.5 pc. For galactic center and anti-center hemispheres, respectively, $f \sim 0.40$ and $f \sim 0.26$. Mean cloud lengths are similar for both hemispheres. The highest values are $f \sim 0.57$, towards α Aql (5 pc) and 61 CygA (3.5 pc), and the lowest values are $f \sim 0.11$ towards χ^1 Ori (8.7 pc), which is ~14° from the downwind direction. The sightline towards Sirius (2.7 pc, 43° from the downwind direction) has $f \sim 0.26$. These filling factors indicate that the neutral ISM does not fill the sightline towards any of the nearest stars, including towards α Cen where $f \sim 0.5$, unless instead the true value for $\langle n(\mathrm{H}^\circ)\rangle$ is much smaller than 0.2 cm^{-3}.

6.3.4 Cloud Temperature, Turbulence, and Implications for Magnetic Pressure

The basic thermal properties of the CLIC are presented in the Redfield and Linsky (2004) survey of ~50 cloudlets towards 29 stars with distances 1–95 pc. Cloudlet temperatures (T) and turbulence (ξ) values are found to be in the range $T = 1,000 - 13,000$ K and $\xi = 0$–5.5 km s^{-1} for clouds within ~100 pc of the Sun. The mean temperature is 6680 ± 1490 K, and the mean turbulent velocity is 2.24 ± 1.03 km s^{-1}. From these values, RL estimate the mean thermal ($P_{\mathrm{Th}}/k = nT$) and mean turbulent ($P_\xi/\mathrm{k} = 0.5\rho\xi^2/\mathrm{k}$) pressures

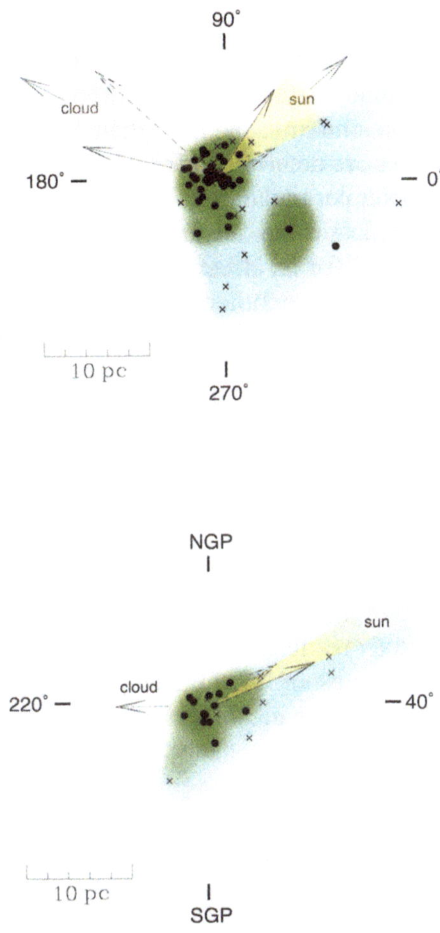

Figure 6.5. The distance to the CLIC edge for $n(\mathrm{H}^{\circ}) \sim 0.2\ \mathrm{cm}^{-3}$ and a continuously distributed ISM. The dots give distances derived from $N(\mathrm{D}^{\circ})$ and $N(\mathrm{H}^{\circ})$ data, and the x's show Ca^{+} distances. The extended H II region found towards λ Sco (§6.3.1, York, 1983) is shown as the origin of the excess cloud length towards the galactic center indicated by the Ca^{+} data. It is interpreted to indicate fully ionized gas near the Sun because the diffuse H II region seen towards λ Sco is at CLIC velocities. The arrows give the motions based on the standard (solid) and Hipparcos (dashed) solar apex motions for both the Sun and CLIC. Top: The edges of the CLIC are shown projected onto the galactic plane. Bottom: The CLIC distribution is shown for a meridian cut 50° wide in longitude, and extending between $l = 40°$ and 220° (the plane of the solar apex motion). The CLIC LSR motion is nearly perpendicular to the solar apex motion.

of 2,280 ± 520 K cm^{-3} and 89 ± 82 K cm^{-3}, respectively, by assuming $n(H^o) = 0.1$ cm^{-3} and $n(e) = 0.11$ cm^{-3}. The thermal pressure calculation includes contributions by H^o, He^o, electrons, protons, and assumes that He is entirely neutral. The pressure will be underestimated by ~15% if He is ~50% ionized as indicated by radiative transfer models (§6.4). For comparison, if these clouds have ionization similar to the LIC, then P_{Th}~2300 cm^{-3} K.

These cloud temperatures are determined from the mass dependence of line broadening using the Doppler parameter, b_D, so spectral data on atoms or ions with a large spread in atomic masses are needed. In practice, observations of the D^o Lα line are required for an effective temperature determination that distinguishes between thermal and nonthermal broadening.

When the star sample is restricted to objects within 10.5 pc, the cloud temperature is found to be anticorrelated with turbulence, and to be correlated with $N(D^o)$ (Fig. 6.6). From a larger sample of components, RL have concluded that the T–ξ anti-correlation is significant. However the likelihood that unresolved velocity structure is present in these UV data allows for the T–ξ anti-correlation to contain some contribution from systematic errors. High resolution optical data show that velocity crowding for interstellar Maxwellian components persists down to component separations below 1 km s^{-1} (§6.1.3), so that the weak positive correlation between T and $N(D^o)$, and negative correlation between T and ξ may result from unresolved component structure.

There are no direct measures of the magnetic field strength in the LIC, but the field strength is presumed to be non-zero based on observations of

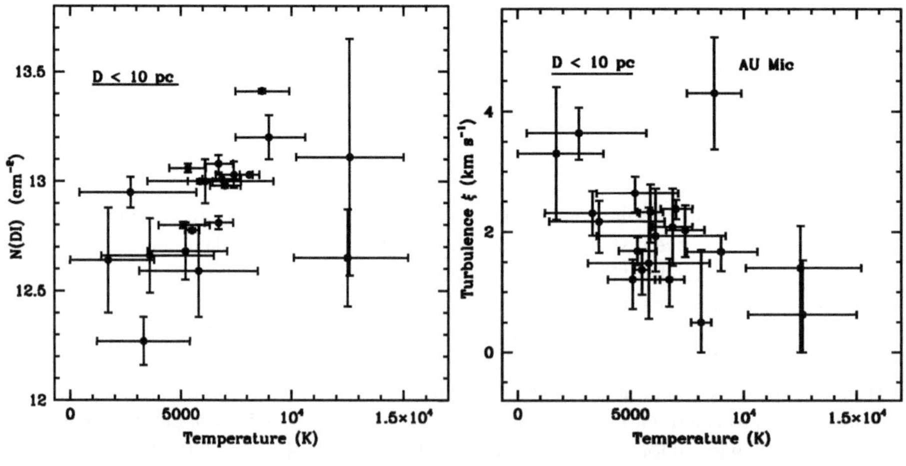

Figure 6.6. Temperature versus turbulence ξ (right) and temperature versus column density $N(D^o)$ (left) for interstellar absorption components seen towards stars within 10 pc of the Sun (based on data in RL).

polarized starlight for nearby stars, which may originate from magnetically aligned grains trapped in interstellar magnetic field lines draped over the heliosphere (Frisch, 2005). The thermal properties of the CLIC have implications for pressure equilibrium and magnetic field strength. The magnetic field strength and density fluctuations can be constrained using equipartition of energy arguments. If clouds in the CLIC are in thermal pressure equilibrium with each other, $P_{\rm Th}/k = nT$, where n is the total number of neutral and charged particles in the gas, and if magnetic field $B = 0$, then the temperature range of $T \sim 10^3$–10^4 K found by RL indicates that densities must vary by an order of magnitude. Since particle number densities vary by a factor of $\sim$2 as the cloud becomes completely ionized, most of the temperature variation must be balanced either by variations in the mass-density or in the magnetic field strength if the cloud is in equilibrium. If the CLIC has a uniform total density, and if thermal pressure variations are balanced by magnetic pressure, $P_{\rm B}/k = B^2/8k\pi$, then magnetic field strengths must vary by factors of $\sim$3 in the CLIC. If the star set is restricted to objects within $\sim$10 pc, a temperature and turbulence range of T = 1,700–12,600 K and $\xi = 0 - 5.5$ km s^{-1} are found, with mean values of $6,740 \pm 2800$ K and $\xi = 1.9 \pm 1.0$ km s^{-1}.

A rough estimate is obtained for the magnetic field strength in the LIC by assuming equipartition between thermal and magnetic energies, and using the results of the RT models that predict neutral and ion densities (see §6.4). The first generation of models gives a LIC thermal energy density of $E_{\rm Th}/k \sim 3600$ cm^{-3} K for $T = 6340$ K, and including H°, H$^+$, e$^-$, He°, and He$^+$. Lower total densities and ionization in the second generation of models reduce the thermal energy density somewhat. Equipartition between thermal and magnetic energy density gives $E_{\rm Th} = E_{\rm B}$ and $E_{\rm B}/k = B^2/8k\pi$. These assumptions then give $B \sim 3.1$–3.8 μG for the LIC.

The interstellar magnetic field strength in the more extended CLIC can be guessed using the RL value for the mean thermal pressure of 2280 cm^{-3} K, and assuming that the mean magnetic and thermal pressures are equal. For this case $B \sim 2.8$ μG. If these clouds are, instead, in pressure equilibrium with the Local Bubble plasma, then $(P_{\rm Th}+P_{\rm B})/k = P_{\rm LB}/k \sim 5 \times 10^3$ K cm^{-3} (§6.2.2), and magnetic field strengths are $\sim$3.1 μG. In contrast, for $P_{\rm B}{\sim}P_\xi$ then $B{\sim}0.6$ μG. Based on equipartition of energy arguments, typical field strengths of $B \sim$2 μG seem appropriate for the CLIC, with possible variations of a factor of 3.

The LIC turbulence appears to be subsonic. Treating the LIC as a perfect gas, the isothermal sound speed is $V_{\rm S}{\sim}\ 0.09\sqrt{T} \sim 7.1$ km s^{-1}, and turbulent velocities are 0.5–2.7 km s^{-1} (Hebrard et al., 1999, Gry and Jenkins, 2001, RL). The Alfven velocity is given by $V_{\rm A}{\sim}\ 2.2B_\mu/\sqrt{n(\rm p)}$, where B_μ is the interstellar magnetic field in μG, $V_{\rm A}$ is in km s^{-1}, and the proton density $n(\rm p)$ is in cm^{-3}. For gas at the LIC temperature (6,300 K), the Alfven velocity exceeds the sound speed for $B_\mu > 1.3$ μG. The velocity of the Sun with respect to the LIC (26.3 km s^{-1}) is both supersonic and super-Alfvenic for interstellar field strengths $B < 3.7$ μG.

6.4 Radiative Transfer Models of Local Partially Ionized Gas

Radiative transfer (RT) effects dominate the ionization level of the tenuous ISM at the Sun. The solar environment is dominated by low opacity ISM, N(H°) $\lesssim 10^{18.5}$ cm^{-2}. In contrast to dense clouds where only photons with $\lambda > 912$ Å penetrate to the cloud interior, the low column density ISM near the Sun is partially opaque to H-ionizing photons and nearly transparent to He-ionizing photons. At 912 Å the cloud optical depth $\tau \sim 1$ for log$N$(H°) $\sim$ 17.2 cm^{-2}, and at the He° ionization edge wavelength of 504 Å, $\tau \sim 1$ for log$N$(H°) $\sim$ 17.7 cm^{-2}. The average H° column and mean space densities for stars within 10 pc of the Sun are $\langle N(\mathrm{H}^\circ)\rangle \sim 10^{18}$ cm^{-2} and $\langle n(\mathrm{H}^\circ)\rangle \sim 0.07$ cm^{-3}, so the heliosphere boundary conditions and the ratio n(H°)/$n(\mathrm{H}^+)$ vary from radiative transfer effects alone as the Sun traverses the CLIC (Fig. 6.8). Warm, $T > 5000$ K, partially ionized gas is widespread near the Sun and is denoted WPIM (§6.3.1). Charged interstellar particles couple to the interstellar magnetic field and are diverted around the heliopause, while coupling between interstellar neutrals and the solar wind becomes significant inside of the heliosphere itself. The density of charged particles in the ISM surrounding the Sun supplies an important constraint on the heliosphere, and this density varies with the radiation field at the solar location, which is now described.

6.4.1 The Local Interstellar Radiation Field

The interstellar radiation field is a key ingredient of cloud equilibrium and ionization at the solar position. This radiation field has four primary components: A. The FUV background, mainly from distant B stars, B. Stellar EUV emission from sources including nearby white dwarfs and B stars (ϵ CMa and β CMa); C. Diffuse EUV and soft X-ray emission from the Local Bubble hot plasma, as we discussed in §6.2.2; and D. Additional diffuse EUV emission thought to originate in an interface between the warm LIC/CLIC gas and the Local Bubble hot plasma (§6.2.2). This last component is required because, although radiative transfer models show that the stellar EUV and Local Bubble emission account for most LIC ionization, it is not sufficient to account for the high He ionization inferred throughout the cloud from the *EUVE* white dwarf data (Cheng and Bruhweiler, 1990, Vallerga, 1996, Slavin, 1989). The spectra of these radiation sources are shown in Fig. 6.7

The interstellar radiation flux at the cloud surface must be inferred from data acquired at the solar location, together with models of radiative transfer effects.

LISM column densities are so small that dust attenuation is minimal, e.g. for the LIC $A_V < 10^{-4}$ mag, and fluxes longwards of $\lambda \sim 912$ Å are similar at the solar location and cloud surface (with the exception of Ly α absorption at 1215.7Å). For wavelengths $\lambda < 912$ Å the situation is different, however, and the spectrum hardens as it traverses the cloud because of the high H$^{\circ}$-ionizing efficiency of 800−912 Å photons. Thus a self-consistent analysis is required to unravel cloud opacity effects, and extrapolate the EUV radiation field observed at the Sun to the cloud surface. The observational constraints on the 200–912 Å radiation field are weak, partly due to uncertainties in N(H$^{\circ}$) towards ϵ CMa and partly due to the difficulty in observing diffuse EUV emission, which allows some flexibility in introducing physical models of the cloud interface.

A thin layer of intermediate temperature ISM is expected to exist in the boundary between the CLIC and the LB hot plasma. This layer will emit radiation with a spectrum and flux dependent on the underlying physical mechanisms. Models for this interface emission indicate it radiates strongly in the energy band E = 20–35 eV, which is important for both He$^{\circ}$ and Ne$^{\circ}$ ionization. The exact physical processes at work in interface regions are unclear, but possibilities include thermal conduction, radiative cooling, and shear flow, which lead to evaporative interface boundaries (Cowie and McKee, 1977), cooling flows (Shapiro and Benjamin, 1991) or turbulent mixing layers (Slavin et al., 1993). All of these boundary types produce intermediate temperature gas ($T \sim 10^{4.5}$–$10^{5.5}$ K) that radiates in the EUV, although ionization levels, and thus the spectrum and intensity of the emission, depend on the detailed physics. Parameters that constrain interface properties include the strength and topology of the interstellar magnetic field, B_{IS}, the hot gas temperature, and the relative dynamics of the hot and warm gas. The magnetic field affects the RT models because B_{IS} reduces the evaporative flow by inhibiting thermal conduction in directions perpendicular to field lines and at the same time supports the cloud by magnetic pressure. Since the total (magnetic + thermal) pressure is roughly constant in an evaporative outflow, the magnitude of the magnetic field is an important factor in determining the pressure in any evaporative flow that might be present. Fig. 6.7 shows examples of the EUV spectrum produced by an evaporative interface and a turbulent mixing layer.

6.4.2 Radiative Transfer Models of the Local Cloud and other Tenuous ISM

Our radiative transfer models are constrained by observations of local ISM towards nearby stars, and *in situ* data from LIC neutrals that have penetrated the heliosphere. The *in situ* data includes direct detection of He$^{\circ}$, observations of solar Lα florescence from H$^{\circ}$ in the heliosphere, and observations of the pickup ion and anomalous cosmic ray populations that are seeded by interstellar neutrals (see Moebius et al., Chapter 8). Generally elements with FIP < 13.6 eV are fully ionized in tenuous clouds, while elements such as H, O, N, Ar, He, and Ne with $13.6 \lesssim$ FIP $\lesssim 25$ eV are partially ionized. Neutrals from these partially ionized species enter the heliosphere, where they seed the pickup ion and anomalous cosmic ray populations measured by instruments on various spacecraft. Column densities towards nearby stars constrain sight-line integrated values, and permit the recovery of $n(H^+)/n(H^o)$ as a function of distance to the cloud surface (Fig. 6.9). While the LIC temperature is determined directly from Ulysses observations of He^o, the densities $n(H^o)$ and $n(H^+)$ at the solar location vary as the Sun moves through the LIC, and must be determined from radiative models.

The detailed attention paid here to LIC radiative transfer models is motivated by the facts that $n(H^o)$ and $n(He^o)$ at the solar location are important boundary conditions of the heliosphere, and that ionization gradients in the CLIC are factors in reconstructing the 3D cloud morphology from data. An extensive study of the LIC ionization has resulted in a series of $\sim$50 radiative transfer models appropriate for tenuous ISM such as the LIC and CLIC (Slavin and Frisch, 2002, Frisch and Slavin, 2003, Frisch and Slavin, 2005). These models use the CLOUDY code (Ferland et al., 1998), an interstellar radiation field based on known sources, and models for the interface expected between the LIC and Local Bubble plasma (also see §6.2.2). Boundary conditions for

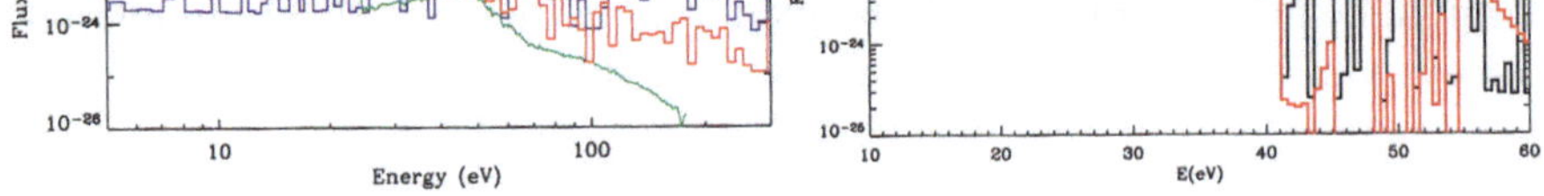

Figure 6.7. Left: Components of the interstellar radiation incident on the local interstellar cloud complex. Contributions from stars show the EUV flux from nearby white dwarf and B stars, after deabsorbing the corresponding $N(H^o)$ to the cloud surface. The cloud boundary flux produced by emission from an evaporative interface between the local gas and the hot gas of the Local Bubble is shown (Slavin and Frisch, 2002). Right: Radiation field from a turbulent mixing layer interface compared to that of a conductive interface. Instabilities at the boundaries of tenuous clouds may lead to quite different radiation fields inside of the clouds, particularly near 584 Å where He^o is ionized.

outer scales are provided by ISM data integrated over the LIC, which can be subject to ionization gradients. On inner scales the boundary conditions are given by *in situ* observations of ISM at the heliosphere. Both sets of constraints are important for evaluating the RT properties of the surrounding ISM where large H^o/H^+ gradients are found. The data sets used here include $n(He^o)$, pickup ions, and anomalous cosmic rays inside of the heliosphere, and absorption line data of the LIC towards the downwind star ϵ CMa (GJ, FS).

These RT models will be updated in the future, as the density, composition, LIC magnetic field, radiation field, and interface regions between the LIC and Local Bubble plasma become better understood. The LIC magnetic field enters indirectly through the contribution of magnetic pressure to the cloud interface. The magnetic pressure in the cloud helps to determine the thermal pressure and thus cooling in the interface, so that further studies of B_{IS} near the Sun will provide insight into the characteristics of the interface radiation. One of the significant results of our study is that the spread in predicted neutral and ion densities demonstrates that low column density ISM can be in equilibrium at a range of ionization levels, and that the ionization itself is highly sensitive to the radiation field, interface characteristics, and other cloud properties.

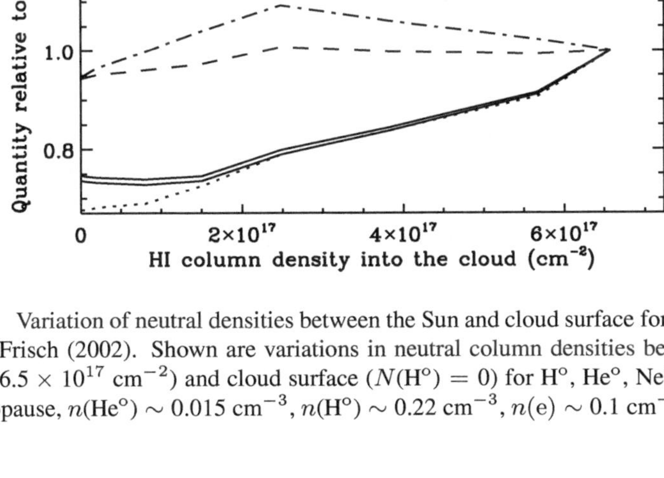

Figure 6.8. Variation of neutral densities between the Sun and cloud surface for Model 2 from Slavin and Frisch (2002). Shown are variations in neutral column densities between the Sun ($N(H^o) = 6.5 \times 10^{17}$ cm^{-2}) and cloud surface ($N(H^o) = 0$) for H°, He°, Ne°, O°, and N°. At the heliopause, $n(He^o) \sim 0.015$ cm^{-3}, $n(H^o) \sim 0.22$ cm^{-3}, $n(e) \sim 0.1$ cm^{-3}.

6.4.3 Results of Radiative Transfer Models

Two generations of radiative transfer (RT) models have been developed, with the focus on matching data for the LIC inside of the heliosphere, and matching data on nearby ISM in the downwind direction where the brightest point sources of EUV radiation (ϵ CMa and β CMa) and low column densities are found. Both sets of models are constrained by LIC data obtained inside of the heliosphere, such as for pickup ions, He^{o}, and anomalous cosmic rays. The PUI and ACR populations are seeded by neutral ISM flowing into the heliosphere, and are subject to filtration losses in the heliosheath regions (Moebius et al., Zank et al., this volume).

The ionization state of the pristine ISM outside of the heliosphere is obtained from the RT models, using slightly different constraints for the first and second generations of models. The first generation of 25 models are constrained by the local ISM towards ϵ CMa, i.e. the sum of the LIC and blue shifted (BC) clouds, and by ISM byproducts inside the heliosphere. The ISM within $\sim$1–2 pc towards ϵ CMa is divided between two clouds, with projected HC velocities of 10 km s^{-1} (the BC) and 18 km s^{-1} (the LIC), and with a total $N(H^{o}) < 10^{18}$ cm^{-2}. The LIC and BC are also observed towards Sirius (α CMa, 2.7 pc). Two models, Model 2 and Model 8, provide a good match to all data except for the cloud temperature inside of the heliosphere. The predicted temperature at the heliosphere is $\sim$3,000 K higher than found from *in situ* He^{o} data, $T = 6{,}400 \pm 340$ K. A possible explanation for this difference is that the abundances of coolants such as C^{+} are incorrect in the models, particularly since $N(C^{+})$ generally has large measurement uncertainties. A second mismatch occurs because predicted $N(Si^{+2})$ values are lower than observed values, which suggests that interface models require additional processes such as a turbulent mixing layer. Models 2 and 8 predict that $n(H^{o}) \sim 0.20$ cm^{-3}, $n(e) \sim 0.1$ cm^{-3}, and ionizations of $\chi(H) \sim 0.30$, and $\chi(He) \sim 0.49$ are appropriate for the solar location (Slavin and Frisch, 2002, Frisch and Slavin, 2003). The mean cloud density to the downwind surface is $\langle n(H^{o}) \rangle_{LIC} \sim 0.17$ cm^{-3}.

The second set of models is constrained by data on the LIC towards ϵ CMa (excluding the BC) and *in situ* ISM. The best of these models are in good agreement with the *in situ* data, including cloud temperature data. However, these models require a high C abundances, possibly in conflict with solar abundances, and fail to predict $N(Mg^{+})/N(Mg^{o})$ in the LIC. The second set of models require that the LIC and BC have different ionization levels, which is similar to the findings of Gry and Jenkins (2001) but remains unexplained. LIC properties predicted by these models, which are still under study, are $n(H^{o}) = 0.19$ cm^{-3}, fractional ionization $\chi(H) = 0.22$, $n(e) = 0.06$ cm^{-3}, $n(He^{o}) = 0.015$ cm^{-3}, and $\chi(He) = 0.37$.

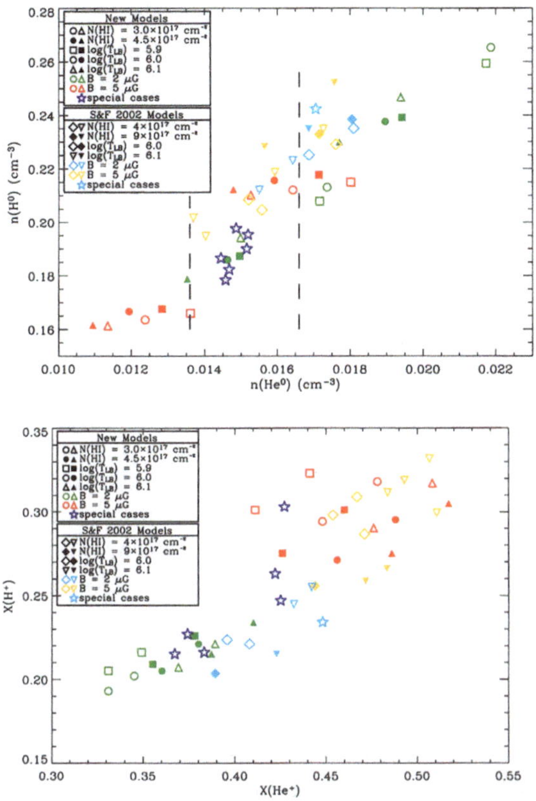

Figure 6.9. Top: H neutral density vs. He neutral density, both at the Solar location for a large set of model calculations. The symbol colors indicate the assumed magnetic field strength in the cloud, the fill indicates the assumed HI column density, and the shape indicates the assumed temperature of the hot gas of the Local Bubble. Stars represent special parameter sets which do not fall on the grid of model parameters, but rather are chosen to better match the data. The density and uncertainties of He° are plotted as vertical lines from Möbius et al., 2004. Note that a range of n(HI) values are consistent with the n(He°) measurements. The radiative transfer models that provide the best agreement with all available data on the ISM also yield a consistent estimate of $n(\mathrm{H}^{\circ}) = 0.19$–$0.21\ \mathrm{cm}^{-3}$. Bottom: Same as left plot but for the H^{+} and He^{+} ion fractions. Note: Some of the differences not explained by $N(\mathrm{H}^{\circ})$, T_{LB}, or B are due to differences in the assumed total H densities.

The results of the RT models that are significant for heliosphere studies are summarized in Figures 6.8 and 6.9. The ionizations of H, He, O, N, and Ne throughout the LIC are shown in Fig. 6.8, with variations of ~25% between the cloud surface and the Sun. Fig. 6.9 shows the extent of ionization states possible under equilibrium conditions in tenuous ISM. Minor variations in the physical assumptions input to the RT models result in a continuum of ionization levels that are in equilibrium for low density clouds like the LIC. Depending on the RT model constraints, H ionization levels are 0.19–0.23, while He ionization levels are 0.32–0.53. *These models show that ionization levels might vary between clouds in the CLIC, so that variations in the heliosphere boundary conditions are expected as the CLIC flows past the Sun (see §6.3).*

6.5 Passages through Nearby Clouds

The transition of the Sun from the near void of the Local Bubble, and into the stream of tenuous ISM flowing away from the Scorpius-Ophiuchus Association, can be probed with the kinematics and column densities of nearby clouds, combined with models of the volume density $n(\mathrm{H^o})$. Specifically, cloud dimensions are assumed to be $\propto N(\mathrm{H^o})/\langle n(\mathrm{H^o})\rangle$. Radiative transfer models of tenuous ISM show that cloud ionization, $\chi(\mathrm{H})$, varies with column density. Since the densities of individual cloudlets are not currently available, $n(\mathrm{H^o})$ determined by LIC radiative transfer models is extrapolated to other nearby clouds (§6.4). With this simple approach, we see that the Sun appears to have entered the CLIC within the past ~150,000 years. The Sun is located close to the downwind edge of the CLIC (Frisch, 1995). CLIC physical properties are discussed in §6.3.

The most predictable passage of the Sun into an interstellar cloud is the epoch at which the Sun entered the LIC, the cloud now surrounding the Sun. The LIC is the only cloud with an accurate 3D velocity vector, which is found from Ulysses measurements of interstellar He° inside of the solar system. The LIC HC velocity is 26.3 ± 0.4 km s^{-1}, and the downwind direction is $\lambda = 74.7 \pm 0.5°$, $\beta = -5.2 \pm 0.2°$ (ecliptic coordinates, Witte, 2004). Several estimates are given for the epoch of the Sun's entry into the LIC. Both the data and radiative transfer models, upon which the estimate relies, need further refinement before answers are conclusive. We test several models of the poorly known LIC shape. The observed cloud velocity and the projected LIC velocity towards the χ^1 Ori, 8.7 pc away and ~15° from the downwind direction, differ by ~2.4 km s^{-1}, suggesting non-LIC gas is contributing (Table 6.4).

The distance to the cloud edge is given by $N(\mathrm{H^o})/\langle n(\mathrm{H^o})\rangle$, where $\langle n(\mathrm{H^o})\rangle$ is the average space density of H° in the cloud. The density of the LIC at the solar location is $n(\mathrm{H^o}) \sim 0.2$ cm^{-3}, and a mean density $\langle n(\mathrm{H^o})\rangle_{\mathrm{LIC}} \sim 0.17$ cm^{-3} is found from radiative transfer models of the LIC in the downwind direction (§6.4). When the cloud structure is not fully resolved, or several velocity components are present, there may be gaps in the cloud that are not incorporated into the adopted value for $\langle n(\mathrm{H^o})\rangle$. We present several estimates below for the entry of the Sun into the LIC, with results that differ by factors of three. However, the answers are consistent with the Sun's entry into the LIC sometime within the past ∼47,000 years for $\langle n(\mathrm{H}^\circ)\rangle = 0.17$ cm^{-3}, and possibly quite recently within the past thousand years.

The difference between the LIC and upwind gas velocities, of $\lesssim 2$ km s^{-1}, including towards the nearest star α Cen, suggest that the properties of the cloud surrounding the Sun will change rather soon, in ∼3,800 years.

6.5.1 First Encounter of Sun with the Local ISM

The projected LIC velocity and N(H°) towards nearby stars in the downwind direction (Table 6.4), indicates that the Sun entered the CLIC sometime within the past 44,000–140,000/$n_{HI,0.2}$ years, where $n_{HI,0.2}$ is the average cloud density in units of 0.2 cm^{-3} (§6.4, Fig. 6.9). The value $n_{HI,0.2} \sim 1$ is reasonable for nearby tenuous ISM, and is consistent with radiative transfer models of the LIC. This transition of the Sun out of the deep vacuum of the Local Bubble, and into the higher density CLIC ISM, would have been accompanied by the appearance of neutral interstellar gas, pickup ions, anomalous cosmic rays, and dust in the heliosphere. The geological record of cosmic ray radioisotopes should have sampled this transition (see the discussions of cosmic rays and the spallation product radioisotopes in Chapters 9, 10, and 12).

The simplest estimate for the date of entry of the Sun into the CLIC is to ignore non-radial motions for all clouds and look only at the nearby stars in the anti-apex direction. ISM towards the three stars α CMaA, α CMaB, and ϵ CMa are useful for this estimate, giving N(H°) $= 5$–7×10^{17} cm^{-2} (based on data and models in Hebrard et al., 1999, Frisch and Slavin, 2003). The distance to the cloud edge in the anti-apex direction is $\propto {n_{HI,0.2}}^{-1}$ pc. The most distant of the two cloudlets seen in these sightlines is a blue-shifted cloud, with a heliocentric radial velocity of 9.2–13.7 km s^{-1} (Table 6.1). We then find that the Sun entered the CLIC within the past (59,000–120,000)/$n_{HI,0.2}$ years.

Using instead the ISM towards the stars χ^1 Ori and α Aur, which are near the heliocentric downwind direction, and for a distance to the cloud edge given by N(H°)/$n_{HI,0.2}$, the observed radial velocities (Table 6.4) indicate that the Sun entered the CLIC (44,000–140,000)/$n_{HI,0.2}$ years ago.

For these estimates, possible non-radial cloud motions and gaps between clouds are ignored. The cloud gas filling factor in the downwind direction is $f \sim 0.26$ in §6.3. If the downwind CLIC consists of a series of wispy clouds with similar velocities, then $n_{HI,0.2} \ll 0.2$ cm^{-3}, and the first solar encounter with the CLIC could have occurred as recently as $\sim 44,000$ years ago based on the χ^1 Ori data.

6.5.2 Entering the Local Cloud

In principle the entry date of the Sun into the CLIC can be determined precisely since the full 3D velocity of the LIC is known from observations of He^{o} inside of the heliosphere (Table 6.1). The LIC shape is uncertain because components at the LIC velocity are not resolved in many sightlines. Although there may be high density contrasts from low speed waves in the LIC, for instance acoustic waves will have velocities $\sim$7 km s^{-1}, we take the approach that local clouds are defined by the component structure and look at several geometric LIC models to estimate the epoch the Sun entered the LIC. ISM data for downwind stars are summarized in Table 6.4.

The simplest estimate (A) uses N(H^{o}) and the LIC velocity towards χ^1 Ori, $\sim$15° from the downwind LIC direction. Estimate (B) is based on the closest observed star in the downwind hemisphere, α CMa, which is $\sim$45° from the downwind direction, combined with the assumption that the normal to the downwind cloud surface is parallel to the LSR LIC velocity. This assumption results in an entry epoch that varies with the assumed LSR (§6.1.2), and ignores data for more distant downwind stars. Estimate (C) approximates the downwind LIC surface as a flat plane, defined by the distance of the LIC edge towards any three stars from Table 6.4, moving through space at the LIC velocity. The final estimate (D) relies on the Colorado LIC model (RL0), which however does not incorporate the most recent data for the LIC, but can be expected to substantially improve in the future.

A. The values for the LIC N(H^{o}) and the projected LIC velocity towards χ^1 Ori° (Table 6.4) indicate that the Sun entered the LIC $\sim$40,000/$n_{HI,0.2}$ years ago, where $n_{HI,0.2}$ is the mean n(H^{o}) to cloud surface in units of 0.2 cm^{-3}. Radiative transfer models indicate $n_{HI,0.2} \sim$ 0.85 (§6.4). The observed cloud velocity towards χ^1 Ori differs from the projected LIC velocity by 2.4 $\pm$ 1.0 km s^{-1}, so that additional cloud gas is present in this sightline that is not at the LIC velocity. In addition, the turbulent contribution to the line broadening is $\xi = 2.38^{+0.15}_{-0.17}$ km s^{-1} towards χ^1 Ori (§6.1.3, Redfield and Linsky, 2004b), indicating unresolved cloudlets may be present.

B. For this estimate we use LIC data towards the nearest star in the downwind direction for which ISM data are available, α CMaAB (Sirius). The LIC downwind surface is assumed to be oriented such that the surface normal is parallel to the LIC LSR velocity, based on the model of Frisch (1994, Figure 6.10). Note that the "downwind" nomenclature is referenced with respect to the Sun, and the "downwind surface" is the leading edge of the LIC as it moves through the LSR. Since the LIC surface is referenced to the LSR, the solar apex motion is a variable (Table 6.1, §6.1.2). The Sun then encountered the LIC within the past $\sim$13,500/$n_{HI,0.2}$ years or $\sim$ 10^3/$n_{HI,0.2}$ years, for the Hipparcos or Standard solar apex motions, respectively. The present distance to the LIC downwind surface is ~0.01–0.3 pc for this model. The velocity of most of the interstellar gas observed towards χ Ori and α Aur differs by a small amount, 0.7–1.4 km s^{-1}, from the projected LIC velocity (Table 6.4), and therefore is located in a separate cloud. Because the LIC also extends less than $\sim10^4$ AU in the upwind direction, this model implies that the LIC is filamentary or sheet-like, similar to global low density ISM (§6.6). This model results in an interstellar magnetic field direction, $B_{\rm IS}$, which is approximately parallel to the cloud surface (perpendicular to the surface normal).

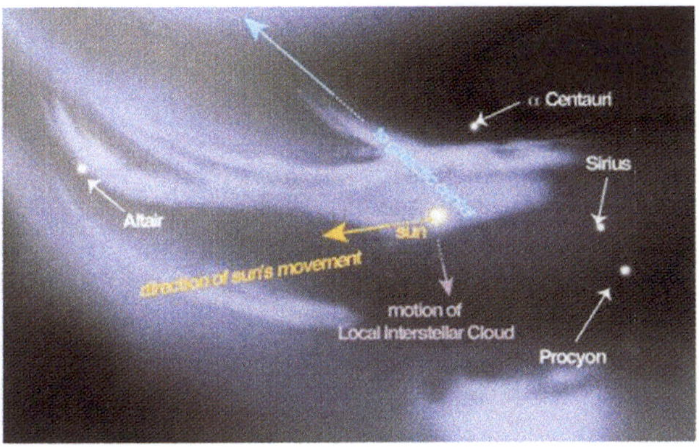

Figure 6.10. The distribution of nearby ISM is shown based on the assumption that the LSR motion of the LIC is parallel to the normal of the cloud surface (model B, §6.5.2). The Sun and LIC motions are in the LSR. Figure from Frisch (2000), courtesy of *American Scientist.*

C. The third estimate models the downwind cloud surface as a locally flat surface moving with the LIC velocity, based on the model in Mueller et al. (2005). We then use any three stars in Table 6.4 to define this plane, and the time at which the Sun crosses this plane marks the entry of the Sun into the LIC. For this model, the Sun entered the LIC between 33,000/$n_{HI,0.2}$ to 36,000/$n_{HI,0.2}$ years ago.

D. The final estimate relies on the Colorado LIC model, based on density $\langle n(\mathrm{H}^\circ)\rangle = 0.1$ cm^{-3} and a LIC velocity determined from UV absorption lines towards nearby downwind stars (RL). This UV LIC vector differs by ~0.6 km s^{-1} in velocity, and ~0.8° in direction, from the LIC velocity given by data on interstellar He$^\circ$ inside of the solar system (Witte, 2004). The Colorado Model Web calculator gives a distance to the downwind surface of 4.5 pc, indicating the Sun entered the LIC ~170,000 years ago for a relative Sun-LIC velocity of 26.3 km s^{-1}, or alternatively ~85,500/$n_{HI,0.2}$ years ago. The column density towards the downwind direction is predicted to be $N(\mathrm{H}^\circ) = 10^{18.14}$ cm^{-2}, however, which differs substantially from recent

Table 6.4. Stars Sampling the Downwind LIC.

Object	ℓ, b (°, °)	Dist. (pc)	$V_{\rm observed}$ (km s^{-1})	$V^{proj}_{\rm LIC}$ (km s^{-1})	log$N$(H°) (cm^{-3})
α Aur	162.6, 4.6	12.9	*21.5±0.5*	23.0±0.3	18.26
LIC He°	183.3, –15.9	0	*26.3±0.3*	*26.3±0.3*	–
χ^1 Ori	188.5, –2.7	8.7	*23.1±0.7*	25.5±0.3	17.80
α CMi	213.7, 13.0	3.5	*20.5±1.0*, 24.0	19.6±0.2	17.90
α CMaA	227.2, –8.9	2.7	13.7, *19.5±0.3*	19.1±0.2	17.25
α CMaB	227.2, –8.9	2.7	11.7, *17.6±1.5*	19.1±0.2	17.63
ϵ CMa	239.8, –11.3	132	9.2, *16.2±1.5*	15.1±0.2	~17.54

Notes: The LIC velocity component is in *italics*. $V^{proj}_{\rm LIC}$ is the LIC velocity projected towards the star. All velocities are heliocentric. The column densities and data are from Witte (2004), Hebrard et al. (1999), Frisch & Welty (2005), RL, GJ, SF, and FS.

values for χ^1 Ori (Table 6.4). We therefore disregard this estimate, but expect further improvements in the Colorado model to significantly increase our understanding of the LIC morphology.

6.5.3 Future Cloud Encounters

The best limits on the distance to the upwind edge of the LIC are found from observations of 36 Oph, 6 pc away and ~16° from the heliosphere nose. The ISM velocity towards 36 Oph is –28.5±0.6 km s^{-1}, which corresponds to the G-cloud velocity (Table 6.3). The limit on N(H°) at the projected LIC velocity of 25.7 km s^{-1} is N(H°) $< 6 \times 10^{16}$ cm^{-2} (Wood et al., 2000b). These numbers indicate that the Sun will exit the LIC sometime within the next ~3700/$n_{HI,0.2}$ years, and that the distance to the upwind LIC surface is <0.10 pc. Support for this result is provided by the ISM velocity towards the nearest star α Cen, which is ~50° from the HC nose direction. Towards α Cen, the observed cloud velocity and the projected LIC velocity differ by ~1 km s^{-1} (Landsman et al., 1984, Linsky and Wood, 1996). Column densities of N(H°) $=$ 17.6–18.0 cm^{-2} were originally found towards α CenAB (Linsky and Wood, 1996), which indicate that $\langle n(\mathrm{H}^\circ)\rangle$ may be similar to the LIC density of ~0.2 cm^{-3}. Later interpretations of the data favor the low end of this column density range (Wood et al., 2005), indicating that the sightline is only partially filled with warm gas.

Possibilities for the next cloud to be encountered by the Sun include the G and Apex clouds (Table 6.3, Frisch, 2003). Both clouds have LSR upwind directions, (ℓ, b) $=$ (348 ± 3°, 5 ± 5°), that are approximately perpendicular to the LSR solar apex motion (§6.1.2). Observations of α Cen and α Aql show that the G-cloud is within 1.3 pc, and the Apex cloud is within 5 pc of the Sun, giving upper limits for an encounter date of ~45,000 and ~175,000 years, respectively. The column densities of the G-cloud are ~3 times larger than Apex cloud column densities, which are $N(\mathrm{H^o}) \sim 10^{17.47}$ cm^{-2} towards 70 Oph and α Aql. Typical G-cloud column densities are $N(\mathrm{H^o}) \sim 10^{17.92}$ cm^{-2}.

Both clouds are inhomogeneous. Only the Apex cloud is seen towards AU Mic (d = 9.9 pc, $\ell = 12.7°$, $b = -36.8°$). Only the G-cloud is seen towards o Ser (d = 13 pc, $\ell = 13.3°$, $b = 9.2°$) and ν Ser (d = 59 pc, $\ell = 10.6°$, $b = 13.5°$). Both clouds are seen towards α Aql (d = 5 pc, $\ell = 47.7°$, $b = -8.9°$) and α Oph (d = 14 pc, $\ell = 35.9°$, $b = 22.6°$).

The G-cloud $\mathrm{Ca^+}$ component ($N(\mathrm{Ca^+}) = 1.58 \times 10^{11}$ cm^{-2}) towards α Oph is extraordinarily strong for a nearby star, and it implies a density of $n(\mathrm{H^o}) > 5$ cm^{-3} in the G-cloud, which is ~15 times the density of the LIC (Frisch, 2003). This estimate is found by using the G-cloud limit of $N(\mathrm{Ca^+})/N(\mathrm{H^o}) < 7.1 \times 10^{-9}$ towards α Cen in order to estimate $N(\mathrm{H^o})$ for the G-cloud in the α Oph sightline, and by assuming the G-cloud is entirely foreground to α Cen, giving a distance limit of 1.3 pc. However, should the G-cloud temperature be colder towards α Oph than towards α Cen, or should a significant amount of the G-cloud be beyond the distances of 36 Oph and α Cen, then the requirement for a high density becomes diminished.

6.6 The Solar Environment and Global ISM

The global properties of ISM in the solar neighborhood offer hints about potential variations in the ISM that might impact the heliosphere.

Over the past ~50 years, radio and UV data have slowly disproved simple ISM models consisting of gravitationally clumped clouds embedded in a substrate of warm or hot low density gas. Evidence supporting a filamentary nature for low column density ISM, with $N(\mathrm{H}) < 10^{21}$ cm^{-2}, is provided by H° 21 cm observations of high latitude H° filaments, and also by UV observations that are interpreted as high space density ($n(\mathrm{H^o}) > 10$ cm^{-3}), low column density ($N(\mathrm{H^o}) < 10^{19}$ cm^{-2} and $N(\mathrm{H^+}) < 10^{18}$ cm^{-2}) nearby gas (Hartmann and Burton, 1997, Welty et al., 1999, Welty et al., 2002). Towards 23 Ori, warm low velocity diffuse gas has densities in the range $n(\mathrm{H^o}) = 15$–20 cm^{-3}, and cloud thicknesses, L, are 0.7–0.9 pc. The densities of cold clouds, with temperatures $\leq$100 K, are slightly lower, $n(\mathrm{H^o}) = 10$–15 cm^{-3}. Diffuse ionized gas is present at intermediate velocities, $|V_{\mathrm{LSR}}| \geq 20$ km s^{-1}, $n(\mathrm{e}) \leq 5$ cm^{-3}, and $L = 0.0006$–0.04 pc. Thin diffuse ionized clouds, with $n(\mathrm{e}) \sim 0.17$ cm^{-3}, $L = 0.2$–2.7 pc, and similar velocities, are observed towards Orion stars spread over >15°, which indicates that these clouds must be thin sheets or filaments.

Winds of massive stars and supernova in young stellar associations eject energy into the surrounding ISM, and evacuate cavities in the ISM that are surrounded by irregular shells of WNM. These shells, which may overlap or be filled with X-ray plasma, are traced by H° 21-cm emission, and explain some of the ISM filamentary structure (MacLow and McCray, 1988, Cox and Smith, 1974, Dickey, 2004, Heiles and Troland, 2003b). The CLIC appears to be a fragment of a superbubble shell from one epoch of star formation in the SCA, and α Oph may be in a direction that is tangential to this shell (Frisch, 1995).

6.6.1 Warm Ionized and Partially Ionized Gas

Interstellar plasma and neutrals interact with the heliosphere in fundamentally different ways, and small variations in the interstellar radiation field convert neutral gas to plasma. Nearby WPIM was discussed in §6.3.1, and we now compare warm interstellar plasma over 100 pc scales and show that the CLIC and global observed diffuse ionized gas are similar. We conclude that dense ionized ISM has not been, and will not be, part of the immediate solar environment over short timescales. However, diffuse WPIM similar to the LIC and CLIC is much more widespread.

Dense fully ionized ISM is not found close to the Sun, although diffuse ionized gas surrounds many hot stars at distances of >150 pc, such as β CMa bordering the giant ionized Gum Nebula, β Cen, which energizes the interior of Loop I, and several stars in Upper Scorpius. Generally H II regions surround hot O, early B, and white dwarf stars that emit strongly in the EUV, $\lambda < 912$ Å. The solar path is unlikely to traverse an H II region surrounding an O–B1 star over timescales of ± 4 Myrs because there are no hot stars or dense H II regions nearby. The closest H II region surrounds the high latitude star α Vir, 80 pc away. White dwarf stars are relatively frequent near the Sun. Over 25 white dwarf stars with surface temperatures $T > 10^4$ K are found within 20 pc, and >40% of these stars are hotter than 15,000 K. Small Stromgren spheres will surround the hottest white dwarf stars, but the densities will be very low for nearby stars (Tat and Terzian, 1999).

Diffuse ionized gas is widespread, in contrast to dense ionized gas. It is traced by weak optical recombination lines such as the Hα line, N II, He I, He II, O I, and Si II (Haffner et al., 2003, Reynolds, 2004), UV absorption lines (Welty et al., 1999, Holberg et al., 1999, Lehner et al., 2003, GJ), and EUV observations of He° and He^+ towards white dwarf stars (§6.3). Pulsar dispersion measures trace the interactions between pulsar wave packets and electrons (e.g. Taylor and Cordes, 1993, Cordes and Lazio, 2002, Armstrong et al., 1995). These data reveal a low density, ionized component, $n(\mathrm{e}) \sim 0.1$ cm^{-3}, which fills ~20% of the volume of a 2-kpc-thick layer around the Galactic disk. The diffuse H^+ has been mapped in the northern hemisphere by WHAM, however the LIC emission measure is an order of magnitude below the WHAM sensitivity (e.g. Reynolds, 2004). About 30% of diffuse H° and H^+ gas are found in spatially and kinematically associated clouds with densities ~0.2–0.3 cm^{-3} (Reynolds et al., 1995). If the H° and H^+ were spatially coincident in these clouds then χ(H) would be $\sim$ 0.4, but Reynolds et al. argue that the H° and H^+ are spatially separated. Despite the evidence from spatial maps and line widths, however, we believe that the existence of some true WPIM (perhaps at lower ionization levels) cannot be ruled out in these regions. These clouds are associated with large filamentary or sheet-like structures, and exhibit cloud densities and hydrogen ionization levels that are similar to LIC values. In other directions, Hα line emission indicates that H is highly ionized in the WIM, χ(H) $\sim$ 1. Emission at 5880 Å from He° shows that for the WIM χ(He) $\sim$ 0.3–0.6. Ratios of χ(He)/χ(H) are ~0.3–0.6 are found, versus ~1.6 for the LIC, implying that the LIC radiation field is harder than for most of the WIM, perhaps because of lower column densities. The WIM has a continuum of temperatures, 5000–10,000 K, and ionization levels, as demonstrated by emission in [N II], [O I], and [S II] lines that trace temperature and ionization.

The interstellar plasma is structured over all spatial scales, and these plasma clumps appear to be similar to the LIC. The plasma is partially opaque to radio emission in the energy range 0.1 to 10 MHz, and as a result synchrotron emission probes the WIM clumpiness (Kulkarni and Heiles, 1988, Peterson and Webber, 2002). Models for the propagation of low frequency synchrotron emission through a clumpy WIM fit the radio data better than do propagation models, that assume a uniformly distributed parallel slab model for the WIM. These clumps have density n(e) ~ 0.2 cm^{-3}, fill 8%–15% of the disk, and have a free-free opacity at 10 MHz that is consistent with WIM temperatures as low as 4500 K. The dispersion, refraction and scintillation of pulsar wave packets indicates that the WIM is turbulent over scale sizes 10^{-2}–10^2 AU, or even larger, ~10^2 pc, if Faraday rotation and plasma density gradients are included (Armstrong et al., 1995).

The WIM properties are consistent with predictions of the radiative transfer models, which find $\chi(H^+)$ = 0.15–0.35, T $\sim$ 5,000–9,000 K, and n(H°) = 0.2–0.3 cm^{-3} for low column density clouds (Fig. 6.9, Slavin and Frisch, 2002). For very low column densities, <10^{18} cm^{-2}, a continuum of ionization and temperature levels are expected from variations in the cloud column density and the EUV radiation field. The WNM and WIM in the CLIC may show similar variations, if the N(H°) $\sim$ 10^{19} cm^{-2} clouds represent clusters of lower column density objects. The uncertain role of turbulence in cloud evolution allows this possibility.

6.6.2 Neutral Interstellar Gas and Turbulence

Radio and UV observations of low $N(\mathrm{H^o})$ cloudlets in the solar vicinity indicate that dense, low column density, cloudlets may be hidden in the CLIC flow. Radio 21-cm data indicate ISM of this type is widespread, and has column densities and kinematics similar to CLIC ISM.

Two populations of interstellar clouds, WNM and CNM, are found based on observations of the collisionally populated H I hyperfine 21-cm line towards radio-continuum sources. The recent Arecibo Millennium Survey used an ON-OFF observing strategy to survey the properties of $\mathrm{H^o}$ clouds towards ∼80 radio continuum sources, including emission and and opacity profiles, which were then fit with ∼375 Gaussian components. The detailed attention paid to sources of noise and line contamination yielded a data set that provides a new perspective on the statistics of low $N(\mathrm{H^o})$ CNM and WNM clouds (Heiles and Troland, 2003a, Heiles and Troland, 2003b).

The CNM and WNM components observed in the Arecibo survey show that ∼25% of the ISM $\mathrm{H^o}$ mass is contained in clouds traveling with velocities ≥10 km $\mathrm{s^{-1}}$ through the local standard of rest. When the solar apex motion of 13–20 km $\mathrm{s^{-1}}$ is included, this means that Sun-cloud encounters with relative velocities larger than 25 km $\mathrm{s^{-1}}$ are quite likely, as we might have guessed because of the LIC heliocentric velocity of –26.3 km $\mathrm{s^{-1}}$.

For 21-cm absorption components with spin temperature T_s, the opacity is $\tau \propto N(\mathrm{H^o})/T_\mathrm{s}$ and the line width is FWHM$\propto\sigma_\mathrm{v}$, where σ_v is the dispersion in cloud velocity. Therefore the turbulent (ΔV_turb) and thermal contributions to σ_v can be distinguished for the CNM. Spin temperatures, T_s, for the CNM are typically ∼20–100 K.

The Arecibo survey discovered that 60% of the diffuse ISM is WNM, and ∼40% is CNM. Median column densities for the WNM and CNM towards sources with $|b| > 10^\circ$ are $N(\mathrm{H^o}) = 1.3 \times 10^{19}$ cm^{-3} and 5.2×10^{19} cm^{-3}, respectively. The finding that CNM column densities are typically lower than WNM values was unexpected. The column density weighted median spin temperature for the CNM is 70 K, although minimum temperatures are 20 K or lower. Upper limits on WNM kinetic temperatures are 15,000 K or higher, with a typical value of 4,000 K. The turbulent velocities derived for CNM components indicate that the turbulence must be highly supersonic, with a turbulent Mach number $M_\mathrm{turb} = 3\Delta V^2_\mathrm{turb}/C^2_\mathrm{s} \sim 3$, where C^2_s is the sound speed and the factor of 3 converts to three-dimensional turbulence (Heiles, 2004b). The densities of the CNM observed towards 23 Ori are >10 cm^{-3} (Welty et al., 1999).

The velocities of CLIC, WNM, and CNM components are compared in Fig. 6.11. Only high latitude WNM and CNM data with $|b| > 25^\circ$ are included,

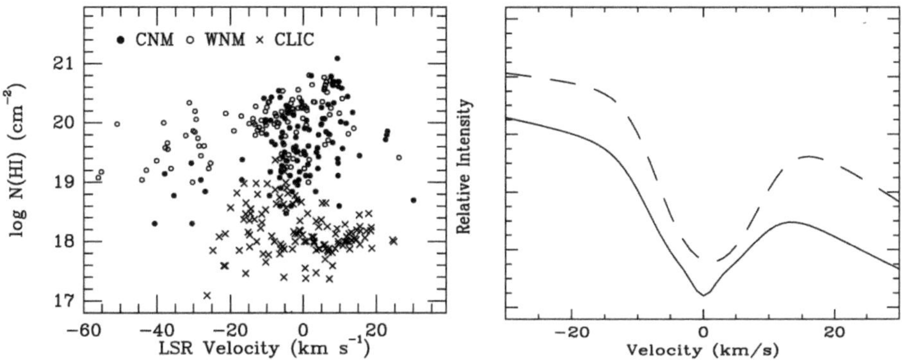

Figure 6.11. Left: CLIC velocities, for d $<$ 50 pc stars, are plotted against CNM and WNM velocities for sightlines with $|b| > 25°$ in the Arecibo Millennium Survey. The Arecibo clouds are within ~425 pc if the H° layer thickness is 185 pc. The CLIC N(H°) is based on D°/H°= 1.5×10^{-5}, Ca$^+$/H°= 8.3×10^{-9}, or Mg$^+$/H°= 6.7×10^{-6}, and LSR$_{Std}$ is used. Right: This composite profile shows that some CNM components can be lost in the WNM at UV resolutions. The solid line shows a D° 1215 Å absorption component formed by two clouds at the same velocity (0 km s^{-1}) and column density (N(D°) $= 10^{13.17}$ cm^{-2}), but different temperatures (50 K and 7,000 K). The dashed line shows WNM only, with N(D°) $= 10^{13.17}$ cm^{-2} and $T = 7000$ K. This theoretical profile incorporates a Gaussian instrument resolution of FWHM 3 km s^{-1} appropriate for UV data. Turbulent broadening is not included.

in order to avoid distant gas near the galactic plane. Some of the CNM and WNM components are found at velocities lower than –25 km s^{-1}, which may be partly due to infalling H° gas at high latitudes. Otherwise, the CLIC kinematics and column densities overlap generic WNM and CNM values.

Of special interest are very low column density cold CNM, or "tiny" clouds, found for a broad range of velocities. An example is a component with T_s = 16.6 K, N(H°)=2 $\times 10^{18}$ cm^{-2} and velocity -40.6 km s^{-1} observed towards 3C 225. Another example is a component with T_s = 43 K, N(H°) = 3×10^{18} cm^{-2}, and velocity –4.6 km s^{-1} observed towards 4C 32.44. In addition, CNM components with N(H°) $\sim 10^{18}$ cm^{-3}, $T_s \sim$ 30–90 K, and velocities –30 to 0 km s^{-1}, are confirmed towards several other sources (Stanimirovic and Heiles, 2005, Braun and Kanekar, 2005).

CNM ISM is detected in the UV towards 23 Ori, with low column densities, weak ionization, and low temperatures: N(H) $\gtrsim$ 2 x 10^{19} cm^{-2}, $T_s \sim$ 60–150 K, and ~1% ionization. Cold clouds with lower column densities may be difficult to resolve at UV resolutions. The densities of these clouds are found from pressure considerations using C I and C II fine structure lines (Jenkins, 2002), and yield $n_H \sim$ 10–15 cm^{-3} and a cloud thickness of ~0.5 pc (Welty et al., 1999). These components are formed in the Orion-Eridanus Bubble shell. Similar features are observed in other Orion stars. This CNM is either filamentary or sheet-like, and it is a plausible template for the CNM in Fig. 6.11.

The evident crowding of absorption components in velocity space may introduce problems with the interpretation of UV data (§6.1.3). CNM and WNM clouds with low $N(\mathrm{H^o})$ and similar velocities would be unresolved at a nominal 3 km s^{-1} UV resolution, as shown by the theoretical composite D^o profile plotted in Fig. 6.11. This profile is synthesized from the expected profiles of two clouds at 0 km s^{-1}, temperatures 7,000 K and 50 K clouds, and each with $N(\mathrm{H^o}) = 10^{18}$ cm^{-2}. An instrumental resolution of 3 km s^{-1} is included, but turbulence is not. Resolution problems are worse for heavy elements, since the line FWHM$\propto m^{-1/2}$. Therefore local low $N(\mathrm{H^o})$ cold cloudlets can not be ruled out, and if present would be traversed by the heliosphere in ∼1–20,000 years.

6.6.3 Magnetic Fields

Interstellar magnetic fields interact strongly with the heliosphere, and modify the propagation of galactic cosmic rays in space. Unfortunately the uniformity and strength of the interstellar magnetic field (B_{IS}) at the Sun are unknown. We showed that magnetic field strengths of $B_{\mathrm{IS}} \sim$ 0.6–3.7 μG would satisfy several assumptions about the equilibrium of local ISM (§6.3.4). However, the only direct data on the nearby magnetic field are the starlight polarization data attributed to weakly aligned interstellar grains in the nearest ∼35 pc (Tinbergen, 1982, Frisch, 1990). More recently, it appears as if these grains are instead trapped in the magnetic wall of the heliosheath (Frisch, 2005). The global nearby B_{IS} thus must be used to obtain insight on B_{IS} near the Sun.

The properties of the interstellar magnetic field, including the weak spatially uniform (B_{u}) and stronger random (B_{r}) components, are known for both CNM and WIM (e.g. Beck, 2001, Crutcher et al., 2003, Heiles, 2004a, Heiles and Crutcher, 2005). The CNM median field strength of 6.0±1.8 μG is determined from data on the Zeeman splitting of the H I 21 cm line (Heiles and Troland, 2005), but may not apply to the CLIC where CNM is not yet detected. Although $B_{\mathrm{IS}} \sim n$ for densities larger than 10^3 cm^{-3}, at lower densities B_{IS} strengths are uncorrelated with cloud density, in violation of simple flux freezing assumptions.

Pulsar data, synchrotron emission, and starlight polarization data indicate that the ratios of the random (B_{r}) and uniform (B_{u}) components of the local field are $B_{\mathrm{u}}/B_{\mathrm{r}} \sim$ 0.6–1.0. Synchrotron emission of external galaxies indicate that the ratio $B_{\mathrm{u}}/B_{\mathrm{r}}$ increases in interarm regions such as surrounding the Sun, but decreases in spiral arm regions as the magnetic fields become more random. Comparisons of magnetic field strengths determined from polarized synchrotron emission and pulsar dispersion measure observations, which interact with $B_{\rm IS}$ and the WIM, give the total local field strength and the relative strengths of $B_{\rm u}$ and $B_{\rm r}$. Synchrotron emission gives a total field strength locally of $B_{\rm IS}\sim$ 6 ± 2 μG, and the uniform component strength $B_{\rm u}\sim$ 4 μG. Faraday rotation from the propagation of the oppositely polarized circular components produces a polarization phase shift, given by the rotation measure RM $=\int_0^L B_{||}(\mathrm{x}) n_{\rm e}(x)dx$, which responds to the parallel component of the magnetic field, $B_{||}$, weighted by the electron density, $n_{\rm e}(x)$, and integrated over pathlength L. $B_{\rm IS}$ is directed away from the observer for RM<0. Wave packet dispersion in the WIM gives the dispersion measure DM = $\int_0^L n_{\rm e}(x)dx$. The mean local field, weighted by $n(\mathrm{e})$, is found from the ratio RM/DM and is $B_{||}\sim$ 6.5 μG, where $B_{\rm u}\sim B_{||}$.

Starlight polarization data show that the local $B_{\rm IS}$ is relatively uniform, and not strongly strongly distorted or tangled by turbulence. Locally the total field, $B_{\rm IS}$, follows the pitch angle, 9.4°, of the local spiral arm, and has a curvature radius of ~7.8 kpc centered on $\ell = 344.6°$. The starlight polarization vectors converge near ℓ~80° and ~270°, where the random component of the field, $B_{\rm r}$, results in a polarization angle that is rotated over all angles.

A large excursion from $B_{\rm u}$ is seen towards within 150 pc, towards Loop I, where both polarized starlight and synchrotron emission indicate a magnetic field conforms to the superbubble shell around the Scorpius-Centaurus Association. Surveys of $B_{||}$ in morphologically distinct objects such as the Loop I, Eridanus, and Orion supershells, show that the magnetic field is strong, $B_{||} > 5$ μG, and ordered throughout these shell features.

6.6.4 Origin of Local ISM

Several origins have been suggested for the CLIC. CLIC kinematics and abundance patterns led to the suggestion that the CLIC is part of a superbubble shell surrounding the Scorpius-Centaurus Association (Frisch, 1981, Frisch and York, 1986, de Geus, 1992, Frisch, 1995). The SCA is surrounded by a well-known supershell formed by epochs of star formation, winds from massive stars and supernovae (Weaver, 1979). Using superbubble expansion models and the star formation rates in the SCA, Frisch modeled the CLIC as a fragment of the shell from the formation of the Upper Scorpius subgroup (~4 Myrs old) of the SCA, which has expanded away from the high-density ISM close to the association to the lower pressure regions around the Sun (Frisch, 1995). A related suggestion attributes the origin of the CLIC to Rayleigh-Taylor instabilities in the shell of the younger Loop I supernova remnant (~0.5 Myrs old), which has propelled ISM towards the Sun (Breitschwerdt et al., 2000). A third origin suggests that the CLIC originates as a magnetic flux tube that is originally captured in the inner wall of the Local Bubble surrounding the Sun, and is driven back into the Local Bubble interior by magnetic tension

(Cox and Helenius, 2003). This scenario requires that the CLIC velocity and magnetic fields are parallel, which appears unlikely because of the asymmetric capture of small interstellar dust grains in the heliosheath (Frisch, 2005). Each of these suggestions is consistent with the CLIC kinematics, but should have different consequences for short-term variations in the solar environment.

6.7 Summary

In this chapter we have tried to present the essential information required to understand short-term variations in the galactic environment of the Sun, and at the same time make an educated guess about the times that the Sun has transitioned between different types of ISM for the past 10^3-10^5 years. It is our hope some day that the geological radio isotope records can be searched for the signatures of these transitions, although earlier searches have been inconclusive (e.g., Frisch, 1981, Sonett et al., 1987, Frisch, 1997).

Both the Sun and ISM move through our local neighborhood of the Milky Way Galaxy, which is described by the LSR rest velocity frame (with its own uncertainties). The Sun has been in the nearly empty space of the Local Bubble interior for several million years. Sometime in the past $\sim$140,000/$n_{HI,0.2}$ years, the Sun exited the nearly empty Local Bubble and entered an outflow of ISM with an upwind direction towards the Scorpius-Centaurus Association. Even more recently, $\sim$40,000/$n_{HI,0.2}$ years ago, the Sun entered the cloud now surrounding us, the LIC. Within $\sim$4,000 years the Sun will exit the LIC. What's next? The flow of local ISM past the Sun is bearing down on the Sun at a relative velocity of $\sim$28 km s^{-1}, the cloudlets in this flow are expected to pass over the Sun more often than roughly once per $\sim$70,000 years. The next cloud the Sun is most likely to traverse is either the G-cloud or the Apex Cloud. This cloud must be either either inhomogeneous or denser by a factor of $\sim$30 than the LIC.

Most of this nearby ISM is warm partially ionized gas, not dissimilar to warm ISM observed elsewhere, except that column densities tend to be lower. The properties of this gas are mainly consistent with predictions of radiative transfer models, which then suggests that the local gas is close to equilibrium. If viewed from elsewhere in the galaxy, the ensemble of ISM close to the Sun could easily pass as warm neutral material such as observed by H I 21 cm data. The general characteristics of nearby ISM are similar to the global warm neutral material, except that column densities are very low so that H^o ionizing photons penetrate to the cloud cores. The physical properties of the CLIC vary both between and within clouds. The low column densities of this ISM lead to ionization variations inside of a cloud. As the Sun passes between clouds, the interstellar ionization, dynamic pressure or the thermal pressure may vary, possibly all at once. As a consequence, the boundary conditions of the heliosphere may experience dramatic variations as the Sun moves through the CLIC. The efforts to understand these variations has just begun.

One of the more intriguing possibilities is that tiny cool ISM cloudlets are undetected in the nearby warm partially ionized gas (WPIM). Such structures, with $N(\mathrm{H^o}) \sim 10^{18}$ cm^{-2}, are seen elsewhere, although infrequently. A new class of cloud models, perhaps based on macro-turbulence, will be required if such structures exist locally.

Most of our conclusions have substantial inherent uncertainties. The most prominent is that all distances, such as those to cloud edges, are based on $N(\mathrm{H^o})$, which is moderately well known for most sightlines discussed here, and an assumption about the mean $\mathrm{H^o}$ space density, $\langle n(\mathrm{H^\circ})\rangle$. We have a very good idea of the value for $\langle n(\mathrm{H^\circ})\rangle$ in the LIC because of extensive development of radiative transfer models for the LIC. Our confidence in the models follows from their ability to predict both ISM densities inside of the heliosphere, including the Ulysses satellite values for $n(\mathrm{He^o})$, and ISM column densities towards nearby stars. Since column densities are too small locally for measurements of the fine-structure lines that yield $n(\mathrm{H^o})$, we have assumed that the LIC value for $\langle n(\mathrm{H^\circ})\rangle$ applies elsewhere in the CLIC. In addition, the transition epochs for the Sun are calculated by assuming that there are no gaps between individual cloudlets, as defined by velocity. The LIC $\langle n(\mathrm{H^\circ})\rangle$ value is probably representative of the downwind CLIC, because the Sun is so close to the downwind edge. However, we can not rule out gaps between the cloudlets, and such gaps would alter estimates of the epoch the Sun entered the CLIC.

One of the most significant improvements will be to make good maps of nearby ISM. Such maps would require a new generation of space instrumentation, capable of high resolution, R $\sim$ 500,000, and high signal-to-noise observations of both bright and faint objects over the spectral range 912–3,000 Å. We also need more precise measurements of the diffuse interstellar radiation field, throughout the full spectral interval of 2000 Å to 0.1 keV. Such data would provide a strong constraint on the interface emission and reduce uncertainties in the $\mathrm{Mg^o}$ photoionization rate.

In this chapter we have reviewed the short-term variations in the galactic environment of the Sun. Our galactic environment affects the heliosphere, and by analogy the astrospheres of nearby stars, so that the space trajectory of a star is a filter for other planetary systems with conditions conducive to stable climates for exoplanets (Frisch, 1993). As other papers in this volume demonstrate, the ISM-modulated heliosphere has a pronounced effect on the cosmic ray flux in the inner heliosphere, which in turn appears to affect the climate. Similar processes will affect the climates of planets outside of the solar system. Such possibilities make the study of short-term variations in the galactic environment of the Sun highly topical.

Acknowledgments: The authors thank NASA for supporting this research through grants NAG5-11005, NAG5-13107 and NAG5-13558.

References

Adams, T. F. and Frisch, P. C. (1977). High-resolution observations of Lyman alpha sky background. *Astrophys. J.*, 212:300–308.

Adams, W. S. (1949). Observations of interstellar H and K, molecular lines, and radial velocities in spectra of 300 O and B stars. *Astrophys. J.*, 109: 354–379.

Armstrong, J. W., Rickett, B. J., and Spangler, S. R. (1995). Electron density power spectrum in the local interstellar medium. *Astrophys. J.*, 443: 209–221.

Bash, F. (1986). Present, Past and Future Velocity of Nearby Stars: The path of the Sun in 10^8 Years. In *Galaxy and the Solar System*, pages 35–46. University of Arizona Press.

Beck, R. (2001). Galactic and Extragalactic Magnetic Fields. *Space Science Reviews*, 99:243–260.

Bohlin, R. C., Savage, B. D., and Drake, J. F. (1978). A survey of interstellar H I from L-alpha absorption measurements. *Astrophys. J.*, 224:132–142.

Braun, R. and Kanekar, N. (2005). Tiny H I clouds in the local ISM. *Astron. & Astrophys.*, 436:L53–L56.

Breitschwerdt, D., Freyberg, M. J., and Egger, R. (2000). Origin of H I clouds in Local Bubble. I. Hydromagnetic Rayleigh-Taylor instability caused by interaction of Loop I and Local Bubble. *Astron. & Astrophys.*, 361:303–320.

Bruhweiler, F. C. and Kondo, Y. (1982). UV spectra of nearby white dwarfs and the nature of the local interstellar medium. *Astrophys. J.*, 259:232–243.

Cheng, K. and Bruhweiler, F. C. (1990). Ionization processes in local interstellar medium - effects of hot coronal substrate. *Astrophys. J.*, 364:573–581.

Cordes, J. M. and Lazio, T. J. W. (2002). NE2001.I. A New Model for Galactic Distribution of Free Electrons and its Fluctuations. *ArXiv Astrophysics e-prints.*

Cowie, L. L. and McKee, C. F. (1977). Evaporation of spherical clouds in a hot gas. I - Classical and saturated mass loss rates. *Astrophys. J.*, 211:135.

Cox, A. N. (2000). *Allen's Astrophysical Quantities*, pages 29–30. AIP Press.

Cox, D. P. (1998). Modeling the Local Bubble. *LNP Vol. 506: IAU Colloq. 166: The Local Bubble and Beyond*, 506:121–131.

Cox, D. P. and Helenius, L. (2003). Flux-Tube Dynamics and Model for Origin of Local Fluff. *Astrophys. J.*, 583:205–228.

Cox, D. P. and Smith, B. W. (1974). Large-scale effects of supernova remnants on Galaxy: Generation and maintenance of a hot network of tunnels. *Astrophys. J. Let.*, 189:L105–L108.

Cravens, T. E. (2000). Heliospheric X-ray Emission Associated with Charge Transfer of Solar Wind with Interstellar Neutrals. *Astrophys. J. Let.*, 532: L153–L156.

Cravens, T. E., Robertson, I. P., and Snowden, S. L. (2001). Temporal variations of geocoronal and heliospheric X-ray emission associated with solar wind interaction with neutrals. *J. Geophys. Res.*, 106:24883–24892.

Crutcher, R., Heiles, C., and Troland, T. (2003). Observations of Interstellar Magnetic Fields. *Lecture Notes in Physics, Berlin Springer Verlag*, 614:155–181.

Crutcher, R. M. (1982). The local interstellar medium. *Astrophys. J.*, 254:82–87.

de Geus, E. J. (1992). Interactions of stars and interstellar matter in Scorpio Centaurus. *Astron. & Astrophys.*, 262:258–270.

Dehnen, W. and Binney, J. J. (1998). Local stellar kinematics from Hipparcos data. *Mon. Not. Roy. Astron. Soc.*, 298:387–394.

Dickey, J. M. (2004). Is the Local Fluff typical? *Adv. Space Res.*, 34:14–19.

Dupuis, J., Vennes, S., Bowyer, S., Pradhan, A. K., and Thejll, P. (1995). Hot White Dwarfs in Local Interstellar Medium: Hydrogen and Helium Interstellar Column Densities and Stellar Effective Temperatures from EUVE Spectroscopy. *Astrophys. J.*, 455:574.

Dutra, C. M. and Bica, E. (2002). A catalogue of dust clouds in the Galaxy. *Astron. & Astrophys.*, 383:631–635.

Ebel, D. S. (2000). Variations on solar condensation: Sources of interstellar dust nuclei. *J. Geophys. Res.*, 105:10363–10370.

Eggen, O. J. (1963). Luminosities, colors, and motions of the brightest A-type stars. *Astron. J.*, 68:689.

Ferland, G. J., Korista, K. T., Verner, D. A., Ferguson, J. W., Kingdon, J. B., and Verner, E. M. (1998). Cloudy 90: Numerical simulation of plasmas and their spectra. *Pub. Astron. Soc. Pac.*, 110:761–778.

Ferlet, R., Lagrange-Henri, A.-M., Beust, H., Vitry, R., Zimmermann, J.-P., Martin, M., Char, S., Belmahdi, M., Clavier, J.-P., Coupiac, P., Foing, B. H., Sevre, F., and Vidal-Madjar, A. (1993). Beta Pictoris protoplanetary system. XIV - Observations of Ca II H and K lines. *Astron. & Astrophys.*, 267: 137–144.

Fitzgerald, M. P. (1968). Distribution of interstellar reddening material. *Astron. J.*, 73:983.

Florinski, V., Pogorelov, N. V., Zank, G. P., Wood, B. E., and Cox, D. P. (2004). On the Possibility of a Strong Magnetic Field in the Local Interstellar Medium. *Astrophys. J.*, 604:700–706.

Frisch, P. (1999). *Galactic Environments of Sun and Cool Stars*, pages 3–10. Editions Frontieres.

Frisch, P. and York, D. G. (1986). Interstellar clouds near the Sun. In *The Galaxy and the Solar System*, pages 83–100. University of Arizona Press.

Frisch, P. C. (1979). Interstellar Material towards Chi Ophiuchi. I - Optical observations. *Astrophys. J.*, 227:474–482.

Frisch, P. C. (1981). The Nearby Interstellar Medium. *"Nature"*, 293:377–379.

Frisch, P. C. (1990). Characteristics of the local interstellar medium. In Grzedzielski, S. and Page, D. E., Eds., *Physics of the Outer Heliosphere*, pages 19–22.

Frisch, P. C. (1993). G-star astropauses - A test for interstellar pressure. *Astrophys. J.*, 407:198–206.

Frisch, P. C. (1994). Morphology and Ionization of Interstellar Cloud Surrounding the Solar System. *Science*, 265:1423.

Frisch, P. C. (1995). Characteristics of Nearby Interstellar Matter. *Space Sci. Rev.*, 72:499–592.

Frisch, P. C. (1997). Journey of the Sun. *http://xxx.lanl.gov/*, page astroph/9705231.

Frisch, P. C. (2000). The Galactic Environment of Sun. *American Scientist*, 88:52–59.

Frisch, P. C. (2003). Local Interstellar Matter: The Apex Cloud. *Astrophys. J.*, 593:868–873.

Frisch, P. C. (2005). Tentative Identification of Interstellar Dust in the Magnetic Wall of Heliosphere. *Astrophys. J. Let.*, 632:L143–L146.

Frisch, P. C., Welty, D. E., York, D. G., and Fowler, J. R. (1990). Ionization in nearby interstellar gas. *Astrophys. J.*, 357:514–523.

Frisch, P. C., Dorschner, J. M., Geiss, J., Greenberg, J. M., Grün, E., Landgraf, M., Hoppe, P., Jones, A. P., Krätschmer, W., Linde, T. J., Morfill, G. E., Reach, W., Slavin, J. D., Svestka, J., Witt, A. N., and Zank, G. P. (1999). Dust in the Local Interstellar Wind. *Astrophys. J.*, 525:492–516.

Frisch, P. C., Grodnicki, L., and Welty, D. E. (2002). Velocity Distribution of Nearest Interstellar Gas. *Astrophys. J.*, 574:834–846.

Frisch, P. C., Jenkins, E. B., Johns-Krull, C., Sofia, U. J., Welty, D. E., York, D. G., and Aufdenberg, J. (2005). Local Interstellar Matter towards η UMa. *in preparation.*

Frisch, P. C., Sembach, K., and York, D. G. (1990). Studies of the local interstellar medium. VIII - Morphology and kinematics of diffuse interstellar clouds toward Orion. *Astrophys. J.*, 364:540–548.

Frisch, P. C. and Slavin, J. D. (2003). Chemical Composition and Gas-to-Dust Mass Ratio of Nearby Interstellar Matter. *Astrophys. J.*, 594:844–858.

Frisch, P. C. and Slavin, J. D. (2005). Heliospheric Implications of Structure in the Interstellar Medium. *Adv.Sp. Res.*, 35:2048–2054.

Frisch, P. C. and Welty, D. E. (2005). CaII Observations of Local Interstellar Material. *In preparation.*

Frisch, P. C. and York, D. G. (1983). Synthesis Maps of Ultraviolet Observations of Neutral Interstellar gas. *Astrophys. J. Let.*, 271:L59–L63.

Frisch, P. C., York, D. G., and Fowler, J. R. (1987). Local interstellar medium. VII - Local interstellar wind and interstellar material towards star alpha Ophiuchi. *Astrophys. J.*, 320:842–849.

Frogel, J. A. and Stothers, R. (1977). Local complex of O and B stars. II - kinematics. *Astron. J.*, 82:890–901.

Gayley, K. G., Zank, G. P., Pauls, H. L., Frisch, P. C., and Welty, D. E. (1997). One- versus Two-Shock Heliosphere: Constraining Models with Goddard High Resolution Spectrograph Ly-alpha Spectra toward alpha Centauri. *Astrophys. J.*, 487:259–270.

Grenier, I. A. (2004). Gould Belt, star formation, and the local interstellar medium. *ArXiv Astrophysics e-prints*.

Gry, C. and Jenkins, E. B. (2001). Local clouds: Ionization, temperatures, electron densities and interfaces, from GHRS and IMAPS spectra of epsilon Canis Majoris. *Astron. & Astrophys.*, 367:617–628.

Haffner, L. M., Reynolds, R. J., Tufte, S. L., Madsen, G. J., Jaehnig, K. P., and Percival, J. W. (2003). Wisconsin Hα Mapper Northern Sky Survey. *Astrophys. J. Supl.*, 149:405–422.

Hartmann, D. and Burton, W. B. (1997). *Atlas of Galactic Neutral Hydrogen.* Cambridge University Press, Cambridge.

Hebrard, G., Mallouris, C., Ferlet, R., Koester, D., Lemoine, M., Vidal-Madjar, A., and York, D. (1999). Ultraviolet observations of Sirius A and Sirius B with HST-GHRS. An interstellar cloud with a possible low deuterium abundance. *Astron. & Astrophys.*, 350:643–658.

Heiles, C. (2004a). Observational Magnetogasdynamics: 21 Years of HI Zeeman Splitting Measurements... and More. *Astrophys. & Space Sci.*, 292: 77–88.

Heiles, C. (2004b). Physical Properties of the Diffuse HI. In *ASP Conf. Ser. 317: Milky Way Surveys: Structure and Evolution of our Galaxy*, pages 323.

Heiles, C. and Crutcher, R. (2005). Magnetic Fields in Diffuse H I and Molecular Clouds. *ArXiv Astrophysics e-prints*.

Heiles, C. and Troland, T. H. (2003a). Millennium Arecibo 21 Cm Absorption-Line Survey. I. Techniques and Gaussian Fits. *Astrophys. J. Supl.*, 145: 329–354.

Heiles, C. and Troland, T. H. (2003b). Millennium Arecibo 21 Cm Absorption-Line Survey. II. Properties of the Warm and Cold Neutral Media. *Astrophys. J.*, 586:1067–1093.

Heiles, C. and Troland, T. H. (2004). Millennium Arecibo 21 Cm Absorption-Line Survey. III. Techniques for Spectral Polarization and Results for Stokes V. *Astrophys. J. Supl.*, 151:271–297.

Heiles, C. and Troland, T. H. (2005). Millennium Arecibo 21 Centimeter Absorption-Line Survey. IV. Statistics of Magnetic Field, Column Density, and Turbulence. *Astrophys. J.*, 624:773–793.

Holberg, J.B., Bruhweiler, F.C., and Dobie, M.A. Barstow P. D. (1999). Far-UV spectra of the white dwarf REJ1032+532. *Astrophys. J.*, 517:841–849.

Jenkins, E. B. (1987). Element abundances in the interstellar atomic material. In *ASSL Vol. 134: Interstellar Processes*, pages 533–559.

Jenkins, E. B. (2002). Thermal Pressures in Neutral Clouds inside the Local Bubble, as Determined from C I Fine-Structure Excitations. *Astrophys. J.*, 580:938–949.

Jenkins, E. B., Oegerle, W. R., Gry, C., Vallerga, J., Sembach, K. R., Shelton, R. L., Ferlet, R., Vidal-Madjar, A., York, D. G., Linsky, J. L., Roth, K. C., Dupree, A. K., and Edelstein, J. (2000). Ionization of the local interstellar medium as revealed by FUSE observations of N, O, and Ar toward white dwarf stars. *Astrophys. J. Let.*, 538:L81–L85.

Jones, A. P., Tielens, A. G. G. M., Hollenbach, D. J., and McKee, C. F. (1994). Grain destruction in shocks in the interstellar medium. *Astrophys. J.*, 433: 797–810.

Kimble, R. A., Davidsen, A. F., Long, K. S., and Feldman, P. D. (1993). Extreme ultraviolet observations of HZ 43 and the local H/He ratio with HUT. *Astrophys. J. Let.*, 408:L41–L44.

Kulkarni, S. R. and Heiles, C. (1988). *Neutral hydrogen and the diffuse interstellar medium*, pages 95–153. Galactic and Extragalactic Radio Astronomy.

Kuntz, K. D. and Snowden, S. L. (2000). Deconstructing the Spectrum of the Soft X-Ray Background. *Astrophys. J.*, 543:195–215.

Lallement, R. and Bertin, P. (1992). Northern-hemisphere observations of nearby interstellar gas - possible detection of the local cloud. *Astron. & Astrophys.*, 266:479–485.

Lallement, R. and Ferlet, R. (1997). Local interstellar cloud electron density from Mg and Na ionization: A comparison. *Astron. & Astrophys.*, 324:1105–1114.

Lallement, R., Vidal-Madjar, A., and Ferlet, R. (1986). Multi-component velocity structure of the local interstellar medium. *Astron. & Astrophys.*, 168: 225–236.

Landsman, W. B., Henry, R. C., Moos, H. W., and Linsky, J. L. (1984). Observations of interstellar hydrogen and deuterium toward Alpha Centauri A. *Astrophys. J.*, 285:801–807.

Lehner, N., Jenkins, E., Gry, C., Moos, H., Chayer, P., and Lacour, S. (2003). FUSE Survey of Local Interstellar Medium within 200 Parsecs. *Astrophys. J.*, 595:858–879.

Leroy, J. L. (1999). Interstellar dust and magnetic field at the boundaries of the Local Bubble. Analysis of polarimetric data in the light of HIPPARCOS parallaxes. *Astron. & Astrophys.*, 346:955–960.

Linsky, J. L. (2003). Atomic Deuterium/Hydrogen in the Galaxy. *Space Science Reviews*, 106:49–60.

Linsky, J. L. and Wood, B. E. (1996). Alpha Centauri line of sight: D/H ratio, physical properties of local interstellar gas. *Astrophys. J.*, 463:254–270.

Lodders, K. (2003). Solar System Abundances and Condensation Temperatures of the Elements. *Astrophys. J.*, 591:1220–1247.

Lucke, P. B. (1978). Distribution of color excesses and interstellar reddening material in the solar neighborhood. *Astron. & Astrophys.*, 64:367–377.

Möbius, E., Bzowski, M., Chalov, S., Fahr, H.-J., Gloeckler, G., Izmodenov, V., Kallenbach, R., Lallement, R., McMullin, D., Noda, H., Oka, M., Pauluhn, A., Raymond, J., Ruciński, D., Skoug, R., Terasawa, T., Thompson, W., Vallerga, J., von Steiger, R., and Witte, M. (2004). Synopsis of the interstellar He parameters from combined neutral gas, pickup ion and UV scattering. *Astron. & Astrophys.*, 426:897–907.

MacLow, M. and McCray, R. (1988). Superbubbles in disk galaxies. *Astrophys. J.*, 324:776–785.

McCammon, D., Almy, R., Apodaca, E., Bergmann Tiest, W., Cui, W., Deiker, S., Galeazzi, M., Juda, M., Lesser, A., Mihara, T., Morgenthaler, J. P., Sanders, W. T., Zhang, J., Figueroa-Feliciano, E., Kelley, R. L., Moseley, S. H., Mushotzky, R. F., Porter, F. S., Stahle, C. K., and Szymkowiak, A. E. (2002). High Spectral Resolution Observation of the Soft X-Ray Diffuse Background. *Astrophys. J.*, 576:188–203.

McClintock, W., Henry, R. C., Linsky, J. L., and Moos, H. W. (1978). Ultraviolet observations of cool stars. VII - Local interstellar H and D Lyman-alpha. *Astrophys. J.*, 225:465–481.

McRae Routly, P. and Spitzer, L., Jr. (1952). A Comparison of components in Interstellar Na and Ca. *Astrophys. J.*, 115:227.

Mihalas, D. and Binney, J. (1981). *Galactic Astronomy*. Freeman, San Francisco.

Morton, D. C. (1975). Interstellar absorption lines in the spectrum of ζ Ophiuchi. *Astrophys. J.*, 197:85–115.

Mueller, H. R., Frisch, P. C., Florinski, V., and Zank, G. P. (2005). Heliospheric Response to Different Possible Interstellar Environments. *submitted to Astrophys. J.*.

Munch, G. and Unsold, A. (1962). Interstellar gas near the Sun. *Astrophys. J.*, 135:711–715.

Oegerle, W. R., Jenkins, E. B., Shelton, R. L., Bowen, D. V., and Chayer, P. (2005). A Survey of O VI Absorption in the Local Interstellar Medium. *Astrophys. J.*, 622:377–389.

Pepino, R., Kharchenko, V., Dalgarno, A., and Lallement, R. (2004). Spectra of the X-Ray Emission Induced in the Interaction between the Solar Wind and the Heliospheric Gas. *Astrophys. J.*, 617:1347–1352.

Perryman, M. A. C. et al. (1997). HIPPARCOS Catalogue. *Astron. & Astrophys.*, 323:L49–L52.

Peterson, J. D. and Webber, W. R. (2002). Interstellar Absorption of the Galactic Polar Low-Frequency Radio Background Synchrotron Spectrum as an Indicator of Clumpiness in the Warm Ionized Medium. *Astrophys. J.*, 575: 217–224.

Pottasch, S. R. (1972). A model of the interstellar medium. Interpretation of the Na/Ca ratio. *Astron. & Astrophys.*, 20:245.

Ratkiewicz, R., Barnes, A., Molvik, G. A., Spreiter, J. R., Stahara, S. S., Vinokur, M., and Venkateswaran, S. (1998). Local interstellar magnetic field and exterior heliosphere. *Astron. & Astrophys.*, 335:363–369.

Redfield, S. and Linsky, J. L. (2000). Three-dimensional Structure of the Warm Local Interstellar Medium. *Astrophys. J.*, 534:825–837.

Redfield, S. and Linsky, J. L. (2001). Microstructure of the Local Interstellar Cloud and the Identification of the Hyades Cloud. *Astrophys. J.*, 551:413–428.

Redfield, S. and Linsky, J. L. (2002). Structure of the Local Interstellar Medium. I. High-Resolution Observations of Fe II, Mg II, and Ca II toward Stars within 100 Parsecs. *Astrophys. J. Supl.*, 139:439–465.

Redfield, S. and Linsky, J. L. (2004a). Structure of the Local Interstellar Medium. II. Observations of D I, C II, N I, O I, Al II, and Si II toward Stars within 100 Parsecs. *Astrophys. J.*, 602:776–802.

Redfield, S. and Linsky, J. L. (2004b). Structure of the Local Interstellar Medium. III. Temperature and Turbulence. *Astrophys. J.*, 613:1004–1022.

Reid, M. J., Readhead, A. C. S., Vermeulen, R. C., and Treuhaft, R. N. (1999). Proper Motion of Sagittarius A*. I. First VLBA Results. *Astrophys. J.*, 524: 816–823.

Reynolds, R. J. (2004). Warm ionized gas in the local interstellar medium. *Adv. Space Res.*, 34:27–34.

Reynolds, R. J., Tufte, S. L., Kung, D. T., McCullough, P. R., and Heiles, C. (1995). A Comparison of Diffuse Ionized and Neutral Hydrogen Away from the Galactic Plane: H alpha -emitting HI Clouds. *Astrophys. J.*, 448: 715–726.

Rogerson, J. B., York, D. G., Drake, J. F., Jenkins, E. B., Morton, D. C., and Spitzer, L. (1973). Results from the Copernicus Satellite. III. Ionization and Composition of the Intercloud Medium. *Astrophys. J. Let.*, 181:L110–L114.

Sanders, W. T., Edgar, R. J., Kraushaar, W. L., McCammon, D., and Morgenthaler, J. P. (2001). Spectra of the 1/4 keV X-Ray Diffuse Background from the Diffuse X-Ray Spectrometer Experiment. *Astrophys. J.*, 554:694–709.

Savage, B. D. (1995). Gaseous Galactic Corona. In *ASP Conf. Ser. 80: Physics of the Interstellar Medium and Intergalactic Medium*, pages 233–250.

Savage, B. D. and Sembach, K. R. (1996). Interstellar and Physical Conditions toward Distant High-Latitude Halo Stars. *Astrophys. J.*, 470:893.

Sfeir, D. M., Lallement, R., Crifo, F., and Welsh, B. Y. (1999). Mapping the contours of the Local Bubble: preliminary results. *Astron. & Astrophys.*, 346:785–797.

Shapiro, P. R. and Benjamin, R. A. (1991). "New results concerning the galactic fountain". *Pub. Astron. Soc. Pac.*, 103:923.

Slavin, J. D. (1989). Consequences of a conductive boundary on the local cloud. I - No dust. *Astrophys. J.*, 346:718–727.

Slavin, J. D. and Frisch, P. C. (2002). Ionization of Nearby Interstellar Gas. *Astrophys. J.*, 565:364–379.

Slavin, J. D., Jones, A. P., and Tielens, A. G. G. M. (2004). Shock Processing of Large Grains in the Interstellar Medium. *Astrophys. J.*, 614:796–806.

Slavin, J. D., Shull, J. M., and Begelman, M. C. (1993). Turbulent mixing layers in the interstellar medium of galaxies. *Astrophys. J.*, 407:83.

Snow, T. P. (2000). Composition of interstellar gas and dust. *J. Geophys. Res.*, 105:10239–10248.

Snow, T. P. and Meyers, K. A. (1979). Interstellar abundances in the zeta Ophiuchi clouds. *Astrophys. J.*, 229:545–552.

Sonett, C. P., Morfill, G. E., and Jokipii, J. R. (1987). Interstellar Shock Waves and 10/BE from Ice Cores. *Nature*, 330:458.

Spitzer, L. (1978). *Physical Processes in the Interstellar Medium*. John Wiley & Sons, Inc., Newrk.

Spitzer, L. J. (1954). Behavior of Matter in Space. *Astrophys. J.*, 120:1–17.

Stanimirovic, S. and Heiles, C. (2005). Thinnest cold HI clouds in the diffuse interstellar medium? *ArXiv Astrophysics e-prints*.

Tat, H. H. and Terzian, Y. (1999). Ionization of the Local Interstellar Medium. *Pub. Astron. Soc. Pac.*, 111:1258–1268.

Taylor, J. H. and Cordes, J. M. (1993). Pulsar distances and the galactic distribution of free electrons. *Astrophys. J.*, 411:674–684.

Tinbergen, J. (1982). Interstellar polarization in the immediate solar neighborhood. *Astron. & Astrophys.*, 105:53–64.

Vallerga, J. V., Vedder, P. W., Craig, N., and Welsh, B. Y. (1993). High-resolution Ca II observations of the local interstellar medium. *Astrophys. J.*, 411:729–749.

Vallerga, John (1996). Observations of the local interstellar medium with the EUVE. *Space Sci. Rev.*, 78:277–288.

Vandervoort, P. O. and Sather, E. A. (1993). On the Resonant Orbit of a Solar Companion Star in the Gravitational Field of the Galaxy. *Icarus*, 105:26–47.

Vergely, J.-L., Egret, D., Freire Ferrero, R., Valette, B., and Koeppen, J. (1997). Extinction in Solar Neighborhood from HIPPARCOS Data. In *ESA SP-402: Hipparcos - Venice '97*, volume 402, pages 603–606.

Vergely, J.-L., Freire Ferrero, R., Siebert, A., and Valette, B. (2001). NaI, HI 3D density distribution in solar neighborhood. *Astron. & Astrophys.*, 366:1016–1034.

Vidal-Madjar, A. and Ferlet, R. (2002). Hydrogen Column Density Evaluations toward Capella: Consequences on the Interstellar Deuterium Abundance. *Astrophys. J. Let.*, 571:L169–L172.

Wargelin, B. J., Markevitch, M., Juda, M., Kharchenko, V., Edgar, R., and Dalgarno, A. (2004). Chandra Observations of the "Dark" Moon and Geocoronal Solar Wind Charge Transfer. *Astrophys. J.*, 607:596–610.

Warwick, R. S., Barber, C. R., Hodgkin, S. T., and Pye, J. P. (1993). EUV source population and the Local Bubble. *Mon. Not. Roy. Astron. Soc.*, 262:289–300.

Weaver, H. (1979). Large supernova remnants as common features of the disk. In *IAU Symp. 84: Large-Scale Characteristics of the Galaxy*, volume 84, pages 295–298.

Welty, D. E. and Hobbs, L. M. (2001). A high-resolution survey of interstellar K I absorption. *Astrophys. J. Supl.*, 133:345–393.

Welty, D. E., Hobbs, L. M., and Kulkarni, V. P. (1994). A high-resolution survey of interstellar Na I D1 lines. *Astrophys. J.*, 436:152–175.

Welty, D. E., Hobbs, L. M., Lauroesch, J. T., Morton, D. C., Spitzer, L., and York, D. G. (1999). Diffuse Interstellar Clouds toward 23 Orionis. *Astrophys. J. Supl.*, 124:465–501.

Welty, D. E., Jenkins, E. B., Raymond, J. C., Mallouris, C., and York, D. G. (2002). Intermediate- and High-Velocity Ionized Gas toward ζ Orionis. *Astrophys. J.*, 579:304–326.

Welty, D. E., Morton, D. C., and Hobbs, L. M. (1996). A high-resolution survey of interstellar Ca II absorption. *Astrophys. J. Supl.*, 106:533–562.

Witte, M. (2004). Kinetic parameters of interstellar neutral He. Review of results obtained during one solar cycle with the Ulysses/GAS-instrument. *Astron. & Astrophys.*, 426:835–844.

Wolff, B., Koester, D., and Lallement, R. (1999). Evidence for an ionization gradient in the local interstellar medium. *Astron. & Astrophys.*, 346: 969–978.

Wood, B. E., Ambruster, C. W., Brown, A., and Linsky, J. L. (2000a). Mg II and Ly-alpha lines of nearby K dwarfs. *Astrophys. J.*, 542:411–420,.

Wood, B. E., Linsky, J. L., Hébrard, G., Vidal-Madjar, A., Lemoine, M., Moos, H. W., Sembach, K. R., and Jenkins, E. B. (2002). Deuterium Abundance toward WD 1634-573: Results from FUSE. *Astrophys. J. Supl.*, pages 91–102.

Wood, B. E., Linsky, J. L., and Zank, G. P. (2000b). Heliospheric, astrospheric, and interstellar Ly-alpha; absorption toward 35 Oph. *Astrophys. J.*, 537:304–311.

Wood, B. E., Redfield, S., Linsky, J. L., Mueller, H., and Zank, G. P. (2005). Stellar Lyman-alpha Emission Lines in the Hubble Space Telescope Archive: Intrinsic Line Fluxes and Absorption from the Heliosphere and Astrospheres. *ArXiv Astrophysics e-prints*.

York, D. G. (1976). A UV picture of the gas in the interstellar medium. *Memorie della Societa Astronomica Italiana*, 47:493–551.

York, D. G. (1983). *Lambda Sco. Astrophys. J.*, 264:172–195.

York, D. G. and Kinahan, B. F. (1979). Alpha Virginis. *Astrophys. J.*, 228: 127–146.

Zank, G. P. and Frisch, P. C. (1999). Consequences of a Change in the Galactic Environment of the Sun. *Astrophys. J.*, 518:965–973.

Chapter 7

VARIATIONS OF THE INTERSTELLAR DUST DISTRIBUTION IN THE HELIOSPHERE

Heliospheric Dust Environment as a Function of the Surrounding Cloud

Markus Landgraf
ESA/ESOC, Robert-Bosch-Str. 5
64293 Darmstadt, Germany
Markus.Landgraf@esa.int

Abstract Cosmic dust is a significant ingredient in the interstellar environment of the Sun. Information about this solid phase of matter in interstellar space comes from the measurement of the extinction, and polarization of starlight, from emission spectra in the infrared, from the analysis of the isotopic composition of presolar grains in primitive meteorites, and from in situ measurements taken by dust detectors on board interplanetary spacecraft. During the more than 10 years of in situ data coverage, our solar system has traveled less than 0.3×10^{-4} pc, only a tiny fraction of the local interstellar cloud that surrounds the Sun. Therefore we do not expect any change in the flux of interstellar dust through the solar system. And, indeed, the observed variation of the local interstellar dust flux can fully be accounted for by the modulation of the interstellar dust stream and by its interaction with the heliospheric magnetic field. Modelling the current interaction between interstellar dust and the heliosphere provides an excellent laboratory for investigating the consequences of an encounter of the Sun with a much denser interstellar cloud that also should contain a higher concentration of dust grains. If the heliosphere is sufficiently compressed such that the planets are exposed to interstellar material, the small interstellar dust grains, which are currently excluded from the heliosphere, would create an increased amount of secondary dust particles when they collide with asteroids and comets. This would lead to a more dusty interplanetary environment with higher accretion rates of dust onto planetary atmospheres.

Keywords: Interplanetary dust, interstellar dust, heliosphere, Kuiper Belt

P. C. Frisch (ed.), Solar Journey: The Significance of our Galactic Environment for the Heliosphere and Earth, 195–208.

Introduction

The discovery that cosmic dust is a major component of the galactic environment was made as early as the 1920s, when the reddening of starlight was attributed to galactic interstellar dust (Schalén, 1929, Trümpler, 1930, Öhman, 1930). It was argued that the small interstellar grains scatter predominantly blue light, thereby removing it from the line of sight between the observer and the star. The red portion of the light is less affected, and consequently the stars appear reddened when a dust cloud is in front of them. The first model of the composition of interstellar grains was created in the 1940s (Oort and Van de Hulst, 1946, Van de Hulst, 1949). It described an icy dust population that consisted of hydrogenated compounds of the most abundant heavy elements (heavier than hydrogen and helium) O, C, and N. Mainly H_2O, CH_4, and NH_3 ice were believed to condense in interstellar clouds and to form conglomerates with typical sizes of about $0.3\ \mu\text{m}$. In order to explain the spectral properties of the reddening, a model grain distribution was introduced. It was found to explain the observed extinction at that time (Struve, 1937, Henyey and Greenstein, 1941, Stebbins and Whitford, 1943), and could be interpreted as an equilibrium state between growth and collisional fragmentation of grains in the interstellar cloud.

Information about the dust temperature in interstellar clouds can be obtained by observing the emitted radiation. Just like interplanetary dust grains, most energy emitted by interstellar dust is in the infrared. The detection of interstellar dust closer to our Sun became possible after infrared astronomy was used more frequently to analyze interstellar phenomena. The spectroscopic analysis of the infrared light also allowed the identification of molecular bands that provided information on the chemical composition of interstellar dust. Prominent silicate (Si–O) features near 10 and $20\,\mu\text{m}$, and the polycyclic aromatic hydrocarbon (PAH) features at 3.3, 6.2, 7.7, and $11.3\ \mu\text{m}$ were detected. It was the European Infrared Space Observatory (ISO) that provided, for the first time, high resolution spectra in the wavelength range from 2 to $200\ \mu\text{m}$. With the ISO data, it became possible to obtain an inventory of molecular ice features of H_2O, CO, CO_2, and CH_3OH in molecular clouds. Already early infrared observations showed, however, that interstellar absorption spectra of lines of sight through the *diffuse* medium lacked the O–H stretching feature of H_2O ice at $3.08\,\mu\text{m}$ that was predicted by the icy dust model (Danielson et al., 1965, Knacke et al., 1969). Consequently, new models of the interstellar dust composition called for refractory grains, mainly consisting of graphite (Hoyle and Wickramasinghe, 1962), SiO_2 (Kamijo, 1963), and SiC (Friedemann, 1969). Modern models that also take into account the cosmic abundance of chemical elements, as well as polarization measurements, suggest grains of various sizes and compositions: small graphite grains, silicate grains, and composite grains containing carbon (amorphous, hydrogenated, or graphitic), silicates, and oxides (Mathis, 1996). Also grains consisting of a silicate core with an organic refractory mantle, as well as nanometer-sized grains, or large molecules, of polycyclic aromatic hydrocarbons are discussed (Li and Greenberg, 1997).

What is the origin of interstellar grains? It is known that evolved stars continuously lose mass. About 90% of the stellar mass loss is provided by cool high-luminosity stars, in particular by asymptotic giant branch (AGB) and post-AGB stars. As the ejected material cools in the expanding stellar wind, solid particles condense out of the gas phase (for reviews see Whittet, 1989, Whittet et al., 1992, and Sedlmayr, 1994). This so-called "stardust" provides the seeds for grains that grow in cool interstellar clouds by accretion of atoms and molecules and by agglomeration (Ossenkopf and Henning, 1994). The laboratory analysis of the isotopic composition of primitive meteorites shows that in addition to stardust from evolved AGB stars there is a significant amount of dust generated in the envelopes of super nova explosions (Hoppe, 2004). From observations of the interstellar medium, it is known that dust is not only created there, but also destroyed. Shocks in the medium are mainly caused by supernova explosions and supersonic stellar winds. In these shocks dust grains are ground down by the plasma and mutual collisions (Jones et al., 1994). Additionally, dust grains lose their volatile constituents when they are exposed to the interstellar UV radiation (Greenberg et al., 1995). Ultimately, an interstellar dust grain can be incorporated and destroyed in a newly forming star, or it might become part of a planetary system. This way cosmic dust is repeatedly recycled through the galactic evolution process (Dorschner and Henning, 1995).

Early attempts to detect interstellar dust in the solar system with in-situ instruments on board the Pioneer 8 and 9 spacecraft failed or were inconclusive (McDonnell and Berg, 1975). The non-detection of interstellar dust by Pioneer 8 and 9 was interpreted to be caused by the electrostatic charge that grains acquire in the heliospheric environment. This charge causes the grains to interact with the solar wind magnetic field that diverts them out of the heliosphere. A more detailed analysis of the proposed diversion effect (Gustafson and Misconi, 1979, Morfill and Grün, 1979) showed that the solar wind magnetic field is not only capable of diverting small interstellar grains, but also of concentrating them towards the plane of the solar magnetic equator, depending on the phase of the solar cycle. More than a decade after these discussions, interstellar dust grains were unambiguously detected in the solar system with in-situ instruments (Grün et al., 1993). Once it became evident that galactic interstellar dust is accessible to in-situ detection and even sample return to Earth, NASA selected the Stardust mission to analyze and return to Earth samples of interstellar dust at asteroid belt distances (Brownlee et al., 1997). While the primary mission of Stardust is the encounter with comet Wild-2 in January 2004, its secondary goal is to collect and detect the interstellar grains that have been discovered by Ulysses. With its time-of-flight mass spectrometer Stardust is capable of analyzing the chemical composition of local interstellar dust. Stardust returns to Earth in January 2006, when it sends a capsule into the Earth's atmosphere that contains the collected dust samples.

7.1 The Contemporary Interstellar Dust Environment of the Heliosphere

Our current understanding of the contemporary interstellar environment of the heliosphere comes mainly from the in-situ data collected with Ulysses. The Ulysses spacecraft circles the Sun on an elliptical orbit that is inclined by 79° with respect to the ecliptic plane. Its closest point to the Sun is at 1.3 AU, and the furthest point from the Sun is at 5.4 AU. Ulysses carries an impact ionization dust detector (Grün et al., 1992b, Grün et al., 1992a), which measures the plasma cloud released on impact of cosmic dust grains onto its sensitive target. The plasma cloud is separated by an electrostatic field, and the charge signals are measured at the electrodes. A detection is confirmed by multiple coincidence of three independent charge signals. Masses and impact speeds of the impactors are determined from the measured amplitudes and rise times of the impact charge signals (Grün et al., 1995). In addition to the mass and speed of the impacting grains, the coarse impact direction can be determined from the time of the impact, since the dust detector has a limited circular field of view of 140° (full angle) that scans directions perpendicular to the spacecraft's spin axis as the spacecraft rotates.

The discovery of interstellar dust grains in the Ulysses data was achieved by observing the dust impact direction, impact speed, and the dependence of the dust impact rate as a function of the ecliptic latitude of the spacecraft. It was found that interstellar dust arrives from the opposite direction than does dust from interplanetary sources like short period comets or asteroids. In addition the impact speeds of dust grains from the retrograde direction have been above the local solar system escape speed, even when radiation pressure effects are neglected. The observation that the impact rate of small dust grains remained in the same order of magnitude after Ulysses left the ecliptic plane (Baguhl et al., 1996), where most interplanetary sources are concentrated, shows that the impacts detected by Ulysses are in fact dominated by interstellar grains (Krüger et al., 1999).

Since Ulysses left the ecliptic plane in February 1992, it monitors the stream of interstellar dust grains through the solar system. It was found that the solar wind filtration causes a deficiency of detected interstellar grains with masses below 10^{-17} kg (Grün et al., 1994). This is illustrated by Figure 7.1, which shows how the heliospheric magnetic field creates a cavity void of small interstellar grains around the Sun. An additional filtration by solar radiation

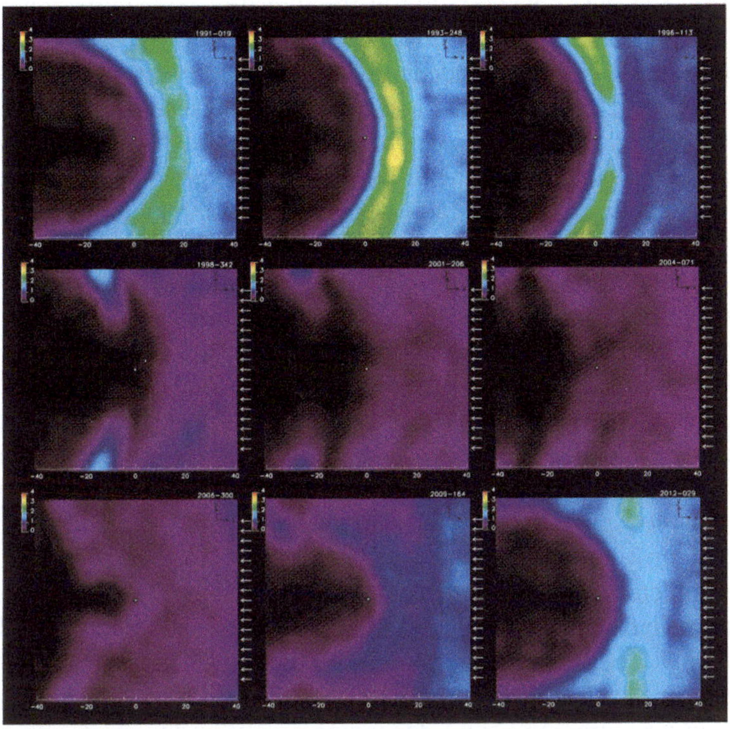

Figure 7.1. Time-lapse sequence (left to right, top to bottom) of the distribution of small ($m < 10^{-17}$ kg) interstellar dust grains in the heliosphere over one solar cycle (Landgraf, 2000). The distributions are shown in a plane that includes the Sun at its center (shown by the white dot) and the initial velocity vector of the dust grains, which is parallel to the x-axis (scale is shown in AU). The colour scales on the left of each panel indicate the spatial concentration relative to the value in the LIC.

pressure, which deflects grains with masses of 10^{-16} kg, was found to be effective at solar distances below 4 AU (Landgraf et al., 1999a). In mid-1996 a decrease of the interstellar dust flux density was observed, from initial values of $1.5 \times 10^{-4}\,\mathrm{m}^{-2}\,\mathrm{s}^{-1}$ to $0.5 \times 10^{-4}\,\mathrm{m}^{-2}\,\mathrm{s}^{-1}$ (Landgraf et al., 1999b). In early 2000 the flux returned to values around $1 \times 10^{-4}\,\mathrm{m}^{-2}\,\mathrm{s}^{-1}$, which it maintained up to end of November 2002 (Landgraf et al., 2003).

What is the cause of the temporal variation of the interstellar dust flux as seen in Figure 7.3? In general there are three possible explanations: (a) a spatial variation of the dust concentration in the heliosphere that translates into temporal variation with spacecraft motion; (b) a variation in the efficiency of the filtration mechanisms at the heliopause and within the heliosphere; and (c) a variation of the dust concentration in the interstellar medium. The spatial variation of the interstellar dust distribution in the heliosphere, as indicated by possibility (a) and the dependence of the measurement on the spacecraft location, must be considered when simulating the measurements (Landgraf, 2000). If the flux variation was solely due to the spatial distribution, however, the flux value would be periodic with Ulysses' orbital period of 6.2 years. This is not the case, as the flux measured in early 1993 was three times greater than in mid-1999.

If an acceptable fit is achieved under the assumption of possibility (b), an efficiency change of the filtration mechanisms, we can conclude that the dust phase of the LIC is homogeneous on scales of a few 10 AU, as the Sun traveled 50 AU relative to the LIC in the period from end-1992 to end-2002. The filtration in the heliopause region will not have a strong effect on the detected dust grains because the dust grains that can be detected by the Ulysses detector are typically larger than 0.2 μm in diameter, and thus their charge to mass ratio is too small for the heliopause to have a significant effect (Linde and Gombosi, 2000).

In order to determine the influence of the varying solar wind magnetic field on the interstellar dust flux, a model of the heliospheric interaction of interstellar grains was constructed (Landgraf, 2000, Landgraf et al., 2003). In this model the grains are assumed to be spherical, with their charge to mass ratio and radiation pressure coefficient depending solely on the grain radius. Figure 7.2 shows the predicted flux at the location and in the frame of reference of the Ulysses spacecraft, normalized to the initial flux, for various grain sizes.

As expected, the simulation shows that the filtration is most efficient for the smallest grains with radii of 0.1 μm. The spatial distribution of these grains is dominated by the interaction of these grains with the solar wind magnetic field resulting from their high charge-to-mass ratio of $q/m = 5.31$ C kg^{-1}. The Lorentz-force created by the solar wind magnetic field sweeping past the grains both repels and focuses the grains, depending on the configuration of the field and thus on the phase of the solar cycle (Grün et al., 1994, Gustafson and Lederer, 1996, Grogan et al., 1996). At the end of the focusing solar cycle, the grains are concentrated in a sheet about the plane of the solar equator. The concentration of small grains is thus expected to be small at high heliographic latitudes not too close to the Sun. Because the solar wind magnetic field moves radially outward with the solar wind, the Lorentz-force on the grains is not perfectly perpendicular to the solar equator, but also contains a small outward directed component. This force decelerates small grains and causes a region of enhanced density upstream of the Sun, which is visible as the arc-shaped feature in Figure 7.1. During the solar minimum phase, when the average interplanetary magnetic field has a strongly diverting effect on small dust grains, the region of high grain density is diverted before it reaches the solar vicinity.

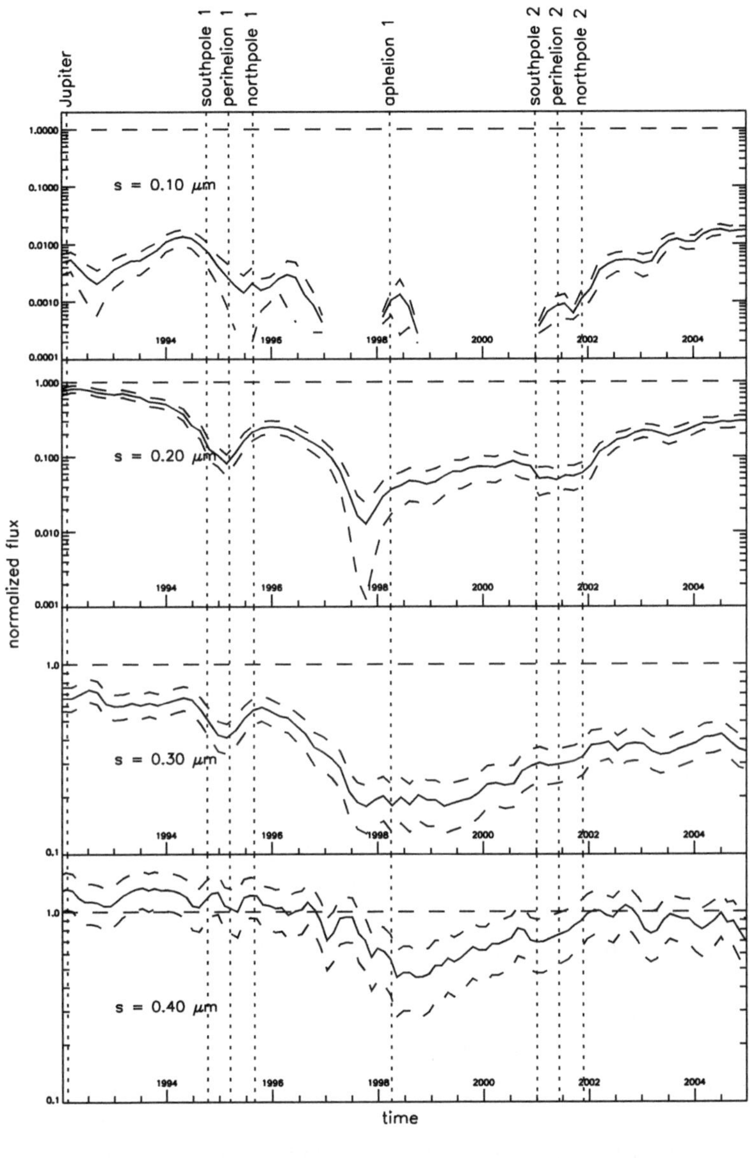

Figure 7.2. Predicted temporal variation of the flux (solid lines) of interstellar dust arriving at the Ulysses spacecraft for four different grain radii. The dashed lines around the flux prediction curves indicate the 1σ uncertainty that results from the small-number-statistics of the Monte-Carlo simulation. The vertical dotted lines mark the milestones along Ulysses' orbit. The flux is normalized to its value at large distances from the Sun.

According to the Monte-Carlo simulation, the flux of grains with radii of $0.1\ \mu\text{m}$ never reached more than 2% of its value at infinity between 1992 and 2005. In the second half of the last solar cycle, between 1997 and 2001, these small grains were even completely removed from the part of the solar system inside 5 AU. Only around aphelion was Ulysses in the right location to detect them. The prediction also shows that the flux of small interstellar grains at the location of Ulysses should be reduced when the spacecraft was over the solar poles, as expected. Already the next bigger grains with a radius of $0.2\ \mu\text{m}$ are much less affected by the solar wind magnetic field. Their flux is close to its value in the interstellar medium at the beginning of the last solar cycle. The strong decrease in flux by one order of magnitude around Ulysses' perihelion is caused by the strong effect of radiation pressure on grains of this size. The simulation predicts the maximum suppression of the flux at Ulysses by two orders of magnitude just before the first aphelion in 1998. Grains with radii of 0.3 and $0.4\ \mu\text{m}$ follow the same trend as the $0.2\ \mu\text{m}$ grains, but exhibit much less variation due to their smaller charge to mass ratios.

Comparisons between the interstellar dust flux variation, as measured at Ulysses, and the Monte-Carlo simulation results allows us to determine which combination of charge-to-mass ratio and radiation pressure coefficient must be dominant in the interstellar dust stream. This combination then gives us information on the typical grain sizes. Figure 7.3 shows the best fit of the simulated flux profiles to the Ulysses data.

The fit results in a dominant contribution from $0.3\ \mu\text{m}$ grains with some minor contribution from $0.4\ \mu\text{m}$ grains. The contribution from 0.1 and $0.2\ \mu\text{m}$ grains is not needed to achieve the fit. The fitted flux profile successfully represents the qualitative features of the measurement: The almost constant flux between 1992 and 1996, the decrease in 1997, and the rebound until 2002. Also, the relatively low value of the normalized χ^2 of 0.1 (or not normalized $\chi_8^2 = 0.9$ for 8 degrees of freedom) shows the good agreement of prediction and measurement. In the period between 1998 and 2000, however, the measured values are systematically higher, indicating an earlier rebound of the dust flux than predicted.

This result, that it is possible to fit the observed variation of the interstellar dust flux in the heliosphere assuming a constant dust concentration in the LIC, lets us conclude that the dust phase of the LIC is homogeneously distributed over length scales of 50 AU, the distance inside the LIC traveled by the Sun between end-1992 and end-2002, which is the period when the data was acquired. From this observation it can be predicted that the flux of interstellar dust onto the Ulysses detector levels off at a constant value just above $1 \times 10^{-4}\ \text{m}^{-2}\,\text{s}^{-1}$ until the end of 2004.

While the Ulysses observations have shown that in the solar vicinity the LIC is homogeneous on scales much smaller than the extent of the cloud itself,

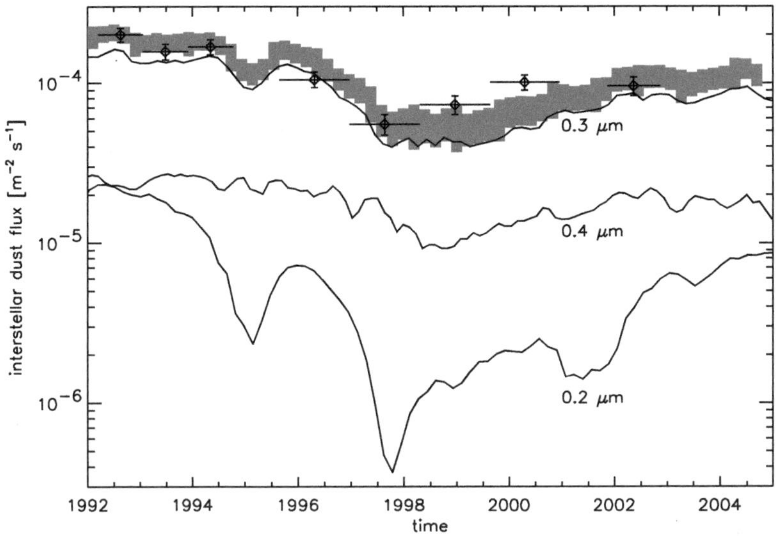

Figure 7.3. Fit of simulated to measured flux. The fit parameters are the relative contribution of grains of sizes between 0.1 and 0.4 μm (The 0.1 μm curve is not shown, because it did not contribute to the fit). The flux measurements by Ulysses are shown as in Figure 7.3. The solid lines show the flux profiles of the simulated grains of various sizes, scaled with their best-fit relative contributions. The shaded region indicates the best-fit total predicted flux, with its vertical extent giving the 1σ uncertainty.

recent results (Price et al., 2001) from UV absorption measurements along lines of sight toward nearby stars indicate that local inhomogeneities on scales of 10 AU exist. Now the question is: how will the interstellar dust population in the heliosphere change if the Sun encounters a different interstellar environment?

7.2 Consequences of a Changing Interstellar Environment

In the past few million years the Sun has traveled through a region of space with very low interstellar gas and dust densities. This region, commonly referred to as the Local Bubble, is located between two spiral arms of the Milky Way galaxy. Our galactic environment changes with time since the Sun moves through space with a velocity of $\approx$16.5 pc every one million years. What will the response of the heliosphere be if the Sun traverses a region 10^4 times more dense than our current environment? The modelling of the hydrodynamic interaction of the solar wind plasma with a much more dense environment (Yeghikyan and Fahr, 2004) shows that the heliosphere can be compressed to an upwind radius of less than 1 AU, which would expose the Earth directly to interstellar material.

Another, probably equally significant, effect will be that the small-grain component of the interstellar dust population is no longer deflected, as the solar wind magnetic field is only effective in the domain of the wind. The depletion of interstellar dust grains smaller than $0.2\ \mu\mathrm{m}$ observed by Ulysses today (see above) keeps the interstellar dust flux in the order of hundreds of impacts per square meter each year. Since the grain size distribution rises steeply as grain size decreases, a compression of the heliosphere will expose the affected regions, including objects in these regions, to interstellar dust fluxes orders of magnitude higher than today. The increase of the interstellar dust mass accreted will not be significant, as the total mass of interstellar dust is dominated by the larger grains (Landgraf et al., 2000), the dynamics of which are not influenced by the presence or non-presence of the heliosphere. There will be however a strong effect on the secondary dust production rate of comets, asteroids, and Kuiper Belt objects, as this parameter is proportional to the cross section flux density that is dominated by the smaller grains.

The Kuiper belt dust production is an especially important factor in the dust equilibrium of the solar system. It is known that about half of the dust produced by the Kuiper Belt is released by interstellar dust impacts. From measurements by the Pioneer 10 meteoroid experiment (Humes, 1980, Landgraf et al., 2002), we know that the Kuiper belt must produce 50 tons per second in order to maintain the equilibrium dust concentration beyond Saturn's orbit. It is difficult to estimate how much this production rate would increase if the inner, dense part of the Kuiper belt were to be exposed to a significantly increased flux of small interstellar grains. The surface properties of Kuiper belt objects are not well known, and thus we cannot know the smallest grain size that will effectively eject dust for an impact velocity of $20\ \mathrm{km\ s^{-1}}$.

In general it can be stated that the compression of the heliosphere will lead to a changed dust environment in interplanetary space. More small interstellar dust grains will be present, and probably also more interplanetary dust. This will change the accretion rates onto the planets. The Earth today accretes about $40,000$ tons of interplanetary dust each year (Love and Brownlee, 1993), which is significant for the chemistry of the high atmosphere and the delivery of exotic elements to the Earth's surface. With an increased dust concentration in interplanetary space due to a compressed heliosphere, the effect on the Earth can only be bigger.

References

Baguhl, M., Grün, E., and Landgraf, M. (1996). In situ measurements of interstellar dust with the Ulysses and Galileo spaceprobes. In von Steiger, R., Lallement, R., and Lee, M. A., editors, *The Heliosphere in the Local*

Interstellar Medium, volume 78 of *Space Science Reviews*, pages 165–172. Kluwer Academic Publishers.

Brownlee, D. E., Tsou, P., Burnett, D., Clark, B., Hanner, M. S., Horz, F., Kissel, J., McDonnell, J. A. M., Newburn, R. L., Sandford, S., Sekanina, Z., Tuzzolino, A. J., and Zolensky, M. (1997). The Stardust mission: Returning comet samples to Earth. *Meteoritics & Planetary Science*, 32:A22.

Danielson, R. E., Woolf, N. J., and Gaustad, J. E. (1965). A search for interstellar ice absorption in the infrared spectrum of μ Cephei. *Astrophysical Journal*, 141:116–125.

Dorschner, J. and Henning, T. (1995). Dust metamorphis in the galaxy. *The Astron. Astrophs. Rev.*, 6:271–333.

Friedemann, C. (1969). Evolution of silicon carbide particles in the atmospheres of carbon stars. *Physica*, 41:139–143.

Greenberg, J. M., Li, A., Mendoza-Gomez, C. X., Schutte, W. A., Gerakines, P. A., and Groot, M. De (1995). Approaching the interstellar grain organic refractory component. *Astrophysical Journal Letters*, 455:L177.

Grogan, K., Dermott, S. F., and Gustafson, B.Å. S. (1996). An estimation of the interstellar contribution to the zodiacal thermal emission. *Astrophysical Journal*, 472:812–817.

Grün, E., Baguhl, M., Fechtig, H., Kissel, J., Linkert, D., Linkert, G., and Riemann, R. (1995). Reduction of Galileo and Ulysses dust data. *Planetary and Space Science*, 43:941–951.

Grün, E., Fechtig, H., Giese, R. H., Kissel, J., Linkert, D., Maas, D., McDonnell, J. A. M., and Morfill, G. E. (1992a). The Ulysses dust experiment. *Astrophysical Supplement Series*, 92:411–423.

Grün, E., Fechtig, H., Hanner, M. S., Kissel, J., Lindblad, B.-A., Linkert, D., Maas, D., Morfill, G. E., and Zook, H. A. (1992b). The Galileo dust detector. *Space Science Reviews*, 60:317–340.

Grün, E., Gustafson, B. Å. S., Mann, I., Baguhl, M., Morfill, G. E., Staubach, P., Taylor, A., and Zook, H. A. (1994). Interstellar dust in the heliosphere. *Astronomy and Astrophysics*, 286:915–924.

Grün, E., Zook, H. A., Baguhl, M., Balogh, A., Bame, S. J., Fechtig, H., Forsyth, R., Hanner, M. S., Horányi, M., Kissel, J., Lindblad, B.-A., Linkert, D., Linkert, G., Mann, I., McDonnell, J. A. M., Morfill, G. E., Phillips, J. L., Polanskey, C., Schwehm, G., Siddique, N., Staubach, P., Svestka, J., and Taylor, A. (1993). Discovery of jovian dust streams and interstellar grains by the Ulysses spacecraft. *Nature*, 362:428–430.

Gustafson, B. Å. S. and Lederer, S. M. (1996). Interstellar grain flow through the solar wind cavity around 1992. In Gustafson, B.Å. S. and Hanner, M. S., editors, *Physics, Chemistry and Dynamics of Interplanetary Dust*, volume 104 of *Astronomical Society of the Pacific Conference Series*, pages 35–39.

Gustafson, B.Å. S. and Misconi, N. Y. (1979). Streaming of interstellar grains in the solar system. *Nature*, 282:276–278.

Henyey, L. G. and Greenstein, J. L. (1941). Diffuse radiation in the galaxy. *Astrophysical Journal*, 93:70–83.

Hoppe, P. (2004). Stardust in meteorites. In Witt, A. N., Clayton, G. C., and T. Draine, B., editors, *Astrophysics of Dust*, volume 309 of *Astronomical Society of the Pacific Conference Series*, pages 265–283.

Hoyle, F. and Wickramasinghe, N. C. (1962). On graphite particles as interstellar grains. *Monthly Notices of the Royal Astronomical Society*, 124:417.

Humes, D. H. (1980). Results of Pioneer 10 and 11 meteoroid experiments: Interplanetary and near-Saturn. *Journal of Geophysical Research*, 85(A/II): 5841–5852.

Jones, A. P., Tielens, A. G. G. M., Hollenbach, D. J., and McKee, C. F. (1994). Grain destruction in shocks in the interstellar medium. *Astrophysical Journal*, 433:797–810.

Kamijo, F. (1963). A theoretical study of the long period variable stars III. Formation of solid or liquid particles in the circumstellar envelope. *Publ. Astron. Soc. Japan*, 15:440–448.

Knacke, R. F., Cudaback, D. D., and Gaustad, J. E. (1969). Infrared spectra of highly reddened stars: A search for interstellar ice grains. *Astrophysical Journal*, 158:151.

Krüger, H., Grün, E., Landgraf, M., Baguhl, M., Dermott, S., Fechtig, H., Gustafson, B.Å. S., Hamilton, D. P., Hanner, M. S., Horányi, M., Kissel, J., Lindblad, B. A., Linkert, D., Linkert, G., Mann, I., McDonnell, J. A. M., Morfill, G. E., Polanskey, C., Schwehm, G., Srama, R., and Zook, H. A. (1999). Three years of Ulysses dust data: 1993 to 1995. *Planetary and Space Science*, 47:363–383.

Landgraf, M. (2000). Modeling the motion and distribution of interstellar dust inside the heliosphere. *Journal of Geophysical Research*, 105(A5):10303–10316.

Landgraf, M., Augustsson, K., Grün, E., and Gustafson, B. Å. S. (1999a). Deflection of the local interstellar dust flow by solar radiation pressure. *Science*, 286:2319–2322.

Landgraf, M., Baggaley, W. J., Grün, E., Krüger, H., and Linkert, G. (2000). Aspects of the mass distribution of interstellar dust grains in the solar system from in-situ measurements. *Journal of Geophysical Research*, 105(A5): 10343–10352.

Landgraf, M., Krüger, H., Altobelli, N., and E. Grün, E. (2003). Penetration of the heliosphere by the interstellar dust stream during solar maximum. *Journal of Geophysical Research (Space Physics)*, 108(A10):5–1.

Landgraf, M., Liou, J.-C., Zook, H. A., and Grün, E. (2002). Origins of solar system dust beyond Jupiter. *The Astronomical Journal*, 123(5):2857–2862.

Landgraf, M., Müller, M., and Grün, E. (1999b). Prediction of the in-situ dust measurements of the Stardust mission to comet 81p/wild 2. *Planetary and Space Science*, 47:1029–1050.

Li, A. and Greenberg, J. M. (1997). A unified model of interstellar dust. *Astronomy and Astrophysics*, 323:566–584.

Linde, T. J. and Gombosi, T. I. (2000). Interstellar dust filtration at the heliospheric interfaces. *Journal of Geophysical Research*, 105(A5):10411–10418.

Love, S. G. and Brownlee, D. E. (1993). A direct measurements of the terrestrial mass accretion rate of cosmic dust. *Science*, 262:550–553.

Mathis, J.S. (1996). Dust models with tight abundance constraints. *Astrophysical Journal*, 472:643.

McDonnell, J. A. M. and Berg, O. E. (1975). Bounds for the interstellar to solar system microparticle flux ratio over the mass range 10^{-11}–10^{-13} g. In *Space Research XV*, pages 555–563. Akademie-Verlag, Berlin.

Morfill, G. E. and Grün, E. (1979). The motion of charged dust particles in interplanetary space – II. interstellar grains. *Planetary and Space Science*, 27:1283–1292.

Öhman, Y. (1930). Spectrophotometric studies of B, A and F-type stars. *Medd. Uppsala*, 48.

Oort, J. H. and Van de Hulst, H. C. (1946). Gas and smoke in interstellar space. *Bull. Astron. Inst. Netherlands*, 10:187.

Ossenkopf, V. and Henning, T. (1994). Dust opacities for protostellar cores. *Astronomy and Astrophysics*, 291:943–959.

Price, R. J., Crawford, I. A., Barlow, M. J., and Howarth, I. D. (2001). An ultra-high-resolution study of the interstellar medium towards Orion. *Monthly Notices of the Royal Astronomical Society*, 328:555–582.

Schalén, C. (1929). Untersuchungen über Dunkelnebel. *Medd. Astron. Obs.*, 58.

Sedlmayr, E. (1994). From molecules to grains. In Jorgensen, U. G., editor, *Molecules in the Stellar Environment VIII*, number 146 in Proceedings of the IAU Colloquium, page 163. Springer, Berlin, Heidelberg, New York.

Stebbins, J. and Whitford, A. E. (1943). Six-color photometry of stars I. The law of space reddening from the colors of O and B stars. *Astrophysical Journal*, 98:20–32.

Struve, O. (1937). On the interpretation of the surface brightness of diffuse galactic nebulae. *Astrophysical Journal*, 85:194–212.

Trümpler, R. (1930). Preliminary results on the distances, dimensions, and space distribution of open star clusters. *Lick Obs. Bull*, 420.

Van de Hulst, H. C. (1949). The solid particles in interstellar space. *Rech. Astron. Obs. Utrecht*, 11(2).

Whittet, D. C. B. (1989). The composition of dust in stellar ejects. In *Interstellar Dust*, number 135 in Proceedings of the IAU Colloquium, page 455.

Whittet, D. C. B., Martin, P. G., Hough, J. H., Rouse, M. F., Bailey, J. A., and Axon, D. J. (1992). Systematic variations in the wavelength dependence of interstellar linear polarization. *Astrphysical Journal*, 386:562–577.

Yeghikyan, A. and Fahr, H. (2004). Effects induced by the passage of the Sun through dense molecular clouds - I. flow outside of the compressed heliosphere. *Astronomy & Astrophysics*, 415:763–770.

Chapter 8

EFFECTS IN THE INNER HELIOSPHERE CAUSED BY CHANGING CONDITIONS IN THE GALACTIC ENVIRONMENT

Eberhard Möbius[1], Maciek Bzowski[2], Hans-Reinhard Müller[3], Peter Wurz[4]

1 Dept. of Physics and Space Science Center, University of New Hampshire, Durham, NH 03824, U.S.A.

2 Space Research Centre, Warsaw, Poland

3 Dept. of Physics and Astronomy, Dartmouth College, Hanover, NH, U.S.A.

4 Physikalisches Institut, Universität Bern, Bern, Switzerland

Abstract Interstellar neutral gas flows through the inner heliosphere because of the Sun's relative motion with respect to the surrounding local interstellar medium. For contemporary medium conditions, interstellar He constitutes the largest neutral gas contributor in interplanetary space at Earth's orbit. This neutral gas is the source of a small, but noticeable population of pickup ions in the solar wind, which subsequently is injected into ion acceleration processes much more efficiently than solar wind ions. In fact, He^+ has been found to be the third most abundant energetic ion species at 1 AU after H^+ and He^{2+}. During its history, the solar system must have encountered a variety of interstellar environments, including hot and dilute bubbles and much denser clouds, and will continue to do so in the future. We have studied the neutral gas and its secondary products in the inner heliosphere, such as pickup ions and energetic particles, for conditions that range from that of a hot bubble to dense clouds with densities up to 100 times the current value, allowing for variation in temperature and bulk flow velocity. We have used multi-fluid models of the global heliosphere and kinetic models of the He flow, to derive the spatial distribution of interstellar neutral H and He in the inner heliosphere. With this

P. C. Frisch (ed.), Solar Journey: The Significance of our Galactic Environment for the Heliosphere and Earth, 209–258.

limitation the termination shock is at 8 AU, an d the Earth remains in the supersonic solar wind. Except for the case when the Sun is immersed in a hot bubble with no neutral gas, interstellar He maintains its dominant role in the inner heliosphere. However, even for the highest densities, the interstellar gas generally does not yet affect the solar wind dynamically at 1 AU, except for the region of the He focusing cone on the downwind side of the interstellar flow. Because dense clouds usually are also cold, a rather narrow cone with a drastically increased density develops, which acts like a huge and stationary comet in the sense that the solar wind would be decelerated to approximately the interstellar flow speed at 1 AU. Under such conditions energetic particles generated from interstellar pickup ions would dominate the energetic particle population at least during solar minimum.

Key words: Neutral particles; pickup ions; energetic neutral atoms; energetic particles; heliosphere-ISM interaction; interstellar cloud variations

8.1 Introduction

Since the late 1960's it has been well known that the Sun's environment cannot be compared with a Strömgren Sphere, as seen around massive and luminous O or B stars. A Strömgren Sphere is basically a spherical HII region around a star in which the interstellar gas is completely ionized by the star's ultraviolet radiation, and the radius of the sphere is determined by an equilibrium between ionization and recombination at the boundary (Strömgren, 1939). Due to the Sun's motion with ≈26 km/s relative to the interstellar medium (ISM) the equilibrium condition is violated because the typical time constant for ionization of H atoms is much longer than the travel time for distances larger than a few astronomical units (e.g. Fahr, 1968a). As a consequence a wind of interstellar neutral gas blows through the entire heliosphere, which is not affected significantly by the solar wind and its embedded magnetic field. In this way, the interstellar medium fills even the inner solar system with extrasolar neutral gas. It forms a recognizable pattern with a small central void that depends on the ionization potential of the species, the interstellar flow speed and temperature, as well as the Sun's gravitation and radiation pressure (see, e.g., reviews by Axford, 1972; Fahr, 1974; Holzer, 1977; Fahr, 2004).

Over the past 10 years, substantial attention has been devoted to potential changes in the overall size, shape and morphology of the heliosphere in response to variations in the physical parameters of the ISM environment of the Sun over time (Frisch 2000; Zank and Frisch, 1999). There has been speculation about potential effects on the Earth, in case it is exposed to the ISM rather than the solar wind (Yeghikyan and Fahr, this volume). This may happen during a very drastic compression of the heliosphere to smaller than 1 AU, which will also substantially alter the modulation of cosmic rays and thus the Earth's exposure to them (Florinski and Zank, this volume). However, less effort has been spent on

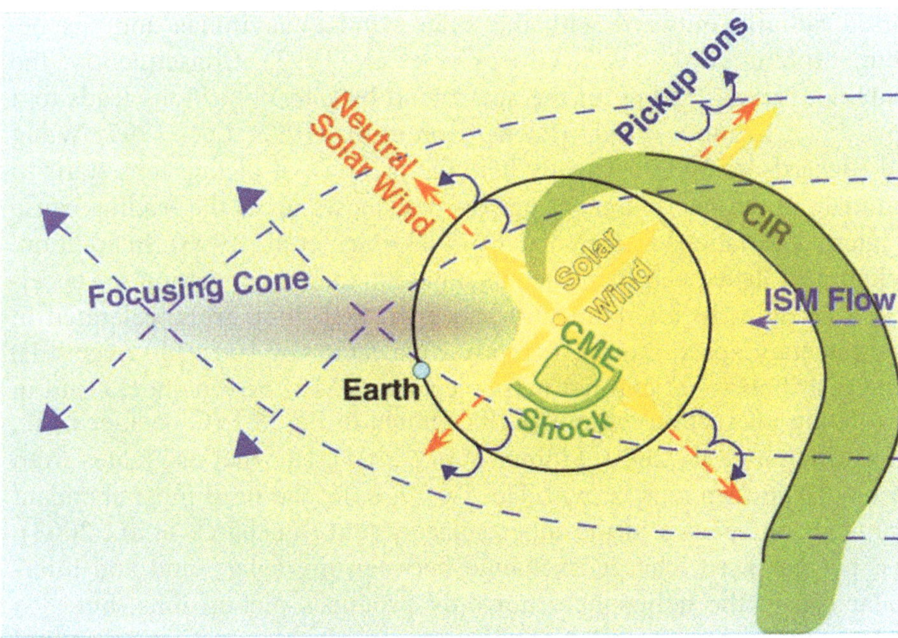

Figure 8.1. Schematic view of the inner heliosphere with the interstellar gas, its secondary particle populations, such as neutral solar wind and pickup ions, as well as their energetic offspring. The decreasing influence of the solar wind with distance from the Sun is shown in yellow and the increasing importance of the ISM in light blue.

related changes of the interstellar neutral gas inventory in the inner heliosphere in response to more moderate variations in the physical parameters of the local galactic environment. These will form the topic of the present chapter.

Throughout the heliosphere, including its innermost portion, the interstellar neutral gas flow is the source for a number of secondary products, which together with the original gas itself are important contributors to the neutral and ionized particle environment in the heliosphere. Figure 8.1 presents a view of the inner heliosphere together with the major interstellar and solar particle sources, including their secondary products and places where energetic particles are generated. Under current conditions the total density of interstellar particles dominates that of solar particles outside 5–10 AU, depending on solar activity. Through ionization by solar UV, charge exchange with solar wind ions, and electron impact ionization, interstellar pickup ions are formed that are transported radially outward with the solar wind (Vasyliunas and Siscoe, 1976; Möbius et al., 1985; Gloeckler et al., 1993). Consequently, the continuous mass-loading of the solar wind by interstellar ions leads to a slow-down beyond 30 AU (Richardson et al., 1995; Lee, 1997; Wang and Richardson, 2003). Already here the pressure of pickup ions starts to dominate in the interplanetary plasma and shows up as the leading cause of internal variations in the solar wind (Burlaga et al., 1996). In addition, pickup ions have been identified as a potent source population for accelerated particles in interplanetary space. In fact, they are accelerated in interplanetary space more efficiently by a factor of 100–200 over solar wind distributions at coronal mass ejection (CME) driven shocks and in co-rotating interaction regions (CIR), shown in Fig. 8.1 (Gloeckler et al., 1994; Chotoo et al., 2000; Möbius et al., 2001). He^+ that originates from interstellar pickup ions is, with $He^+/He^{2+} \approx 0.06$, the third most abundant energetic ion species in the inner solar system (Kucharek et al., 2003). Last but not least, charge exchange between the solar wind and interstellar gas in the heliosphere not only produces pickup ions, but also leads to the generation of a small yet recognizable fraction of neutral solar wind (e.g. Fahr, 1968b; Gruntman, 1994; Gruntman, 1997; Collier et al., 2001). Increase of the neutral gas population in the inner heliosphere due to changes in density, velocity, and temperature of the surrounding medium will, therefore, lead to an increase in the importance of each of the aforementioned secondary products. The influence of interstellar pickup ions on the solar wind dynamics will be pushed closer to the Sun. The increased density of pickup ions is mirrored in an increased importance of the relevant energetic ion populations. Finally, neutral solar wind is not blocked by the Earth's magnetic field and, thus, will have direct access to interaction with the upper layers of the Earth's atmosphere. If the neutral solar wind flux reaches a certain level, substantial effects on the particle and energy balance in the upper atmosphere are expected (Bzowski et al., 1996; Yeghikyan and Fahr, 2004). In extreme cases the interaction with the neutral solar wind may look more like the direct exposure of the Venusian or Martian atmosphere to the solar wind (Breus et al., 1989), except for the magnetic pile-up.

The purpose of this chapter is not to go into detail about the effects of all these secondary products on the Earth and the Earth's magnetosphere, which are covered in the chapters by Yeghikyan and Fahr (this volume) and by Parker (this volume). It will rather describe how the primary neutral gas distributions and the related secondary products change in response to variations in the local interstellar medium over a significant and reasonable range in interstellar density, temperature, and velocity. For the purpose of this book we will include the bandwidth of current solar activity in the range of parameters, but potential changes in the activity level with the age of the Sun go beyond the scope of this chapter. We will start the discussion with a summary of the current understanding of the interstellar inventory in the heliosphere, including knowledge about its secondary products, based on observations and heliospheric modeling. In the following section we will model changes in the hydrogen (H) and helium (He) inventory for similar sets of parameters. These two key species are kept separate because of their genuinely different interaction with the heliosphere. Helium passes through the heliospheric boundary unimpeded and is the dominant interstellar species at 1 AU because of its high ionization potential. Hydrogen is substantially filtered in the heliospheric interface region, leading to a reduction in density, slowdown, and heating. As the majority component in the ISM, H plays the dominant role in the determination of the overall size and shape of the heliosphere. Finally, after establishing the ISM neutral distribution under these circumstances, we will discuss the effects of their variation on all the secondary products.

8.2. Observations and Modeling of Neutrals in the Contemporary Heliosphere

As has been pointed out in the introduction, neutral interstellar gas flows through most of the heliosphere. This gas flow is responsible for a number of secondary products, which can affect the environment of the Earth if their fluxes reach a high enough level. On the other hand, the presence of the neutral gas flow enables the use of powerful in situ diagnostic tools to study the gas and its products in the inner solar system, from Earth orbiting and interplanetary spacecraft. In the following we will summarize our knowledge of the interstellar neutral gas and its products under current interstellar boundary conditions.

8.2.1 Modeling of Interstellar Neutral H and He Gas in the Heliosphere

The most abundant element both in the solar wind and in the local interstellar medium is hydrogen, which consequently also determines the overall dynamics between both domains. The interaction of the solar wind plasma with the ionized component of the interstellar wind creates the heliosphere with its characteristic heliospheric boundaries, namely, the termination shock and the heliopause. The ISM plasma experiences the heliosphere as an obstacle and flows around it, the heliopause being the separation surface between the solar wind and interstellar plasma. Before the supersonic solar wind reaches the heliopause, it is decelerated and heated at the termination shock and then turned into a tailward flow relative to the interstellar neutral gas (ISN) flow. Other species of the neutral gas mostly follow the bulk flow, but are dynamically decoupled because the mean free path length for collisions exceeds or is of the order of the size of the heliosphere.

Close to the Sun the ISN that flows into the heliosphere is depleted due to photo-ionization by solar UV radiation, charge exchange with solar wind ions, and electron impact ionization. The newborn ions are picked up by the interplanetary magnetic field (hence pickup ions) and are swept away with the solar wind. They will be discussed in detail further down in this section. The local loss rate of neutral particles at 1 AU is

$$\begin{aligned} -\frac{dn_{ISN}}{dt} &= n_{ISN} n_{SW} v_{rel} \sigma_p(v_{rel}) + n_{ISN} n_{SW} v_{rel} \sigma_e(v_{rel}) + n_{ISN} \beta_{phot} \\ &= n_{ISN}\left(\beta_p + \beta_e + \beta_{phot}\right) = n_{ISN} \beta_A \end{aligned} \tag{8.1}$$

where n_{ISN} is the ISN density, n_{SW} is the solar wind plasma density ($n_p \approx n_e \approx n_{SW}$), and v_{rel} is the relative speed between the interacting particles. The individual losses of neutral particles (β_p, β_e, and β_{phot}) are combined into a total loss rate β_A for each species A. The resulting change in density of the interstellar neutrals at distance r from the Sun is then given as

$$\frac{d\, n_{ISN}}{dl} = -n_{ISN} \beta_A \frac{1}{v_{ISN}} \left(\frac{r_E}{r}\right)^2 \tag{8.2}$$

where v_{ISN} is the velocity of the interstellar gas in the heliosphere and β_A is the total ionization rate for species A at Earth's orbit, i.e., at distance r_E. An inverse quadratic dependence with distance from the Sun is implemented in Eq. 8.2, which is adequate for photo-ionization and charge exchange, while electron impact usually is only a small contribution. For a first estimate the remaining fraction of neutral interstellar gas at distance r from the Sun results from integrating the ionization along the particle trajectory, based on a cold model of the ISN gas (Vasyliunas and Siscoe, 1976).

$$n_{A,ISN}(r,\theta) = n_{A,\infty} \exp\left(-\frac{\beta_A r_E^2}{v_{ISN} r} \frac{\theta}{\sin\theta}\right) \tag{8.3}$$

θ is the angle of position relative to the upwind direction and $n_{A,\infty}$ is the interstellar gas density of species A outside the influence of the Sun (i.e., the neutral density at "infinity"). It should be noted that the use of a cold model is justified in the upwind direction, while the ISN distribution on the downwind side is substantially influenced by the gas temperature. We present simple analytical relations based on the cold model here so that the reader may use them for order of magnitude estimates of the global effects of interstellar environment variations. Equation 8.3 contains the assumption that v_{ISN} does not vary with distance and the interstellar gas density does not change other than through ionization. A constant v_{ISN} approximately applies to the hydrogen flow, for which $\mu \approx 1$ (Bzowski, 2001) (μ being the ratio between radiation pressure and gravitation). Using the cold model approach of Eq. 8.3 and a total ionization rate for H in the range $_{H}=(3.5 \quad 8.0)\bullet 10^{\ 7} s^{\ 1}$ (a typical variation with solar activity) leads to a remaining fraction of ISN hydrogen in the upwind direction ($\theta = 0$) at Earth's orbit ranging from 0.092 to 0.0042 and from 0.62 to 0.34 at Jupiter's orbit under current interstellar medium conditions. Towards the downwind direction the remaining neutral fractions become very small inside a few AU, and the depleted volume of interstellar gas is often referred to as the ionization cavity of the Sun, although the evacuation of interstellar atoms from the ionization cavity is not complete.

The effective inflow velocity of H is lowered to ≈ 22 km/s from the original ISM value of 26 km/s because of the strong interaction of H at the heliospheric interface, as observed through the Doppler shift of the backscattered Ly α radiation (Clarke et al., 1998; Quémerais et al., 1999). In the decelerated plasma just upwind of the heliopause, charge exchange creates a density enhancement of slow neutral H, the so-called hydrogen wall, also seen in absorption in high-resolution spectra of neighboring stars (e.g., Wood et al. 2000). Neutral H atoms that enter the hot heliosheath (the region between the heliopause and the termination shock) experience charge exchange losses (Ripken and Fahr, 1983; Fahr, 1991; Osterbart and Fahr, 1992; Baranov and Malama, 1993; Izmodenov et al., 1999; Müller et al., 2000; Izmodenov et al., 2004). Charge exchange in this region provides both a cooling mechanism in the heliosheath, and heating for the region upwind of the heliopause through secondary charge exchange of the neutrals created in the heliosheath. This also produces a hot neutral H distribution with a low bulk velocity, which leaks into the heliosphere and mixes with the remainder of the original ISN H flow to form a combined decelerated and heated neutral gas distribution. Overall the neutral H density inside the heliopause is substantially depleted from its value in the pristine ISM, known as filtration of the incoming neutral gas. Therefore, inferring the ISM H (and O) parameters from heliospheric observations is strongly model dependent.

Heavier atoms experience gravitational acceleration when approaching the Sun, as radiation pressure is negligible for them and thus $\mu \approx 0$ applies. Making use of energy conservation along the trajectory, a relation that is modified from Eq. 8.3 is derived.

$$n_{A,ISN}(r) = n_{A,\infty} \exp\left(-\frac{\beta_A r_E^{2}}{v_{ISN}} \delta\left(\sqrt{1+\tfrac{2}{\delta r}} - 1\right) \right) \text{ where} : \delta = \tfrac{v_{ISN}^2}{M_S G} \qquad (8.4)$$

Here M_S is the mass of the Sun and G the gravitational constant. The validity of Eq. 8.4 is restricted solely to the upwind axis. Using as a reasonable range over solar activity for the He ionization rate of $_{He} = (6 \quad 20)\bullet 10^{\;8}\, s^{\;1}$ (McMullin et al., 2004), the fractional He density at 1 AU in the ISN upwind direction is 0.79 – 0.45 at Earth's orbit and 0.95 –0.90 at Jupiter's orbit. The values for $_{He}$ are variable also on shorter time scales; they change with solar UV flux as well as with solar wind parameters (e.g. Rucinski et al., 1996; Bzowski 2001; Lallement et al., 2004b; McMullin et al., 2004; Witte et al., 2004), but it is sufficient to use averaged values here.

All species except H form a region of increased interstellar gas density, the focusing cone, downwind of the Sun, which is a result of gravitational bending of the particle trajectories, as indicated in Fig. 8.1. H does not show such focusing because its gravitational attraction is essentially compensated by radiation pressure and $\mu \approx 1$, or even $\mu > 1$ for high solar activity. This focusing cone has been extensively studied for interstellar He because significant density enhancements are observed already at Earth's orbit (Gloeckler et al., 2004; Möbius et al., 1995, 2004). Heavier atoms are also gravitationally focused by the Sun, but at larger distances (Gloeckler and Geiss, 2001).

A more detailed model that takes into account the random motion in a hot gas is used, from this point on, for the evaluation of the neutral He gas variations in the inner heliosphere. In particular the focusing cone cannot be adequately described with a simple cold gas approximation. In such models the density of the interstellar helium gas at a given point in space, described by the heliocentric distance r and the angular distance from the upwind axis θ, is calculated as an integral of the local distribution function $f_{ISN}(r,\theta,\mathbf{v})$ of the ISN gas. In a first step, an intermediate distribution function $f_{ISN0}(\mathbf{v}(r,\theta))$ is computed by applying Liouville's theorem along Keplerian trajectories to the distribution function $f_{\infty}(\mathbf{v}_0)$ of the unperturbed gas outside the heliosphere (assumed as a Maxwellian shifted by the bulk flow vector $\mathbf{v}_{ISN}$ of the ISN).

$$f_{\infty}(\mathbf{v}_0)=\left(\frac{m}{2\pi\,kT}\right)^{3/2}\exp\left[-\frac{m\left(\mathbf{v}_0-\mathbf{v}_{ISN}\right)^2}{2kT}\right] \tag{8.5}$$

To account for ionization losses on the way, $f_{ISN0}(\mathbf{v}(r,\theta))$ is multiplied by the survival probability $E(r,\theta,\mathbf{v})$ according to

$$f_{ISN}(r,\theta,\mathbf{v})=f_{ISN0}(\mathbf{v}(r,\theta))\,E(r,\theta,\mathbf{v})\,. \tag{8.6}$$

The survival probability is computed as

$$E(r,\theta,\mathbf{v})=\exp\left[-\int_t^{\infty}\beta_{He}(r_E)\left(\frac{r_E}{r'(r,\theta,\mathbf{v},t')}\right)^2 dt'\right]. \tag{8.7}$$

$r'(r,\theta,\mathbf{v},t)$ is the heliocentric distance of an individual atom following a Keplerian trajectory from outside the heliosphere to a location (r, θ). $\beta_{He}(r_E)$ is the total ionization rate of the He atoms at 1 AU. The velocity vector of an individual atom $\mathbf{v}_0(r,\theta)$ is calculated by tracing the trajectories backwards from the point of observation.

The modeling shown here was performed assuming a time-invariable ionization rate that depends on the heliocentric distance as $1/r^2$ and is equal to 10^{-7} s^{-1} at 1 AU, a value representative of the average present-day photo-ionization rate (McMullin et al., 2004). Apart from the EUV ionization rate, close to the Sun (inside ≈2 AU) another ionization mechanism, electron impact (Rucinski and Fahr, 1989, 1991), needs to be taken into account. The dependence of its rate on the heliocentric distance has usually been described by $1/r^{\alpha}$, where $\alpha \neq 2$ (Rucinski and Fahr, 1991). Departure from the $1/r^2$ law is mainly due to the variation of the electron temperature with heliocentric distance (Pilipp et al., 1987a,b; Marsch et al., 1989), which scales as $T_e \propto r^{-\gamma}$, with the limiting values $\gamma = 0$ and 4/3. The exact value of γ seems to vary with heliocentric distance and also with the regime (fast or slow) of the solar wind (McMullin et al., 2004) and consequently with the phase of the solar cycle. However, contributions from electron ionization remain less than that of photo-ionization, even inside 1 AU.

The EUV ionization rate varies over the solar cycle by a factor of ≈3 (McMullin et al., 2004), which leads to He density variations whose amplitude decreases with heliocentric distance and depends on the offset angle from the upwind direction (Rucinski et al., 2003). Outside the cone, this amplitude is ≈ 25% at 1 AU and ≈ 10% at 5 AU. On the cone axis the variations are much stronger, with amplitudes of ≈80% at 1 AU and ≈ 25% at 5 AU.

8.2.2 Observation of Interstellar Neutral H and He Gas in the Heliosphere

Interstellar He was first observed in the inner heliosphere through resonant backscattering of the solar He I 58.4 nm line with rocket-borne (Paresce et al., 1974) and satellite-borne (Weller and Meier, 1974) instrumentation. Density, bulk flow speed, and temperature could be deduced from the total intensity of the glow, as well as the relative intensity and width of the focusing cone, after this structure was modeled with a hot gas distribution (e.g. Fahr et al., 1978; Wu and Judge, 1979). Over many years the derived parameters varied substantially from observation to observation and carried large uncertainties (see compilations by Chassefière et al., 1986; Möbius et al., 1993). The results from the He UV observations were significantly improved when EUVE, with its low

Earth orbit, provided the opportunity to measure the He velocity distribution by using the Earth's exosphere as a natural gas absorption cell (Flynn et al., 1998). In addition, the discovery of interstellar He pickup ions at 1 AU (Möbius et al., 1985) introduced a first in situ method to probe interstellar particles and to deduce the interstellar He flow parameters (Möbius et al., 1995). Meanwhile, helium is probably the best-measured species of the inflowing interstellar gas since it is detected directly with suitable instrumentation, i.e., the GAS instrument on the Ulysses mission (Rosenbauer et al., 1983). The most recent and most advanced analysis of the GAS data gives for the bulk speed $v_{He\infty} = 26.3 \pm 0.4$ km/s (He is accelerated when approaching the Sun, $v_{He,\infty} = v_{ISN}(r = \infty)$), the flow direction in ecliptic longitude $\lambda_\infty = 74.7° \pm 0.5°$ and ecliptic latitude $\beta_\infty = -5.2° \pm 0.2°$, and the temperature $T_{He\infty} = 6300 \pm 340$ K (Witte, 2004). Since He does not interact at the boundary of the heliosphere, this set of physical parameters is considered valid for all interstellar gas species outside the influence of the Sun. The interstellar He density of $n_{He\infty} = 0.015$ cm^{-3} has been derived from Ulysses/GAS neutral He data (Witte, 2004) and from Ulysses/SWICS He^{2+} pickup ion observations (Gloeckler et al., 2004), with a smaller (10%) uncertainty based on pickup ions, because this method does not require an absolute instrument calibration. Similar flow vector and temperature values are derived for He from observations of backscattered solar He I 58.4 nm line emission (e.g. Lallement et al., 2004b; Vallerga et al., 2004). For He these different methods are compared by Möbius et al. (2004) and show good agreement in the derived parameters.

For interstellar hydrogen so far no direct neutral gas observations are available. The Ulysses/GAS instrument is not sensitive to H. Suitable instruments for the direct detection of interstellar H have only been developed recently (Wieser et al., 2005), and are currently being implemented for a space mission (McComas et al., 2004). However, starting with the analysis of backscattered solar Lyman α intensity sky maps (Bertaux and Blamont, 1971; Thomas and Krassa, 1971), the parameters of ISN H became accessible. Based on early modeling using a cold interstellar gas flow (e.g. Blum and Fahr, 1970; Holzer and Axford, 1971; Axford, 1972; Fahr, 1974), the general flow direction and an order of magnitude estimate for the density could be deduced. Through the Doppler effect, high-resolution profiles of backscattered Ly α from Copernicus provided a first reasonable value for the H bulk speed inside the heliosphere (≈22 km/s) and constraints on the temperature (Adams and Frisch, 1977). Substantial progress towards the kinetic parameters was made with the use of hydrogen absorption cells, which provide the best information on the H flow velocity vector ($v_{H\infty}$ = 18–22 km/s, λ_∞ = 74° and β_∞ = –7°) and the temperature ($T_{H\infty}$ = 8000 – 12000 K) inside the heliosphere (Bertaux et al., 1985; Ajello et al., 1994; Lallement, 1996; Clarke et al., 1998; Quemerais et al., 1994, 1999; Costa et al., 1999). The most recent determination of the parameters for hydrogen using hydrogen absorption cells have been reported by Lallement et al. (2005) where $v_{H\infty}$ = 22 ± 1 km/s, $T_{H\infty}$ = 11500 ± 1000 K, λ_∞ = 72.5±0.5° and β_∞ = –8.8±0.5° are found. The flow direction of the interstellar H inside the heliosphere is offset from that of He by 4±1° (Lallement et al., 2005) and close to the velocity vector of the local interstellar cloud, λ_{LIC} = 74.5° and β_{LIC} = –7.8°, as observed by HST (Lallement, 1996). The observed difference between the H and He flow direction in the inner heliosphere is probably due to the deflection of interstellar plasma by the interstellar magnetic field and the strong neutral gas plasma interaction of hydrogen. In this connection it should be noted that potential experimental evidence for a second, more energetic, neutral particle stream possibly related to the interstellar gas has been presented (Collier et al., 2004; Wurz et al., 2004). However, its interpretation is still rather controversial, and improved observations are needed to test the hypothesis of such a second stream.

As a consequence of the interaction with the plasma H is also substantially filtered at the heliospheric boundary. Therefore, the density outside the heliosphere has to be inferred through modeling. Pickup ion observations (discussed in more detail in Section 8.2.4) provide an estimate for the hydrogen density at the termination shock of $n_{H,TS}$ = 0.095±0.01 cm^{-3} (Gloeckler and Geiss, 2004), which agrees well with $n_{H,TS}$ = 0.09±0.02 cm^{-3} as derived from the slowdown of the solar wind through mass loading with H^+ pickup ions (Wang and Richardson, 2003). Combining the measured ISN H density inside the heliosphere with a pressure balance between solar wind and ISM and using a kinetic model for the filtration, Izmodenov et al. (1999, 2004) deduced a neutral H density $n_{H,ISM}$ = 0.18 cm^{-3} and a proton density n_{H^+ISM} = 0.06 cm^{-3} for the local ISM. UV instrumentation is also part of the Pioneer and Voyager spacecraft. These observations provide insight into the interstellar neutrals at larger distances from the Sun and have been used to infer H and H^+ densities in the surrounding ISM (Gangopadhyay et al., 2002; 2004). In their most recent work, which includes a Monte Carlo radiative transfer model and a global heliosphere model (Baranov and Malama, 1993), Gangopadhyay et al. (2004) report a neutral density $n_{H,ISM}$ = 0.15 cm^{-3} and a fraction of ionized hydrogen $n_p/(n_p + n_{H,ISM}) = 0.3$. Their best fit to the UV data for the H density at the termination shock suggests 0.08 cm^{-3}. Considering the rather difficult absolute calibrations for UV observations, these results are in good agreement with the values derived from the in situ particle observations. The results imply that only $\approx$50% of the H atoms pass through the heliospheric boundary.

8.2.3 Neutral Solar Wind

Neutral solar wind (NSW) originates from solar wind ions, which become neutralized during their travel from the Sun through interplanetary space. Consequently, NSW moves in the anti-Sunward direction with solar wind velocity, i.e., 300–800 km s^{-1}. The mechanisms suggested for the creation of NSW are charge exchange with neutral gas (e.g. the inflowing interstellar gas or the exosphere of a planet) or interaction of solar wind ions with interplanetary dust particles. The dominant portion of dust particles stems from the fragmentation of comets and asteroids. Kuiper belt dust and interstellar dust are only a small fraction of the dust population in the inner heliosphere (Mann et al., 2004). In addition, it has been speculated that neutral atoms in the solar wind could originate directly from ejections of solar matter (Akasofu 1964a,b; Illing and Hildner, 1994). Fahr (1968b) suggested that charge exchange between solar protons and neutral hydrogen near the Sun would create NSW that would penetrate into the Earth's thermosphere and could raise the temperature and density. Holzer (1977) examined the consequences of interplanetary atomic H on the solar wind and concluded that charge transfer would produce a NSW fraction of $3 \cdot 10^{-5}$ at 1 AU. Gruntman (1994) estimated the NSW fraction observed near Earth as $10^{-5} - 10^{-4}$, depending on the time of the year. Such small fractions have a negligible effect on the Earth's upper atmosphere (Yeghikyan and Fahr, 2004).

The charge exchange process between an ISN atom, A_{ISN}, and a solar wind ion, B^+_{SW}, is described as

$$A_{ISN} + B^+_{SW} \xrightarrow[\sigma_{CE}]{} A^+_{ISN} + B_{SW} \tag{8.8}$$

with a charge exchange cross section σ_{CE}. The resulting ion A^{+}_{ISN} is a newly created pickup ion, and B_{SW} is a NSW atom. Typically one electron is exchanged between the collision partners, which means that NSW created by this process will be composed almost entirely of H. The most important charge exchange processes occur between H and solar wind protons,

$$H_{ISN} + H^{+}_{SW} \xrightarrow[\sigma_{CE}]{} H^{+}_{ISN} + H_{SW} \tag{8.8a}$$

with a cross section $\sigma_{CE} = 1.7 \cdot 10^{\ 15}$ cm^2 (Maher and Tinsley, 1977), and between He and solar wind alpha particles

$$He_{ISN} + He^{++}_{SW} \xrightarrow[\sigma_{CE}]{} He^{++}_{ISN} + He_{SW} \tag{8.8b}$$

with a cross section $\sigma_{CE} = 2.4 \cdot 10^{\ 16}$ cm^2 (Barnett et al., 1990). The highly charged heavy ions (e.g. C^{5+}, O^{6+} or Fe^{10+}) may reduce their charge states as a result of collisions with neutral gas atoms, but will not end up as neutral atoms. The total NSW flux Φ^{ISN}_{NSW} from charge exchange with the ISN gas is derived by integrating this process from the Sun to the location of the observer.

$$\Phi^{ISN}_{NSW}(r,\theta) = \int_0^r n_{SW} \left(\frac{r_E}{r'} \right)^2 v_{SW}\, \sigma_{CE}\, n_{ISN}(r',\theta)\, dr' \tag{8.9}$$

n_{SW} is the solar wind density at r_E = 1 AU, which scales quadratically with distance from the Sun. With the interstellar gas density from Eq. 8.3 this yields

$$\begin{aligned} \Phi^{ISN}_{NSW}(r,\theta) &= n_{SW} v_{SW} \frac{\sigma_{CE}\, v_{ISN}}{\beta_A} \frac{\sin\theta}{\theta} n_\infty \exp\left(-\frac{\beta_A r_E^{\ 2}}{v_{ISN} r} \frac{\theta}{\sin\theta} \right) \\ &= n_{SW} v_{SW} \frac{\sigma_{CE}\, v_{ISN}}{\beta_A} \frac{\sin\theta}{\theta} n(r,\theta) \end{aligned} \tag{8.10}$$

Neutralization of solar wind ions by dust in the inner heliosphere can occur through two processes. Firstly, solar wind ions are implanted into the dust grains, saturate the outer layer with gas over time, and are released later via natural outgassing of the dust particle (Banks, 1971; Fahr et al., 1981). Similarly, atoms and a small fraction of ions are released from the surface of dust particles into space through sputtering by impinging solar wind ions. Fortuitously, the solar wind energy is about 1 keV/nuc, which is about the energy for the maximum sputter yield, and the total sputter yield of solar wind is approximately 0.15 atoms per incident ion, assuming a typical mix of solar wind ion species. The sputtered neutrals and ions contribute material from the dust grains to the environment. Together, these processes establish a neutral gas source close to the Sun that results in the charge exchange reactions as described by Eq. 8.8. Fahr et al. (1981) calculated that dust-generated neutral H dominates over ISM H inside 0.5 AU. NSW created in this way, again, will be mostly H. In a second process solar wind ions traverse sufficiently small dust particles and emerge with a lower charge state or even neutralized (Collier et al., 2003; Wimmer-Schweingruber and Bochsler, 2003). With typical energies of 1 keV/nuc solar wind ions will penetrate dust grains in the size range 100 to 300 Å or the corresponding range at the edge of larger grains. The second process also allows for NSW atoms other than H. However, upon traversal through a dust grain the particles will lose a substantial fraction of their energy, and part of the resulting NSW will be significantly slower than the ionized solar wind. Ions that are generated out of the dust-generated neutral gas and the sputtered ions are referred to as "inner source" pickup ions (Geiss et al., 1996). Therefore, also solar wind ions that have assumed a low charge state and end up with very low speed after penetration of dust grains may contribute to the "inner source" ions (Wimmer-Schweingruber and Bochsler, 2003).

NSW produced by dust via the first process is given by Banks (1971)

$$\Phi_{NSW}^{dust,1}(r) = \frac{\sigma_{CE} \cdot (n_{SW} v_{SW})^2}{\beta} r_E \Omega_1 \left(\frac{r_E}{r}\right)^2 \ln\left(\frac{r}{r_{\min}}\right) \tag{8.11}$$

where Ω_1 is the total integrated cross-section of the dust from the Sun to the observer located at r_E and β is the rate for charge exchange between solar wind ions and dust-generated neutral gas at 1 AU. For small distances, Ω_1 can be imagined as the probability per unit length that a particle will be absorbed by dust.

$$\Omega_1 = \int_{s_{\min}}^{s_{\max}} f(s,r) \pi s^2 \, ds \tag{8.12}$$

Equation 8.12 represents the integral over the entire size distribution $f(s,r)$ of dust particles of size s, assuming spherical dust grains (Banks, 1971), where $f(s,r)ds$ is the local number density of dust grains. For simplicity, a size distribution $f(s,r) = f_E(s) \bullet (r_E / r)$ is assumed, which does not vary in shape, but whose density falls off linearly with distance from the Sun, as is considered valid outside $0.1\ r_E$ (Mann et al., 2004). Previously, radial dependencies were reported that are close to linear: $(r_E/r)^{1.25}$ (Levasseur et al., 1991) and $(r_E/r)^{1.3}$ (Richter et al., 1982). Note that the solar wind flux enters as a quadratic term in Eq. 8.11, because dust particles are saturated with solar wind and thus the outgassing neutral gas flux is also proportional to the solar wind flux into the particle. In addition, there is an inverse quadratic radial dependence that confines this process close to the Sun.

NSW that arises from interaction with dust via the second process is given by

$$\Phi_{NSW}^{dust,2}(r) = p_0\ n_{SW}\ v_{SW} r_E\ \Omega_2 \left(\frac{r_E}{r}\right)^3 \tag{8.13}$$

with p_0 being the probability that the emerging particle is neutral and Ω_2 the integrated cross-section for the dust penetration process at r_E. The cubic radial dependence in Eq. 8.13 is a combination of the quadratic solar wind flux decrease and the linear dust density decrease with distance from the Sun. Thus, NSW from both processes is created preferentially close to the Sun, with an even stronger concentration for the latter process. The integral of the total dust cross-section, Ω_2, based on this second process may be smaller because the interaction with larger grains is limited to their edge regions, and small dust particles could be depleted close to the Sun because of magnetic and drag forces (e.g. Ragot and Kahler, 2003; Mann et al., 2004). Therefore, the contribution of the second process to the generation of NSW is potentially weaker than that of the first process, and may be solar cycle dependent, but a number of questions still remain to be answered.

It should be emphasized that the dust that contributes to the generation of NSW is solely of interplanetary origin. Although a stream of interstellar dust through the solar system has been detected with Ulysses (Grün et al., 1994), these particles constitute only a small fraction of the dust in interplanetary space with a very specific velocity distribution. Compared with gas the contribution of dust is already small in the ISM. For typical ISM clouds, gas to dust mass ratios of 100 to 170 are found (e.g. Spitzer, 1978), and more recently a value of 180 ± 3 has been deduced for the local neighborhood of the heliosphere (Frisch et al., 1999; Landgraf, 2000; Frisch and Slavin, 2003). In addition, the small dust grains are strongly filtered by the interplanetary magnetic field (Landgraf et al., 2003). Therefore, the importance of interstellar over interplanetary dust is rather small in the inner heliosphere (Mann et al., 2004). Although during the passage of denser interstellar clouds the relative importance of interstellar dust will increase (Landgraf, this volume), the high gas to dust ratio in the interstellar material will lead to dominant NSW production from the neutral gas component.

NSW created close to the Sun may be ionized again on its way out. The probability of an atom born at distance r_0 from the Sun to reach a point further out in the heliosphere at distance r is

$$p_S(r,r_0)=\exp\left(-\frac{\beta_A r_E^{\,2}}{v_{SW}}\left(\frac{1}{r_0}-\frac{1}{r}\right)\right) \tag{8.14}$$

Since NSW propagates with the solar wind speed, the relative velocity between solar wind ions and NSW is close to zero and the probability for charge exchange is negligible (see Eq. 8.1). In addition, this would not change the NSW flux, since two co-moving particles would exchange their charges. However, NSW may still be photo-ionized, but these rates are lower than the charge exchange rates between ISM H and solar wind (Rucinski et al., 1996) and NSW travels at solar wind speed, much faster than ISN. Thus most of the NSW will escape into the interplanetary medium. Only NSW created inside ≈ 0.5 AU will be noticeably affected.

Observational evidence of NSW thus far is very scarce because of the difficulty in measuring neutral particle fluxes several orders of magnitude lower than the solar wind flux, while simultaneously being exposed to full sunlight. The only observation thus far was made with LENA on IMAGE, from which a NSW fraction of $\sim 10^{-4}$ was derived (Collier et al., 2001). Enhancements in the NSW flux due to charge exchange in the geocorona were observed, with the NSW fraction increasing to a few times 10^{-4} (Collier et al., 2001). Since the NSW fraction outside the geocorona was found to be roughly constant along the Earth's orbit for large intervals, it was concluded that the neutralizing agent cannot be the ISN gas flow (Collier et al., 2003), which should have a pronounced seasonal modulation (Bzowski et al., 1996). Only in the upwind direction will charge exchange with the ISN gas give a NSW fraction of ~10^{-4} (Wurz et al., 2004). Therefore, interaction with dust was suggested, which is distributed evenly in the inner heliosphere (Richter et al., 1982). From the observed NSW fluxes the total dust cross section was determined to be $\Omega < 6 \cdot 10^{\ 19}$ cm$^{\ 1}$ at 1 AU (Collier et al., 2003).

Obviously under current conditions NSW is mainly generated by interplanetary dust, because the fluxes of ISN gas into and the available gas densities in the inner heliosphere are very small. Therefore, the production of NSW by charge exchange is insignificant when evaluated for effects on planetary atmospheres. If the ISN velocity were substantially increased and/or the ISN gas density were much higher, a NSW fraction in the range of a few percent, or even beyond, may result as will be discussed below.

8.2.4 Pickup Ions

Through charge exchange with the solar wind, which also produces NSW, as well as through photo and electron impact ionization pickup ions are generated out of the neutral interstellar gas. These pickup ions constitute an important component of the interplanetary plasma population throughout the heliosphere. They were first detected for He in 1984 (Möbius et al., 1985) and for H in the 1990's (Gloeckler et al., 1993). Because the interstellar gas flow is slow compared with the solar wind, the initial velocity at ionization, which equals the neutral gas velocity, is usually neglected and pickup ions are modeled starting with $v = 0$. In the solar wind frame, they are injected with negative solar wind velocity $-v_{SW}$, which leads to a gyration about, and motion along, the interplanetary magnetic field, with the perpendicular and parallel component of the injection velocity distributed according to the orientation of the interplanetary magnetic field (IMF). In velocity space seen in the solar wind frame, the initial distribution of pickup ions in polar coordinates is $f_i(\mathbf{v}) = f_0\delta(v \quad v_{SW})\delta(\theta \quad \theta_o)$, where θ_o is the initial pitch angle of the pickup ions, which is equal to the angle of the IMF relative to the solar wind.

In the standard picture, pitch angle scattering due to Alfvén waves embedded in the IMF redistributes the pickup ions into a spherical shell in velocity space, and subsequently the shell shrinks due to adiabatic cooling during the radial expansion of the solar wind (e.g. Vasyliunas and Siscoe, 1976; Möbius et al., 1988). Observations have shown that the pickup ion distribution very often deviates drastically from this simple spherically symmetric shape due to inefficient pitch angle scattering (Gloeckler et al., 1995; Möbius et al., 1998; Oka et al., 2002) and that their flux and spectra are highly variable in space and time (Gloeckler et al., 1994; Schwadron et al., 1999; Saul et al., 2003). However, in order to evaluate the large-scale contribution to the particle population in the inner heliosphere by particles produced out of the interstellar gas flow, we can safely integrate over such variations and thus neglect them, as has been successfully done for the determination of the local interstellar parameters from pickup ions (Gloeckler et al., 1996; Gloeckler and Geiss, 2001; Gloeckler et al., 2004; Möbius et al., 2004). In this standard view the pickup ion distribution function $f(v)$ describes their production along the spacecraft-Sun line by

$$f(v)4\pi v^2 dv = n_{ISN}(r)\cdot \beta_A dt \qquad (8.15)$$

Assuming adiabatic cooling of the pickup ions in the solar wind like an ideal gas and using the fact that both the pickup ion density of already accumulated ions and the ionization rate decrease as $1/r^2$ with distance from the Sun, the relation

$$f(v) = \frac{3n_{ISN}(r)\cdot \beta_A \cdot r_{obs}}{8\pi\, v_{SW}^4}\left(\frac{v}{v_{SW}}\right)^{-3/2} \qquad (8.16)$$

emerges for the interstellar pickup ion distribution. In this relation β_A represents the ionization rate at the location of the observer. Recognizing that

$$r = r_{obs}\cdot\left(\frac{v}{v_{SW}}\right)^{3/2} \qquad (8.17)$$

describes the radial dependence of the adiabatic cooling, $f(v)$ represents a mapping of the radial distribution $n_{ISN}(r)$ of the interstellar neutrals inside the observer location into the observed velocity distribution of the pickup ions.

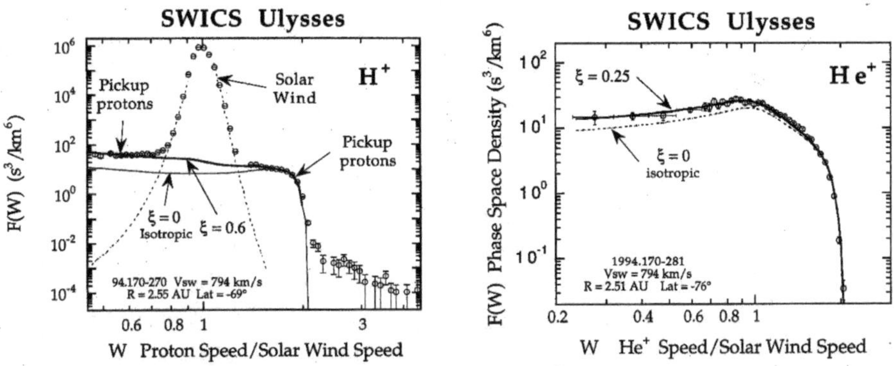

Figure 8.2. Phase space density spectra of H^+ (left) and He^+ (right) pickup ions (Gloeckler and Geiss, 1998). The H^+ distribution includes a suprathermal tail above $W > 2$, which is invisible for He^+, because the dynamic range in this plot is more limited.

Figure 8.2 shows energy spectra of the pickup ion phase space density for H and He (Gloeckler and Geiss, 1998). Here the observations are shown as a function of the ratio W of the pickup ion speed v to the solar wind speed v_{SW} to be able to integrate over times of varying solar wind speeds. The phase space density spectrum of H^+ pickup ions appears much flatter than that of He^+ between $1 < W < 2$, which reflects the fact that the H ISM density falls off much faster towards the Sun than that of He. Observations of the He^+ spectra over the course of one year in Earth's orbit, or at the Lagrange point L1, contain the distinct longitudinal structure of the He focusing cone (Möbius et al., 1995; Gloeckler et al., 2004), as can be seen in Fig. 8.3, which shows a combination of data from the 1980's with AMPTE SULEICA and data obtained after 1997 with ACE SWICS. The substantial variation of the cone, with maximum height during solar activity minima around 1985 and 1997 and reduced height during solar maximum, becomes obvious. The prominence of the He cone as the most important spatial feature of the interstellar gas flow through the inner heliosphere is a combined function of ionization rate, interstellar flow speed and temperature (Fahr et al., 1979; Wu and Judge, 1979) and thus depends on solar and interstellar conditions.

The knowledge of the radial distribution of the neutrals and the local ionization rate are sufficient to calculate the pickup ion velocity distribution and their absolute density. Under current conditions He^+ pickup ions make up $\approx 8 \cdot 10^{-4}$ of the solar wind He flux at 1 AU outside the focusing cone and $\approx (4–8) 10^{-3}$ in the cone center, depending on the solar cycle phase. Compared with the bulk solar wind density, this fraction is a factor of ≈16 lower, because the average abundance of He^{2+} is ≈6% in the solar wind. Therefore, pickup ions at 1 AU constitute a mere test particle population in interplanetary space. While interstellar He is present deep inside Earth's orbit and dominates the pickup ion population at 1 AU, the main component of the ISM, hydrogen, is noticeable only outside ≈3 AU. Dust-related "inner source" pickup ions dominate close to the Sun (Gloeckler et al., 2000). Because the dust population falls off rapidly with distance from the Sun as discussed above, the inner source pickup ion velocity distribution is already strongly cooled at 1 AU and will continue to do so at larger distances. As the solar wind accumulates interstellar pickup ions on its way outward from the Sun, they become increasingly more important at larger distances. Because interstellar pickup ions are continually injected into the solar wind even at large distances, they form a genuinely hot population and thus dominate the kinetic pressure in the solar wind outside ≈30 AU. Here they lead to so-called pressure balance structures (Burlaga et al., 1996). Outside these distances the implantation of pickup ions also leads to a slowdown of the solar wind due to mass-loading (Richardson et al., 1995; Wang and Richardson, 2003) and substantial pressure modifications (Fahr and Rucinski, 1999, 2001). An extensive review of mass-loading and its effects is given by Szegö et al. (2000). Still, a substantial reduction in solar wind speed due to the accumulation of pickup ions is not felt until close to the solar wind termination shock. In this chapter we will restrict ourselves to effects of the ISM neutrals and related secondaries in the inner heliosphere, while the bulk interaction that leads to the termination of the solar wind is described in the chapter by Zank et al. (this volume).

8.2.5 Singly Ionized Energetic Ions

Over the past decade it has become evident that interstellar pickup ions are a formidable source population for energetic particles in the heliosphere. The initially puzzling observation of a substantial fraction of He^+ in interplanetary energetic ion populations by Hovestadt et al. (1984a), has found its natural explanation with the detection of interstellar pickup He^+ in the solar wind (Möbius et al., 1985). Gloeckler et al. (1994) have identified interstellar pickup He^+ as the main constituent of energetic He ions at energies up to 60 keV in a CIR at 4.5 AU. At 1 AU He^+ also contributes substantially to the suprathermal (up to 300 keV)

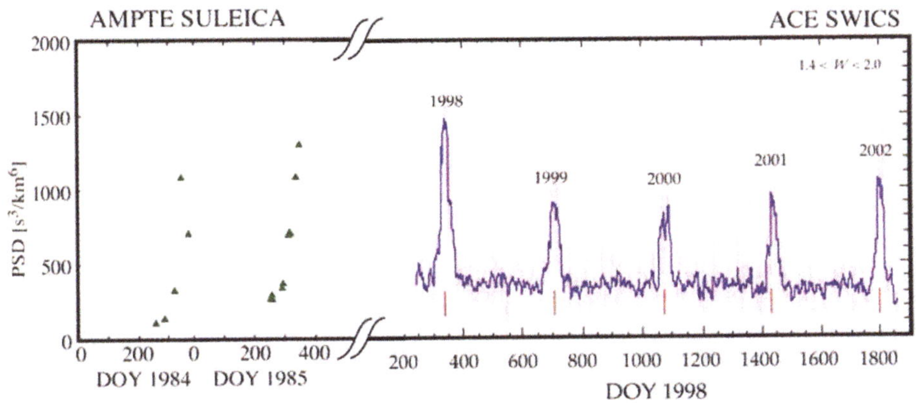

Figure 8.3. Helium focusing cone as seen in pickup ions (adapted from Gloeckler et al., 2004) as observed with AMPTE SULEICA (left) and ACE SWICS (right). The SULEICA data are selected 8–10 hour averages when the satellite was in the solar wind. The SWICS data are shown as 1-day averages (pink) and 10-day sliding averages (blue), indicating the strong intrinsic pickup ion variability.

He population in CIRs (Hilchenbach et al., 1999; Chotoo et al., 2000). Möbius et al. (2002) confirmed this result for more energetic ions (0.25– 0.8 MeV/n) and showed that also interstellar Ne^+ is present in CIRs. Yet, with $He^+/He^{2+} \approx 0.25$, He^+ is substantially less abundant at 1 AU than in CIRs at 4–5 AU with its dominance over He^{2+}. Moreover, Morris et al. (2001) demonstrated that the He^+/He^{2+} ratio increases as the observing spacecraft is magnetically connected to the CIR at increasing distances from the Sun, when the CIR structure sweeps across the observer due to the Sun's rotation. The observed increase of the energetic He^+/He^{2+} ratio with time, and thus with distance to the CIR along the magnetic field line, reflects the increasing importance of interstellar He as source material with increasing distance from the Sun.

These results have led to the suggestion that pickup ions are an important source of ions, which can be accelerated very efficiently at interplanetary shocks (Gloeckler, 1999). In a survey of the abundance of He^+ and He^{2+} (shown in Fig. 8.4) Kucharek et al. (2003) have demonstrated that He^+ is in fact the third most abundant species in the energetic ion population in the inner heliosphere, after H^+ and He^{2+}, with $He^+/He^{2+} \approx 0.06$ and $He^{2+}/H^+ \approx 0.03$ (Hovestadt et al., 1984b). Over and above this average ratio, He^+ is substantially enhanced in CIRs as well as at interplanetary shocks. Bamert et al. (2002) and Kucharek et al. (2003) show that the overwhelming majority of the He^+ accelerated at these shocks cannot stem from cold prominence material, even in associated CMEs, since the substantial enhancement in the energetic population occurs outside the region that contains He^+ in the solar wind. Generally, the He^+/He^{2+} ratio is substantially (by a factor of 50–200) enhanced in the energetic ion population (0.25 – 0.8 MeV/n) over its value in the source population, represented by the ratio of pickup He^+ over solar wind He^{2+}. Based on the observed increased efficiency of pickup ions to be injected into the process for acceleration to higher energies, Gloeckler (1999) has argued that the "inner source" of pickup ions may also contribute substantially to the energetic particle population in CIRs. However, in the same study that revealed interstellar He^+ and Ne^+ in the energetic CIR population, Möbius et al. (2002) did not find any evidence for the singly charged C, O, or Mg, expected from the inner source.

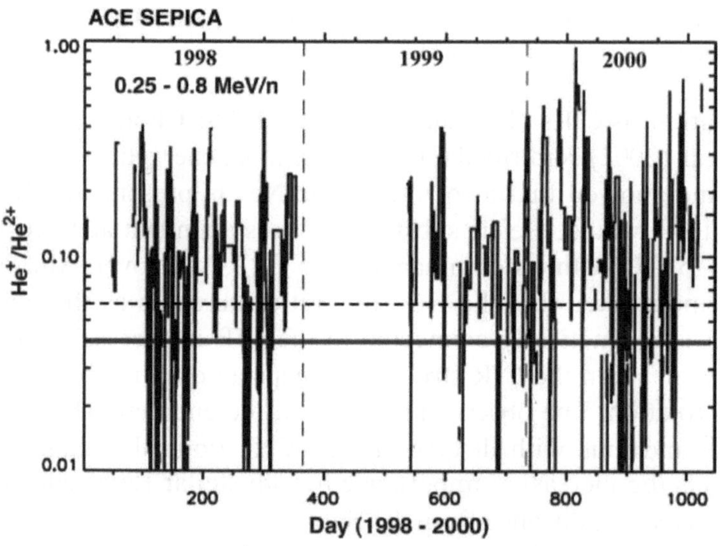

Figure 8.4. Survey of the He^+/He^{2+} ratio in energetic particles for the years 1998 to 2000. An average of 0.06 (indicated by the dashed line) is found with a large variability with values up to 1. The full line indicates the detection limit for He^+ in each individual sampling interval that is determined by a fixed number of He counts.

Because of their advantage in the injection/acceleration process, interstellar pickup ions are a major source of energetic particles in the inner solar system under the current conditions, and their relative importance increases with distance from the Sun. If the density of the neutral gas in the inner heliosphere is boosted by an increased inflow of interstellar gas, both pickup ion density and the abundance of singly-charged energetic ions will increase accordingly. Such an increase is likely to continue approximately linearly with the increase in the neutral gas density. This increase may be reduced or stopped when mass-loading of the solar wind leads to a substantial slowdown and thus a decrease in the free energy that is available for acceleration and/or when the further acceleration of pickup ions consumes a substantial fraction of the total kinetic energy of the solar wind.

In their interaction with the interstellar neutral gas flow, the suprathermal and energetic particle populations in the heliosphere also generate energetic neutral atoms (ENAs). Gruntman et al. (2001) have shown that the most important source of such ENAs are the energetic particles that are accelerated at the termination shock and the heated solar wind distribution in the heliosheath. As all the ENA populations will roughly scale linearly with the generating ISN density it can be expected that the ENAs from the outer heliosphere remain the dominant contributors also under a wide variety of heliospheric conditions.

8.3 Interstellar Neutral Gas and its Secondary Products under Varying Interstellar Conditions

After discussing the interstellar particle populations in the inner heliosphere, as they are observed in our contemporary heliosphere, we want to set up a grid of models with modified ISM conditions for the discussion of all related modifications in the heliosphere. The size and shape of the heliosphere and thus the related inventory of interstellar neutral gas in the heliosphere react to changes of the total H density, the ionization degree, as well as the ISM temperature and relative velocity. In addition, the heliosphere reacts also to the interstellar magnetic field and cosmic ray environment, but these parameters will influence mainly the influx of cosmic rays into the heliosphere. We will concentrate our discussion in this chapter on substantial changes in the interstellar neutral gas inventory, and we will discuss the consequences for the solar wind, the production of secondary products of the interstellar neutrals, and their influence on the energetic particle populations. However, we neglect the potential subtle modifications of the neutral gas flow due to magnetic field related variations in the interaction between the neutrals and plasma at the heliospheric interface.

A systematic variation of all four relevant parameters would lead to a four-dimensional matrix of cases to be modeled, with an unrealistic amount of effort. However, some of the parameters are linked in their variations in realistic interstellar cloud environments, and not all parameter combinations provide conditions that allow neutral gas influx into the heliosphere. Therefore, we have limited our evaluation to a meaningful subset of cases that we have modeled and that we will discuss in the following sections. The parameter combinations for these cases are compiled in Table 8.1 together with a few descriptive remarks concerning the importance of the choice. Of course, the heliosphere under contemporary conditions is used as a baseline (case 1 in Table 8.1). It is immediately obvious that a variation in the ISM density plays a major role, both for the size of the heliosphere and the resulting densities in the inner heliosphere. Because the effects on the Earth's environment under more drastic changes, when the Earth is either in the outer heliosphere or even outside its boundary, are described in contributions by Zank et al. (this volume) and Fahr et al. (this volume), we restrict the discussion in this chapter to a parameter range for which the termination shock stays at 10 AU, i.e., the Earth remains in the inner heliosphere. As we shall see in the next section, the limiting total density of the ISM under this restriction is $\approx 15\ \mathrm{cm}^{-3}$ (featured as case 6). With the reasonable assumption of cosmic abundance for He this brings the maximum He density to $n_{\mathrm{He}} = 1.5\ \mathrm{cm}^{-3}$.

Via the ram pressure balance, the ISM velocity also controls the size of the heliosphere. In connection with the pertinent ionization rate the velocity determines the radial density profile of the neutral gas in the inner heliosphere according to Eq 8.4. As a reasonable range for the velocity of the solar system relative to the ISM, half and twice the current velocity are chosen as bounding values for our discussion. This covers a typical range of peculiar velocities for so-called intermediate velocity gas clouds in the solar neighborhood (Welty et al., 1999). For H the full range of velocities is explored for the current warm ISM (cases 1–3) and only the high-velocity case for a cold ISM cloud with current density (case 4). High ram pressure situations are added for cold and dense clouds with cases 6 and 7. The focusing cone of He is determined by a combination of the ISN velocity and temperature. Therefore, He is modeled with all velocities for the current temperature and for a cold cloud of 10 K (cases 1–6), while the main variations for H are covered with only two velocities at 10 K. For He the density distribution is modeled relative to the density in the ISM, and the He distributions can be scaled linearly to any density value. Therefore, the density values given in Table 8.1 are only relevant for H. It should be noted that dense interstellar clouds are usually cold and that they have most likely a rather low fractional ionization, with no need of individual variation of these parameters. For H we capitalize on the availability of model heliospheres calculated in the chapter by Zank et al. (this volume) instead of modeling all of our cases listed in Table 8.1. As a trade-off, we incur a slight temperature mismatch for cases 1–3, where we use a LISM H temperature of 7000 K instead of 6000 K for cases 1 and 3, and an H temperature of 3000 K for the low-speed case 2. As these cases have a supersonic LISM, the consequences of temperature deviations, both for the morphology and especially for the neutral hydrogen content in the inner heliosphere, are negligible, and the H models 1–3 presented in detail below are good proxies for the cases 1–3 in Table 8.1. Finally, the case of the hot local bubble environment (case 8) has only been included for completeness, as the almost full ionization precludes the penetration of neutral interstellar gas into the heliosphere. Thus no modeling of the He was performed for this case.

Table 8.1. Environmental conditions for the heliospheric models.

Case #	n_H (cm^{-3})	n_H^+/n_H	T (K)	v_{ISN} (km/s)	Species	Comments
1	0.26	0.18	6000	26	H, He	Contemporary heliosphere
2	0.28	0.14	6000	13	H, He	Warm cloud, different speed
3	0.24	0.42	6000	52	H, He	Warm cloud, different speed
4	0.28	0.14	10	52	H, He	Cold cloud, high speed
5	0.28	0.14	10	13	He	Cold cloud, different speed
6	15	0.01	10	26	H, He	Cold and very dense cloud; He density scales
7	1	0.04	10	52	H	Fast, cold, and dense cloud
8	0.005	1	$1.25 \cdot 10^6$	13	H	Hot bubble; no He inflow

8.3.1 Variation of the interstellar H distribution in the inner solar system

An overview of the ISN gas distributions in the heliosphere and their modeling has been given in Section 8.2.1. It is important to note here that the mean-free-path for charge exchange is mostly larger than typical heliospheric distances, so that neutral H is in effect not in equilibrium in and around the heliosphere. This holds for interstellar environments over a range of different densities up to at least ten times the current ISM density, because the size of the heliosphere varies with the density of the surrounding medium. Therefore, models of neutral H throughout the heliosphere tend to be complex and need to account at least for the behavior of each of the different neutral gas components. Following in detail the evolution of all the neutral populations in space, energy, and time, would call for a full kinetic treatment of the neutral component (e.g. Baranov and Malama, 1993; Izmodenov et al., 2001), which is very computer intensive. An alternate, computationally less expensive approach, which captures the spatial distribution of the neutral gas flow throughout the heliosphere with sufficient accuracy for the purpose of this discussion, is to approximate the neutral distribution with multiple fluids. Therefore, the modeling used in this section draws from a three-fluid approach by Zank et al. (1996). In this model the neutral hydrogen of the ISM (either flowing in from afar or born outside the heliopause) is labeled component 1 (fluid one), neutral H born through charge exchange in the hot heliosheath is labeled component 2, and neutral H products created by charge exchange in the supersonic solar wind (i.e., upstream of the termination shock) are labeled component 3. Component 2 neutrals have a very hot distribution with a mean velocity of ~100 km/s as they are characterized by the underlying million degree heliosheath plasma. Component 3 neutrals are cold and fast, and represent the NSW that is generated in the interaction with the incoming interstellar neutrals. The NSW contribution created by dust interactions is not included in this model, as it does not contain any dust component. However, the dust related NSW is not important in most of our model cases, except for the hot Local Bubble (case 8), when no interstellar neutrals reach the inner heliosphere.

The balance between solar wind ram pressure and interstellar pressure determines the size and shape of the heliosphere. The termination shock distance can be calculated according to:

$$r_{TS} = C\sqrt{P_{1AU} / P_{ISM}} \tag{8.18}$$

where $P_{1AU} \approx \rho_{1AU} v_{SW}^2$ is the solar wind ram pressure at 1 AU. Similarly the total ISM pressure is defined as $P_{ISM} \approx m_H n_H v_{ISM}^2 + P_{therm}$, where P_{therm} is the thermal pressure in the surrounding ISM. The constant C in Eq. 8.18 is of order 1 AU following Zank et al. (this volume). Assuming an unchanging solar wind pressure, it can be expected from pressure balance that the distances to the termination shock and heliopause, respectively, scale inversely with the interstellar velocity, and inversely with the square root of the interstellar density.

For cases 1–7 in Table 8.1 the pressure balance is dominated by interstellar ram pressure, whereas for case 8 the ISM is subsonic and thermal pressure takes over. Case 8 represents the era (thought to be several million years ago) when the Sun was in the interior of the so-called Local Bubble, a hot, tenuous ISM (Frisch and Slavin, this volume). This scenario provides for a relatively uninteresting picture. A total density $n_\infty = 5 \cdot 10^{-3}$ cm^{-3} and a temperature of $T_\infty = 1.25 \cdot 10^6$ K surprisingly lead to a heliosphere of similar size as today. In spite of the much-reduced density the total outside pressure remains comparable to current conditions because of the high temperature. In this case the speed of the Sun relative to the local environment is not even important, because the interaction is strongly subsonic. Therefore, the heliosphere will be axisymmetric along the interstellar magnetic field, unlike today (Parker, 1963).

Because of the high temperature the interstellar medium will be almost fully ionized, and no interstellar neutral gas will enter the solar system. This removes the possibility of creating any significant amount of secondary particle populations inside the heliosphere, such as neutral solar wind, pickup ions, and their more energetic brethren. ACRs will also be almost absent. The generation of any secondary particles will be restricted to the interaction with dust, of interplanetary and interstellar origin, inside the heliosphere. Because the interstellar dust density is typically very low with gas to dust mass ratios of 100 to 170 (Spitzer, 1978) and the total gas density of the ISM is extremely low in case 8, there will be almost no interstellar dust in the heliosphere. Therefore, the amount of dust-related interactions will be controlled completely by the interplanetary dust component and will probably not be much different from what is observed under current conditions for the neutral solar wind (Collier et al., 2001, 2003) and inner-source pickup ions (Geiss et al., 1996; Gloeckler et al., 2000). Under such circumstances there would certainly be no possibility to deduce the surrounding interstellar conditions from in situ observations inside the heliosphere, as is possible today. Astronomers would have to resort solely to remote sensing techniques to obtain just average line-of-sight information about the surrounding medium.

Starting with the contemporary heliosphere (case 1) in the upper left, Figure 8.5 shows four different models, representing a change in interstellar pressure by a factor of 16. It can be seen that both density increases and velocity increases lead to smaller heliospheres. Shown is the plasma temperature in color-coding, which illustrates the location of critical boundaries, such as the termination shock, and significant neutral gas – plasma interaction through increased values. Case 6 represents the smallest heliosphere modeled here, with the upwind termination shock only 8 AU from the Sun and the heliopause at 12 AU. Note that this means that Saturn dips periodically into the hot heliosheath, crossing the termination shock, and Jupiter's orbit goes through regions of heated supersonic solar wind close to the upstream region of the termination shock.

Generally speaking, the morphology of the heliosphere is preserved even in these extreme cases. Going to higher interstellar speeds does however increase the termination shock upwind/downwind asymmetry. Also, the presence of more neutral ISM H inside the termination shock leads to more pickup ion heating, as is evident from the increased plasma temperatures in Figure 8.5.

The amount of interstellar H that penetrates into the heliosphere inside the termination shock is not only a function of the interstellar neutral density at infinity, but also depends on the interstellar velocity. This is evident in the density profiles along the stagnation axis as displayed in the left panel of Figure 8.6, where the red profile corresponds to the contemporary heliosphere. The dark blue curve represents the high-density case 6, the green ones are high-speed cases, and the cyan profile is a low-speed case. Note that the modeled cases can have component 1 neutral densities that are two orders of magnitude higher at 1 AU than in the contemporary configuration. This higher neutral H density naturally results in a higher charge-exchange rate, and the by-products are component 3 neutrals whose densities are shown in the right panel of Figure 8.6. Again, at 1 AU, the absolute NSW densities of case 6 are two orders of magnitude higher than in the contemporary heliosphere. Even in this cold high-density case the neutral H density at 1 AU does not exceed 0.3 cm^{-3} and thus is at least one order of magnitude below the average solar wind density. As we will see below, all secondary products can still be considered as test particles, and they do not contribute significantly to the solar wind dynamics. Even on the upwind side and crosswind, the density of interstellar He is higher by a factor of 3–5 than that of H.

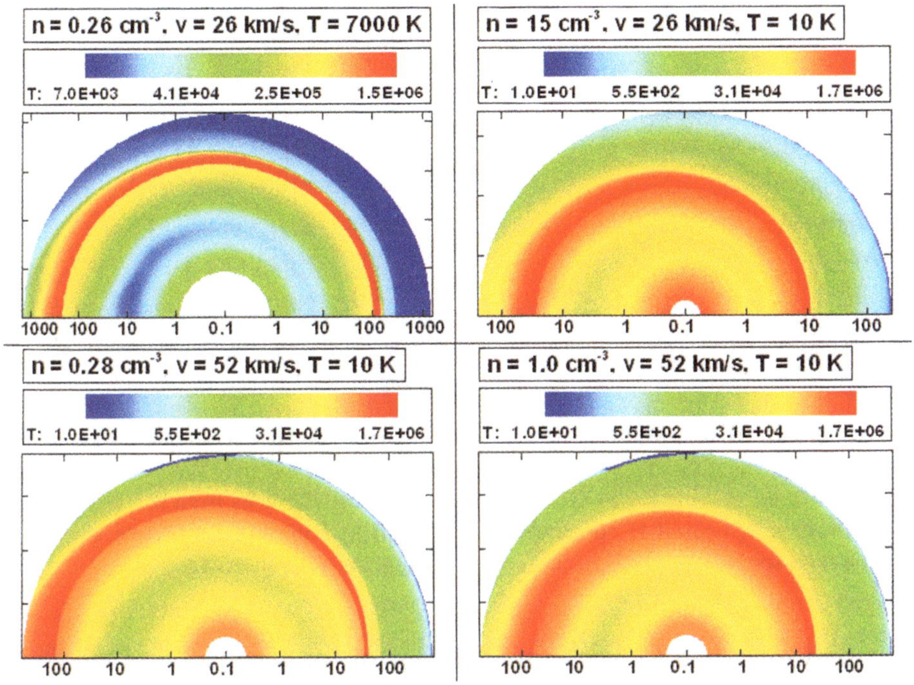

Figure 8.5. Two-dimensional maps of plasma temperature for four different models, counterclockwise from upper left: case 1, case 4, case 7, and case 6. The distance scale is logarithmic in AU, with stagnation axis angle preserved. In each map, the ISM comes in from the right; the first temperature rise is the bow shock, the thin red feature is the heliosheath/heliotail between heliopause and termination shock. The supersonic solar wind is heated by the pickup ions.

8.3.2 Variation of the interstellar He distribution in the inner solar system

Under contemporary conditions He is the dominant interstellar species in the inner solar system, inside approximately 3 AU, because of its rather high ionization potential. It is expected that He will remain the dominant species for all situations with a substantial neutral gas flow through the heliosphere. The density of interstellar helium in the inner heliosphere scales linearly over a wide range with the density of the inflowing gas. Therefore, it will suffice to model the conditions for a single neutral gas density. For total gas densities below $\approx 100\ \mathrm{cm}^{-3}$ the gas can be considered collisionless, which allows for a simple scheme with

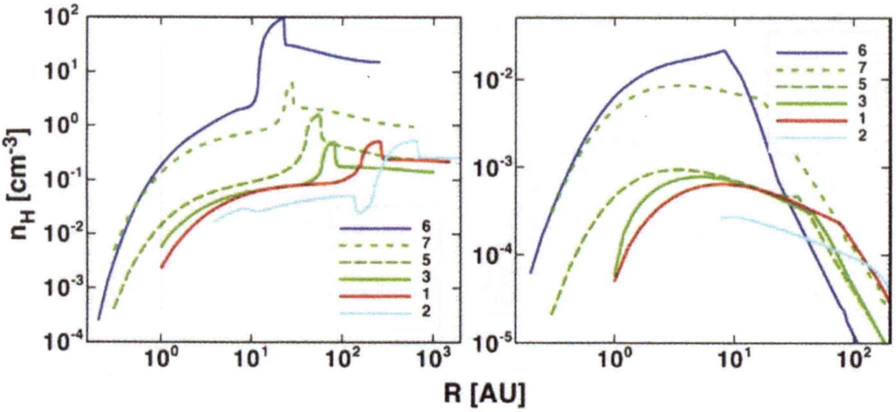

Figure 8.6. Left Panel: Profiles of neutral H density (interstellar + hydrogen wall component), in cm^{-3}, along the upwind stagnation axis, in AU for cases 1–3 and 5–7. The hydrogen walls are evident, as is photo-ionization for small distances from the Sun. Right Panel: Profiles of neutral H component 3 density (i.e., born through charge exchange with the supersonic solar wind), in cm^{-3}, along the upwind stagnation axis.

the atoms following Keplerian trajectories for the calculation of the density and higher moments of its distribution function. We have restricted ourselves to densities even below 15 cm^{-3}, because for densities exceeding this number the Earth will not be in the inner heliosphere, since the termination shock will move inside 10 AU (Fahr et al., Zank et al., this volume).

Because the interplay between the He flow, the Sun's gravitation, and ionization leads to a characteristic focusing cone structure in the inner heliosphere, which depends on the flow velocity and temperature of the gas as well as on the ionization rate, a range of velocities and temperatures need to be explored. We restrict ourselves to typical ionization rates for the contemporary heliosphere (see Rucinski et al., 1996; McMullin et al., 2004) since the emphasis in this book is on the response of the heliosphere to external conditions and not to solar activity. There is no point in calculating a model for a hot interstellar medium with T_∞ = 10^6 K because the medium is almost completely ionized and no noticeable neutral gas flow will reach the inner heliosphere. Calculation of "warm" heliospheres (T_∞ = 6000 K) with different velocities does not differ from the contemporary case since the gas has a thermal character (its thermal velocity is comparable to the bulk speed). For these cases a hot model (Wu and Judge 1979, Fahr 1979) was used.

The calculation of "cold" heliospheres (T_∞ = 10 K, cases 4–6) can in principle be performed with the use of the same model, even the same numerical code. In practice, however, the hot model approach becomes computationally costly because of the extremely narrow peaks of the distribution functions related to the very small thermal spread. Fortuitously, for such a low temperature the analytical cold gas approximation (Axford 1972) can be used except close to the cone center, where the thermal spread of the gas, however small it is, must be taken into account to avoid singularities in the computational scheme around the downwind axis in the cold model.

We are using a stationary model for the He distribution here, although it is known that the temporal variation of the ionization rate influences the spatial He distribution inside the heliosphere. However, this simplified approach is justified because for the discussions in this chapter we are interested in the average and extreme values of the neutral gas densities in the inner heliosphere, and not so much their temporal evolution or their detailed spatial pattern. To illustrate the differences between the contemporary warm interstellar gas environment and a typical cold gas cloud, Figure 8.7 shows a 2-dimensional cut through the He density distribution that includes the cone for a warm and a cold interstellar environment. The color-coding has been adjusted to the corresponding maximum density in the cone for each case and does not have any bearing on absolute densities. Compared is the contemporary heliosphere for interstellar gas with a flow velocity of 26 km/s and a temperature of 6000 K with a fast and cold interstellar gas flow of v_∞ = 52 km/s and T_∞ = 10 K. It is obvious that the fast and cold flow produces an extremely narrow focusing cone that is concentrated along the downwind axis. The inset shows the same density distribution, stretched along the y-axis for better resolution of the cone structure. The full density enhancement is concentrated at much less than 0.1 AU around the cone axis. Otherwise the He distribution is very evenly distributed.

To be more quantitative, Fig. 8.8 shows longitudinal density profiles of the neutral interstellar He density at 1 AU normalized to the He density at infinity for the current conditions (case 1), along with cases 2 and 3 in the upper panel, and for a cold interstellar cloud at T_∞ = 10 K with three different relative velocities (cases 4 – 6) in the lower panel. For the cold cloud conditions with an average ionization rate for He of $\beta_{He} = 10^{-7}$ s^{-1} at 1 AU the density enhancement in the peak of the cone exceeds a

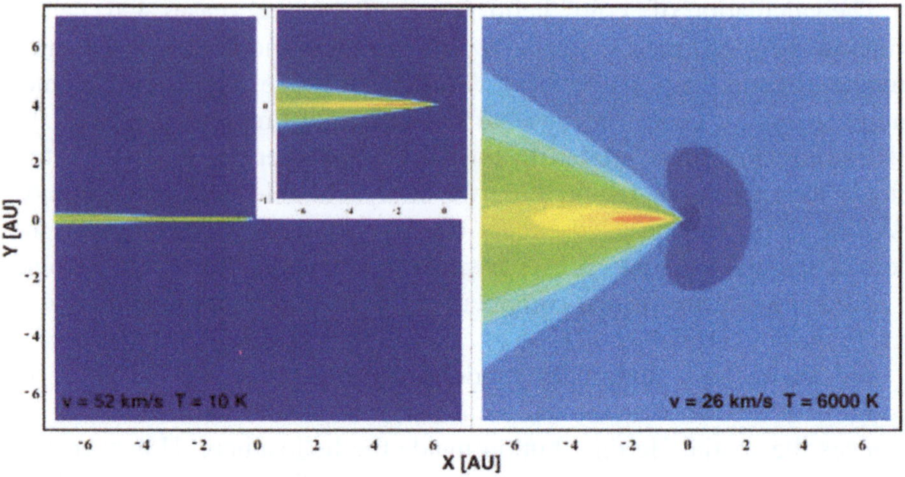

Figure 8.7. Two-dimensional cuts of the neutral interstellar He density distribution in the inner heliosphere in a color-coding based on logarithmic scaling. The colors are adjusted to the maximum density in the cone and do not reflect any absolute calibration. Compared are a contemporary heliosphere (on the right) and a fast cold interstellar gas (on the left). Because of the extremely narrow cone the inset shows a version that is stretched along the y-axis.

factor of 80. Taking the low ionization rate of $\beta_{He} = 6 \cdot 10^{-8}$ s^{-1} at solar minimum, or the high rate of $\beta_{He} = 2 \cdot 10^{-7}$ s^{-1} at solar maximum, the cone is higher by a factor of ≈ 1.5 or lower by a factor of ≈ 3, respectively. In addition, the cones are very concentrated. In order to make the central cone visible for the cold clouds, the angle range is stretched by a factor of 6 in the inset. For the fastest velocity chosen (52 km/s), the cone is extremely narrow, but for 13 km/s the half width of the cone at 1 AU reaches 5°, which is equivalent to almost 0.1 AU, still a substantial width. It should be noted though that even for the highest density in the cone, the mean free path for collisions is ≈ 0.25 AU and thus noticeably larger than the cone width. Therefore, the collisionless approximation used in the simulations is still valid. For both 13 and 26 km/s the density enhancement is equal or stronger over the entire angle range as compared with contemporary conditions. Taking into account the increased overall density in the cold dense cloud environment, i.e., $n_{He\infty} = 1.5$ cm^{-3}, the He density will be substantially higher at all locations.

It should be noted that the present interstellar flow vector is inclined about 6° relative to the ecliptic plane (Witte, 2004; Möbius et al., 2004). Therefore, the bulk of such a narrow cone would mostly miss the Earth's

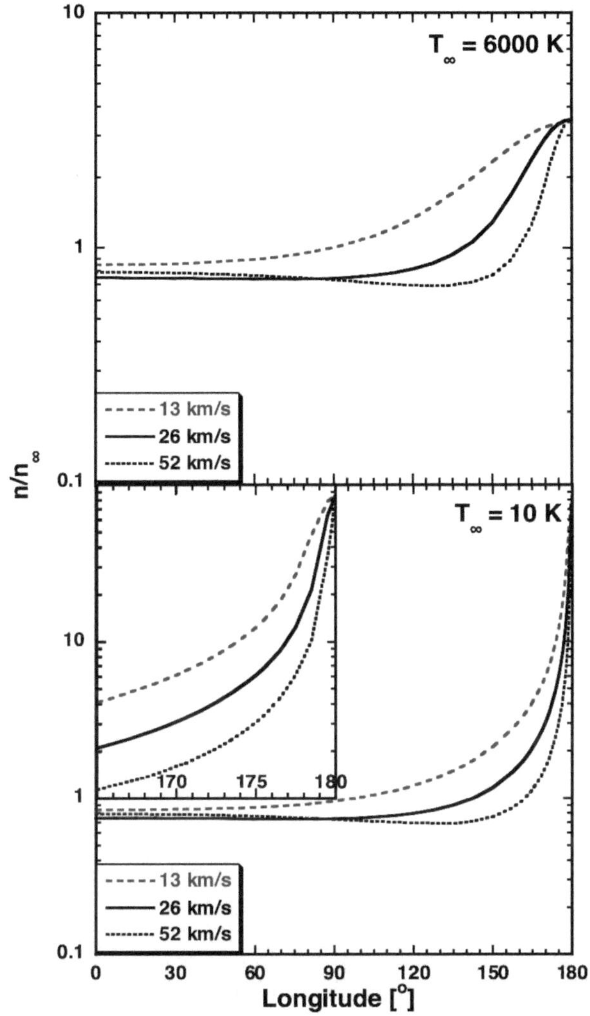

Figure 8.8. Longitudinal density profile of interstellar He at 1 AU exactly across the gravitational focusing cone for ISN gas with $T_\infty = 6000$ K and 13, 26, and 52 km/s flow, including the contemporary heliosphere (upper panel). Same set of velocities for the cold gas with $T_\infty = 10$ K (lower panel). Inset: tailward detail of the cold cones (Möbius et al., 2005).

orbit for the current flow direction, and for the majority of interstellar clouds the cone may not directly influence the Earth's environment. If the orientation of the velocity vector were completely random, there is still an $\approx 5\%$ probability for the Earth's orbit cutting through the half-width of the cone for cold conditions with a 13 km/s relative velocity.

8.3.3 Secondary Particles in the Inner Heliosphere

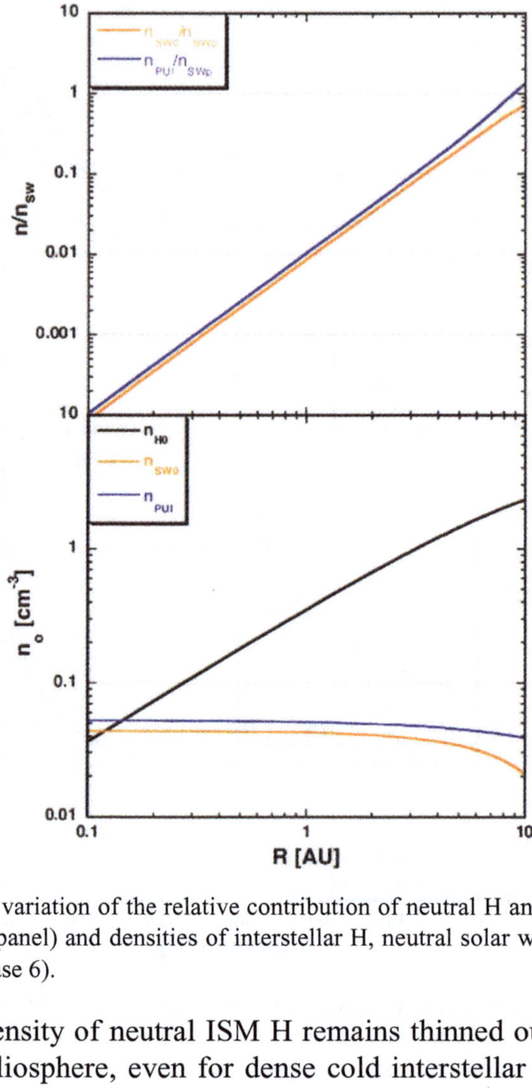

Figure 8.9. Radial variation of the relative contribution of neutral H and pickup H^+ to the solar wind (upper panel) and densities of interstellar H, neutral solar wind H, and pickup H^+ (lower panel, case 6).

Since the density of neutral ISM H remains thinned out substantially in the inner heliosphere, even for dense cold interstellar clouds, the related secondary products, neutral solar wind H, and H^+ pickup ions, constitute a tracer population. Figure 8.9 shows their radial distribution along the upwind axis for the cold and dense ISM (case 6). Since the neutral gas is more depleted in any other direction, it is sufficient for the purpose of this chapter only to consider the upwind direction.

As can be seen from Fig. 8.9, the absolute density of neutral solar wind H, n_{SWO}, and pickup H^+, n_{PUI}, remain below 0.1 cm^{-3} at all distances. Their contribution to the solar wind only becomes substantial outside 1 AU and is comparable near the termination shock at about 10 AU. At 1 AU both species make up 1% of the solar wind. Although this is substantially more than under current conditions, the solar wind environment will not be changed significantly at Earth's orbit for a cold dense cloud of 15 cm^{-3} H density.

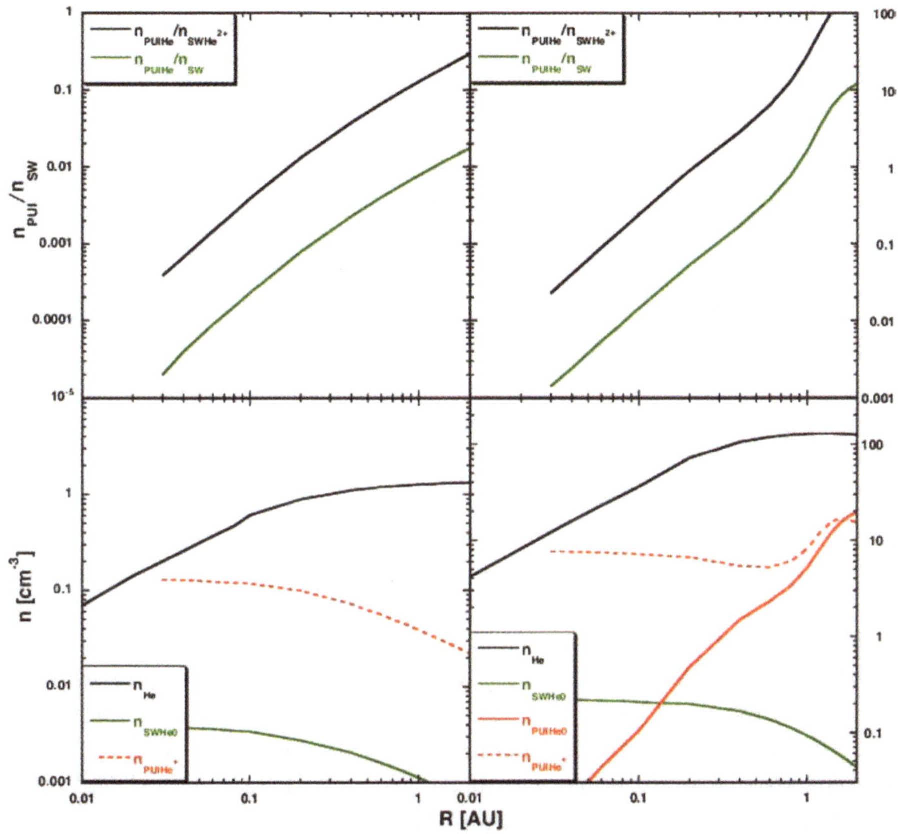

Figure 8.10. Densities of neutral interstellar He (n_{He}), neutral solar wind He (n_{SWHe0}), He^+ pickup ions (n_{PUIHe^+}), and pickup ions turned neutrals (n_{PUIHe0}) (lower panels) as well as the density ratio of He^+ pickup ions versus solar wind He^{2+} and H^+ (upper panels) as a function of distance from the Sun (case 5). Upwind is shown on the left and downwind on the right.

As discussed above, neutral He is the dominant interstellar species in the inner heliosphere under all interstellar conditions, and it features the well-known gravitational focusing cone with enhanced density on the downwind side of the interstellar flow. The dominance of this spatial structure is reflected in the NSW and the pickup ions. Figure 8.10 shows the radial variation of the neutral interstellar He, NSW He, and pickup ion He^+ density in its lower panels. Shown are both radial profiles, along the upwind (left panel) and downwind (right panel) axis. On the downwind side the density of neutrals is also included that stem from charge exchange of pickup ions with the ISN gas. The upper panels contain the fraction of pickup He^+ relative to solar wind protons and alphas. The 13 km/s ISM flow (case 5) has been chosen because it produces the widest cone structure. To compute the secondary products charge exchange cross sections for a 400 km/s solar wind and an average ionization rate of 10^{-7} s^{-1} for He were used. The secondary particles are accumulated starting at 0.03 AU, leaving out the innermost portion of the region where the solar wind is still accelerating and where the neutral densities are negligible. As far as secondary products are concerned, He^+ pickup ions are most important under all circumstances. They exceed the neutral He in the solar wind, typically by a factor of 30, because photo-ionization is the prevalent production process for He^+ ions (Rucinski et al., 1996; McMullin et al., 2004). NSW hydrogen, produced by the interaction with interstellar He, is even rarer because of the extremely low charge exchange cross section between protons and He, which is lower by about two orders of magnitude compared with double charge exchange between He^{2+} and He. Therefore, NSW hydrogen produced by interstellar He is not even shown here.

Surprisingly, a tertiary neutral particle product becomes very important on the downwind axis. Through charge exchange with the interstellar He neutrals that are strongly enhanced in the cone, He^+ pickup ions, which are also strongly enhanced there, are turned into neutral He with the velocity distribution of the pickup ions. Because the enhancement in the cone translates quadratically into the density of this product, its density is enhanced by about four orders of magnitude compared with the upwind direction. To compute the radial profile of these pickup-ion-generated neutrals, we accumulated their production along the downwind axis, since they also follow the solar wind on average. However, these neutrals will also disperse from the cone region according to their pickup ion inherited velocity distribution. This effect is modeled through a free expansion of the cone structure of these neutrals in two dimensions, using an equivalent thermal speed of the generating pickup ion distribution according to Burlaga et al. (1996). For simplicity, the density profile across the cone is assumed to be Gaussian.

With regard to neutral particles that the Earth's atmosphere may be exposed to, the neutral interstellar He is the dominant constituent with its density outside the heliosphere of $\approx 1.5\ cm^{-3}$ for cases 4 – 6 chosen here. In the focusing cone on the downwind side of the ISN flow the He density even exceeds 100 cm^{-3}. In terms of energy flux delivered to the upper atmosphere, the neutral solar wind is more important because of its much higher kinetic energy. On the upwind side both H and He contribute, with He containing more energy because of its higher mass. The total energy flux in the neutrals reaches almost 5% of that of the solar wind ions. On the downwind axis the dominant fraction of the energy flux is in the neutrals that are generated by charge exchange of He^+ pickup ions with interstellar He neutrals. It delivers approximately the full solar wind energy flux directly into the upper atmosphere.

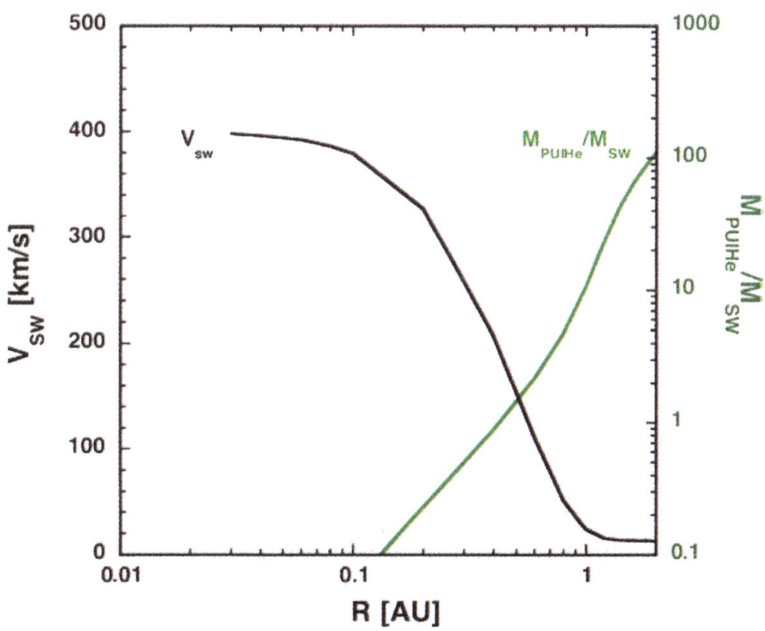

Figure 8.11. Radial evolution of the effective solar wind speed and the mass-loading by He^+ pickup ions on the downwind cone axis for case 5, assuming a homogeneous interaction between neutral He and the solar wind (Möbius et al., 2005).

On the upwind side the abundance of pickup He^+ reaches about 10% of the solar wind alpha particles and $\approx 1\%$ of the solar wind protons at 1 AU, still not important for the solar wind dynamics. However, on the cone axis pickup He^+ reaches the average solar wind density at 1 AU, with obvious impact on the solar wind dynamics. As discussed already above, the density of a tertiary product, pickup ions that turned neutral, becomes comparable with that of pickup ions here. These neutrals also carry away momentum from the solar wind, because they were originally accelerated on average to the solar wind speed and thus contributed to its mass-loading. Since both the interstellar neutral He and pickup ion densities are two orders of magnitude lower outside the cone, the neutrals generated from the pickup ions are negligible everywhere else. To demonstrate the order of magnitude of the solar wind slowdown due to the interaction with interstellar He in the density enhancement of the cone, Fig. 8.11 shows the radial evolution of the mass-loading ratio ($M_{PUI,He^+}/M_{SW}$) and the effective solar wind speed (v_{SW}), assuming a homogenous interstellar density. The mass-loading ratio has reached almost 10, and the solar wind has slowed down dramatically at 1 AU, as discussed by Möbius et al. (2005). At a distance of $\approx$3 AU from the Sun, the solar wind blends smoothly into the interstellar gas flow by approaching its speed of 13 km/s. To compute this estimate the loss of momentum to neutral solar wind and neutrals, which originate from He^+ pickup ions, was included, since they completely decouple dynamically from the solar wind flow. This momentum loss becomes important already at $\approx$ 0.3 AU. It should be noted though that the overall solar wind slowdown is overestimated, because the density enhancement in the cone is limited to a region that ranges from less than 1° to approximately 5° in both directions, depending on the interstellar gas flow speed. Through magnetic tension, the solar wind in the cone is still dynamically coupled to the region outside the cone, where the mass-loading is substantially weaker. As long as the solar wind speed outside the cone still exceeds the Alfvén velocity, the coupling to the surrounding medium is weak, because the information of the slowdown can only travel at the Alfvén velocity. Because this condition is mostly fulfilled at 1 AU outside the cone, the estimate given here for the solar wind slowdown in the cone is roughly representative of what can be expected under the conditions chosen here. For a more detailed evaluation, the cone needs to be treated like a comet in the solar wind (e.g. Schmidt and Wegmann, 1980; Schmidt et al., 1993). In a nutshell, the gravitational focusing cone in a dense cold interstellar gas cloud scenario will behave like a giant stationary cometary coma in the inner solar system (Möbius et al., 2005).

8.3.4 Energetic Particles of Interstellar Origin in the Inner Heliosphere

As has been discussed for the contemporary conditions in Section 8.2.5, pickup ions constitute a very important source population for energetic particles in the heliosphere. They are injected into further acceleration with an efficiency that exceeds by more than a factor of 100 that of solar wind ions. On average the abundance of interstellar He^+ is already 6% of the total energetic He population at 1 AU under current conditions (Kucharek et al., 2003). Even for the dense interstellar cloud conditions considered here (cases 4 – 6), He^+ pickup ions constitute only 1% of the total solar wind ion density, and they can still be treated mostly as test particles in the solar wind. Therefore, it is justified to estimate the expected energetic particle density that arises from acceleration of these pickup ions through linear scaling based on the increased pickup ion density. For an interstellar cloud with a He density of 1.5 cm^{-3} the average He^+/He^{2+}ratio in the energetic particles should grow to $n_{He^+}/n_{He^{2+}} \approx 6$, making He^+ dominant over He^{2+} in the energetic particle population.

Due to its enhancement in the focusing cone, the abundance of He will probably exceed that of H and become the dominant species of interplanetary accelerated particles on the downwind side of the ISM flow. Because pickup ions and, in particular, energetic particles are spatially much more widely distributed than their source, the energetic particle enhancement will spread well beyond the cone and become noticeable in Earth's orbit, even if the interstellar flow vector is not well aligned with the ecliptic plane.

During solar minimum, co-rotating interaction regions are the dominant generators of energetic particles and they accelerate interstellar pickup ions very efficiently (Gloeckler et al., 1996; Möbius et al., 2002). In CIRs $He^+/He^{2+} \approx 0.2$–0.25 on average at 1 AU, and about 20–25 times more He^+ than He^{2+} is expected under cold dense cloud conditions. Therefore, the flux of He^+ would exceed that of H^+ in CIRs even outside the focusing cone. If one includes the compounding effects of the focusing cone, pickup ions will play the dominant role as a source of energetic particles during solar minimum conditions. Assuming that the solar wind and solar activity conditions remain similar to what is observed in the contemporary heliosphere, the abundance of interstellar pickup ions will boost the total flux of energetic particles in general, at least under solar minimum conditions. Since also the intensity of ACRs is substantially increased throughout the entire heliosphere (Fahr et al., this volume) and modulation of galactic cosmic rays will be significantly reduced because of the smaller heliosphere dimensions (Florinski and Zank, this volume), the total energetic particle fluxes around the Earth will be substantially enhanced over the entire energy range. Therefore, the change in the energetic particle spectra and fluxes around Earth is probably one of the most important effects of a moderately enhanced ISM density on the Earth's environment.

Acknowledgments: The authors would like to thank the International Space Science Institute (ISSI) for hosting a coordination meeting for the work towards the manuscript. E. M. would like to thank the staff at ISSI and at the Physics Institute of the University of Bern during part of the preparation of this chapter and thankfully acknowledges support by the Hans-Sigrist Stiftung. The authors thank M.A. Lee and the reviewer for carefully reading the manuscript and helpful suggestions. The work by E. M. was supported under NASA grants NAG5-12929 and NAG5-10890, H.-R. M. was supported under NASA grants NAG5-12879 and NAG5-13611, and M. B. under the Polish SCSR grant 1P03D00927.

REFERENCES

Adams, T.F., and Frisch, P.C. (1977). High-resolution observations of the Lyman-α sky background. *Astrophys. J.*, 212:300–308.

Ajello, J.M., Pryor, W.R., Barth, C.A., Hord, C.W., Stewart, A.I.F., Simmons, K.E., and Hall, D.T. (1994). Observations of interplanetary Lyman-alpha with the Galileo Ultraviolet Spectrometer: Multiple scattering effects at solar maximum. *Astron. Astrophys.*, 289:283–303.

Akasofu, S.-I., (1964a). A source of the energy for geomagnetic storms and auroras. *Planet. Space Sci.*, 12:801–833.

Akasofu, S.-I, (1964b). The neutral hydrogen flux in the solar plasma flow. *Planet. Space Sci.*, 12: 905–913.

Axford, W.I., (1972). The interaction of the solar wind with the interstellar medium, In Sonnett, E.P., Coleman, P. J., Wilcox, J. M. Editors, *Solar Wind*, NASA SP-308, pages 609–660.

Bamert, K., Wimmer-Schweingruber, R.F., Kallenbach, R., Hilchenbach, M., Klecker, B., Bogdanov, A., and Wurz, P., (2002). Origin of the May 1998 suprathermal particles: Solar and heliospheric observatory/Charge, Element, and Isotope Analysis System/(Highly) Suprathermal Time of Flight results, *J. Geophys. Res.,* 107:1130, 10.1029/2001JA900173.

Banks, P.M., (1971). Interplanetary hydrogen and helium from cosmic dust and the solar wind. *J. Geophys. Res.,* 76:4341–4348.

Baranov, V.B., and Malama, Yu.G., (1993). Model of the solar wind interaction with the Local Interstellar Medium: numerical solution of self-consistent problem. *J. Geophys. Res.,* 98:15157–15163.

Barnett, C.F., Hunter, H.T., Kirkpatrick, M.I., Alvarez, I., Cisneros, C., and Phaneuf, R.A., (1990). Collisions of H, H_2, He and Li atoms and ions with atoms and molecules, In: *Atomic data for fusion*. Oak Ridge National Laboratories, ORNL-6086/V, Oak Ridge, Tenn.

Bertaux, J.-L., and Blamont, J.E., (1971). Evidence for a source of an extraterrestrial hydrogen Lyman-alpha emission: the interstellar wind. *Astron. Astrophys.,* 11:200–217.

Bertaux, J. L., Lallement, R., Kurt, V. G., and Mironova, E. N., (1985). Characteristics of the local interstellar hydrogen determined from PROGNOZ 5 and 6 interplanetary Lyman-alpha line profile measurements with a hydrogen absorption cell. *Astron. Astrophys.*, 150:1–20.

Breus, T.K., Bauer, S.J., Krymskii, A.M., and Mitnitskii, V.Ya., (1989). Mass loading in the solar wind interaction with Venus and Mars. *J. Geophys.*, Res., 94:2375–2382.

Burlaga, L.F., Ness, N.F., Belcher, J.W., and Whang, Y.C., (1996). Pickup protons and pressure-balanced structures from 39 to 43 AU: Voyager 2 observations during 1993 and 1994, *J. Geophys. Res.*, 101:15523–15254.

Bzowski, M., Fahr, H.J., and Rucinski, D. (1996). Interplanetary neutral particle fluxes influencing the Earth's atmosphere and the terrestrial environment, *Icarus* 124:209–219.

Bzowski, M., (2001). Time-dependent radiation pressure and time dependent, 3D ionisation rate for heliospheric modelling, In Scherer, K., Fichtner, H., Fahr, H.J., Marsch E., Editors, *The Outer Heliosphere: The Next Frontier*, COSPAR Coll. Ser., 11:69–72, Elsevier, Pergamon.

Chassefière, E., Bertaux, J.-L., Lallement, R., and Kurt, V.G. (1986). Atomic hydrogen and helium densities of the interstellar medium measured in the vicinity of the sun. *Astron. Astrophys.*, 160:229–242.

Chotoo, K., Schwadron, N.A., Mason, G.M., Zurbuchen, T.H., Gloeckler, G., Posner, A., Fisk, L.A., Galvin, A.B.; Hamilton, D.C. and Collier, M.R., (2000). The suprathermal seed population for corotating interaction region ions at 1 AU deduced from composition and spectra of H^+, He^{++}, and He^+ observed on Wind. *J. Geophys. Res.,* 105:23107–32122.

Clarke, J.T., Lallement, R., Bertaux, J.L., Fahr, H.-J., Quémerais, E., and Scherer, H. (1998). HST/GHRS observations of the velocity structure of interplanetary hydrogen. *Astrophys. J.,* 499:482–488.

Collier, M.R., Moore, T.E., Ogilvie, K.W., Chornay, J.D., Keller, J.W., Boardsen, S., Burch, J.L., El Marji, B., Fok, M.-C., Fuselier, S.A., Ghielmetti, A.G., Giles, B.L., Hamilton, D.C., Peko, B.L., Quinn, J.M., Stephen, T.M., Wilson, G.R., and Wurz, P., (2001). Observations of neutral Atoms from the Solar Wind. *J. Geophys. Res.,* 106:24893–24906.

Collier, M.R., Moore T.E., Ogilvie, K., Chornay, D.J., Keller, J., Fuselier, S., Quinn, J., Wurz, P., Wuest, M., and Hsieh, K.C., (2003). Dust in the wind: The dust geometric cross section at 1 AU based on neutral solar wind observations, Solar Wind X. *American Institute Physics.,* 679:790–793.

Collier, M.R., Moore, T.E., Simpson, D., Roberts, A., Szabo, A., Fuselier, S., Wurz, P., Lee, M.A., and Tsurutani, B., (2004). An unexplained 10°– 40° shift in the location of some diverse neutral atom data at 1 AU. *Adv. Space Res.,* 34:166–171.

Costa, J., Lallement, R., Quémerais, E., Bertaux, J.-L., Kyrölä, E., and Schmidt, W., (1999). Heliospheric interstellar H temperature from SOHO/SWAN H cell data. *Astron. Astrophys.,* 349:660–672.

Fahr, H.-J., (1968a) On the Influence of Neutral Interstellar Matter on the Upper Atmosphere, *Astrophys. Space Sci.*, 2:474–495.

Fahr, H.J., (1968b). Neutral corpuscular energy flux by charge-transfer collisions in the vicinity of the sun. *Astrophys. Space Sci.*, 2:496–503.

Fahr, H.J., (1974). The extraterrestrial UV-background and the nearby interstellar medium. *Space Sci. Rev.,* 15:483–540.

Fahr, H.J., Lay, G. and Wulf-Mathies, C. (1978). Derivation of interstellar helium gas parameters

Fahr, H.J., (1979). Interstellar hydrogen subject to a net repulsive solar force field *Astron. Astrophys.,* 77:101–109.

Fahr, H.J., Ripken, H.W., and Lay, G., (1981). Plasma-Dust Interactions in the Solar Vicinity and their Observational Consequences. *Astron. Astrophys*., 102:359–370.

Fahr, H.J. (1991). Local interstellar oxygen in the heliosphere: its analytic representation and observational consequences, *Astron. Astrophys.*, 241:251–259.

Fahr, H.J., and Rucinski, D. (1999). Neutral interstellar gas atoms reducing the solar wind Mach number and fractionally neutralizing the solar wind, *Astron. Astrophys.* 350:1071–1078.

Fahr, H.J., and Rucinski, D. (2001). Modification of properties and dynamics of distant solar wind due to its interaction with neutral interstellar gas, *Space Sci. Rev.*, 97:407–412.

Fahr, H.-J. (2004). Global structure of the heliosphere and interaction with the local interstellar medium: three decades of growing knowledge, Adv. Space Sci., 34:3–13.

Fahr, H.-J., Fichtner, H., Scherer, K., and Stawicki, O. (2005). Variable terrestrial particle environments during the galactic orbit of the sun, In: P. C. Frisch Editor, *Solar Journey: The Significance of Our Galactic Environment for the Heliosphere and Earth.* This Volume, Springer, Dordrecht.

Florinski, V., and Zank, G. P. (2005). Variations in the galactic cosmic ray intensity in the heliosphere in response to variable interstellar environments, In: P. C. Frisch Editor, *Solar Journey: The Significance of Our Galactic Environment for the Heliosphere and Earth*. This Volume, Springer, Dordrecht.

Flynn, B., Vallerga, J., Dalaudier, F., and Gladstone, G.R. (1998). EUVE measurements of the local interstellar wind and geocorona via resonance scattering of solar He I 584-Å line emission, *J. Geophys. Res*., 103:6483–6494.

Frisch, P.C., Dorschner, J.M., Greenberg, J.M., Grün, E., Landgraf, M., Hoppe, P., Jones, A.P., Krätschmer, W., Linde, T.J., Morfill, G.E., Reach, W., Slavin, J.D., Svestka, J., Witt, A.N., Zank, G.P., (1999). Dust in the Local Interstellar Wind, *Astrophys. J.*, 525:492–516.

Frisch, P.C., and Slavin, J. (2003). The Chemical Composition and Gas-to-Dust Mass Ratio of Nearby Interstellar Matter. *Astrophys. J.,* 594:844–858.

Frisch, P.C., (2000). The galactic environment of the Sun. *J. Geophys. Res.,* 105:10279–10290.

Frisch, P.C., and Slavin, J. (2005) Short-term variations in the galactic environment of the Sun, In: P. C. Frisch Editor, *Solar Journey: The Significance of Our Galactic Environment for the Heliosphere and Earth.* This Volume, Springer, Dordrecht.

Gangopadhyay, P., Izmodenov, V., Gruntman, M., and Judge, D.L. (2002). Interpretation of Pioneer 10 Ly-β based on heliospheric interface models: Methodology and first results, *J. Geophys. Res.,* 107:doi:10.1029/2002JA009345.

Gangopadhyay, P., Izmodenov, V.V., Quemerais, E., Gruntman, M.A., and Judge, D.L., Interpretation of Pioneer 10 and Voyager 2 Ly alpha data: first results, *Adv. Space Res.,* 34:94–98.

Geiss, J., Gloeckler, G., and von Steiger, R., (1996). Origin of C^+ Ions in the Heliosphere. *Space Sci. Rev.,* 78:43–52.

Gloeckler, G., Geiss, J., Balsiger, H., Fisk, L.A., Galvin, A.B., Ipavich, F.M., Ogilvie, K.W., von Steiger, R., and Wilken, B., (1993). Detection of interstellar pick-up hydrogen in the Solar System. *Science,* 261:70–73.

Gloeckler, G., Geiss, J., Roelof, E.C., Fisk, L.A., Ipavich, F.M., Ogilvie, K.W., Lanzerotti, L.J., von Steiger, R., and Wilken, B., (1994). Acceleration of interstellar pickup ions in the disturbed solar wind observed on Ulysses. *J. Geophys. Res.,* 99:17637–17643.

Gloeckler, G., Jokipii, J.R., Giacalone, J., and Geiss, J., (1994). Concentration of interstellar pickup H^+ and He^+ in the solar wind. *Geophys. Res. Lett.,* 21:1565–1568.

Gloeckler, G., Schwadron, N.A., Fisk, L.A., and Geiss, J., (1995). Weak pitch angle scattering of few MV rigidity ions from measurements of anisotropies in the distribution function of interstellar pickup H^+. *Geophys. Res. Lett.,* 22:19:2665–2668.

Gloeckler, G., (1996). The abundance of atomic ^1H, ^4He and ^3He in the local interstellar cloud from pickup ions observations with SWICS on Ulysses. *Space Sci. Rev.,* 78:335–346.

Gloeckler, G., and Geiss, J., (1998). Interstellar and inner source pickup ions observed with SWICS on Ulysses. *Space Sci. Rev.,* 86:127–159.

Gloeckler, G., (1999). Observation of injection and pre-acceleration processes in the slow solar wind. *Space Sci. Rev.,* 89:91–104.

Gloeckler, G., Fisk, L.A., Geiss, J., Schwadron, N.A., and Zurbuchen, T.H., (2000). Elemental composition of the inner source pickup ions. *J. Geophys. Res.,* 105:7459–7463.

Gloeckler, G., and Geiss, J., (2001). Heliospheric and interstellar phenomena deduced from pickup ions. *Space Sci. Rev.,* 97:169–181.

Gloeckler, G., Möbius, E., Geiss, J., Bzowski, M., Rucinski, D., Terasawa, T., Noda, H., Oka, M., McMullin, D.R., Skoug, R., Chalov, S., Fahr, H., von Steiger, R., Yamazaki, A., and Zurbuchen, T., (2004). Observations of the Helium Focusing Cone with Pickup Ions. *Astron. Astrophys.,* 426:845–854.

Grün, E., Gustafsson, B.A.S., Mann, I., Baguhl, M., Morfill, G., Staubach, P., Taylor, A., and Zoog, H.A. (1994). Interstellar dust in the heliosphere. *Astron. Astrophys.,* 286:915–924.

Gruntman, M.A. (1994). Neutral solar wind properties: Advance warning of major geomagnetic storms. *J. Geophys. Res.*, 99:19213–19227.

Gruntman, M.A. (1997). Energetic neutral atom imaging of space plasmas. *Rev. Sci. Instr.,* 68:3617–3656.

Gruntmann, M.A., Roelof, E.C., Mitchell, D.G., Fahr, H.J., Funsten, H.O., and McComas, D.J. (2001). Energetic neutral atom imaging of the heliospheric boundary region. J. Geophys. Res., 106:15767–15782.

Hilchenbach, M., Grünwaldt, H., Kallenbach, R., Klecker, B., Kucharek, H., Ipavich, F.M. and Galvin, A.B., (1999). Observation of suprathermal helium at 1 AU: Charge states in CIRs, in: *Solar Wind Nine*. ed. Habbal et al., 605–608.

Holzer, T.E., (1977). Neutral hydrogen in interplanetary space. *Rev. Geophys. and Space Phys.,* 15:467–490.

Holzer, T.E., and Axford, I. (1971). Interaction between interstellar helium and the solar wind, *J. Geophys. Res.*, 76:6965-6970.

Hovestadt, D., Gloeckler, G., Klecker, B., and Scholer, M., (1984a). Ionic charge state measurements during He^{+}-rich solar energetic particle events. *Astrophys. J.,* 281:463–467.

Hovestadt, D., Klecker, B., Scholer, M., Gloeckler, G., and Ipavich, F.M., (1984b), Survey of He^{+}/He^{2+} abundance ratios in energetic particle events, *Astrophys. J.*, 282:L39–L42.

Illing, R.M.E., and Hildner, E., (1994) Neutral hydrogen in the solar wind at 1 AU, *Eos Trans. AGU,* 75(16):261.

Izmodenov, V.V., Geiss, J., Lallement, R., Gloeckler, G., Baranov, V.B., and Malama, Yu.G. (1999). Filtration of interstellar hydrogen in the two-shock heliospheric interface: Inferences on the LIC electron density. *J. Geophys. Res.*, 104:4731–4741.

Izmodenov, V.V., Gruntman, M.A., and Malama, Yu.G., (2001). Interstellar hydrogen atom distribution function in the outer heliosphere. *J. Geophys. Res.,* 106:10681–10689.

Izmodenov, V., Malama, Y.G., Gloeckler, G. and Geiss, J. (2004). Filtration of interstellar H, O, N atoms through the heliospheric interface: Inferences on local interstellar abundances of the elements, *Astron. Astrophys.*, 414:L29–L32.

Kucharek, H., Möbius, E., Li, W., Farrugia, C.J., Popecki, M.A., Galvin, A.B., Klecker, B., Hilchenbach, M., and Bochsler, P. (2003). On the source and the acceleration of energetic He^{+}: Long term observations with ACE/SEPICA. *J. Geophys. Res.,* 108:8030, doi:10.1029/2003.JA009938.

Lallement, R., (1996). Relations between ISM inside and outside the heliosphere. *Space Sci. Rev.*, 78:361–374.

Lallement, R., Raymond, C.J., Vallerga, J., Lemoine, M., Dalaudier, F., and Bertaux, J.L., (2004a). Modeling the interstellar-interplanetary helium 58.4 nm resonance glow: Towards a reconciliation with particle measurements. *Astron. Astrophys.,* 426:875–884.

Lallement, R., Raymond, J.C., Bertaux, J.-L., Quémerais, E., Ko, Y.-K., Uzzo, M., McMullin, D., and Rucinski, D., (2004b). Solar cycle dependence of the helium focusing cone from SOHO/UVCS observations. *Astron. Astrophys.,* 426:867–874.

Lallement, R. Quémerais, E. Bertaux, J.L. Ferron, S. Koutroumpa, D. and Pellinen, R. (2005). Deflection of the Interstellar Neutral Hydrogen Flow Across the Heliospheric Interface, *Science*, 307:1449–1451.

Landgraf, M. (2000). Modeling the motion and distribution of interstellar dust inside the heliosphere. *J. Geophys. Res.*, 105:10303–10316.

Landgraf, M., Krüger, H., Altobelli, N., and Grün, E. (2003). Penetration of the heliosphere by the interstellar dust stream during solar maximum. *J. Geophys. Res.*, 108:8030 doi:10.1029/2003JA009872.

Landgraf, M. (2005). Variations in interstellar dust distribution in the heliosphere, In: P. C. Frisch Editor, *Solar Journey: The Significance of Our Galactic Environment for the Heliosphere and Earth*. This Volume, Springer, Dordrecht.

Lee, M.A. (1997). Effects of cosmic rays and interstellar gas on the dynamics of a wind, in: *Cosmic Winds and the Heliosphere*, eds. J.R. Jokipii, C.P. Sonett, and M.S. Giampapa, pp. 857–886.

Levasseur-Regourd, A.C., Renard, J.B., and Dumont, R., (1991). The zodiacal cloud complex, In: A.C. Levasseur-Regourd and Hasegawa, H. Editors, *Origin and Evolution of Interplanetary Dust*. Pages 131–138, Kluwer Academic Publishers.

Maher, L.J., and Tinsley, B.A., (1977). Atomic hydrogen escape due to charge exchange with hot plasmaspheric ions. *J. Geophys. Res.,* 82:689–695.

Mann, I., Kimura, H., Biesecker, D.A., Tsurutani, B.T., Grün, E., McKibben, B., Liou, J.-C., MacQueen, R.M., Mukai, T., Guhathakurta, L., and Lamy, P., (2004). Dust near the Sun. *Space Sci. Rev.,* 110:269–305.

Marsch, E., Pilipp, W.G., Thieme, K.M., and Rosenbauer, H. (1989). Cooling of solar wind electrons inside 0.3 AU. *J. Geophys. Res.,* 94:6893–6898.

McComas, D., Allegrini, F., Bochsler, P., Bzowski, M., Collier, M., Fahr, H., Fichtner, H., Frisch, P., Funsten, H., Fuselier, S., Gloeckler, G., Gruntman, M., Izmodenov, V., Knappenberger, P., Lee, M., Livi, S., Mitchell, D., Moebius, E., Moore, T., Reisenfeld, D., Roelof, E., Schwadron, N., Wieser, M., Witte, M., Wurz, P., and Zank, G., (2004). The Interstellar Boundary Explorer (IBEX). *AIP Conference Proceedings*. 719:162–181.

McMullin, D., McMullin, D.R., Bzowski, M., Möbius, E., Pauluhn, A., Skoug, R., W.T., Thompson, Witte, M., von Steiger, R., Rucinski, D., Banaszkiewicz, M., and Lallement, R., (2004). Heliospheric conditions that affect the interstellar gas inside the heliosphere. *Astron. Astrophys.,* 426:885–895.

Möbius, E., Hovestadt, D., Klecker, B., Scholer, M., G.,Gloeckler and Ipavich, F.M., (1985). Direct observation of He^+ pick-up ions of interstellar origin in the solar wind. *Nature*. 318:426–429.

Möbius, E., Klecker, B., Hovestadt, D., and Scholer, M., (1988). Interaction of interstellar pick-up ions with the solar wind. *Astrophys. Space Sci.*, 144:487–505.

Möbius, E. (1993). Gases of Non-Solar Origin in the Solar System. In: *Landoldt-Börnstein, Numerical Data and Functional Relationships in Science and Technology*, VI/3A Chapter 3.3.5.1, pp. 184–188.

Möbius, E., Rucinski, D., Hovestadt, D., and Klecker, B., (1995). The helium parameters of the very local interstellar medium as derived from the distribution of He^+ pickup ions in the solar wind. *Astron. Astrophys.,* 304:505–519.

Möbius, E., Rucinski, D., Lee, M.A., and Isenberg, P.A., (1998). Decreases in the antisunward flux of interstellar pickup He^+ associated with radial interplanetary magnetic field. *J. Geophys. Res.* 103:257–265.

Möbius, E., Morris, D., Popecki, M.A., Klecker, B., Kistler, L.M., and Galvin, A.B., (2002). Charge States of Energetic Ions Obtained from a Series of CIRs in 1999 – 2000 and Implications on Source Populations. *Geophys. Res. Lett.,* 29, 10.1029/2001GL013410.

Möbius, E., Bzowski, M., Chalov, S., Fahr, H.-J., Gloeckler, G., Izmodenov, V., Kallenbach, R., Lallement, R., McMullin, D., Noda, H., Oka, M., Pauluhn, A., Raymond, J., Rucinski, D., Skoug, R., Terasawa, T., Thompson, W., Vallerga, J., von Steiger, R., and Witte, M., (2004). Synopsis of the interstellar He parameters from combined neutral gas, pickup ion and UV scattering observations and related consequences. *Astron. Astrophys.*, 426:897–907.

Möbius, E., Bzowski, M., Müller, H.-R., and Wurz, P., (2005). Impact of Dense Interstellar Gas Clouds on the Neutral Gas and Secondary Particle Environment in the Inner Heliosphere, in: Proceedings of the Solar Wind 11/SOHO 16 Conference, T. Zurbuchen and B. Fleck eds., *ESA Special Publication*, SP-592, 367–370.

Morris, D., Möbius, E., Lee, M,A., Popecki, M.A., Klecker, B., Kistler, L.M., and Galvin, A.B., (2001). Implications for Source Populations of Energetic Ions in Co-Rotating Interaction Regions from Ionic Charge States, in: Solar and Galactic Composition, *AIP Conference Proceedings*, 598:201–204.

Müller, H.-R., Zank, G. P., and Lipatov, A. S. (2000). Self-consistent hybrid simulations of the interaction of the heliosphere with the local interstellar medium. *J. Geophys. Res.*, 105:27419–27438.

Oka, M., Terasawa, T., Noda, H., Saito, H., and Mukai, T., (2002). Torus Distribution of Interstellar Pickup Ions: Direct Observation. *Geophys. Res. Lett.*, 29:1612–1615.

Osterbart, R., and Fahr, H.-J. (1992). A Boltzmann-kinetic approach to describe the entrance of neutral interstellar hydrogen into the heliosphere, *Astron.Astroph ys.,* 264:260–269.

Paresce, F., Bowyer, S., and Kumar, S. (1974). Observations of He I 584 □ nighttime radiation; evidence for an interstellar source of neutral helium. *Astrophys. J.*, 187:633–639.

Parker, E.N. (1963) Interplanetary Dynamical Processes, *Interscience Monographs and Texts in Physics and Astronomy*, Wiley & Sons, New-York, Vol. VIII, pp. 126–127.

Parker, E.N. (2005) Interstellar conditions and planetary magnetospheres, In: P.C. Frisch Editor, *Solar Journey: The Significance of Our Galactic Environment for the Heliosphere and Earth.* This Volume, Springer, Dordrecht.

Pilipp, W.G., Miggenrieder, H., Montgomery, M.D., Mühlhäuser, K.-H., Rosenbauer, H., and Schwenn, R. (1987a). Characteristics of electron velocity distribution functions in the solar wind derived from the Helios plasma experiment. *J. Geophys. Res.,* 92:1075–1092.

Pilipp, W.G., Miggenrieder, H., Montgomery, M.D., Mühlhäuser, K.-H., Rosenbauer, H., and Schwenn, R. (1987b). Variations of electron distribution functions in the solar wind. *J. Geophys. Res.,* 92:1103–1118.

Quemerais, E., Bertaux, J.-L., Sandel, B. R., and Lallement, R., (1994). A new measurement of the interplanetary hydrogen density with ALAE/ATLAS 1. *Astron. Astrophys.* 290:941–955.

Quemerais, E.; Bertaux, J.L., Lallement, R., Berthé, M., Kyrölä, E., and Schmidt, W. (1999). Interplanetary Lyman a line profiles derived from SWAN/SOHO hydrogen cell measurements: Full-sky vector field. *J. Geophys. Res.*, 104:12585–12603.

Ragot, B.R., and Kahler, S.W. (2003). Interactions of dust grains coronal mass ejections and solar cycle variations of the F-corona brightness. *Astrophys. J.*, 594:1049 –1059.

Richardson, J.D., Paularena, K.I., Lazarus, A.J., and Belcher, J.W. (1995). Evidence for a solar wind slowdown in the outer heliosphere? *Geophys. Res. Lett.*, 22:1469–1472.

Ripken, H.W., and Fahr, H.-J. (1983). Modification of the local gas properties in the heliospheric interface. *Astron. Astrophys.*, 122:181–192.

Rosenbauer, H., Fahr, H.J., Keppler, E., Witte, M., Hemmerich, P., Lauche, H., Loidl, A., and Zwick, R., (1983). ISPM Interstellar Neutral Gas Experiment. *ESA SP-1050*, pp. 125–139.

Richter, I., Leinert, C., and Planck, B., (1982). Search for the variations of Zodiacal Light and optical detection of interplanetary plasma clouds. *Astron. Astrophys.*, 110:115–120.

Rucinski, D., Cummings, A.C., Gloeckler, G., Lazarus, A.J., Möbius, E., and Witte, M., (1996). Ionization processes in the heliosphere – Rates and methods of their determination. *Space Sci. Rev.,* 78:73–84.

Rucinski, D., and Fahr, H.J., (1989). The influence of electron impact ionization on the distribution of interstellar helium in the inner heliosphere; possible consequences for determination of interstellar helium parameters. *Astron. Astrophys.*, 224:290–298.

Rucinski, D., and Fahr, H.J., (1991). Nonthermal ions of interstellar origin at different solar wind conditions. *Ann.Geophys.*, 9:102–110.

Rucinski, D., Bzowski, M., and Fahr, H.J., (2003). Imprints from the solar cycle on the helium atom and helium pick-up ion distributions. *Ann.Geophys.*, 21:1315 –1330.

Saul, L., Möbius, E., Smith, C.W., Bochsler, P., Grünwaldt, H., Klecker, B., and Ipavich, F.M., (2004). Observational Evidence of Pitch Angle Isotropization by IMF Waves. *Geophys. Res. Lett.*, 31:L05811 10.1029/2003GL019014.

Schmidt, H. U., and Wegmann, R., (1980). MHD–Calculations for Cometary Plasmas. *Comp. Phys. Comm.*, 19:309–326.

Schmidt, H. U.; Wegmann, R., and Neubauer, Fritz M., (1993). MHD modeling applied to Giotto encounter with comet P/Grigg-Skjellerup. *J. Geophys. Res.*, 98:21, 009–21,016.

Schwadron, N.A., Zurbuchen, T.H., Fisk, L.A., and Gloeckler, G., (1999). Pronounced enhancements of pickup hydrogen and helium in high-latitude compressional regions. *J. Geophys. Res.,* 104:535–548.

Spitzer, L. (1978). *Physical Processes in the Interstellar Medium*. John Wiley & Sons, New York.

Strömgren, E. (1939). The physical state of interstellar hydrogen, *Astrophys. J.*, 89:526–547.

Szegö, K., Glassmeier, K.-H., Brinca, A., Cravens, T., Fischer, C., Fisk, L., Gombosi, T., Haerendel, G., Lee, M.A., Mazelle, C., Möbius, E., Motschmann, U., Isenberg, P., Sauer, K., Schwadron, N., Shapiro, V., Tsurutani, B., and Zank,, G., (2000). Physics of Mass Loaded Plasmas. *Space Sci. Rev.,* 94:429 –671.

Thomas. G.E., and Krassa, R.F., (1971). OGO 5 measurements of the Lyman Alpha sky background. *Astron. Astrophys.* 11:218–233.

Vallerga, J., Lallement, R., Lemoine, M., Dalaudier, F., and McMullin, D., (2004). EUVE observations of the helium glow: Interstellar and solar parameters. *Astron. Astrophys.,* 426:855–865.

Vasyliunas,V.M., and Siscoe, G.L., (1976). On the flux and the energy spectrum of interstellar ions in the solar system. *J. Geophys. Res.* 81:1247–1252.

Wang, C., and Richardson, J.D., (2003). Determination of the solar wind slowdown near solar maximum. *J. Geophys. Res.* 108:1058 10.1029/2002JA009322.

Weinberg, J.L., (1964). The zodiacal light at 5300 Å, *Ann. Astrophys.* 27: 718–738.

Weller, C.S., and Meier, R.R. (1974). Observations of helium in the interplanetary /interstellar wind: the solar-wake effect. *Astrophys. J.,* 193:471–476.

Welty, D.E., Hobbs, L.M., Lauroesch, J. T., Morton, D. C., Spitzer, L., and York, D. G. (1999). The Diffuse Interstellar Clouds toward 23 Orionis. *Astrophys. J. Suppl.*, 124:465–501.

Wieser, M., Wurz, P., Bochsler, P., Moebius, E., Quinn, J., Fuselier, S.A., Ghielmetti, A., DeFazio, J., Stephen, T.M., and Nemanich, R.J. (2005). NICE: An Instrument for Direct Mass spectrometric Measurement of Interstellar Neutral Gas. *Meas. Sci. Technol.* 16: 1667–1676.

Wimmer-Schweingruber, R. F., and Bochsler, P. (2001). Lunar Soils - An Archive for the Galactic Environment of the Solar System? In Wimmer-Schweingruber, R., Editor, *Solar and Galactic Composition, AIP Conference Proceedings*, 598:399–402.

Wimmer-Schweingruber, R. F., and Bochsler, P., (2003). On the origin of inner-source pickup ions. *Geophys. Res. Lett*., 30:1077–1080.

Witte, M., (2004). Kinetic parameters of interstellar neutral helium. *Astron. Astrophys.,* 426:835–844.

Witte, M., Banaszkiewicz, M., Rosenbauer, H., and McMullin, D., (2004). Kinetic parameters of interstellar neutral helium: updated results from the ULYSSES/GAS-instrument. *Adv. Space Res.,* 34:61–65.

Wood, B.E., Linsky, J. L., and Zank, G.P. (2000). Heliospheric, astrospheric, and interstellar Ly-α absorption toward 36 Ophiuchi, *Astrophys. J.*, 537:304–311.

Wu, F.M., and Judge, D.L. (1979). Temperature and velocity of the interplanetary gases along solar radii. *Astrophys. J.*, 231:594–605.

Wurz, P., Collier, M.R., Moore, T.E., Simpson, D., Fuselier, S., and Lennartson, W. (2004). Possible Origin of the Secondary Stream of Neutral Fluxes at 1 AU. *AIP Conference Proceedings* 719:195–200.

Yeghikyan, A., and Fahr, H.-J. (2004). Terrestrial atmospheric effects induced by counterstreaming dense interstellar cloud material. *Astron. Astrophys.,* 425:1113–1118.

Yeghikyan, A., and Fahr, H.-J. (2005) Accretion of interstellar material into the heliosphere and onto Earth, In: P. C. Frisch Editor, *Solar Journey: The Significance of Our Galactic Environment for the Heliosphere and Earth*. This Volume, Springer, Dordrecht.

Zank, G.P., and Frisch, P.C. (1999). Consequences of a Change in the Galactic Environment of the Sun. *Astrophys. J.*, 518:965–997.

Zank, G.P., et al. (2005). Variations in the heliosphere in response to variable interstellar material, In: P. C. Frisch Editor, *Solar Journey: The Significance of Our Galactic Environment for the Heliosphere and Earth*. This Volume, Springer, Dordrecht.

Chapter 9

VARIABLE TERRESTRIAL PARTICLE ENVIRONMENTS DURING THE GALACTIC ORBIT OF THE SUN

Hans J. Fahr
Institut für Astrophysik und Extraterrestrische Forschung, Universität Bonn, Auf dem Hügel 71, 53121 Bonn, Germany
hfahr@astro.uni-bonn.de

Horst Fichtner
Institut für Theoretische Physik IV: Weltraum- und Astrophysik, Ruhr-Universität Bochum, 44780 Bochum, Germany
hf@tp4.rub.de

Klaus Scherer
Institut für Astrophysik und Extraterrestrische Forschung, Universität Bonn, Auf dem Hügel 71, 53121 Bonn, Germany
ks@tp4.rub.de

Olaf Stawicki
Unit for Space Physics, School of Physics, North-West University, Campus Potchefstroom, 2520 Potchefstroom, South Africa
os@tp4.ruhr-unibochum.de

Abstract The effects of variable terrestrial environments of the Sun during its galactic orbit are discussed. Emphasis is put on the energetic particle populations, like pick-up ions, anomalous and galactic cosmic rays. After putting the analysis briefly into the context of the ongoing debate regarding a cosmic ray-terrestrial climate link, we study the structure of the heliosphere in various interstellar environments, particularly in dense clouds of atomic hydrogen. We continue

P. C. Frisch (ed.), Solar Journey: The Significance of our Galactic Environment for the Heliosphere and Earth, 259–279.

with a computation of the energy spectra of both anomalous and galactic cosmic rays for galactic regions located in clouds and/or in galactic spiral arms. In the light of these results we discuss the consequences for the terrestrial and planetary environments, as well as for small solar system bodies.

Keywords: Pick-up ions, energetic neutral atoms, anomalous cosmic rays, dynamic heliosphere

9.1 Introductory Remarks on Cosmic Rays and Climate

The Earth in space represents a thermodynamical non-equilibrium system because it maintains a pronounced exchange of energies with its space environment. The terrestrial electromagnetic and particle radiation environments are changing in time due to solar as well as galactic variabilities. Particularly, the terrestrial climate is a result of a very complicated interactions between open thermodynamic subsystems, like different cloud layers and atmospheric and oceanic circulation patterns on top of a rotating solid body. Consequently, the whole climate system is highly structured with many nonlinear thermal couplings between the lithosphere (i.e. solid Earth crust), the hydrosphere (i.e. the oceans), the atmosphere, and perhaps the biosphere (see, e.g., Cuntz et al., 2003, von Bloh et al., 2003).

It has been argued (Lean and Rind, 1999, Fröhlich and Lean, 2002) that the prime external climate driver on short time scales (Schwabe and Hale cycle, i.e. decades) and on intermediate time scales (Gleisberg cycle and grand activity minima, i.e. centuries and millenia) is solar irradiation, i.e. the integrated electromagnetic energy input into the Earth's upper atmosphere. However, various terrestrial climate drivers, like temperature and cloud coverage of the Earth. correlate well with energetic particle irradiation (Svensmark and Friis-Christensen, 1997, Svensmark, 1998, Svensmark, 2000, Carslaw et al., 2002) on short time scales. For intermediate time scales, the cosmogenic isotope concentration during the Wolf-, Spoerer-, Maunder- and Dalton-Minimum of solar activity indicate an increased cosmic ray flux as recently modelled by Caballero et al. (2004) and Scherer and Fichtner (2004), suggesting that this correlation also holds, because these grand minima coincide with cold periods.

While it cannot yet be determined whether solar or particle irradiation is the prime external climate driver on short and intermediate time scales, this is different for long time scales (galactic arm crossings, i.e. millions of years). For long time scales, the cosmic ray flux varies quasi-periodically during the solar galactic orbit due to enhanced fluxes within spiral arms (Shaviv, 2003), and correlates with long-term temperature variations on Earth (Shaviv and Veizer, 2003). As one should not expect the solar irradiance to correlate with spiral arm crossings, these results are a strong argument for the cosmic rays as climate drivers.

The discussion so far, however, has neither taken into account the variable shielding effect of the heliosphere during a solar galactic orbit nor the permanent presence of an additional energetic particle component, namely anomalous cosmic rays (ACRs). To study both we first investigate the coupling of the interstellar medium (ISM) with the solar wind. An essential role is played by neutral atoms that become ionized within the heliosphere, i.e. the so-called pick-up ions (PUIs), as they have a significant influence on the dynamics of the solar wind and they are the seed population of ACRs. In order to derive the energetic particle spectra at Earth we, secondly, need to include the fluxes of galactic cosmic rays (GCRs), which we estimate from both observations and theoretical considerations. This allows us, finally, to compute the total cosmic ray spectra at Earth.

9.2 The Heliosphere in Different Interstellar Environments

In contrast to the present-day situation, where the heliosphere is located in a dilute ISM (Frisch, 1998, Lallement et al., 2003, Freyberg and Breidtschwerdt, 2003, Breitschwerdt, 2004), a spiral arm crossing provides a dense interstellar environment (DISE). Therefore, we consider typical cases of DISEs in the following. Note, however, that we exclude molecular clouds providing environments of very high interstellar density ($n_H > 1000\,\mathrm{cm}^{-3}$), because the Bonn model (Fahr et al., 2000, Fahr, 2000) we use assumes atomic rather than molecular hydrogen (for the case of molecular clouds see Yeghikyan and Fahr, 2004a). The computations are performed using the five-fluid Bonn model that describes the dynamics of the heliosphere in the presence of solar wind and interstellar protons, interstellar hydrogen, PUIs, ACRs, and GCRs, all of which are mutually interacting (Fahr et al., 2000). The PUIs behave thermodynamically as an individual fluid, which is convected with the solar wind. In turn, the solar wind gets decelerated by momentum loading from these suprathermal particles influencing the compression ratios of the termination shock (TS). Depending on the local strength of the TS, a certain fraction of PUIs is injected at the TS into the process of diffusive shock acceleration, thereby constituting the local ACR source.

9.2.1 The Physics in the Bonn model

In the following we briefly describe the dynamical and thermodynamical coupling of low- and high-energy plasma components as modelled in the Bonn code. The first-order interaction of the solar wind with the interstellar plasma is a hydrodynamic proton-proton interaction. In addition to protons, however, also interstellar H-atoms are flying into the heliosphere, which upon ionization in the region of the supersonic solar wind produce H-pick-up ions (PUIs). Due to effective pitch-angle scattering, they are rapidly isotropized in velocity space and are comoving with the solar wind, thereby representing a keV-energetic ion load of the expanding solar wind. These PUIs are treated in the Bonn model as an additional ion fluid, separated from the solar wind proton fluid by its number density and its pressure. A certain fraction of these PUIs can be injected into the Fermi-1 diffusive acceleration process at the termination shock (Chalov and Fahr, 1997) and act as a seed for the ACR population, which in the multifluid model is also treated by its relevant moment, i.e. the energy density or pressure. This high energy fluid interacts with the solar wind plasma flow via its pressure gradient and, for instance, decelerates it in the region near and upstream of the termination shock, forming a precursor. For stationary conditions the physics in this precursor for the convected low energy plasma species is treated by the following continuity equations (Chalov and Fahr, 1994, Chalov and Fahr, 1995, Chalov and Fahr, 1997, Zank, 1999):

$$\nabla \cdot (\rho_i \vec{u}) = Q_{\rho,i} \tag{2.1}$$

$$\nabla \cdot \left(\rho \vec{u}\vec{u} + (P_{SW} + P_{PUI} + P_{ACR} + P_{GCR})\overset{\leftrightarrow}{I} \right) = \vec{Q}_p \tag{2.2}$$

$$\nabla \cdot \left(\rho \vec{u} \left(\frac{U^2}{2} + \frac{\gamma}{\gamma - 1} \frac{1}{\rho} (P_{SW} + P_{PUI}) \right) + \vec{F}_{CR} \right) = Q_e \tag{2.3}$$

where $\rho = \rho_{SW} + \rho_{PUI}$ denotes the total mass density of the mixed fluids, $\vec{u}$ the solar wind velocity, $\overset{\leftrightarrow}{I}$ the unit tensor, and the quantities P_{SW}, P_{PUI}, P_{ACR}, P_{GCR} the pressures of the solar wind protons, the PUIs, the ACRs and the GCRs, respectively. The polytropic index $\gamma = 5/3$ is taken to be valid for protons and PUIs as well, and the quantities $Q_{\rho,i}$, $\vec{Q}_p$, Q_e denote mass-, momentum-, and energy exchange rates per unit volume and time ($i = SW, PUI$). The contribution to the energy flow of the coupled ACRs and GCRs is separately denoted by $\vec{F}_{CR} = \vec{F}_{ACR} + \vec{F}_{GCR}$ and is itself given by the following equations:

$$\vec{F}_{ACR} = \frac{\gamma_{ACR}}{\gamma_{ACR} - 1} \vec{u} P_{ACR} - \frac{\kappa_{ACR}}{\gamma_{ACR} - 1} \nabla P_{ACR} \tag{2.4}$$

$$\vec{F}_{GCR} = \frac{\gamma_{GCR}}{\gamma_{GCR} - 1} \vec{u} P_{GCR} - \frac{\kappa_{GCR}}{\gamma_{GCR} - 1} \nabla P_{GCR} \tag{2.5}$$

where $\gamma_{ACR,GCR} = 4/3$ and $\kappa_{ACR,GCR}$ are the polytropic indices of the high-energy CR-components and the energy-averaged CR spatial diffusion coefficient, respectively. Both ACRs and GCRs are treated as massless fluids by separate energy-averaged particle transport equations. The coupling of these massless fluids occurs because of convective interactions of the CR fluids with the background plasma, inducing a modification of the plasma flow by ACR and GCR pressure gradients.

The required consistency within this model must also include the dynamical and thermodynamical coupling of the mentioned four species to one more

Table 9.1. The heliocentric distances r_i and the compression ratios s_i of the TS in the upwind (u), crosswind (c), and downwind (d) direction ($i = \{u, c, d\}$) for four LISM number densities n_H. The radial velocity profile is approximated by $v(r) \approx v_0 \left(\frac{r_0}{r}\right)^\alpha$ such that at the inner boundary r_0 the radial solar wind speed $v_0 = 400\,\text{km/s}$. v_u is the velocity just upstream from the TS in upwind direction.

n_H cm^{-3}	r_0 AU	v_u km/s	α	r_u AU	s_u	r_c AU	s_c	r_d AU	s_d	r_u/r_d
0.1	5.1	366.3	0.03	78.3	3.5	107.6	2.2	148.2	2.9	0.53
1.0	5.1	258.4	0.22	32.6	1.6	44.0	1.4	63.6	1.6	0.51
50.0	0.5	286.8	0.12	9.1	1.7	12.5	1.5	17.4	2.1	0.52
100.0	0.5	348.6	0.07	4.0	2.3	6.7	2.1	11.8	3.3	0.34

species, namely the H-atoms. In the Bonn model we treat the H-atoms as an additional hydrodynamical fluid coupled by charge exchange reactions to protons and PUIs. The complete modelling within a two-dimensional multifluid simulation is explained in detail by Fahr (2000) and Fahr et al. (2000), the so-called five-species Bonn model.

9.2.2 Results Obtained with the Bonn Model

In the five-species Bonn model, various parameters have to be specified, for a thorough discussion see Scherer & Fahr (2003). It would go far beyond the scope of this article to study the behavior of the heliosphere under the variation of all parameters. We will rather restrict our modelling to a series of values for the hydrogen density n_H being representative for DISEs along the solar galactic orbit. While it is not quite realistic to keep all other parameters constant, particularly the temperature, it nonetheless is a reasonable approximation, because for all considered model runs the external ram pressure exceeds the thermal pressure significantly. Therefore, reducing the interstellar temperature to ensure a pressure equilibrium in the ISM has no significant consequences for the dynamics of the TS and the heliopause. We consider the values for n_H listed in the first column of Table 1 will refer to the case $n_H = 0.1\,\text{cm}^{-3}$ as the "standard model".

In Fig. 9.1 we show the resulting radial profiles of the radial proton speed, the hydrogen number density, the PUI number density, and the energy density of the ACRs in the upwind direction.

Inspection of Fig. 9.1(d) illustrates how the TS moves inward with increasing interstellar hydrogen density. For the upwind, crosswind, and downwind direction the resulting heliocentric TS distances are listed in Table 1, along with the corresponding compression ratios. At a first glance, astonishingly, the compression ratio is not monotonously decreasing, but increases again for very high

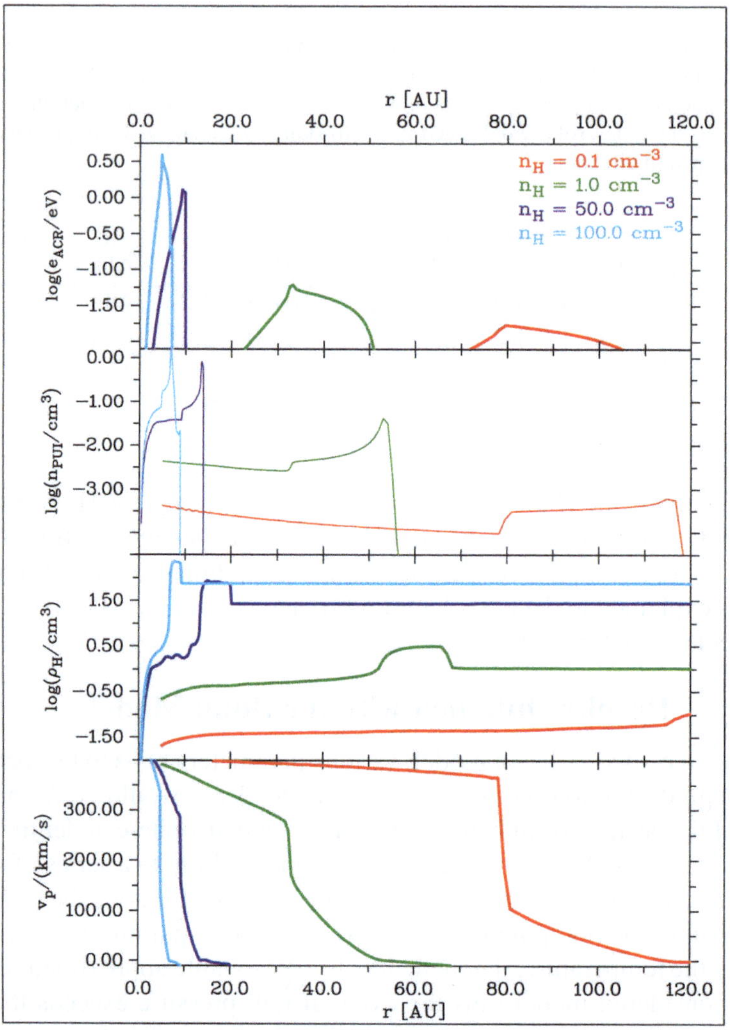

Figure 9.1. Shown from top to bottom are: (a) the ACR energy density, the number densities of the PUIs (b), and the number densities of the neutral hydrogen (c), as well as (d) the proton radial velocity. The red, green, dark blue, and light blue lines represent hydrogen number densities of $n_H = 0.1$, 1.0, 50 and 100 cm^3, respectively.

n_H. This is consistent with the same behaviour of the upstream speed. The reason for this counter-intuitive increase of both is that the net decrease in speed is limited by the short distance to the TS. Therefore, despite the strong momentum loading-induced deceleration resulting from the high hydrogen number density, the solar wind remains fast upstream causing a stronger shock.

Figure 9.1(c) shows that, as expected, the hydrogen density increases at all distances with the hydrogen wall moving inward. Note that the relative strength of the hydrogen wall roughly remains unchanged. It is defined as the maximum value of the hydrogen density inside the bow shock divided by the value at infinity, which is identical to that at the integration boundary and varies by a factor of two between the maximum ($n_H = 0.5\,\mathrm{cm}^{-3}$) and the minimal case ($n_H = 0.1\,\mathrm{cm}^{-3}$), while the cases $n_H = 50$ and $100\,\mathrm{cm}^{-3}$ are close to the value at $n_H = 0.5\,\mathrm{cm}^{-3}$. As a consequence of the increased neutral atom density, the atmospheres of the planets are not only exposed to an enhanced hydrogen influx (Bzowski et al., 1996, Yeghikyan and Fahr, 2004a, Yeghikyan and Fahr, 2004b), but could be located outside the heliopause (see also Begelmann and Rees, 1976, Scherer, 2000),

The heliopause can be nicely seen in panel (b) of Fig. 9.1 at the position where the PUI number density shows a cutoff towards the interstellar medium. Obviously, the PUI density is increased for DISEs. Astonishingly, the ACR energy density shown in panel (a) follows the PUI density increase, although the compression ratio of the TS and, thus, the injection efficiency of PUIs into the process of diffusive shock acceleration varies significantly. The ACR energy density (normalized to that of the standard model) is displayed in Fig. 9.2 and reveals a nonlinear increase with increasing hydrogen number density.

Evidently, the ACR energy density increases by a factor of about 250 from the model with the lowest hydrogen number density compared to that with the highest. Moreover, the peak of the ACR energy density migrates inwards with increasing n_H to about 4 AU.

With the compression ratio, the velocity field, and the size of the heliosphere representing the modulation region (see Table 1), we can compute the associated heliospheric cosmic ray spectra as is done in the following section.

9.3 Cosmic Ray Spectra

Ideally, one would need a self-consistent 'hybrid model' that, on the one hand, describes the (hydro-)dynamics of the heliosphere embedded in the local interstellar medium and, on the other hand, allows for a kinetic treatment of the cosmic ray transport. Several models have been presented that treat either the cosmic ray transport or the heliopheric structure in terms of rather over-simplifying approximations (see, e.g. Myasnikov et al., 2000, Alexashov et al.,

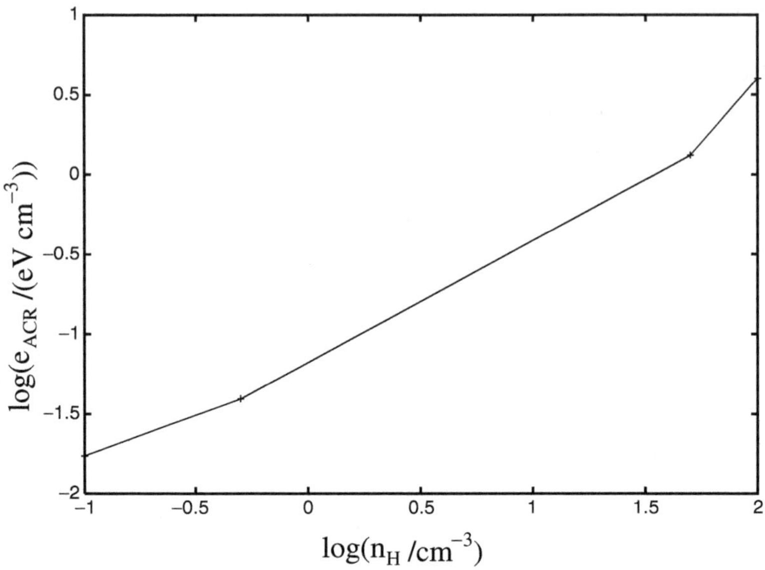

Figure 9.2. The logarithm of normalized ACR energy density e_{ACR} at the TS as a function of the logarithm of the interstellar hydrogen density.

2004), who developed a hybrid description regarding the neutral atoms but used a constant diffusion coefficient, or Florinski et al. (2004), who approximated the heliosheath by a spherical shell.

There are just a few approaches that incorporate both the cosmic ray transport and heliospheric structure in a more advanced fashion. Predecessors to hybrid models in the above sense were studied by Fahr et al. (2000), Scherer & Fahr (2003), and Sreenivasan & Fichtner (2001). While the first two studies are self-consistent two-dimensional fluid models that describe hydrodynamically all considered species (solar and interstellar plasma, neutral gas, pick-up ions, anomalous and galactic cosmic rays), the third study is a first attempt to solve the (steady-state) cosmic ray transport equation including drifts and anisotropic diffusion in a realistic three-dimensionally structured heliosphere.

The first hybrid models describing the cosmic ray transport kinetically everywhere in a "realistic" heliosphere were presented by Florinski et al. (2003), Ferreira et al. (2004), and Ferreira & Scherer (2004). While Florinski et al. concentrated on a partially self-consistent dynamics but neglected drifts, Ferreira et al. and Ferreira and Scherer included fully the drift and diffusion of the cosmic rays, but assumed a substantially simplified dynamics of the heliosphere.

So far, we are still missing a truly dynamical 3-D model for the large-scale heliosphere that includes self-consistently a sophisticated cosmic ray transport comprising anisotropic diffusion and drifts. However a 2-D version has been developed very recently by Scherer & Ferreira (2005). Because this code is still under development, we follow here a different strategy to compute the cosmic rays spectra.

To determine the latter, we employ the semi-analytic model developed by Stawicki et al. (2000), with which, for given solar wind velocity profile ($v(r) \sim r^{-\alpha}$), compression ratio s, and heliocentric distance of the modulation boundary, we estimate the modulated cosmic ray spectra. Note, however, that this is only used to estimate the CR spectra, while the full CR pressure is self-consistently included in the Bonn model as described above.

9.3.1 Boundary Spectra

In order to solve the complete modulation problem, i.e. to compute the spectra of ACRs and GCRs at any location in the heliosphere, we need the ACR and GCR spectra at the boundary of the modulation region. For simplicity, we assume this boundary to be identical with the TS (see Table 1), i.e. neglecting any so far unobserved (but presently discussed) modulation in the inner heliosheath between the TS and the heliopause.

Anomalous Cosmic Rays. It is the general consensus that ACRs result from diffusive acceleration of PUIs at the TS. Therefore, the accelerated spectra are power laws $f \sim p^{-q}$ in the relativistic particle momentum $p = \gamma\, m\, v$, where $q = 3s/(s-1)$ is a function of the compression ratio s, and v is the particle speed. These spectra are normalized such that at low energies they match the PUI spectrum ensuring that the associated energy density

$$e_{ACR} = \frac{p_{ACR}}{\gamma_{ACR} - 1} = \frac{4\pi}{3(\gamma_{ACR} - 1)} \int_0^\infty f\, v\, p^3 dp \qquad (3.1)$$

with $4/3 \leq \gamma_{ACR} \leq 5/3$ is consistent with that computed with the Bonn model (see Fig. 9.2). At high energies there is an exponential cut-off due to the curvature of the TS. The ACR spectra resulting for the four cases listed in Table 1 are shown in Fig. 9.3, where the ACR spectra in the region dominated by the present-day GCR contribution are indicated by the dotted lines. For the present-day interstellar GCR spectrum (see the dotted line below 1 MeV and the solid line above 100 MeV) we have used a standard representation, for a compilation see, e.g., Mori (1997) or Stawicki et al. (2000).

The ACR spectra have a steepening slope as expected for decreasing compression ratio s (see Table 1). At the same time the differential intensity is increased linearly with the interstellar neutral atom number density n_H. The underlying picture here is that the PUI flux is proportional to n_H and that the ACR spectra match the PUI spectra at 1 keV, so that the energy densities

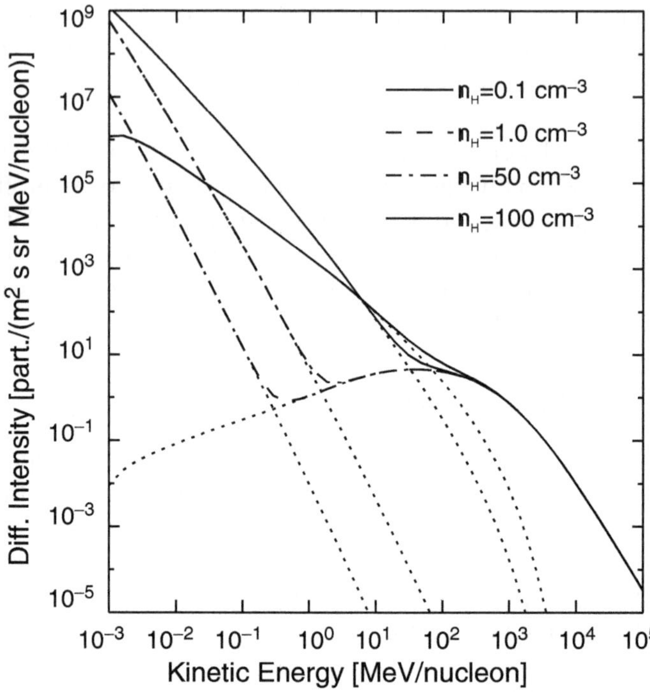

Figure 9.3. The unmodulated ACR spectra at the boundary of the modulation region for present-day conditions (solid line) or inside DISEs. The dotted lines indicate the ACR spectra in the region where GCRs dominate the total flux. For the latter the present-day spectrum is assumed.

corresponding to the computed ACR spectra correspond to those shown in Fig. 9.1. Obviously, only in the case of very dense interstellar environments can the intensity decrease (resulting from the steepening of the spectra) be (over-) compensated by the general increase due to the increased PUI flux: for the two cases $n_H = 1$ and $50\,cm^{-3}$ the ACR intensity dominates the total spectrum only below 1 and 10 MeV, respectively. As modulation will further decrease these fluxes significantly, the resulting intensity at 1 AU is even below the present-day intensity (see Fig. 9.4). From the latter figure, it is evident that only interstellar neutral atom densities of $n_H > 50\,cm^{-3}$ lead to a noticeable intensity increase at 1 AU, i.e. that the decrease of the heliospheric modulation region does not compensate for the steepening of the spectra for lower values. This finding is in agreement with the study by Florinski et al. (2003), who studied the case of a tenfold increase of the ACR intensities and showed that, at Earth, they remain below the GCR intensities at all energies. Therefore, in the following we concentrate on the two cases $n_H = 0.1\ cm^{-3}$ ("present-day") and $n_H = 100\ cm^{-3}$ ("DISE").

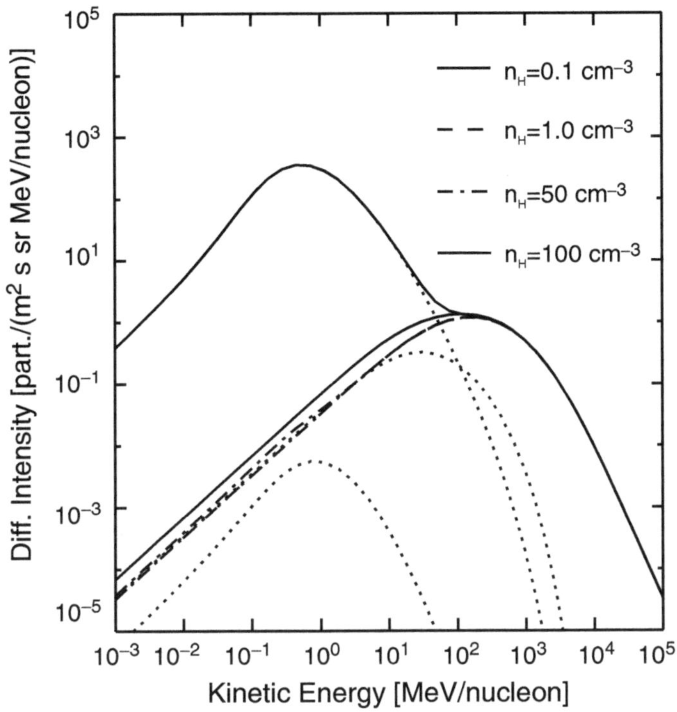

Figure 9.4. The modulated ACR spectra at 1 AU for present-day conditions (solid line) and for DISEs. The dotted lines indicate the ACR spectra in the region where GCRs dominate the total flux. For the latter the present-day spectrum is assumed. Note that the vertical scale has been changed compared to Fig. 9.3.

Galactic Cosmic Rays. While the present-day unmodulated interstellar GCR spectrum in the vicinity of the heliosphere can be derived from observations (for a compilation, see Mori, 1997), the GCR spectra elsewhere can only be obtained from theoretical considerations. It has been demonstrated that there is, at a given position in the Galaxy, a time variation in the spectra of various GCR species resulting from randomly distributed supernova explosions (Büsching et al., 2003). These supernovae represent localized acceleration sites of GCRs with different characteristics essentially resulting in a variation of the spectral indices of the GCR energy spectra at these sources. In addition to this “statistical” variation, one should expect a “regular” variation of the cosmic ray spectra due to the structure of the Galaxy. One of its main features are the spiral arms.

In general, the arms of spiral galaxies are considered to be a result of density waves driven in the galactic plane by an inner bar-like perturbation of the galactic gravitational field, i.e. of the responsible galactic mass distribution. For the Galaxy it is expected that four symmetrically placed spiral arms are developed. They extend from the inner (4/1) and the outer (1/4) Lindblad resonances of Keplerian periods with the corotation period of the bar-like field perturbation (Binney and Tremaine, 1988, Taylor and Cordes, 1993). Not only are these spiral arm regions of condensed galactic (i.e. interstellar matter) often appearing as frequent and dense interstellar clouds, but they also represent regions of an enhanced star formation, of enhanced average electron densities because of abundant HII regions, and of enhanced magnetic fields with a pattern essentially conformal with the density pattern.

The notion that enhanced star formation, for a typical initial mass function, naturally leads to enhanced production rates of massive O-, and B-stars, which then result in supernova events after comparatively short main sequence periods (about 10 to 15 Myr according to Taylor and Cordes, 1993), suggests that spiral arms constitute sites of enhanced supernovae activity and, thus, of enhanced GCR production. Besides, this structure in the source distribution of GCRs, it is also plausible that the diffusive propagation of GCRs is different inside galactic spiral arms as a consequence of both the enhanced magnetic fields, as well as the enhanced turbulence generation created by supernova shock waves. This led to the idea to study the variation of the GCR flux along the galactic orbit of the Sun through arms and interarm regions (Shaviv, 2003). A comparison of the GCR intensities inside and outside spiral arms, envisioning these arm regions as sites of increased star formation, density, and magnetic fields, results in the finding that the GCR flux at the heliospheric modulation boundary exhibits a regular variation with a quasi-period of about 140 Myr, on the basis of an energy-independent spatial diffusion model applied in the well known leaky-box approximation. A study of the cosmic ray exposure age of iron meteorites, as well as the ^{18}O to ^{16}O-isotope ratio in fossils formed in tropical oceans, seem to confirm that such a quasi-period really exists (Shaviv, 2003, Shaviv and Veizer, 2003). The latter correlation, in particular, establishes the strong cosmic ray-climate link mentioned above.

As we are mainly interested in the ACRs, we postpone a more detailed investigation of the GCR flux variation and adopt the general finding that the GCR flux can increase up to a factor of five compared to the present-day value. Therefore, we select two typical cases: first, the established present-day spectrum (for suitable parametrizations, see le Roux & Fichtner (1997) or Stawicki et al. (2000) and, second, a spectrum with five times higher spectral intensities being representative for the inside of a galactic spiral arm (Shaviv, 2003).

All unmodulated total boundary spectra under consideration are displayed in Fig. 9.5. The solid line gives the present-day total spectrum and the dash-dotted line shows the result for an increased ACR intensity in a DISE within a spiral

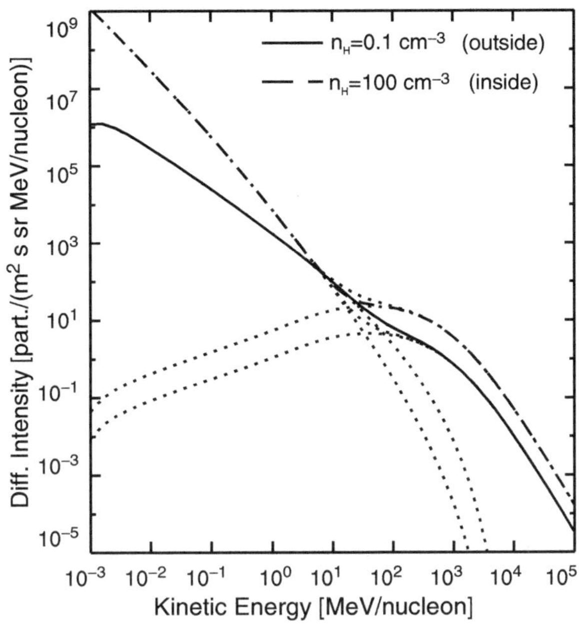

Figure 9.5. The unmodulated ACR and GCR spectra at the boundary of the modulation region for present-day conditions outside clouds or spiral arms and for conditions inside both.

arm leading to a higher GCR flux. The remaining combinations are indicated by the dotted lines.

9.3.2 Modulated Total Spectra at 1 AU

The differential intensity $j = f(\vec{r}, \vec{p}, t)\, p^2$ of cosmic rays with momentum $\vec{p}$ at a location $\vec{r}$ inside the modulation boundary can be estimated using a semi-analytical solution of their phase-space transport equation

$$\frac{\partial f}{\partial t} = \nabla \cdot (\overleftrightarrow{\kappa} \nabla f) - \vec{u} \cdot \nabla f + \frac{1}{3}\,(\nabla \cdot \vec{u})\frac{\partial f}{\partial \ln p} + S(\vec{r}, \vec{p}, t) \qquad (3.2)$$

where $\overleftrightarrow{\kappa}$ and $\vec{u}$ denote the diffusion tensor of energetic particles and the solar wind velocity, respectively, and $S(\vec{r}, \vec{p}, t)$ is an injection source. Drift motions are neglected.

By assuming spherical symmetry, and assuming power laws for the dependence of the spatial diffusion coefficient on both heliocentric distance and momentum, as well as for the dependence of the solar wind speed on heliocentric distance (see the caption of Table 1), we can derive the general steady-state solution in the following form (Stawicki et al., 2000):

$$j(r,p) = p^2 f(r,p) = \frac{3}{b}\int dr_0 \int dp_0 \frac{S(r_0,p_0)}{u(r_0)}\frac{p_0 y_0}{F}$$

$$\times (r_0/r)^{\frac{1+\beta}{2}} (p_0/p)^{\frac{3\beta-4\alpha-5}{2(2+\alpha)}} \exp\left(-y_0(1+h^2)/F\right)$$

$$\times I_{\frac{1+\beta}{1+\alpha-\beta}}(2y_0 h/F) \tag{3.3}$$

which requires a numerical evaluation for non-trivial, i.e. physically realistic sources $S(r_0, p_0)$. The quantity y_0 denotes the so-called modulation parameter, $F = 1 - (p/p_0)^{\frac{3\nu}{2+\alpha}}$; $h = (r/r_0)^{\frac{1+\alpha-\beta}{2}} (p/p_0)^{\frac{3}{2b}}$, I_n is a modified Bessel function of the first kind, $b = (2+\alpha)/(1+\alpha-\beta)$ and $\nu = (1+\alpha-\beta+\frac{2+\alpha}{3}\eta)$. The variables α and β are parameters describing the radial dependence of the solar wind speed and spatial diffusion coefficient, respectively, and η is the dependence of the latter on particle momentum.

While the assumed spherical symmetry of the solution limits its applicability to regions close to a given direction, which we chose as the upwind direction, it nevertheless allows us to estimate the basic changes in the cosmic ray flux levels as function of heliospheric distance r. This way, we obtained the 1 AU spectra displayed in Fig. 9.6.

As in Fig. 9.5 the solid line gives the present-day total spectrum, the dash-dotted line shows the result for an increased ACR intensity in a DISE within a spiral arm, leading to a higher GCR flux, while the other combinations are indicated by the dotted lines. One can clearly see that an n_H-increase associated with a DISE affects the total spectrum for energies below about 50 MeV, while for higher energies a DISE-associated change of the interstellar GCR spectrum is not very important. Note that this conclusion will change for very dense interstellar environments with $n_H > 1000\,\mathrm{cm}^{-3}$, which are typical for molecular clouds. Models for such cases are not as straightforward as, under these conditions, the surrounding medium is dominated by molecular rather than atomic hydrogen, and consequently the interaction scenario might be substantially changed (Yeghikyan and Fahr, 2004a).

9.4 Consequences of Variable Particle Environments

9.4.1 Outer Heliosphere

Obviously, the structure of the whole heliospheric interface adapts to an altered interstellar environment. While nowadays valuable information about this interface region can be extracted from the fluxes of energetic neutral atoms

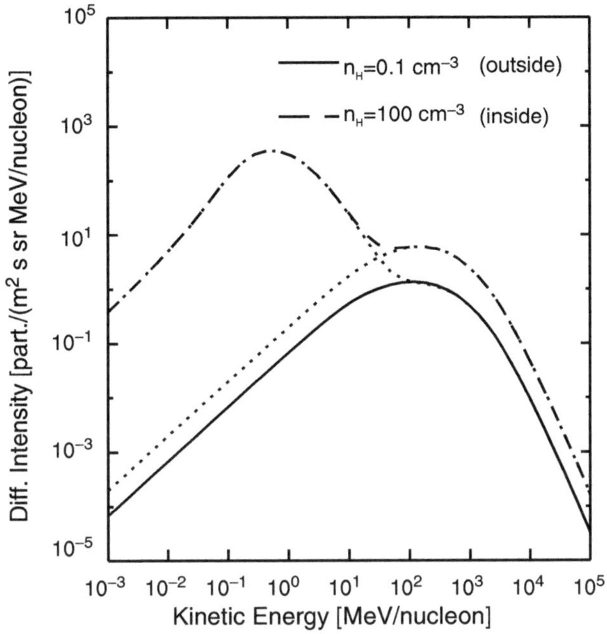

Figure 9.6. The modulated total spectra at 1 AU for $n_H = 0.1\,\mathrm{cm}^{-3}$ and $n_H = 100\,\mathrm{cm}^{-3}$. Note that the vertical scale has been changed as compared to Fig. 5.

(ENAs), which are produced by charge exchange processes between hydrogen atoms and solar wind protons, as well as from PUI protons (Hsieh et al., 1992, Gruntman, 1997), this information would not necessarily be available in cases with different inner and outer boundary conditions. With the Interstellar Boundary Explorer (IBEX), a new space mission is coming up (McComas et al., 2004) that uses the fact that valuable details of the heliospheric structures are reflected in spectral ENA fluxes at Earth for present-day conditions.

In fact, it can be stated that the most valuable information about the present structure of the outer heliosphere comes from the ENA fluxes, which are produced downstream of the termination shock by charge exchange processes of H-atoms with shocked solar wind protons at comparatively low energies ($<$5 keV). In a highly dynamic heliosphere, as characteristic for the present one with its 11-year solar activity cycle, the contribution to the ENA fluxes produced by pick-up protons upstream and downstream of the termination shock are significant only at higher energies greater than 10 keV (Fahr and Scherer, 2004). There is no ENA flux observable at Earth connected with particles originating from upstream solar wind protons, since their velocity vectors responsible for the corresponding ENA velocity vector is almost always directed away from the Sun (supersonic solar wind).

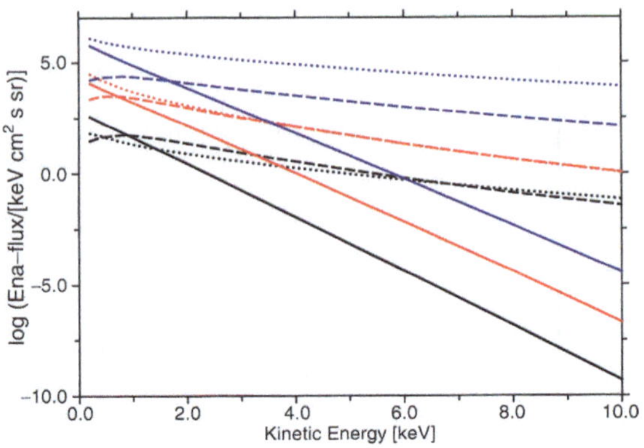

Figure 9.7. The ENA fluxes at 1 AU for $n_H = 0.1,5.0$, and $100\,\mathrm{cm}^{-3}$ produced by the downstream solar wind protons (solid lines) and by the downstream (dashed lines) and upstream (dotted lines) PUI protons. The lowest/highest line in each set corresponds to the lowest/highest hydrogen density in the LISM.

During extended periods of low solar activity (like extended grand minima) the situation is of course different. For quasistatic heliospheric conditions, Figure 9.7 shows the ENA fluxes of the three relevant contributing seed populations (shocked solar wind protons, upstream and downstream PUIs), for some of the DISE cases given in Table 1, and computed for a quasi-static heliosphere from the expression:

$$\Psi(\vec{r},\vec{v}) = \sum_{i=1}^{3} \Psi_i(\vec{r},\vec{v}) = \sum_{i=1}^{3} |n_{\mathrm{i}} f_{\mathrm{i}}(\vec{v}) n_{\mathrm{H}} \sigma_{\mathrm{ex}}(v_{\mathrm{rel}}) v_{\mathrm{rel}}(v)|_{\vec{r}} \tag{4.1}$$

Here, n_{i} and n_{H} are the number densities of the local protons (i $=$ 1: SW downstream; i $=$ 2: PUIs downstream; i $=$ 3: PUIs upstream) and of the H-atoms, respectively, $f_{\mathrm{i}}(\vec{v})$ is the velocity distribution function, σ_{ex} is the charge exchange cross section, and $v_{\mathrm{rel}}(v)$ is the mean relative velocity between H-atoms and the ions with velocity $\vec{v}$. For further details see Fahr & Scherer (2004).

Evidently, the ENA flux produced by the solar wind protons downstream of the termination shock increases with increasing hydrogen number density (see the solid lines in Fig. 9.7), which also holds true for the ENA fluxes due to the other two seed populations. Closer inspection of Eq. 4.1, however, reveals that these increases are nonlinear. Different from the dynamic present-day situation, the latter two dominate the total flux for all hydrogen number densities at nearly all energies. Under these circumstances the sensitivity of the total ENA flux to the interface structure depends on the relative contribution from upstream and downstream PUIs. Note, however, that the situation will change in a dynamical heliosphere governed by variable solar wind momentum flows, triggered by the solar activity cycle as described by Fahr and Scherer (2004).

9.4.2 Solar System Bodies

In the following we discuss the consequences of an increase of the cosmic ray intensities below about 50 MeV due to the enhanced production of ACRs.

Earth. The terrestrial atmosphere is influenced mostly by cosmic ray particles in the energy range 0.1-1 GeV (see the discussion in Scherer, 2000, Scherer et al., 2002), because for lower energies the particles cannot penetrate deep enough through the Earth' magnetosphere into the atmosphere and for higher energies the total flux is too low. Figure 9.6 reveals that, at least for DISE with $n_H \leq 100\,\mathrm{cm}^{-3}$, one should not expect any significant influence of ACRs on the atmosphere and, in particular, on the terrestrial climate as is discussed for GCRs (Svensmark, 1998, Svensmark, 2000 or Shaviv and Veizer, 2003). As mentioned above, this might change for very dense interstellar environments that, however, cannot be modelled with the current model.

As an aside, we mention that the discussion of whether, and if, cosmic rays at all energies could influence the terrestrial climate has just begun. It might turn out that the upper layer of the stratosphere plays an essential role in the long-term behavior of the terrestrial climate through its high altitude clouds. Also, the lower energetic cosmic rays like ACRs can penetrate to such high altitudes. Relevant work on this whole new context is just under discussion, and a more quantitative discussion should be possible in the near future (for example in the forthcoming special issue of Adv. Space Research on the "Influence of the Sun's Radiation and Particles on the Earth's Atmosphere and Climate", edited by J. Pap).

Planets. While the cosmic ray-climate connection might not be affected for cases of a heliosphere immersed in a DISE, there is still another potentially important effect. Within a DISE, not only do PUI or ACR fluxes increase but, of course, also the direct flux of low energy neutral atoms (LENA's). As hydrogen easily chemically reacts with most constituents of planetary atmospheres, these atmospheres can be significantly affected by such a flux increase (see Bzowski et al., 1996, Yeghikyan and Fahr, 2004a, Yeghikyan and Fahr, 2004b). This effect can be amplified for the outer planets, which might be located outside the reduced dimensions of the heliosphere and, thus, be directly in contact with the LISM. In that case the structure of the planetary magnetospheres will also be affected.

Small bodies. Especially those objects which do not have an atmosphere or magnetosphere are directly influenced by the ACR radiation, potentially dominating the fluxes at intermediate and low energies. The top layer of the surfaces of the outer planetary moons can be chemically influenced, as well as the objects from the Kuiper Belt. In DISEs, Kuiper Belt objects will move most likely in orbits located beyond the termination shock, and are thus immersed in the local interstellar medium, while the planetary moons are exposed to an enhanced ACR flux. The cosmic ray bombardment of their surfaces alters the matrix of the minerals and is an additional explanation for the different albedo of these objects. The same holds true for cometary dust and Polycyclic Aromatic Hydrocarbons (PAHs).

Unfortunately, the meteorites found on Earth undergo an ablation process during their passage through the atmosphere and lose parts of their surface. Therefore, they are not good tracers for the cosmic ray radiation history. The surface of the Moon may also be a poor tracer, because the signatures of cosmic rays have to be disentangled from those of the solar wind and of solar energetic particles. With future missions, it might become possible to detect meteorites on the Moon that are not ablated, so that their surface mineralogy may serve as a tracer for the cosmic history of the solar system.

References

Alexashov, D. B., Chalov, S. V., Myasnikov, A. V., Izmodenov, V. V., Kallenbach, R., The dynamical role of anomalous cosmic rays in the outer heliosphere, Astron. Astrophys. 420, 729, 2004

Begelmann, M. C. and Rees M. J., Can cosmic clouds cause climatic catastrophes?, Nature 261, 298–299, 1976

Binney, J., Tremaine, S., *Galactic Dynamics*, Princeton University Press, 1988

Breitschwerdt, D., Self-consistent Modelling of the interstellar medium, Astrophys. Space Sci. 289, 489–498, 2004

Büsching, I., Kopp, A., Pohl, M., Schlickeiser, R., First results of a new cosmic ray propagation code, 28th Int. Cosmic Ray Conf., Tsukuba, Japan, OG, 1981–1984, 2003

Bzowski, M., Fahr, H. J., Rucinski, D., Interplanetary Neutral Particle Fluxes Influencing the Earth's Atmosphere and the Terrestrial Environment, Icarus, 124, 209–219, 1996

Caballero-Lopez, R. A., Moraal, H., McCracken, K. G., McDonald, F. B., The heliospheric magnetic field from 850 to 2000 AD, submitted

Carslaw, K. S., Harrison, R. G., Kirkby, J., Cosmic Rays, Clouds, and Climate, Science 298, 1732–1737, 2002

Chalov, S. V. and Fahr, H. J., A two-fluid model of the solar wind termination shock modified by shock-generated cosmic rays including energy losses, Astron. Astrophys., 288, 973–980, 1994

Chalov, S. V. and Fahr, H. J., The Multifluid Solar Wind Termination Shock and its Influence on the Three-dimensional Plasma Structure Upstream and Downstream, Space Sci. Rev., 72, 237, 1995

Chalov, S. V. and Fahr, H. J., The three-fluid structure of the particle modulated solar wind termination shock, Astron. Astrophys., 326, 860–869, 1997

Cuntz, M., von Bloh, W., Bounama, C., Franck, S., On the possibility of Earth-type habitable planets around 47 UMa, Icarus 162, 214–221, 2003

Fahr, H.-J., Formation of the heliospheric boundaries and the induced dynamics of the solar system: a multifluid view, In: The outer heliosphere: beyond the planets. Based on the spring school "Die äußere Heliosphäre - Jenseits der Planeten," Bad Honnef, Germany, 12-16 April 1999, Eds.: K. Scherer, H. Fichtner, E. Marsch, Katlenburg-Lindau, Germany, Copernicus-Gesellschaft., 67–89, 2000

Fahr, H. J., Scherer, K., Energetic neutral atom fluxes from the heliosheath varying with the activity phase of the solar cycle, ASTRA, 1, 3–15, 2004

Fahr, H. J., Kausch, T., Scherer, H., A 5-fluid hydrodynamic approach to model the solar system-interstellar medium interaction, Astron. Astrophys., 357, 268–282, 2000

Ferreira, S. E. S., Scherer, K., Modulation of Cosmic-Ray Electrons in the Outer Heliosphere, Astrophys. J., 616, 1215–1223, 2004

Ferreira, S. E. S., Potgieter, M. S., Scherer, K., Modulation of Cosmic-Ray Electrons in a Nonspherical and Irregular Heliosphere, Astrophys. J., 607, 1014, 2004

Florinski, V, Zank, G. P., Axford, W. I., The Solar System in a dense interstellar cloud: Implications for cosmic-ray fluxes at Earth and 10Be records, Geophys. Res. Lett., 30, 2206, 2003

Florinski, V., Zank, G. P., Jokipii, J. R., Stone, E. C., Cummings, A. C., Do Anomalous Cosmic Rays Modify the Termination Shock?, Astrophys. J., 611, 1169, 2004

Freyberg, M. J., Breitschwerdt, D. XMM-Newton local bubble and galactic halo survey, Astron. Nach. 324, 162, 2003

Frisch, P. C., Interstellar Matter and the Boundary Conditions of the Heliosphere, Space Sci. Rev., 86, 107–126, 1998

Fröhlich, C., Lean, J., Solar irradiance variability and climate, Astron. Nach., 323, 203–212, 2002

Gruntman, M., Energetic neutral atom imaging of space plasmas, Rev. Sci. Inst., 68, 3617–3656, 1997

Hsieh, K. C., Shih, K. L., Jokipii, J. R. and Grzedzielski, S., Probing the heliosphere with energetic hydrogen atoms, Astrophys. J., 393, 756–763, 1992

Lallement, R., Welsh, B. Y., Vergely, J. L., Crifo, F., Sfeir, D., 3D mapping of the dense interstellar gas around the Local Bubble, Astron. Astrophys., 411, 447–464, 2003

Lean, J., Rind, D., Evaluating Sun-climate relationships since the Little Ice Age, J. Atmos. Sol.-Terr. Phys., 61, 25–36, 1999

le Roux, J. A., Fichtner, H., A self-consistent determination of the heliospheric termination shock structure in the presence of pick-up, anomalous, and galactic cosmic ray protons, J. Geophys. Res., 102, 17365–17380, 1997

McComas, D., Allegrini, F., Bochsler, P., Bzowski, M., Collier, M., Fahr, H., Fichtner, H., Frisch, P., Funsten, H., Fuselier, S., Gloeckler, G., Gruntman, M., Izmodenov, V., Knappenberger, P., Lee, M., Livi, S., Mitchell, D., Mï¿¿bius, E., Moore, T., Reisenfeld, D., Roelof, E., Schwadron, N., Wieser, M., Witte, M., Wurz, P., Zank, G. P., The Interstellar Boundary Explorer (IBEX) in: Physics of the Outer Heliosphere, AIP Conf. Proc., 719, February 2004, Riverside, California. Eds.: V. Florinski, N. V. Pogorelov, G. P. Zank, American Institute of Physics, 162–181, 2004

Mori, M., The galactic diffuse gamma-ray spectrum from cosmic ray proton interactions, Astrophys. J., 478, 225–232, 1997

Myasnikov, A. V., Izmodenov, V. V., Aleksashov, D. B., Chalov, S. V., Self-consistent model of the solar wind interaction with two-component circumsolar interstellar cloud: Mutual influence of thermal plasma and galactic cosmic rays, J. Geophys. Res., 105, 5179, 2000

Scherer, K., Variations of the heliospheric shield, In: *The Outer Heliosphere: Beyond the Planets*, Eds.: Scherer, K., Fichtner, H., Marsch, E., Copernicus-Gesellschaft., 327–355, 2000

Scherer, K., Fichtner, H., Stawicki, O., Shielded by the wind: the influence of the interstellar medium on the environment of Earth, J. Atmos. Sol.-Terr. Phys. 64, 795–804, 2002

Scherer, K., Fahr, H. J., Solar cycle induced variations of the outer heliospheric structures, Geophys. Res. Lett. 30, 17-1, CiteID 1045, DOI 10.1029/2002GL016073

Scherer, K., Ferreira, S. E. S., A heliospheric hybrid model: hydrodynamic plasma flows and kinetic cosmic ray transports, ASTRA, 1, 17–27

Scherer, K., Fichtner, H., Constraints on the heliospheric magnetic field variation during the Maunder Minimum from cosmic ray modulation modelling, Astron. Astrophys. 413, L11–L14, 2004

Shaviv, N. J., The spiral structure of the Milky Way, cosmic rays, and ice age epochs on Earth, New Astron., 8, 39–77, 2003

Shaviv, N. J., Veizer, J., Celestial driver of Phanerozoic climate?, GSA Today 13, 4–10, 2003

Sreenivasan, S. R., Fichtner, H., ACR modulation inside a non-spherical modulation boundary, in: The Outer Heliosphere: The Next Frontiers, Edited by

K. Scherer et al., COSPAR Colloquia Series, Vol. 11, Amsterdam: Pergamon Press, 207, 2001
Stawicki, O., Fichtner, H., Schlickeiser, R., The Parker propagator for spherical solar modulation, Astron. Astrophys., 358, 347–352, 2000
Svensmark, H., Influence of Cosmic Rays on Earth's Climate, Phys. Rev. Lett., 81, 5027–5030, 1998
Svensmark, H., Cosmic rays and Earth's climate, Space Sci. Rev., 93, 175–185, 2000
Svensmark, H., Friis-Christensen, E., Variation of cosmic ray flux and global cloud coverage–a missing link in solar-climate relationships, J. Atm. Sol. Terr. Phys., 59, 1225–1232, 1997
Taylor, J. H., Cordes, J. M., Pulsar distances and the galactic distribution of free electrons, Astrophys. J. 411, 674–684, 1993
von Bloh, W., Cuntz, M., Franck, S., Bounama, C., On the Possibility of Earth-Type Habitable Planets in the 55 Cancri System, Astrobiology 3, 681–688, 2003
Yeghikyan, A. G., Fahr, H. J., Effects induced by the passage of the Sun through dense molecular clouds. I. Flow outside of the compressed heliosphere Astron. Astrophys. 415, 763–770, 2004a
Yeghikyan, A. G., Fahr, H. J., Terrestrial atmospheric effects induced by counterstreaming dense interstellar cloud material, Astron. Astrophys., 425, 1113–1118, 2004b
Zank, G. P., Interaction of the solar wind with the local interstellar medium: a theoretical perspective, Space Sci. Rev., 89, 413, 1999
Zank, G. P., Frisch, P. C., Consequences of the change in the galactic environment of the Sun, Astrophys. J., 518, 965–973, 1999

Chapter 10

THE GALACTIC COSMIC RAY INTENSITY IN THE HELIOSPHERE IN RESPONSE TO VARIABLE INTERSTELLAR ENVIRONMENTS

Vladimir Florinski
Institute of Geophysics and Planetary Physics, University of California, Riverside, CA 92521, USA
vflorins@ucr.edu

Gary P. Zank
Institute of Geophysics and Planetary Physics, University of California, Riverside, CA 92521, USA
zank@ucr.edu

Abstract Indirect indices of cosmic ray intensity at Earth, preserved in cosmogenic isotope records, show strong variability on a wide range of timescales. Some of the variability may be attributed to heliospheric phenomena, such as changes in solar activity and in the geomagnetic field strength. Another possibility is changes in the factors external to the solar system, such as supernova explosions and the variability in the interstellar environment. Here we present a first systematic study of the dependence of cosmic-ray intensity in the inner and the outer heliosphere on the changing galactic environment of the Sun. Three different scenarios (tenuous Local Bubble, Local Interstellar Cloud and diffuse cold interstellar cloud) are investigated as background conditions for galactic cosmic ray propagation. Because cosmogenic isotope production is strongly biased toward higher energies of the primary particles, we compare two models with different levels of turbulence in the energy-containing part of the spectrum. It is shown that the amount of modulation that cosmic rays experience depends on the interplay between turbulence level variability due to pickup ions and the changes in the extent of the heliosheath. Our calculations show that cosmic-ray intensities at intermediate rigidities (1 GeV) vary by more than a factor of four between extreme cases. Cosmogenic isotopes show a similar response giving a

P. C. Frisch (ed.), Solar Journey: The Significance of our Galactic Environment for the Heliosphere and Earth, 281–316.

range of variation between a 25% decline in low-density environments and an increase in excess of 300% in high density clouds.

Keywords: Cosmic rays, solar wind, interstellar clouds, ^{10}Be records

10.1 Introduction

Galactic cosmic rays (GCRs) are highly energetic charged particles believed to be produced by the process of diffusive acceleration at supernova remnant shocks. As GCRs encounter the solar system, they experience a process called modulation. Modulation is caused by cosmic-rays interacting with plasma of the solar wind (SW) and the embedded magnetic field. Modulation may be represented as a superposition of a "regular" component, varying on large scales ($\gg r_g$, the cosmic-ray gyroradius) that enables particles to drift across large-scale magnetic fields, and a small scale chaotic or turbulent component. Particle propagation inside the heliosphere is a combination of four phase-space modes of transport: (a) convection with the mean plasma velocity, (b) drift motion in a nonuniform magnetic field, (c) stochastic motion (diffusion) in turbulent magnetic fields, and (d) adiabatic cooling in expanding geometries due to a prevalence of overtaking interactions with the ambient fluctuations. The interplay between the four processes results in a complicated dependence of GCR intensities and spectra on the properties of the heliosphere at any given location and phase of the solar cycle (e.g., Potgieter, 1998; Jokipii and Kota, 2000; Potgieter et al., 2001). Generally heliospheric modulation affects particles with rigidities of 10 GeV or less. The sensitivity of these modulation processes to the variations in the interstellar environment of the solar system is the subject of this chapter.

While direct measurements of cosmic-ray intensities became possible only a few decades ago, indirect records are available for much earlier times. Past history of cosmic-ray intensity is preserved in cosmogenic isotope records found in both terrestrial (Beer, 1997) and extraterrestrial (Lal, 1972) samples. Because variations in the interstellar environment are expected to occur on relatively long time scales ($>10^4$–10^5 years), isotopes with a longer half-life are preferred. These include ^{10}Be, ^{26}Al, ^{36}Cl, ^{53}Mn and ^{81}Kr. Beryllium-10, with a half-life of 1.5×10^6 years is particularly useful to study past cosmic-ray variability because it is produced in relative abundance in particle showers caused by the passage of energetic cosmic-ray ions through the Earth's atmosphere. High resolution measurements of ^{10}Be are available from polar ice samples (Raisbeck et al., 1985; Wagner et al., 2000) and marine sediments (Cini Castagnoli et al., 1998; McHargue et al., 2000). The available samples provide records of cosmic-ray variability extending more than three million years into the past (Aldahan and Possnert, 2003).

All ^{10}Be records display significant variability on virtually all timescales (Figure 10.1, see also Chapter 12 in this book). The most widely publicized features are two prominent events occurring 35,000–40,000 and 60,000–65,000 years ago when the flux of cosmic rays reaching Earth exceeded the present value by a factor of $\sim$2 for two thousand years. The increases could be attributed to any number of factors pertaining to cosmic ray production and transport from their place of origin in supernova remnant shocks to their "demise" in air showers. These can generally be divided into three categories: (1) increases in production due to a recent supernova explosion within a relatively short distance from the Sun, (2) changes in the strength of heliospheric modulation due to either a change in the solar activity level or in the interstellar environment, and (3) variations in the geomagnetic field that prevents lower energy cosmic rays from reaching the atmosphere at low latitudes.

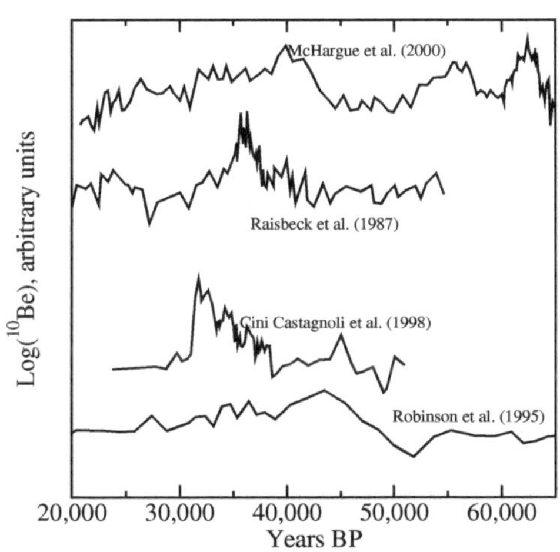

Figure 10.1. ^{10}Be measurements in ice and marine sediment samples from different sources. Note the prominent enhancement at 35,000–40,000 BP.

The supernova hypothesis has been investigated in Sonett et al. (1987) and Kocharov (1994). This scenario involves the passage of a blast wave past the heliosphere carrying with it an enhanced flux of accelerated particles. The Geminga pulsar at a distance of 160 parsec has been suggested as a remnant that caused some of the observed ^{10}Be anomalies (Ellis et al., 1996). Such events are expected to be extremely rare. Axford (1981) estimated that the upper limit on the rate of supernova shock passage of any given point in space is about $4 \times 10^{-7}\,\mathrm{yr}^{-1}$ for shocks that are capable of producing a twofold increase in GCR fluxes. The actual rate will likely be even smaller owing to shock weakening by the cosmic rays. There is currently no substantial evidence for a nearby supernova occurring after the sequence of explosions that formed the Local Bubble and Loop I (see below) some 10^6 years ago (Smith and Cox, 2001).

The effects of solar activity variations on cosmic-ray fluxes in the past are easily identified in records dating back less than 1000 years (Beer, 2000). Prominent increases correspond to solar activity minima, such as the well known Maunder minimum, during which the SW was extremely quiet. Under these conditions both the tilt of the neutral sheet and the amount of turbulence in the SW remain small, which leads to easier GCR access to the inner heliosphere via drift and diffusive transport (Jokipii, 1991). Recorded minima occur every 200 years, on average (Beer, 2000), and are probably responsible for most of the short-term variability in cosmogenic isotopes (Jokipii and Marti, 1986). There is currently no evidence that solar activity minima can span thousands of years or longer.

Over intermediate timescales, changes in the Earth's magnetic dipole strength are believed to be responsible for most of the variability. The geomagnetic field undergoes complete reversals every 250,000 years, on average, as well as more frequent excursions, when the field recovers with the original polarity. During such periods the geomagnetic field can remain weak for thousands of years thus offering no protection against cosmic radiation. Analysis of magnetization levels in marine sediments show that they are reasonably well correlated with the isotopic data (McHargue et al., 2000; Wagner et al., 2000; Aldahan and Possnert, 2003), although discrepancies were also reported (Cini Castagnoli et al., 1998). The geomagnetic hypothesis relies on the assumption that ^{10}Be is being distributed evenly across the globe via, e.g., atmospheric transport, because otherwise the theory could not explain large variations in the polar regions where the magnetic shielding is inefficient. McCracken (2004) analyzed several mixing models and found that a relatively modest amount of latitudinal mixing is sufficient to explain the records.

Changes in the interstellar conditions as a source of cosmic-ray variability have not been considered until recently, when the general understanding of the morphology of the local interstellar medium began to emerge. The Sun and surrounding stars are located inside a low density hot ($T \sim 10^6$ K) Local Bubble (LB) believed to have been formed by supernova explosions some 10^6 years ago. Embedded in the LB are denser partially ionized interstellar clouds such as the Local Interstellar Cloud (LIC) where the Sun is presently located (Ferlet, 1999; Frisch, 2000; Lallement, 2001). The properties of the LIC and the types of interstellar environment that the Sun is likely to encounter during its journey through space are discussed in Chapter 5 and 6 of this book. As far as cosmic rays are concerned, Zank and Frisch (1999) were the first to recognize the possibility that cosmogenic isotope anomalies could have been caused by the solar system passing through a relatively dense interstellar cloud (30 times denser than the LIC in their paper). As shown in Zank and Frisch (1999) and Müller et al. (2001) (see also Chapter 2), the heliosphere shrinks in size by a factor of 5–10 with the termination shock (TS) moving in to 10–20 AU. Further estimates of cosmic-ray intensities under these conditions (Florinski et al., 2003a) based on a 1D transport model revealed that the flux of GCR protons in the energy range between 100 MeV and 1 GeV reaching Earth could be 1.5–3 times higher than at present, mostly as a result of the reduced amount of heliosheath diffusion. Here we present a more systematic investigation of the response of cosmic ray fluxes at 1 AU and in the outer heliosphere to changes in the local interstellar environment. Three models are evaluated ranging from the hot and tenuous LB to a diffuse cold neutral interstellar cloud 30 times denser than the LIC (Table 10.1).

Table 10.1. Summary of interstellar environments investigated in this work. Here $N_{\mathrm{t}\,\infty}$, T_∞, and V_∞ are the total hydrogen density (H+H$^+$), temperature and speed of the cloud relative to the Sun, respectively, and η is the ionization ratio. For the LB, values marked with an asterisk (*) were used in the multifluid numerical model (Müller et al., in preparation) to calculate the positions of the interface boundaries, but only the LISM velocity condition was used here. See Section 10.5 for more details.

Model	$N_{\mathrm{t}\,\infty}$, cm^{-3}	η	T_∞, K	V_∞, km/s
Local Bubble (LB)	0.005*	1.0	$1.2 \times 10^{6*}$	12.5
Local Interstellar Cloud (LIC)	0.21	0.33	7000	25
Diffuse Cloud (DC)	10	0.0	200	25

Another new and intriguing terrestrial aspect of cosmic ray research worth mentioning here is the possibility that GCRs could directly affect the Earth's climate. Svensmark and Friis-Christensen (1997) suggested that cosmic rays may influence the Earth's cloud cover thus modulating the amount of solar radiation received by the surface. The link between GCRs and clouds has not been confirmed yet (Wagner et al., 2001), although some supporting evidence is available (Christl et al., 2004). The physical mechanism that would cause clouds to form as a result in GCR intensity increases also has not yet been identified. One possibility is that cosmic-ray-induced ionization affects the formation of aerosols that facilitate cloud condensation (Marsh and Svensmark, 2000). If the climate connection is confirmed, it would also establish a curious relationship between interstellar and terrestrial clouds via cosmic rays.

10.2 Transport Properties of the Heliospheric Interface

Before we proceed to study each individual model, we would like to reiterate the general properties of the heliospheric interface as a medium for cosmic-ray transport. As we shall see, the general structure of the heliosphere is qualitatively similar in all interstellar environments under consideration, therefore we will discuss its properties using the LIC as the best known example. Figure 10.2 shows the schematic view of the present day heliosphere. The principal regions are the supersonic SW region, enclosed inside the TS, the inner heliosheath region that extends into the heliotail in the anti-apex direction and is separated from the interstellar plasma by the heliopause (HP), and the local interstellar medium (LISM) region that contains the bow shock (BS). Properties of the plasma and neutral gas in the interface have been investigated extensively (see Zank, 1999a for a review). Here we will only focus on the properties of the interface that are important for energetic particle transport and on the methods used to include them in modulation models.

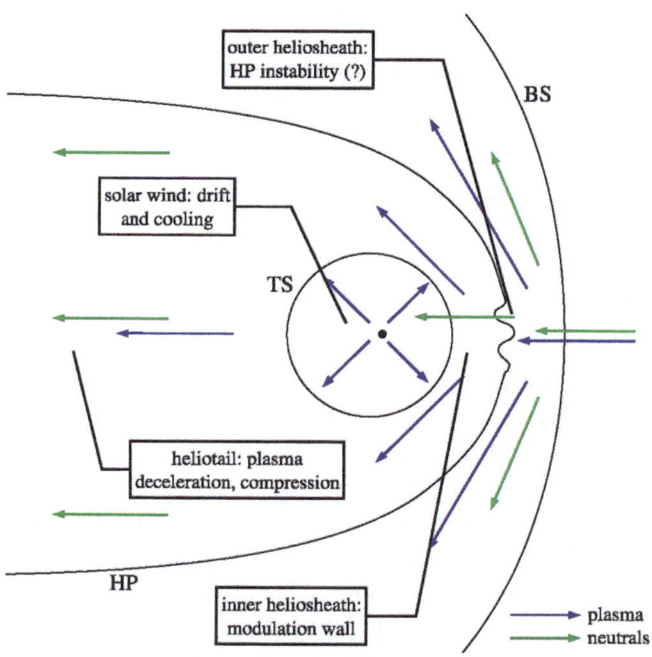

Figure 10.2. Schematic view of the modulation regions in the heliosphere. Region 1 is the SW, Region 2 is the heliosheath and heliotail and Region 3 is the LISM.

10.2.1 Solar Wind (Region 1)

Galactic particle propagation in the solar wind is reasonably well understood. During solar minima, drift motion dominates the transport with positive ions entering at high latitudes during periods of positive field polarity, when the magnetic field is directed away from the Sun in the northern hemisphere and toward the Sun in the southern hemisphere, and through the warped neutral sheet during negative periods, when the magnetic field direction is just the opposite. During periods near solar maxima, drifts are less effective and transport is controlled by diffusion and changes in the magnetic field intensity and topology including transient propagating structures (Jokipii and Kota, 1995; Potgieter et al., 2001). Adiabatic cooling operates at about the same rate during minima and maxima and ensures that low energy spectra are always populated due to cooling at high energies.

At low and intermediate latitudes the SW magnetic field geometry is essentially that of the Archimedian spiral (Parker, 1958), which is in agreement with the Voyager observations (Burlaga et al., 1998). At high latitudes the Parker model predicts a nearly radial field, which is inconsistent with Ulysses measurements of GCR latitudinal gradients that imply that cosmic rays do not have an unimpeded access to the inner heliosphere, but can apparently propagate in latitude with relative ease (Heber, 2001). To account for this discrepancy, two modifications to the high latitude field are commonly made (Fisk and Jokipii, 1999). The first mechanism involves a fluctuating random transverse field component produced from footpoint motion on the Sun (Jokipii and Kota, 1989). To model this effect we include an extra component B_m to the system of equations for the magnetic field, as described in Florinski and Jokipii (1999) (see Equation 3.3 below). This modification reduces both the radial mean free path and drift velocity in the polar regions. The second mechanism (Fisk, 1996) derives a "regular" latitudinal magnetic field component from differential rotation of the solar photosphere and nonradial expansion of wind in the corona, which establishs direct field-line connections between different latitudes. To take this effects into account we adopt a simple approach, commonly used in modulation studies (e.g., Burger et al., 2000; Parhi et al., 2004), of increasing the latitudinal component of the cosmic-ray diffusion tensor as calculated from a theoretical model (see Equation 3.9 below) by a fixed ratio.

Diffusive cosmic ray transport crucially depends on the properties of the interplanetary turbulence. A useful theory for the evolution of turbulence in the SW has been developed by the Bartol group (Zhou and Matthaeus, 1990; Zank et al., 1996a; Matthaeus et al., 1999). In this theory, turbulent fluctuations are described in terms of small-scale Elsässer fields $\mathbf{z}^{\pm} = \delta\mathbf{u} \pm \delta\mathbf{B}/(4\pi\rho)^{1/2}$, where the '$+$' or '$-$' sign implies a sense of propagation parallel or antiparallel to the mean magnetic field $\mathbf{B}$, which satisfy the condition of local incompressibility. Here $\mathbf{u}$ is the solar wind velocity and ρ is the density. It is assumed that the fluctuations are axially symmetric about the direction of $\mathbf{B}$. Two kinds of fluctuations are present in the SW: those with wavevectors $\mathbf{k}$ along $\mathbf{B}$ (slab component) and those with wavevectors perpendicular to $\mathbf{B}$ (2D component). In what follows we will use the terms $k_{\parallel}$ and $k_{\perp}$ for the wavevector magnitudes of the two components, respectively. In both cases, fluctuating fields are in the directions orthogonal to both $\mathbf{k}$ and $\mathbf{B}$. Observations show that most of the energy (75% to 95%) resides in the 2D component (Bieber et al., 1996). The basic theory does not distinguish between slab and 2D turbulence and describes both components in terms of the total fluctuation energy $Z^2 = \langle \mathbf{z}^+ \cdot \mathbf{z}^+ + \mathbf{z}^- \cdot \mathbf{z}^- \rangle / 2 = \langle \delta \mathbf{u}^2 \rangle + \langle \delta \mathbf{B}^2 \rangle / (4\pi\rho)$, and the associated correlation length $l_c = \int \langle \mathbf{z}^+ \cdot \mathbf{z}^{+'} + \mathbf{z}^- \cdot \mathbf{z}^{-'} \rangle dx / Z^2$, where the prime indicates that a quantity is taken at a spatially lagged position and the integration is performed along the direction of variance (parallel to $\mathbf{B}$ for slab and perpendicular to $\mathbf{B}$ for 2D turbulence). Using a single correlation length requires that the cross-helicity $\langle \mathbf{z}^+ \cdot \mathbf{z}^+ - \mathbf{z}^- \cdot \mathbf{z}^- \rangle$ is small, a condition that is well satisfied in the SW, at least for low latitudes at heliocentric distances beyond 2 AU (Roberts et al., 1987).

Under the assumptions of a steady radial superalfvenic solar wind, the equations describing the radial evolution of Z^2 and l_c may be written as (Zank et al., 1996a)

$$u_r \frac{dZ^2}{dr} + \left(\frac{u_r}{r} + \frac{1}{2} \frac{du_r}{dr} \right) Z^2 - \frac{\Gamma u_r}{r} Z^2 = -\frac{Z^3}{l_c} + S_{\text{sh}} + S_{\text{PI}} \tag{2.1}$$

$$u_r \frac{dl_c}{dr} + \frac{\Gamma u_r l_c}{r} = \frac{Z}{2} - \frac{l_c}{2Z^2} (S_{\text{sh}} + S_{\text{PI}}). \tag{2.2}$$

Here Γ is a parameter related to the symmetry properties of the turbulence, $S_{\text{sh}} = C_{\text{sh}} u_r \cos\theta Z^2 / r$ describes driving by large scale compression and shear, and $S_{\text{PI}} = f v_a u_r \dot{n}_{\text{PI}} / n$ is the energy supplied to the turbulence by the process of pickup ion scattering by slab Alfvén waves (Williams and Zank, 1994). In the above expressions $\dot{n}_{\text{PI}}$ is the rate of pickup ion (PUI) production in charge exchange collisions, v_a is the Alfvén speed, $n = \rho / m_{\text{p}}$ is the number density (m_{p} is the proton mass), and C_{sh} is a phenomenological constant. The second term in Equation 2.1 includes the effect of SW slowdown by charge exchange. We use $\Gamma = 0.1$ and $C_{\text{sh}} = 2.0$ in our simulations that make the theory consistent with the turbulence measurements by the Voyager 2 and Pioneer 11 spacecraft in the inner and the outer heliosphere (cf. Smith et al., 2001; Parhi et al., 2004). We also included a latitude dependence factor in the shear driving term, to reflect the fact that large scale structures (shocks and corotating interaction regions) are less common at high latitudes. The factor

f in the PUI driving term implies that only a fraction of the energy difference between the initial ring-beam and the final stable phase space distributions is available in the form of resonantly generated waves. The introduction of this factor is prompted by the work of Smith et al. (2001) who found that the amount of turbulent heating of the solar wind at large heliocentric distances is inconsistent with the values of $f \sim 0.5$–1 predicted by the conventional theory (Williams and Zank, 1994). Isenberg et al. (2003) explain this discrepancy by noting that PUIs can resonantly interact with both Alfvén and whistler waves, and that the initial PUI ring distribution will be spread in energy due to ambient fluctuations that result in a nearly isotropic final distribution, while Chalov et al. (2004) suggest that self-generated waves are immediately absorbed by the PUIs and do not contribute to the cascade to small wavenumbers (dissipation range) where the SW heating occurs. Note that the density of interstellar neutral hydrogen is the principal source of variability (Table 10.1) and thus PUI turbulence driving is of crucial importance in studies of the galactic environment effects on cosmic-ray propagation.

10.2.2 Inner Heliosheath (Region 2)

The Voyager 1 spacecraft is now positioned to cross the TS and enter the inner heliosheath. Evidence for modulation in this region comes from Voyager observations of intensities and radial gradients of GCRs in the outer heliosphere (Webber and Lockwood, 2001a; Webber and Lockwood, 2001b; McDonald et al., 2004), and from the time history of particle recovery following a passage of a global interplanetary disturbance that continues to produce a decrease in GCR intensity after crossing into the heliosheath (McDonald et al., 2000). It is estimated that during positive solar minima, where magnetic fields are north-pole positive, 50–80% of the modulation occurs beyond the TS. During solar maxima the amount of heliosheath modulation inferred is up to 90% at 0.6 GeV and is somewhat less during negative solar cycles. Nevertheless, strong caution must be exercised when interpreting these observations because of a large uncertainty in our knowledge of the interstellar GCR intensities. For example, little can be inferred about the population of lower-energy cosmic rays (below 100 MeV) in interstellar space because these particles are unable to penetrate into the SW and the observed populations are produced in Region 1 by adiabatic cooling from higher energies.

Drift and diffusion in both the SW and the heliosheath are important in producing the global modulation pattern (Florinski et al., 2003b; Ferreira and Scherer, 2004). MHD models show that the Parker spiral extends into the heliosheath where it is wound more tightly as a result of shock compression and flow deceleration (Washimi and Tanaka, 1996; Linde et al., 1998; Pogorelov et al., 2004). On approach to the HP magnetic ridges are formed where the field is strongly compressed. Because of the low plasma speed in the heliosheath, the region preserves an imprint of several preceding solar cycles in a compressed form. The field topology is expected to consist of envelopes with opposite field polarity interspersed with regions with a strongly mixed field as a result of solar rotation and dipole tilt (Nerney et al., 1995). Because the field is so disorganized, large-scale drifts in the heliosheath are likely to be suppressed (Langner et al., 2003).

The observed pattern of solar modulation is qualitatively in agreement with the concept that drifts in the SW control GCR propagation under solar minimum activity conditions (i.e., Jokipii et al., 1993; Florinski and Jokipii, 1999). Eliminating drift motion in the heliosheath has an insignificant impact on modulation (Langner et al., 2003). However, the relative amount of GCR attenuation in the heliosheath increases significantly if the downstream diffusive mean free path decreases by a factor larger than the shock compression ratio (Caballero-Lopez et al., 2004). As we will show, such behavior is expected from theoretical predictions (Section 10.4). Consequently, the inner heliosheath may be filtering a significant part of the incoming GCR flux during the entire solar cycle.

Little has been done to study turbulence evolution in the heliosheath. It is not presently known if any of the driving mechanisms operating in the supersonic SW will be effective here. PUIs produced by charge exchange with the LISM neutrals, for example, will have a velocity distribution that is not too different from the hot background plasma. Consequently, PUIs are probably not a significant source of turbulence in the heliosheath. However, interaction of the TS with large scale solar wind structures is likely to populate the heliosheath with waves and discontinuities (Story and Zank, 1995; Story and Zank, 1997; Zank and Pauls, 1997) that would in turn contribute to turbulent driving.

Due to a lack of suitable theory of turbulence transmission across a shock we use an estimate based on Alfvén wave amplification (McKenzie and Westphal, 1969, see also Chalov and Fahr, 2000)

$$\langle \delta B_2^2 \rangle = \frac{s(s+1)}{2} \langle \delta B_1^2 \rangle, \tag{2.3}$$

where $\langle \delta B_1^2 \rangle$ and $\langle \delta B_2^2 \rangle$ are magnetic variances in front and behind the TS with a compression ratio s. The above expression is valid when the Alfvén velocity in the direction of the shock normal is small compared with the flow speed both upstream and downstream of the shock. Here we treat the TS as quasi-perpendicular at all latitudes in view of a presence of a large transverse Jokipii–Kota component in the polar region. Because $B_2 \simeq sB_1$, the ratio $\langle \delta B^2 \rangle / B^2$ does not change significantly across a quasi-perpendicular shock and we may take $\langle \delta B_2^2 \rangle / B_2^2 = \langle \delta B_1^2 \rangle / B_1^2$. Furthermore, we assume $\langle \delta B^2 \rangle / B^2 = \mathrm{const}$ in the heliosheath. Slab waves preserve their wavenumber $k_\parallel$ on crossing a perpendicular shock, and the downstream spectrum will be an enhanced version of the upstream spectrum with same correlation length l_c. Because we use a single correlation length for both the slab and 2D components, the effect of a possible increase in $k_\perp$ across the TS is necessarily ignored.

10.2.3 Outer Heliosheath (Region 3)

The outer heliosheath is part of the LISM and is therefore filled with plasma of interstellar origin. Florinski et al. (2003b) estimated the diffusion coefficient in this region and concluded that it should cause no additional modulation because the canonical spatial scale of interstellar turbulence ($l_c \sim 10^{17}$–10^{18} cm) is too large and there is consequently little fluctuating power available at typical cosmic-ray gyro-resonance scales (Armstrong et al., 1995). Nevertheless it is possible (see, e.g., Liewer et al., 1996; Zank et al., 1996b) that turbulence can be generated locally as a result of hydrodynamic instabilities operating in the vicinity of the nose of the HP. The common instability in the proximity of the stagnation point is of the Rayleigh–Taylor type, driven by charge exchange between the plasma and the neutral gas (Zank, 1999b; Florinski et al., 2005). This instability may serve as a driver for turbulence on shorter length-scales ($\sim 10^{14}$ cm), which would subsequently transport turbulent energy to even smaller scales by means of a turbulent cascade. It is possible that the diffusion coefficient will be relatively small in the outer heliosheath, which would lead to compressional re-acceleration of cosmic rays in this region. Still, given the lack of theoretical and observational data on the outer heliosheath, we will for the present ignore any possible contributions to global modulation from this region.

10.3 Cosmic Ray Transport Model

Galactic cosmic rays have large energies and their anisotropies are accordingly small. Particle transport may be then described via their pitch-angle averaged distribution function $f(\mathbf{r}, p)$ using Parker's equation (Parker, 1965; Gleeson and Axford, 1967; Skilling, 1975)

$$\frac{\partial f}{\partial t} + \mathbf{u} \cdot \nabla f + \nabla \cdot (\kappa \cdot f) = \frac{\nabla \cdot \mathbf{u}}{3} \frac{\partial f}{\partial \ln p}, \tag{3.1}$$

where κ is the spatial diffusion tensor. While the latter formally contains drift effects in the antisymmetric part, we neglect drift motions here in view of the specific geometry of the problem (see below). The model we use to calculate $\mathbf{u}$ and κ represents a major departure from traditional (local) models of solar modulation (see reviews by Potgieter, 1998 and Jokipii and Kota, 2000). We employ a global heliospheric model (see, e.g., Baranov and Malama, 1993; Pauls et al., 1995; Zank et al., 1996b; Müller et al., 2000; Fahr et al., 2000;

Myasnikov et al., 2000 and Chapters 2 and 3 of this book) that calculates $\mathbf{u}$ and $\mathbf{B}$ self-consistently for the entire region formed by the interaction between the fully ionized solar wind and the partially ionized interstellar gas (note that some of the models mentioned above did not include the magnetic field). For the Local Bubble environment a different approach, based on an analytic approximation for the velocity field is used (Section 10.5). Once the two vector fields are computed, turbulence parameters and diffusion coefficients can be calculated at every grid point as described below. The transport Equation 3.1 is then solved using a time-dependent Alternating Direction Implicit (ADI) numerical scheme.

Our current model is two-dimensional, i.e., axisymmetric about the direction of the interstellar flow (the z-axis). MHD plasma and gas-dynamic neutral conservation laws are solved in the azimuthal half-plane $\phi = 0$ using a second order Total Variation Diminishing (TVD) numerical scheme. For the SW we used the same 1 AU boundary values for all three models, namely, plasma number density $n_0 = 5\,\mathrm{cm}^{-3}$, radial velocity $u_{r0} = 500\,\mathrm{km\,s}^{-1}$, and temperature $T_0 = 10^6$ K. Note that a 2D model allows only limited MHD phenomena to be included in the LISM region. The only permissible configuration is that of a LISM magnetic field coaligned with the plasma flow velocity vector. Some (Frisch, 2003b) argue that the magnetic field in the LIC lies in the galactic plane, which is inclined by 60° to the ecliptic, while others (Cox and Helenius, 2003) suggest that the field should be co-aligned. Both configurations are consistent with currently available theoretical and observational results pertaining to the magnetic field in the LIC (Florinski et al., 2004). Whereas a realistic MHD description of the heliospheric field requires a 3D model, a simpler kinematic approach allowed Florinski et al. (2003b) to include the fully 3D field by ignoring magnetic forces on the plasma. This approximation becomes invalid only near the HP where the flow approaches stagnation (Linde et al., 1998; Pogorelov et al., 2004) and in the vicinity of the neutral sheet (Opher et al., 2004), while being reasonably accurate elsewhere. The transport equations describing the evolution of the heliospheric magnetic field are (Florinski et al., 2003b)

$$\frac{\partial \mathbf{B}_{x,z}}{\partial t} + \nabla_{x,z} \times (\mathbf{u} \times \mathbf{B}_{x,z}) = -\mathbf{u}(\nabla_{x,z} \cdot \mathbf{B}_{x,z}) \tag{3.2}$$

$$\frac{\partial B_{\phi,m}}{\partial t} + \nabla_{x,z} \cdot (\mathbf{u} B_{\phi,m}) = 0, \tag{3.3}$$

where B_m is the magnitude of the Jokipii–Kota fluctuating field. The Parker field solution (Parker, 1958) is assumed at the inner boundary.

The diffusion tensor κ can be determined once the magnetic field topology (Equations 3.2 and 3.3) and the turbulence parameters (Equations 2.1 and 2.2) are known from the transport model, by rotating from the field-aligned coordinate system (where the diffusion tensor is diagonal with components equal to $\kappa_{\|}$ and $\kappa_{\perp}$ in the directions parallel and perpendicular to $\mathbf{B}$, respectively) to obtain

$$\kappa_{ij} = \kappa_{\perp} + (\kappa_{\|} - \kappa_{\perp})\frac{B_i B_j}{B^2}. \tag{3.4}$$

Here $i = x, z$, and B^2 is calculated as the sum of the mean field plus the Jokipii–Kota component, by choosing an appropriate model of particle diffusion parallel and perpendicular to the mean magnetic field. The former can be obtained from the Fokker–Planck pitch-angle diffusion coefficient $D_{\mu\mu}$ as (Jokipii, 1966; Schlickeiser, 1989)

$$\kappa_{\|} = \frac{w^2}{8}\int_{-1}^{1}\frac{(1-\mu^2)^2}{D_{\mu\mu}}d\mu, \tag{3.5}$$

where w is the particle's velocity and μ is the pitch angle cosine. The corresponding mean free path is $\lambda_{\|} = 3\kappa_{\|}/w$.

The pitch-angle diffusion coefficient depends critically on the properties of the interplanetary turbulence (see Dröge, 2000 for a review of popular models of parallel diffusion). The basic physical process that governs pitch-angle diffusion is resonant scattering by slab fluctuations. Whereas at low rigidities wave damping effects are important (Schlickeiser and Achatz, 1993; Bieber et al., 1994), transport of high rigidity particles may with a high degree of accuracy be described by taking the turbulence to be magnetostatic. In quasi-linear theory (QLT) the Fokker–Planck coefficient takes on a simple form (Jokipii, 1966)

$$D_{\mu\mu} = \frac{\pi w(1-\mu^2)}{r_g^2 B^2|\mu|}P_{xx}\left(\frac{1}{|\mu| r_g}\right), \tag{3.6}$$

where $P_{xx}(k_{\|})$ refers to spectral power in any one slab turbulence component so that the total power in slab turbulence is $\langle\delta B^2_{x,sl}\rangle = \int P_{xx}dk_{\|}$. The argument of the power spectrum distribution expresses the resonance condition with $r_g = pc/eB$ being the particle cyclotron radius. Formally one can write

$$\langle\delta\mathbf{B}^2\rangle = \frac{\langle\mathbf{Z}^2\rangle}{4\pi\rho}(1 + r_A), \tag{3.7}$$

where r_A is the constant Alfvén ratio, i.e., the ratio of kinetic and magnetic turbulent energies (Zank et al., 1996a). In practice, it is difficult to use this relationship because the validity of the turbulence transport model, described by Equations 2.1 and 2.2, is limited to fluctuations at the high end of the energy range. Smith et al. (2001) report the average value in one perpendicular component $Z_x^2 = 350\,\mathrm{km}^2/\mathrm{s}^2$ at 1 AU using a variety of spacecraft measurements, a value we adopt as the inner boundary condition. The calculated value of Z_x^2 is then used as a proxy for the total magnetic variance $\langle\delta B_x^2\rangle$ using the normalization value of $\langle\delta B_x^2\rangle = 13.2\,\mathrm{nT}^2$ at 1 AU (Bieber et al., 1994). We use a constant partition coefficient between slab and 2D components $\langle\delta B_{x,sl}^2\rangle/\langle\delta B_x^2\rangle = 0.1$. Such a small ratio is predicted by the theory of weakly compressible turbulence (Zank and Matthaeus, 1993) and is well within the measured limits (0.05–0.25, Bieber et al., 1996).

The power spectrum measured in the solar wind at low latitudes consists of a flat energy-containing range populated by large scale fluctuations ($kl_c \leq 1$), followed by an inertial range ($kl_c \geq 1$) characterized by spectral transfer of energy from large to small scales with a classical Kolmogoroff spectral slope $P_{xx} \sim k^{-5/3}$ (e.g., Bieber et al., 1994). At the same time, some observational evidence points to a k^{-1} dependence at high latitudes (Horbury and Balogh, 2001). The low end of the turbulence spectrum is likely to be populated by structures (shocks and discontinuities) that are generally removed from the data reported (Smith et al., 2001). However, these structures may be responsible for the modulation of very high energy (above 1 GeV) particles. Accordingly, we use two forms of the power spectrum, identical in the inertial range (with a spectral index of $5/3$), but having a different spectral index in the energy-containing spectral region: 0 (Model 1) and -1 (Model 2). Note that the spectrum cannot be normalized in the second model, and we assume that the k^{-1} dependence is simply an excess over the flat spectrum. On assuming that the change in slope of the spectrum occurs at $k = l_b^{-1}$, the parallel mean free path is easily calculated from Equations 3.5 and 3.6 to be

$$\lambda_\parallel = \frac{135}{14\pi}\frac{r_g^{1/3} l_b^{2/3}}{A_{sl}^2}, \qquad r_g < l_b$$

$$\lambda_\parallel = \frac{15}{8\pi}\frac{r_g^2}{l_b A_{sl}^2}\left[\frac{1}{2} + 5\left(\frac{l_b}{r_g}\right)^2 - \frac{5}{14}\left(\frac{l_b}{r_g}\right)^4\right], \qquad r_g > l_b \text{ (Model 1)} \tag{3.8}$$

$$\lambda_\parallel = \frac{15}{2\pi}\frac{r_g}{A_{sl}^2}\left[\frac{1}{3} + \frac{l_b}{r_g} - \frac{1}{21}\left(\frac{l_b}{r_g}\right)^3\right], \qquad r_g > l_b \text{ (Model 2)},$$

where $A_{sl}^2 = \langle\delta B_{sl}^2\rangle/B^2$ is the slab turbulent ratio. The second form of the diffusion coefficient was also used by Burger and Hattingh (1998) and Burger et al. (2000). This form provides a flatter rigidity and magnetic field dependence for the high-energy cosmic rays than the first form. Note that the correlation length (as defined for Model 1) is $l_c = \pi l_b/5 = 0.628 l_b$.

Diffusion perpendicular to the mean magnetic field is caused by field line separation. For convenience, we define the perpendicular "diffusive path" as $\lambda_\perp = 3\kappa_\perp/w$. This does not have the same physical meaning as the parallel mean free path, which can be viewed as the mean distance traveled parallel to $\mathbf{B}$ between successive interactions with magnetic irregularities, because particles follow helical trajectories and do not have a well-defined perpendicular velocity. QLT specifically assumes that particles follow the field lines in the slab geometry, so that $\lambda_\perp \sim A_{sl}^2 l_c$ for particles of sufficiently high energies (Jokipii, 1971; Forman et al., 1974). Bieber and Matthaeus (1997) extended the theory by introducing a more general concept of "orbit decorrelation" rate. Under the assumption that field line diffusion is the source of decorrelation, the Bieber and Matthaeus (1997) expression for $\lambda_\perp$ (called the BAM model) is formally similar to the QLT prediction, however, the effects of the 2D turbulence component could now be accommodated. Nevertheless, numerical simulations of particle motion in turbulent magnetic fields (Giacalone and Jokipii, 1999) were in considerable disagreement with both the QLT and BAM predictions.

The failure of the theories to predict the outcome of numerical simulations was traced to the fact that the theories assumed free streaming in the direction of the mean field, rather than a random walk (diffusion). When parallel diffusion was included in the simulations, perpendicular displacement turned out to be subdiffusive in the slab geometry because the particles tend to retrace their trajectories along similar field lines (Qin et al., 2002a). However, perpendicular diffusion is recovered if the field possesses sufficient transverse structure, i.e., in a composite (slab+2D) geometry (Qin et al., 2002b), although the diffusion rate differs significantly from the QLT prediction. Theory is now converging with the simulations results, because of the introduction of nonlinear guiding center theory, or NLGC (Matthaeus et al., 2003). In this theory, particle gyrocenters follow field lines, but the particles are nevertheless able to sample the transverse magnetic field structure by following stochastic trajectories, rather than the unperturbed orbits of the QLT. The result is that $\lambda_\perp$ depends on both $\lambda_\parallel$ and itself in a complicated fashion through an integral equation. Approximate forms for $\lambda_\perp$ were derived by Zank et al. (2004) and Shalchi et al. (2004). The expression becomes particularly simple at high energies, as appropriate for GCRs. In this limit the contribution of the slab turbulence component to perpendicular transport may be neglected and the perpendicular diffusive path is given by Zank et al. (2004) as

$$\lambda_\perp = \left(\frac{\sqrt{3}\pi a^2}{5}\right)^{2/3} A_{2D}^{4/3} l_b^{2/3} \lambda_\parallel^{1/3}, \tag{3.9}$$

where a is a constant of order 1.

Our model does not include drift transport because of the geometry of the problem. The solar wind magnetic field is axially symmetric about the solar rotation axis, which is also the symmetry axis for 2D modulation models. In our model the interstellar flow direction is the symmetry axis and there is no neutral sheet or a distinct polar axis direction. The problem of global modulation is inherently three-dimensional and a fully 3D model would be required to simulate GCR interaction in all three regions of the heliospheric interface.

10.4 Cosmic Ray Modulation in the Global Heliosphere: Local Cloud Environment

Cosmic-ray propagation in the global heliosphere was first investigated by Florinski et al. (2003b), and the discussion in this section is based on their results. The present work is an extension of our earlier model, which did not include turbulence transport, but rather assumed that both $A^2 = A_{sl}^2 + A_{2D}^2$ and l_c were constant throughout the heliosphere. The introduction of a turbulence model was necessary to properly take into account variations in PUI-driven fluctuation levels caused by the changes in neutral hydrogen density in the LISM. Despite quantitative differences in results obtained with the old and new models, the qualitative features of the modulation process remain the same. Below we summarize the principal findings of our earlier investigation of GCR propagation in the outer heliosphere. Detailed differences between turbulence levels and transport parameters between the three interstellar environments are discussed in the next section. We will also only discuss Model 1 results in this section because this model more closely resembles the diffusion model we used previously (Florinski et al., 2003b). Note that the present model is most appropriate for modulation near positive solar minima, when particle access to the inner heliosphere is dominated by polar transport.

The heliospheric magnetic field intensity in the LIC model is shown in the top panel of Figure 10.3. This result was obtained with the help of a two-fluid MHD-neutral numerical code (Florinski et al., 2003b) using the interstellar parameters that are considered typical of the LIC (Frisch, 2000, Table 10.1). Generally particle mean free paths are small in regions of high field intensity (the heliosheath near the HP) and large in regions where the field is weak (the SW at high latitudes). The most striking difference between "local" modulation models and the global approach is the presence of the "modulation wall", which is a region with very small diffusive mean free path on the inside of the HP (red in color in the top panel of Figure 10.3). The wall is marked by a sharp decrease in radial mean free path, as seen in Figure 10.6a (red lines). The mean free path first decreases across the TS, as a result of azimuthal magnetic field B_ϕ compression by the shock, and then continues to decrease on approach to the stagnation point in the upstream direction. Note

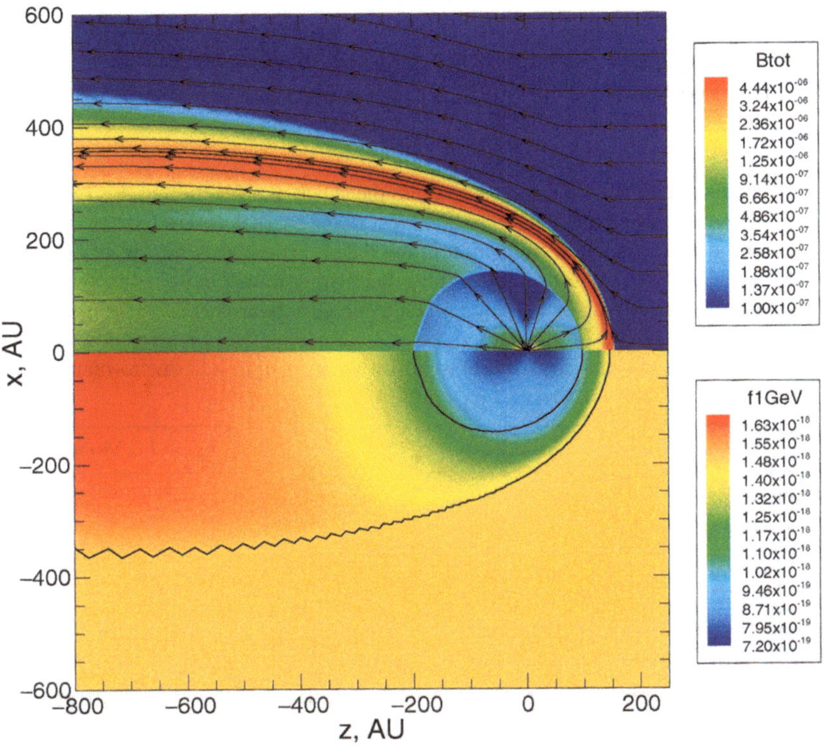

Figure 10.3. Top: Heliospheric magnetic field strength, calculated as $(B_x^2+B_y^2+B_\phi^2+B_\mathrm{m}^2)^{1/2}$ (logarithmic scale, legend marked in microGauss) and plasma streamlines in both the SW and LISM in the LIC environment. Bottom: 1 GeV GCR proton intensity, calculated using diffusion Model 1 (arbitrary units). Positions of the TS and the HP are also shown.

that in Model 1 the decrease in diffusive mean free paths for high-energy particles is larger than the shock compression ratio s, which is approximately 3 in the LIC model. To understand this result we first note that according to Equation 3.7, $\lambda_\| \sim B^{-2}$ in the energy range, which means that the decrease in $\lambda_\|$ across the quasi-perpendicular TS is $\sim (B_1/B_2)^2 \sim s^{-2}$, rather than s^{-1}. Additionally, if the radial diffusion coefficient is dominated by parallel diffusion, i.e., $\kappa_{rr} = \kappa_\perp + (\kappa_\| - \kappa_\perp)B_r^2/B^2 \simeq \kappa_\| B_r^2/B^2$ (see Equation 3.4), there is an additional reduction of the order $(B_{r2}/B_{r1})^2(B_1/B_2)^2 \sim s^{-2}$, which is absent when perpendicular diffusion dominates. Table 10.2 lists all possible limiting case scenarios for radial diffusive mean free path jump conditions across a quasi-perpendicular shock. The actual decrease in κ_{rr} at high energies will not be as large because the radial diffusion will almost always be governed by $\kappa_\perp$ downstream, even though the opposite is usually true in the upstream region. As discussed in Caballero-Lopez et al. (2004), heliosheath modulation becomes important when the GCR diffusive length κ_{rr}/u is smaller downstream of the TS than upstream. One can see that such behavior is exactly what is to be expected from theoretical predictions.

Table 10.2. Radial mean free path decrease across the shock as a function of the TS compression ratio s. Turbulence Models 1 and 2 are identical in the inertial spectral range, and consequently predict the same amount of reduction in the radial diffusion coefficient at low particle energies ($r_g \ll l_c$). Model 2, featuring declining fluctuating power in the energy range, predicts a less dramatic reduction in κ_{rr} at high particle energies than Model 1, with its flat power spectrum, because a decrease in fluctuating power at resonance wavenumber downstream of the shock partially compensates the general trend for a decrease in κ_{rr} due to magnetic field compression.

	$r_g \ll l_c$ (all models)	$r_g \gg l_c$ (Model 1)	$r_g \gg l_c$ (Model 2)
$\kappa_\parallel$ dominates	$s^{-7/3}$	s^{-4}	s^{-3}
$\kappa_\perp$ dominates	$s^{-1/9}$	$s^{-2/3}$	$s^{-1/3}$

Because the heliosheath flow convects the field to higher latitudes (see the streamline pattern in the top panel of Figure 10.3), it follows that the wall confines the entire apex-facing portion of the heliosheath. The bottom panel of Figure 10.3, which shows 1 GeV proton intensity calculated using the first diffusion model, clearly shows a strong decrease across the modulation wall. Because the latter strongly inhibits access of particles with energies below several tens of MeV to the inner heliosphere, their population can only be replenished from adiabatic cooling of more energetic cosmic rays in the expanding SW. The barrier also significantly attenuates higher energy (GeV) particles, so that the amount of modulation inside this barrier is comparable to or exceeds that in the unshocked SW. This effect may be seen from the spectra shown in Figure 10.4 by comparing the left panel, which shows heliosheath modulation only, with the right panel, which shows the combined Region 1 and 2 contribution, i.e., the total amount of modulation from the LISM to 1 AU.

Our calculations also show that the GCRs are generally weakly coupled to the plasma in the heliosphere because diffusion coefficients are large. This, of course, is strictly true in the outer heliosheath, which is a part of the LISM. We do not expect GCRs to modify the structure of the bow shock or even the TS. A possible exception is the inner heliosheath region, since it has relatively large radial gradients in the GCR pressure. This region may be more strongly influenced by cosmic-ray dynamics.

Our third major result concerns the latitudinal asymmetry of the GCR distribution. While the calculated GCR intensity in the heliosphere is highly asymmetric at large distances, the asymmetry disappears at smaller distances due to a specific diffusion pattern in a Parker field, which is dominated by polar transport during positive sunspot minima. It transpires that the interplanetary magnetic field imposes a strong symmetry about the heliospheric polar axis on the GCR distribution, effectively erasing the information about the asymmetry of GCRs from their propagation through the inner heliosheath. This effect is clearly seen in the bottom panel of Figure 10.3, where the GCR intensity pattern inside the TS shows a characteristic symmetric lobe structure. This feature makes it difficult to use GCRs to directly probe the structure of the outer heliosphere when deep inside the TS, requiring spacecraft observations at large heliocentric distances.

Finally, we discovered that a possibility exists for the re-acceleration of GCRs in the decelerated flow in the heliotail. Notice that the intensity in the tail at 1 GeV is higher than the interstellar intensity. Compressive re-acceleration is a result of solar wind slowdown in the heliotail. The flow in this region is decelerated and cooled as a result of charge exchange with the interstellar atoms that cross into the tail from the sides. This effect is noticeably not present when charge exchange processes are ignored. We should point out, however, that the heliotail probably extends to tens of thousands of AU in the downstream direction (Izmodenov and Alexashov, 2003) and a much larger simulation box may be required to properly study the re-acceleration effect.

10.5 Interface Variability Driven by Interstellar Environment Changes: Cosmic Ray Response

As the density and dynamic pressure of the interstellar medium surrounding the solar system changes, the heliosphere expands or contracts relative to its present size (Chapter 2). Some of the most important properties of the interface in the three environments are summarized in Table 10.3. The overall structure of the interface is the same in all three cases, the only exception being the absence of a bow shock in the strongly subsonic LB interstellar wind. There are, however, significant variations in sizes and PUI intensities. A thicker heliosphere may be expected to be a better shield against galactic radiation. The situation, however, is complicated by the fact that turbulence production by the pickup process is less efficient in low density environments, which means that the SW magnetic field will be less "tangled" in the LB than in denser clouds. To answer the question of how these two effects influence GCR propagation one must first understand the problem of heliosheath modulation. This subject is only now beginning to be addressed and no theory of turbulence or diffusion in Region 2 exists yet. Here we use two different diffusion models in order to cover a range of possible scenarios. Model 1 favors stronger heliosheath modulation than Model 2, because the decrease in the mean free path across the TS at high energies is larger by a factor of s (Table 10.2). Conversely, Region 1

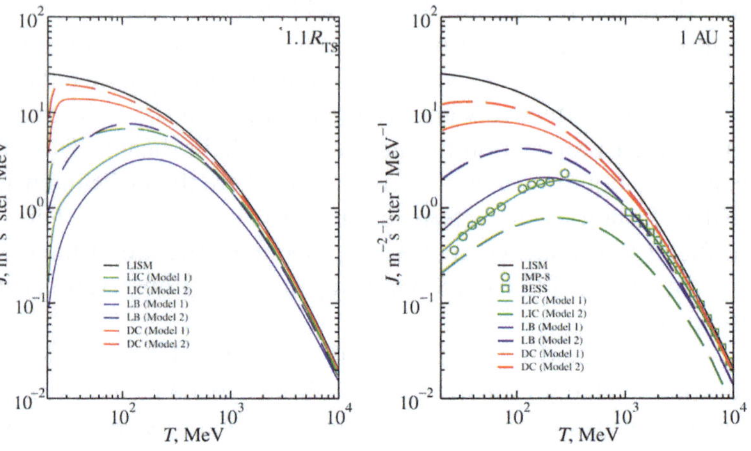

Figure 10.4. Left: spectra of galactic cosmic ray protons at 1.1 the distance to the TS in the upwind direction for the three interstellar environments described in the text. Solid lines correspond to Model 1 results, while Model 2 data is plotted with dashed lines. The black curve is the hypothetical LISM spectrum (Ip and Axford, 1985). Right: same for 1 AU (except the LB model where the results are for the internal boundary at 1.5 AU). Solar minimum data at low energies (IMP-8, McDonald, 1998) and high energies (BESS, Sanuki et al., 2000) are also shown. Notice a general decreasing trend in the modulation at 1 AU from the LB to the DC.

modulation is more important in Model 2. This is most clearly seen in the LIC case by observing the behavior of the radial mean free path in Figure 10.6a, and by comparing the amount of modulation in the heliosheath and the SW (the left and the right panels in Figure 10.4, respectively). Below we discuss the differences between modulation patterns in the low density (LB) and relatively high density (DC) environments in comparison with the present (LIC) conditions in the context of the two modulation scenarios.

Table 10.3. Properties of the heliospheric interfaces formed as a result of SW interaction with the LIC, the LB and the DC. Here $r_{\rm TS}$ is the distance to the TS, $h = (r_{\rm HP} - r_{\rm TS})/r_{\rm TS}$ is the relative thickness of the heliosheath, and s is the TS compression ratio.

Model	$r_{\rm TS}$, AU	h	s
Local Interstellar Cloud (LIC)	100	0.5	3.1
Local Bubble (LB)	100	1.8	4.0
Diffuse Cloud (DC)	12	1.0	1.6

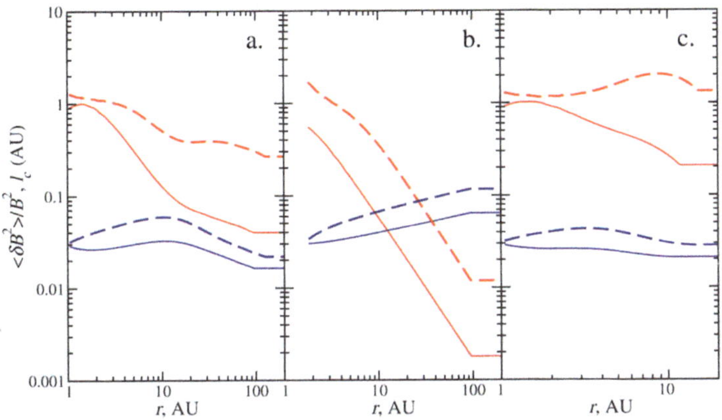

Figure 10.5. Turbulent ratio $\langle \delta B_x^2 \rangle / B^2$ (red lines) and the correlation length l_c in AU (blue lines) in the ecliptic (solid) and polar (dashed) regions. The three plots a, b, and c correspond to the LIC, LB and DC environments, respectively.

10.5.1 Hot Local Bubble (LB) Environment

Some 10^5 years ago the solar system galactic environment was very different from today. The heliosphere was embedded in the Local Bubble, a region of space some $\sim$100 pc in size with an anomalously low density and high temperature ($T > 10^6$ K). The LB is likely to be fully ionized, a fact that significantly simplifies the heliospheric model since there is no need to calculate neutral atom distributions. Plasma parameters inside the LB that serve as external boundary conditions for the heliosphere are shown in Table 10.1. Unlike the two cloud models, which assume that a cloud moves in the opposite direction to the Sun's velocity vector in the "local standard of rest" (LSR) frame, here a stationary background is assumed with the only contribution to the velocity difference between the heliosphere and the LISM being the motion of the Sun relative to the LSR (Frisch, 2000). Because the interstellar flow is substantially subsonic (Mach number $M \sim 0.1$), it is expected to behave as a nearly incompressible fluid. Numerical calculations using a multifluid model (Müller et al., in preparation) performed on a relatively large (several thousand AU) domain reveal that the plasma density is indeed nearly constant in both the heliosheath and the LISM. These preliminary results also show that it is virtually impossible to eliminate the influence of the external (outflow) boundary because of the acoustic waves propagating upstream and modifying the solution in the heliotail. Fortunately, the subsonic regime can be adequately described by an analytic model developed for an incompressible potential flow (Suess and Nerney, 1990). In the spherical coordinate system the components of the plasma velocity in both subsonic regions (2 and 3) may be calculated from

$$u_r = u_\infty \cos\theta \left[\left(\frac{r_s}{r}\right)^3 - 1 \right] + u_0 \frac{\gamma - 1}{\gamma + 1} \left(\frac{r_s}{r}\right)^2 , \tag{5.1}$$

$$u_\theta = u_\infty \sin\theta \left[\frac{1}{2} \left(\frac{r_s}{r}\right)^3 + 1 \right] , \tag{5.2}$$

where r_s is the distance to the TS and $\gamma = 5/3$ is the specific heat ratio. The TS itself is assumed to be spherical and have a compression ratio of $(\gamma + 1)/(\gamma - 1) = 4$ (strong shock) as expected in the absence of charge exchange. The Region 1 flow is assumed to have a constant radial velocity u_0. We set $r_s = 100$ AU, which is essentially the same as in the LIC and in agreement with the multifluid model results, and calculate **B** numerically from the kinematic model using Equations 3.2 and 3.3.

The heliospheric structure in the LB environment is shown in Figure 10.7. The heliosheath is almost three times thicker and the TS is stronger than in the LIC (Table 10.3). The HP is located at 280 AU on the symmetry axis. From Figure 10.5b one can see that the absence of PUIs has a large impact on A^2. Unlike the LIC case, the only sources of turbulence are stream interactions, which are ineffective beyond several AU. The result is that the wave intensity decreases asymptotically as $\langle \delta B_x^2 \rangle \sim r^{-7/2}$, while the correlation length increases as $r^{1/4}$, in agreement with the theory (Zank et al., 1996a). Recalling that $|B| \sim 1/r$ at virtually all latitudes in the presence of the modified field component, we obtain $A^2 \sim r^{-3/2}$. This is quite different from the LIC case, where the fluctuation levels fall off much slower because of PUI driving. The fast decay of A^2 results in the GCR mean free path being larger in Region 1 compared with the LIC (Figure 10.6b). However, the drop across the TS is also larger and the heliosheath diffusion coefficient is about the same in the two cases. The larger decrease is a consequence of both a stronger shock and the dominance of parallel diffusion in the LB case. Generally, a large parallel diffusion implies smaller ratios of $\kappa_\perp/\kappa_\parallel$, because the $\kappa_\perp$ depends on $\kappa_\parallel$ only as a cubic root and tends to become smaller when turbulence levels decrease. On the contrary parallel diffusion is fastest in the absence of fluctuations that cause scattering, thus inhibiting particle propagation in the parallel direction. This also explains why the change in κ_{rr} is larger at high latitudes, where diffusion mean free paths are generally larger as well. As before, the drop across the shock is less in Model 2 in agreement with the general trend (Table 10.2).

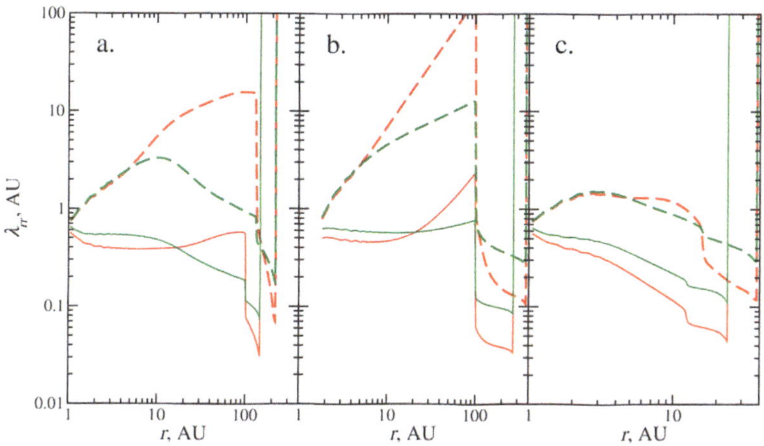

Figure 10.6. Radial diffusive mean free path of 1 GeV protons in the ecliptic (solid lines) and at high latitudes (dashed lines). Red lines are Model 1 and green lines are Model 2 results. The three plots a, b, and c correspond to the LC, LB and DC environments, respectively.

The bottom half of Figure 10.7 shows strong modulation in the magnetic wall region. The GCR intensity distribution inside the TS is nearly spherically symmetric. No compressional acceleration in the tail is visible, as expected in the absence of neutrals. Spectra (blue lines in Figure 10.4) confirm that heliosheath modulation is more dominant than in the LIC (85% and 54% of the total amount of modulation at 1 GeV for Models 1 and 2, respectively, compared with 52% and 25% for the LIC). Unlike the LIC case, Model 2 actually gives a higher GCR intensity at 1 AU than Model 1. The latter is a consequence of diffusion coefficients being quite large in Region 1 in both models, whereas in the LIC Model 2 predicts small SW diffusion. It follows then that if the diffusion coefficient obeys the Model 2 prescription, GCR fluxes at Earth would be larger than at present, while if Model 1 is correct, the fluxes of particles with energies above 300 MeV would be smaller in the Local Bubble environment.

10.5.2 Diffuse Interstellar Cloud (DC) Environment

We now proceed to investigate the properties of the GCR distribution in the heliosphere during the periods when the Sun enters a moderately dense interstellar cloud (Table 10.1). There is some evidence, for example, that the so-called G Cloud, which lies approximately in the path of the Sun's travel,

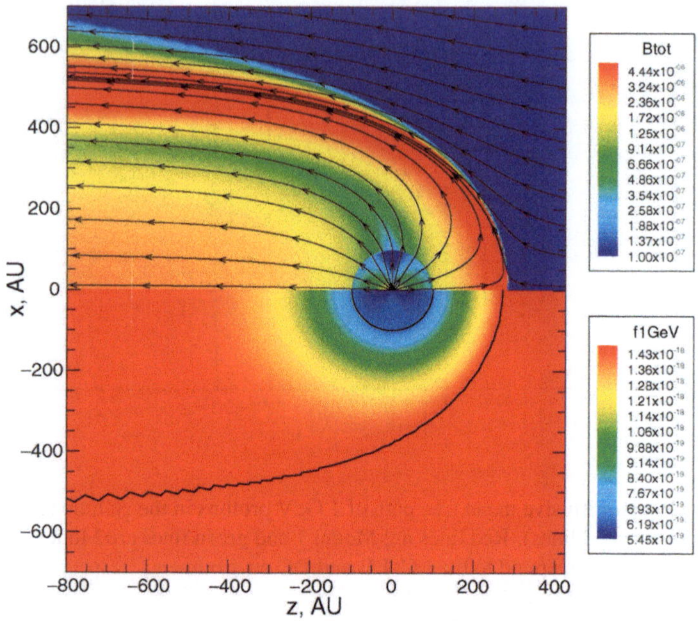

Figure 10.7. Same as Figure 10.3, but for the Local Bubble environment.

may be significantly denser than the LIC. Frisch (2003) (see also Chapter 6) estimates that the number density in the G Cloud could be $>5\,\text{cm}^{-3}$, which is 15–30 times that of the LIC, and that this cloud could become the next galactic environment of the solar system in 10^3 to 10^5 years. Should that be the case, the structure of the heliosphere will change dramatically.

Earlier studies of the SW–cloud interactions were conducted by Zank and Frisch (1999), Müller et al. (2001), and Florinski et al. (2003a) for a cloud similar to the DC, and by Scherer et al. (2002) for a less dense, but strongly ionized cloud with $N_{\text{t}\,\infty} = 1\,\text{cm}^{-3}$. Encounters with very high density molecular clouds ($N_{\text{t}\,\infty} = 100 - 1000\,\text{cm}^{-3}$) were investigated by Yeghikyan and Fahr (2003) (also, Chapter 11 of this book), although they only included neutral gas interactions in their analysis. Cosmic-ray attenuation in such dense clouds is discussed in Chapter 9. The cloud we study here is cold (T = 200 K) in order for it to be in approximate pressure equilibrium with the surrounding warm low density medium. Consequently, the hydrogen ionization ratio in the cloud is low. In fact, we can assume it to be identically zero at the boundary because the cloud is ionized locally by solar Ly α radiation, which is included in the computer model. The model itself is a modification of our earlier study (Florinski et al., 2003a) in that we now include three neutral hydrogen populations instead of one. This "multifluid" model is similar to Zank et al. (1996) and Florinski et al. (2004). The multifluid results for the plasma and neutral distribution are somewhat different from those of Florinski et al. (2003a), because the single fluid approach overestimates neutral atom filtration by the hydrogen wall. In the new model more neutral hydrogen atoms penetrate into the inner heliosphere, causing stronger solar wind deceleration by charge transfer. The SW is slowed down to 250 km/s by the time it reaches the TS, located at 12 AU (i.e., just past the orbit of Saturn) in the upstream direction. The SW is strongly heated by the PUIs and the TS is weak with a compression ratio of only 1.6 (Table 10.3). The HP is located at 24 AU and the thickness of the heliosheath is about equal to the radial extent of Region 1.

Figure 10.5c demonstrates the effect of enhanced turbulent driving by the PUIs. The PUIs are produced in abundance compared to the LIC, and the additional wave generation significantly slows down the decrease of the energy density of the fluctuations with radial distance, which is caused by expansion and turbulent decay. The correlation length is approximately a constant. Proton mean free paths in Region 2 are smaller than in the LIC and decrease with r. The jump across the shock is quite small, however, because perpendicular diffusion is relatively large and the TS compression ratio is small. As a result, the mean free path in the heliosheath is actually larger than in the LIC (cf. Figures 10.6a and 10.6c).

The combined effect of a relatively large heliosheath κ_{rr}, and a much smaller extent of the modulation cavity, is a significant reduction in the heliospheric shielding of GCRs. Figure 10.4 (red lines) shows that the intensity between 300 MeV and 1 GeV is increased by a factor of 1.4–2.4 in Model 1 and by a factor of 4.1–7.6 in Model 2. The Model 1 results are in good agreement with those of Florinski et al. (2003a). The relative contribution of the heliosheath modulation is slightly less here than in the LIC case. Figure 10.8 shows little if any compressional acceleration at 1 GeV despite a significant slowdown of the heliosheath flow in the tail. Because the diffusive mean free path of GeV particles is relatively large, it appears there is not enough time for compressive re-acceleration during the passage through the heliotail that is less than 100 AU in diameter.

To conclude the discussion in this section, we would like to point out that the GCR intensity inside a diffuse cloud need not be identical to that of the ambient interstellar medium. Inside dense clouds, cosmic rays experience ionization and pion production losses that tend to produce a net flux of high-energy particles into the cloud. The resulting anisotropy, suggested by Skilling and

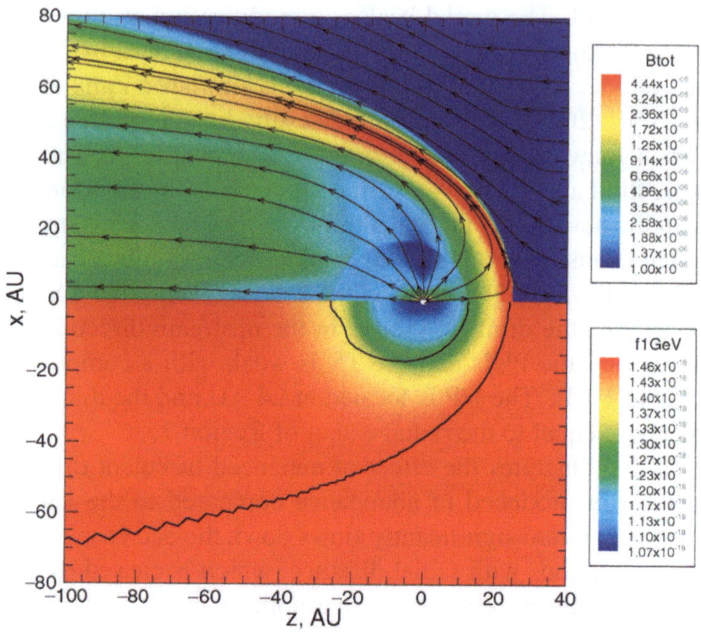

Figure 10.8. Same as Figure 10.3, but for the Diffuse Cloud environment.

Strong (1976),would resonantly excite Alfvén waves, which may in turn inhibit particle diffusion parallel to the ISM magnetic field, thus excluding them from the cloud interior. Alfvén wave damping by ion-neutral friction (Kulsrud and Pierce, 1969) imposes an upper limit on the energy of cosmic rays that can excite waves because the streaming flux decreases with energy and the wave growth rate becomes less than the rate of dissipation. However, Skilling and Strong (1976) neglected the effect of magnetic field enhancement inside a dense cloud, which tends to compress cosmic rays and inhibit inward streaming. When this effect is included, cloud shielding becomes effective only for protons with energies below 50 MeV (Cesarsky and Völk, 1978). Note that the effect of shielding is only relevant for giant molecular clouds with neutral hydrogen column densities greater than $4 \times 10^{22}\,\mathrm{cm}^{-2}$, which is more than 10^4 times larger than the column density of the cloud considered here and in Florinski et al. (2003a). We conclude then that our use of the same background interstellar spectrum for all three environments is justified.

10.6 Cosmogenic Isotope Response

As high-energy galactic cosmic rays strike the Earth's atmosphere, they produce a shower of secondary particles that can interact with the atmospheric nucleons to produce long-lived radioactive isotopes. The contribution of the primary particles to ^{10}Be production is negligible and most of the isotope is generated in spallation reactions between secondary neutrons and N and O nuclei. Modeling this process involves complex Monte-Carlo simulations, following random trajectories of a large number of particles that take into account all possible combinations of targets and product nuclei. Cross sections for ^{10}Be production can be found in Masarik and Beer (1999), which show that secondary neutrons below 100 MeV are the most efficient producers in nitrogen reaction, while those with energies above 100 MeV interact more readily with oxygen. Because the number of secondary particles increases with the energy of the primary proton, the process favors higher energy cosmic rays. However, since there are fewer of these particles in the incident spectrum, the production rate actually peaks at about 2 GeV. The total rate of cosmogenic isotope production is calculated as

$$P = \int_0^\infty S(T) \frac{dJ}{dT} dT, \tag{6.1}$$

where $S(T)$ is the specific yield of the product isotope from the primary particle with a kinetic energy T, integrated over the depth of the atmosphere and dJ/dT is the differential intensity at 1 AU plotted in Figure 10.4. The yields for ^{10}Be were conveniently calculated for primary protons with energies between 300 MeV and 300 GeV by Webber and Higbie (2003) (see their Figure 5). Note that our simulations only include protons up to 10 GeV. However, the error introduced by neglecting higher energy particles is only of the order of 10%, which is probably much smaller than the uncertainties involved in estimates of the diffusion coefficient. Because we are only interested in the relative increase or decrease of cosmogenic isotope production rates in different interstellar environment as compared to the LIC, the exact shape of the GCR spectra assumed for the LISM is not important. In this study we are also ignoring the energy cutoff effect produced by the Earth's magnetic field.

The results of this calculation are presented in Table 10.4. Model 1 predicts a relatively modest amount of variability. The LB environment shows a modest reduction in the ^{10}Be production rate, while the DC shows a somewhat larger increase. The variations are significantly larger in Model 2 results. This is mostly a consequence of a large amount of total modulation predicted for the LIC by this model. Both the LB and the DC exhibit less modulation, and ^{10}Be production rates are increased by a factor of 2–3. The last line in Table 10.4 corresponds to a hypothetical case of an extremely dense cloud where the size of the heliosphere is reduced to such small scales that the modulation is effectively zero. It shows that the largest possible increase in the rate of cosmogenic isotope production, due to galactic environment variability, is of the

Table 10.4. Beryllium-10 production rate relative to the present rate as a function of the interstellar environment and diffusion model.

Galactic environment	Model 1	Model 2
Local Interstellar Cloud (LIC)	100%	100%
Local Bubble (LB)	77%	219%
Diffuse Cloud (DC)	134%	330%
No modulation	175%	407%

order 75–300%. This is certainly enough to explain the variability observed in ^{10}Be records (Figure 10.1).

10.7 Conclusion

This work represents a first systematic study of the effects of a variable interstellar environment on the process of GCR modulation by the heliospheric interface. In contrast to more traditional (local) modulation models, a global heliospheric model is used to calculate cosmic-ray transport parameters from a background formed by a self-consistent interaction between a stellar wind with a partially ionized interstellar flow. Three contrasting environments are considered: the Local Interstellar Cloud presently surrounding the solar system, the tenuous and hot environment of the Local Bubble, and a diffuse cloud of purely neutral hydrogen. We employ two different diffusion models, one favoring more modulation by the heliosheath region and the other giving preference to the solar wind modulation. It is demonstrated that dense clouds are less efficient in shielding the inner heliosphere from highly energetic cosmic radiation. The situation is less certain for the LB, which can either reduce or enhance the intensity of GCRs reaching Earth. Intensity variations would be recorded in cosmogenic isotope records, which could provide valuable information about the types of environments encountered by the Sun during its journey around the center of the Galaxy. Such an identification, however, would require first separating the signals from geomagnetic field polarity excursions and solar activity variations, and this may be difficult at present. Another difficulty is our relatively poor understanding of the properties of the heliospheric interface (and the heliosheath in particular), as a medium for high-energy (GeV) cosmic ray transport. This situation, however, may soon improve when the Voyager 1 spacecraft finally crosses the termination shock and enters the heliosheath. In another 15 years the first Voyager spacecraft will cross the heliopause and the uncertainties in the interstellar cosmic-ray spectrum will hopefully be eliminated.

Acknowledgments: V. F. would like to thank N. V. Pogorelov, H.-R. Müller, W. I. Axford, P. C. Frisch, G. Cini Castagnoli, and K. G. McCracken for valuable discussions. This research was supported, in part, by NSF grant ATM-0296114 and NASA grants NAG5-11621 and NAG5-12903. Numerical computations were performed on the IGPP/UCR "Lupin" cluster.

References

Aldahan, A., and Possnert, G. (2003). Geomagnetic and climatic variability reflected by ^{10}Be during the Quaternary and late Pliocene, *Geophys. Res. Lett.*, 30:1301.

Armstrong, J. W., Rickett, B. J., and Spangler, S. R. (1995). Electron density power spectrum in the local interstellar medium, *Astrophys. J*, 443:209–221.

Axford, W. I. (1981). Acceleration of cosmic rays by shock waves. In *Proc. 17 Intl. Cosmic Ray Conf.*, Vol. 12, pages 155–204.

Baranov, V. B., and Malama, Y. G. (1993). Model of the solar wind interaction with the local interstellar medium: Numerical solution of self-consistent problem, *J. Geophys. Res.*, 98:15,157–15,163.

Beer, J. (1997). Cosmogenic isotopes as a tool to study solar and terrestrial variability. In Cini Castagnoli, G., and Provenzale, A., editors, *Past and Present Variability of the Solar–Terrestrial System: Measurements, Data Analysis and Theoretical models*, Proc. of the Intl. School of Physics "Enrico Fermi" CXXXII, pages 25–36.

Beer, J. (2000). Long-term indirect indices of solar variability, *Space Sci. Rev.*, 94:53–66.

Bieber, J. W., and Matthaeus, W. H. (1997). Perpendicular diffusion and drift at intermediate cosmic-ray energies, *Astrophys. J*, 485:655–659.

Bieber, J. W., Matthaeus, W. H., Smith, C. W., Wanner, W., Kallenrode, M.-B., and Wibberenz, G. (1994). Proton and electron mean free paths: the Palmer consensus revisited, *Astrophys. J.*, 420:294–306.

Bieber, J. W., Wanner, W., and Matthaeus, W. H. (1996). Dominant two-dimensional solar wind turbulence with implications for cosmic ray transport, *J. Geophys. Res.*, 101:2511–2522.

Burger, R. A., and Hattingh, M. (1998). Towards a realistic diffusion tensor for galactic cosmic rays, *Astrophys. J.*, 505:244–251.

Burger, R. A., Potgieter, M. S., and Heber, B. (2000). Rigidity dependence of cosmic ray proton latitudinal gradients by the Ulysses spacecraft: Implications for the diffusion tensor, *J. Geophys. Res.*, 105:27,447–27,455.

Burlaga, L. F., Ness, N. F., Wang, Y.-M., and Sheeley Jr., N. R. (1998). Heliospheric magnetic field strength out to 66 AU: Voyager 1, 1978–1996, *J. Geophys. Res.*, 103:23,727–23,732.

Caballero-Lopez, R. A., Moraal, H., and McDonald, F. B. (2004). Galactic cosmic ray modulation: Effects of the solar wind termination shock and the heliosheath, *J. Geophys. Res.*, 109:A05105.

Chalov, S. V., Alexashov, D. B., and Fahr, H.-J. (2004). Reabsorption of self-generated turbulent energy by pick-up protons in the outer heliosphere, *Astron. Astrophys.*, 416:L31–L34.

Chalov, S. V., and Fahr, H.-J. (2000) Pick-up ion acceleration at termination shock and post-shock pick-up ion energy distribution, *Astron. Astrophys.*, 360: 381–390.

Christl, M., Mangini, A., Holzkämper, S., and Spötl, C. (2004). Evidence for a link between the flux of galactic cosmic rays and Earth's climate during the past 200,000 years, *J. Atmos. Solar Terr. Phys.*, 66: 313–322.

Cesarsky, C. J., and Völk, H. J. (1978). Cosmic ray penetration into molecular clouds, *Astron. Astrophys.*, 70:376–377.

Cini Castagnoli, G., Bonino, G., Taricco, C., and Lehman, B. (1998). Cosmogenic isotopes and geomagnetic signals in a Mediterranean sea sediment at 35,000 y BP, *Il Nuovo Cimento*, 21:243–246.

Cox, D. P., and Helenius, L. (2003). Flux-tube dynamics and a model for the origin of the local fluff, *Astrophys. J.*, 583:205–228.

Dröge, W. (2000). Particle scattering by magnetic fields, *Space Sci. Rev.*, 93: 121–151.

Ellis, J., Fields, B. D., and Schramm, D. N. (1996). Geological isotope anomalies as signatures of nearby supernovae, *Astrophys. J.*, 470:1227–1236.

Fahr, H.-J., Kausch, T., and Scherer, H. (2000). A 5-fluid hydrodynamic approach to model the solar system–interstellar medium interaction, *Astron. Astrophys.*, 357:268–282.

Ferlet, R. (1999). The local interstellar medium, *Astron. Astrophys. Rev.*, 9: 153–169.

Ferreira, S. E. S., and Scherer, K. (2004). Modulation of cosmic-ray electrons in the outer heliosphere, *Astrophys. J.*, 616:1215–1223.

Fisk, L. A. (1996). Motion of the footpoints of heliospheric magnetic field lines at the Sun: Implications for recurrent energetic particle events at high heliographic latitudes, *J. Geophys. Res.*, 101:15,547–15,553.

Fisk, L. A., and Jokipii, J. R. (1999). Mechanisms for latitudinal transport of energetic particles in the heliosphere, *Space Sci. Rev.*, 89:115–124.

Florinski, V., and Jokipii, J. R. (1999). A two-dimensional, self-consistent model of galactic cosmic rays in the heliosphere, *Astrophys. J.*, 523: L185–L188.

Florinski, V., Pogorelov, N. V., Zank, G. P., Wood, B. E., and Cox, D. P. (2004). On the possibility of a strong magnetic field in the local interstellar medium, *Astrophys. J.*, 604:700–706.

Florinski, V., Zank, G. P., and Axford, W. I. (2003a). The Solar System in a dense interstellar cloud: Implications for cosmic-ray fluxes at Earth and ^{10}Be records, *Geophys. Res. Lett.*, 30:2206.

Florinski, V., Zank, G. P., and Pogorelov, N. V. (2003b). Galactic cosmic ray transport in the global heliosphere, *J. Geophys. Res.*, 108:1228.

Florinski, V., Zank, G. P., and Pogorelov, N. V. (2005). Heliopause stability revisited: dispersion analysis and numerical simulations, *Adv. Space Res.*, in press.

Forman, M. A., Jokipii, J. R., and Owens, A. J. (1974). Cosmic-ray streaming perpendicular to the mean magnetic field, *Astrophys. J.*, 192:535–540.

Frisch, P. C. (2000). The galactic environment of the Sun, *J. Geophys. Res.*, 105:10,279–10,289.

Frisch, P. C. (2003a). Local interstellar matter: The Apex cloud, *Astrophys. J.*, 593:868–873.

Frisch, P. C. (2003b). Boundary conditions of the heliosphere, *J. Geophys. Res.*, 108:8036.

Giacalone, J., and Jokipii, J. R. (1999). The transport of cosmic rays across a turbulent magnetic field, *Astrophys. J.*, 520:204–214.

Gleeson, L. J., and Axford, W. I. (1967). Cosmic rays in the interplanetary medium, *Astrophys. J.*, 149:L115–L118.

Heber, B. (2001). Modulation of galactic and anomalous cosmic rays in the inner heliosphere, *Adv. Space Res.*, 27:451–460.

Horbury, T. S., and Balogh, A. (2001). Evolution of magnetic field fluctuations in high-speed solar wind streams: Ulysses and Helios observations, *J. Geophys. Res.*, 106:15,929–15,940.

Ip, W.-H., and Axford, W. I. (1985). Estimates of galactic cosmic ray spectra at low energies, *Astrophys. J.*, 149:7–10.

Isenberg, P. A., Smith, C. W., and Matthaeus, W. H. (2003). Turbulent heating of the distant solar wind by interstellar pickup protons, *Astrophys. J.*, 592:564–573.

Izmodenov, V. V., and Alexashov, D. B. (2003). A model of the tail region of the heliospheric interface, *Astron. Lett.*, 29:58–63.

Jokipii, J. R. (1966). Cosmic ray propagation. I. Charged particles in a random magnetic field, *Astrophys. J.*, 146:480–487.

Jokipii, J. R. (1971). Propagation of cosmic rays in the solar wind, *Rev. Geophys. Space Phys.*, 9:27–87.

Jokipii, J. R. (1991). Variations of the cosmic-ray flux with time. In Sonett, C. P., Giampapa, M. S., and Matthews, M. S., editors, *The Sun in Time*, University of Arizona Press, Tucson, pp. 205–220.

Jokipii, J. R., and Kota, J. (1989). The polar heliospheric magnetic field, *Geophys. Res. Lett.*, 16:1–4.

Jokipii, J. R., and Kota, J. (1995). Three-dimensional cosmic-ray simulations: heliographic latitude and current-sheet tilt, *Space Sci. Rev.*, 72:379–384.
Jokipii, J. R., and Kota, J. (2000). Galactic and anomalous cosmic rays in the heliosphere, *Astrophys. Space Sci.*, 274:77–96.
Jokipii, J. R., Kota, J., and Merenyi, E. (1993). The gradient of galactic cosmic rays at the solar-wind termination shock, *Astrophys. J.*, 405:782–786.
Jokipii, J. R., and Marti, K. (1986). Temporal variations of cosmic rays over a variety of time scales. In Smoluchowski, R., Bahcall, J. N., and Matthews, M. S., editors, *The Galaxy and the Solar System*, University of Arizona Press, Tucson, pp. 116–128.
Kocharov, G. E. (1994). On the origin of cosmic rays, *Astrophys. Lett. Comm.*, 29:227–232.
Kulsrud, R., and Pearce, W. P. (1969). The effect of wave-particle interactions on the propagation of cosmic rays, *Astrophys. J.*, 156:445–469.
Lal, D. (1972). Hard rock cosmic ray archaeology, *Space Sci. Rev.*, 14:3–102.
Lallement, R. (2001). Heliopause and astropauses, *Astrophys. Space Sci.*, 277: 205–217.
Langner, U. W., Potgieter, M. S., and Webber, W. R. (2003). Modulation of cosmic ray protons in the heliosheath, *J. Geophys. Res.*, 108:8039.
Liewer, P. C., Karmesin, S. R., and Brackbill, J. U. (1996). Hydrodynamic instability of the heliopause driven by plasma-neutral charge-exchange interactions, *J. Geophys. Res.*, 101:17,119–17,127.
Linde, T. J., Gombosi, T. I., Roe, P. L., Powell, K. G., and DeZeeuw, D. L. (1998). Heliosphere in the magnetized local interstellar medium — Results of a three-dimensional MHD model, *J. Geophys. Res.*, 103:1889–1904.
Marsh, N. D., and Svensmark, H. (2000). Low cloud properties influenced by cosmic rays, *Phys. Rev. Lett.*, 85:5004–5007.
Masarik, J., and Beer, J. (1999). Simulation of particle fluxes and cosmogenic nuclide production in the Earth's atmosphere, *J. Geophys. Res.*, 104:12,099–12,111.
Matthaeus, W. H., Zank, G. P., Smith, C. W., and Oughton, S. (1999). Turbulence, spatial transport, and heating of the solar wind, *Phys. Rev. Lett.*, 82:3444–3447.
Matthaeus, W. H., Qin, G., Bieber, J. W., and Zank, G. P. (2003). Nonlinear collisionless perpendicular diffusion of charged particles, *Astrophys. J.*, 590:L53–L56.
McCracken, K. G. (2004). Geomagnetic and atmospheric effects upon the cosmogenic ^{10}Be observed in polar ice, *J. Geophys. Res.*, 109:A04101.
McDonald, F. B. (1998). Cosmic-ray modulation in the heliosphere — A phenomenological study, *Space Sci. Rev.*, 83:33–50.
McDonald, F. B., Cummings, A. C., Stone, E. C., Heikkila, B. C., Lal, N., and Webber, W. R. (2004). The energetic particle populations of the distant

heliosphere. In Florinski, V., Pogorelov, N. V., and Zank, G. P., editors, *Physics of the Outer Heliosphere*, AIP Conf. Proc. Vol. 719, pages 139–149.

McDonald, F. B., Heikkila, B., Lal, N., and Stone, E. C. (2000). Relative recovery of galactic and anomalous cosmic rays at 1 AU: Evidence for modulation in the heliosheath, *J. Geophys. Res.*, 105, 1–8.

McHargue, L. R., Donahue, D. J., Damon, P. E., Sonett, C. P., Biddulph, D., and Burr, G. (2000). Geomagnetic modulation of the late Pleistocene cosmic-ray flux as determined by ^{10}Be from Blake Outer Ridge marine sediments, *Nucl. Instr. Methods Phys. Res. B*, 172:555–561.

McKenzie, J. F., and Westphal, K. O. (1969). Transmission of Alfvén waves through the Earth's bow shock, *Planet. Space Sci.*, 17:1029–1037.

Müller, H.-R., Zank, G. P., and Frisch, P. C. (2001). Effect of different possible interstellar environments on the heliosphere: a numerical study. In Scherer, K., Fichtner, H., Fahr, H.-J., and Marsch, E., editors, *The Outer Heliosphere: the Next Frontiers*, COSPAR Coll. Ser. Vol. 11, pages 329–332.

Müller, H.-R., Zank, G. P., and Lipatov, A. S. (2000). Self-consistent hybrid simulations of the interaction of the heliosphere with the local interstellar medium, *J. Geophys. Res.*, 105:27,419–27,438.

Myasnikov, A. V., Alexashov, D. B., Izmodenov, V. V., and Chalov, S. V. (2000). Self-consistent model of the solar wind interaction with three-component circumsolar interstellar cloud: Mutual influence of thermal plasma, galactic cosmic rays, and H atoms, *J. Geophys. Res.*, 105:5167–5177.

Nerney, S., Suess, S. T., and Schmahl, E. J. (1995). Flow downstream of the heliospheric terminal shock: Magnetic field line topology and solar cycle imprint, *J. Geophys. Res.*, 100:3463–3471.

Opher, M., Liewer, P. C., Velli, M., Bettarini, L., Gombosi, T. I., Manchester, W., DeZeeuw, D. L., Toth, G., and Sokolov, I. (2004). Magnetic effects at the edge of the Solar System: MHD instabilities, the de Laval nozzle effects, and an extended jet, *Astrophys. J.*, 611:575–586.

Parhi, S., Bieber, J. W., Matthaeus, W. H., and Burger, R. A. (2004). Heliospheric solar wind turbulence model with implications for ab initio modulation of cosmic rays, *J. Geophys. Res.*, 109:A01109.

Parker, E. N. (1958). Dynamics of the interplanetary gas and magnetic fields, *Astrophys. J.*, 128:664–676.

Parker, E. N. (1965). The passage of energetic charged particles through interplanetary space, *Planet. Space Sci.*, 13:9–49.

Pauls, H. L., Zank, G. P., and Williams, L. L. (1995). Interaction of the solar wind with the local interstellar medium, *J. Geophys. Res.*, 100:21,595–21,604.

Pogorelov, N. V., Zank, G. P., and Ogino, T. (2004). Three-dimensional features of the outer heliosphere due to coupling between the interstellar and interplanetary magnetic fields. I. Magnetohydrodynamic model: interstellar perspective, *Astrophys. J.*, 614:1007–1021.

Potgieter, M. S. (1998). The modulation of galactic cosmic rays in the heliosphere: theory and models, *Space Sci. Rev.*, 83:147–158.

Potgieter, M. S., Burger, R. A., and Ferreira, S. E. S. (2001). Modulation of cosmic rays in the heliosphere from solar minimum to maximum: a theoretical perspective, *Space Sci. Rev.*, 97:295–307.

Qin, G., Matthaeus, W. H., and Bieber, J. W. (2002a). Subdiffusive transport of charged particles perpendicular to the large scale magnetic field, *Geophys. Res. Lett.*, 29:1048.

Qin, G., Matthaeus, W. H., and Bieber, J. W. (2002b). Perpendicular transport of charged particles in composite model turbulence: Recovery of diffusion, *Astrophys. J.*, 758:L117–L120.

Raisbeck, G. M., Yiou, F., Bourles, D., and Kent, D. V. (1985). Evidence for an increase in cosmogenic ^{10}Be during a geomagnetic reversal, *Nature*, 315:315–317.

Roberts, D. A., Goldstein, M. L., Klein, L. W., and Matthaeus, W. H. (1987). Origin and evolution of fluctuations in the solar wind: Helios observations and Helios–Voyager comparisons, *J. Geophys. Res.*, 92:12,023–12,035.

Sanuki, T., Motoki, M., Matsumoto, H., Seo, E. S., Wang, J. Z., Abe, K., Anraku, K., Asaoka, Y., Fujikawa, M., Imori, M., Maeno, T., Makida, Y., Matsui, N., Matsunaga, H., Mitchell, J., Mitsui, T., Moiseev, A., Nishimura, J., Nozaki, M., Orito, S., Ormes, J., Saeki, T., Sasaki, M., Shikaze, Y., Sonoda, T., Streitmatter, R., Suzuki, J., Tanaka, K., Ueda, I., Yajima, N., Yamagami, T., Yamamoto, A., Yoshida, T., and Yoshimura, K. (2000). Precise measurements of cosmic-ray proton and helium spectra with the BESS spectrometer, *Astrophys. J.*, 545:1135–1142.

Scherer, K., Fichtner, H., and Stawicki, O. (2002). Shielded by the wind: the influence of the interstellar medium on the environment of Earth, *J. Atmos. Sol. Terr. Phys.*, 64:795–804.

Schlickeiser, R. (1989). Cosmic-ray transport and acceleration. I. Derivation of the kinetic equation and application to cosmic rays in static cold media, *Astrophys. J.*, 336:243–263.

Schlickeiser, R., and Achatz, U. (1993). Cosmic-ray particle transport in weakly turbulent plasmas. I. Theory, *J. Plasma Phys.*, 49:63–77.

Shalchi, A., Bieber, J. W., and Matthaeus, W. H. (2004). Analytic forms of the perpendicular diffusion coefficient in magnetostatic turbulence, *Astrophys. J.*, 600:675–686.

Skilling, J. (1975). Cosmic ray streaming I. Effect of Alfvén waves on particles, *Mon. Not. R. Astron. Soc.*, 172:557–566.

Skilling, J., and Strong, A. W. (1976). Cosmic ray exclusion from dense molecular clouds, *Astron. Astrophys.*, 53:253–258.

Smith, C. W., Matthaeus, W. H., Zank, G. P., Ness, N. F., Oughton, S., and Richardson, J. D. (2001). Heating of the low-latitude solar wind by dissipation of turbulent magnetic fluctuations, *J. Geophys. Res.*, 106:8253–8272.

Smith, R. K., and Cox, D. P. (2001). Multiple supernova remnant models of the Local Bubble and the soft X-ray background, *Astrophys. J. (Supp.)*, 134:283–309.

Sonett, C. P., Morfill, G. E., and Jokipii, J. R. (1987). Interstellar shock waves and ^{10}Be from ice cores, *Nature*, 330:458–460.

Story, T. R., and Zank, G. P. (1995). The response of a gasdynamical termination shock to interplanetary disturbances, *J. Geophys. Res.*, 100:9489–9501.

Story, T. R., and Zank, G. P. (1997), Response of the termination shock to interplanetary disturbances 2. MHD, *J. Geophys. Res.*, 102:17,381–17,394.

Suess, S. T., and Nerney, S. (1990). Flow downstream of the heliospheric terminal shock 1. Irrotational flow, *J. Geophys. Res.*, 95:5403–6412.

Svensmark, H., and Friis-Christensen, E. (1997). Variation of cosmic ray flux and global cloud coverage — a missing link in solar-climate relationship, *J. Atmos. Solar Terr. Phys.*, 59:1225–1232.

Wagner, G., Livingstone, D. M., Masarik, J., Muscheler, R., and Beer, J. (2001). Some results relevant to the discussion of a possible link between cosmic rays and the Earth's climate, *J. Geophys. Res.*, 106:3381–3387.

Wagner, G., Masarik, J., Beer, J., Baumgartner, S., Imboden, D., Kubik, P. W., Synal, H.-A., and Suter, M. (2000). Reconstruction of the geomagnetic field between 20 and 60 kyr BP from cosmogenic radionuclides in the GRIP ice core, *Nucl. Instr. Methods Phys. Res. B*, 172:597–604.

Washimi, H., and Tanaka, T. (1996). 3-D magnetic field and current system in the heliosphere, *Space Sci. Rev.*, 78:85–94.

Webber, W. R., and Higbie, P. R. (2003). Production of cosmogenic Be nuclei in the Earth's atmosphere by cosmic rays: Its dependence on solar modulation and the interstellar cosmic ray spectrum, *J. Geophys. Res.*, 108:1355.

Webber, W. R., and Lockwood, J. A. (2001a). Voyager and Pioneer spacecraft measurements of cosmic ray intensities in the outer heliosphere: Toward a new paradigm for understanding the global modulation process 1. Minimum solar modulation (1987 and 1997), *J. Geophys. Res.*, 106:29,323–29,331.

Webber, W. R., and Lockwood, J. A. (2001b). Voyager and Pioneer spacecraft measurements of cosmic ray intensities in the outer heliosphere: Toward a new paradigm for understanding the global modulation process 2. Maximum solar modulation (1990–1991), *J. Geophys. Res.*, 106:29,333–29,340.

Williams, L. L., and Zank, G. P. (1994). Effect of magnetic field geometry on the wave signature of the pickup of interstellar neutrals, *J. Geophys. Res.*, 99:19,229–19,244.

Yeghikyan, A. G., and Fahr, H. J. (2003). Consequences of the Solar System passage through dense interstellar clouds, *Ann. Geophys.*, 21:1263–1273.

Zank, G. P. (1999a). Interaction of the solar wind with the local interstellar medium: A theoretical perspective, *Space Sci. Rev.*, 89:413–688.

Zank, G. P. (1999b). The dynamical heliosphere. In Habbal, S. R., Esser, R., Hollweg, J. V., and Isenberg, P. A. editors, *Solar Wind Nine*, AIP Conf. Proc. Vol. 471, pages 783–786.

Zank, G. P., and Frisch, P. C. (1999). Consequences of a change in the galactic environment of the Sun, *Astrophys. J.*, 518:965–973.

Zank, G. P., Li, G., Florinski, V., Matthaeus, W. H., Webb, G. M., and le Roux, J. A. (2004). The perpendicular diffusion coefficient for charged particles of arbitrary energy, *J. Geophys. Res.*, 109:A04107.

Zank, G. P., and Matthaeus, W. H. (1993). Nearly incompressible fluids. II. Magnetohydrodynamics, turbulence, and waves, *Phys. Fluids A*, 5:257–273.

Zank, G. P., Matthaeus, W. H., and Smith, C. W. (1996a). Evolution of turbulent magnetic fluctuation power with heliocentric distance, *J. Geophys. Res.*, 101:17,093–17,107.

Zank, G. P., and Pauls, H. L. (1997). Shock propagation in the outer heliosphere 1. Pickup ions and gasdynamics, *J. Geophys. Res.*, 102:7037–7049.

Zank, G. P., Pauls, H. L., Williams, L. L., and Hall, D. T. (1996b). Interaction of the solar wind with the local interstellar medium: A multifluid approach, *J. Geophys. Res.*, 101:21,639–21,655.

Zhou, Y., and Matthaeus, W. H. (1990). Transport and turbulence modeling of solar wind fluctuations, *J. Geophys. Res.*, 95:10,291–10,311.

Chapter 11

ACCRETION OF INTERSTELLAR MATERIAL INTO THE HELIOSPHERE AND ONTO EARTH

Ararat Yeghikyan
Byurakan Astrophysical Observatory, Armenia
arayeg@mail.ru

Hans Fahr
Institute of Astrophysics and Space Research, University of Bonn, Germany
hfahr@astro.uni-bonn.de

Abstract The solar system must have passed through interstellar molecular clouds many times in the past. These events would have pushed the heliosphere inward to the region of the terrestrial planets, bringing the Earth into immediate contact with interstellar matter, provided that the number density of the cloud's material was the order of, or greater than, 1000 cm^{-3}, and that relative velocities of about 20 km/s prevailed. A simple two-fluids treatment of the incoming flow of interstellar material (ISM) is proposed here. We consider ISM that is ionized only by the solar UV, and then assess the amount of neutral interstellar hydrogen that is accreted by the Earth's atmosphere during a single passage through a dense cloud. The behavior of the flow variables is investigated by a 2D-hydrodynamic approach to model the interaction processes, taking into account both the photoionization and the gravity of the Sun. As we show, the resulting strongly increased neutral hydrogen fluxes, ranging from 10^9 to 10^{11} $cm^{-2}s^{-1}$, cause substantial changes in the terrestrial atmosphere. In that case hydrogen acts as a chemical agent to remove oxygen atoms and to cause ozone concentration reductions above 50 km by a factor of 1.5 at the stratopause, to about a factor of 1000 and more at the mesopause. Thus, depending on the specific encounter parameters, the high mixing ratio of hydrogen in the Earth's atmosphere may substantially decrease the ozone concentration in the mesosphere and may trigger an essential climatic change of relatively long duration.

P. C. Frisch (ed.), Solar Journey: The Significance of our Galactic Environment for the Heliosphere and Earth, 317–348.

Keywords: ISM: clouds, interplanetary physics: heliopause and solar wind termination; solar wind - hydrodynamics, Earth: accretion of interstellar matter

Introduction

The interstellar medium (ISM) in our Galaxy, a mixture of gas and dust, is known to be highly inhomogeneous with respect to density and temperature. It tends to concentrate near the Galactic plane and along the spiral arms, and is composed of a dense, neutral and molecular material, embedded in an ambient warm and tenuous medium (for an overview of the composition of the ISM, see McKee, 1995). Roughly half the ISM mass is confined to discrete clouds occupying only $\sim 1 - 2\%$ of the Galactic disk volume. The dense component consists of cold atomic HI clouds ($T < 100$ K) and giant molecular clouds (GMC, $T \sim$ 10–20 K) with clumpy substructure. These clouds have irregular shapes and overall dimensions and densities ranging from 0.01–50 pc and from 10–100 cm^{-3} (HI) and about of 1000 cm^{-3} (GMC clouds) respectively. Cores of GMC and clumps within GMC are denser (more than 10^3 and 10^5 cm^{-3} correspondingly), but they occupy only ~ 0.1 and $\sim 1\%$ of the GMC volume, respectively (Allen, 2000).

From time to time, the solar system on its galactic itinerary encounters various galactic objects, e.g. spiral arms (Leitch and Vasisht, 1998; Shaviv, 2003), star clusters and associations (Innanen, 1996), galactic diffuse clouds (HI), and giant molecular clouds (Talbot and Newman, 1977), etc. Although all encounter probabilities are finite, only a few of them are high enough to make it worthwhile to consider them. All of these mentioned events correspond to different mean travel times of the solar system between consecutive encounters with the corresponding objects, e.g. depending on their distributions in the galactic plane, their sizes and their peculiar velocities.

Looking into the ancient past and upcoming future of the solar system and its planets, it may be of eminent importance to inquire what are the imprints of encounters with dense interstellar gas clouds on the heliosphere and the planetary environments. Of course the point of termination of the supersonic solar wind by an outer shock, which in the upwind direction presently is expected at about 90 AU (see e.g. Krimigis et al., 2003), will drastically move inwards for increasing densities of the interstellar wind that actually blows over the solar system. As already emphasized in an early paper by Fahr (1980), the outer termination shock (see below, Eq. 11.8) should be located at distances smaller than 1 AU for interstellar gas densities larger than $\mathrm{n}_{H,\infty} \geq 100$ cm^{-3}, as shown in Figure 11.1. The Earth should, under these conditions, be directly exposed to a pristine interstellar medium that mainly consists of hydrogen in atomic and molecular form. This also means that the Earth's atmosphere is swept over by the interstellar wind, with all of the inherent atmospheric and climatologic implications.

As already pointed out by Talbot and Newman (1977), the solar system, since its existence, has probably traversed several dense interstellar clouds (IC) while orbiting the galactic center. Talbot and Newman (1977) estimated that 135 encounters have occurred with clouds of a gas density larger than 10 cm^{-3} and 16 encounters with clouds of densities larger than 1000 cm^{-3}. As a consequence of such encounters, it has been envisaged by these authors that a) interstellar matter is accreted to planets; b) interstellar elements are accreted to the solar photosphere; and c) the solar wind becomes drastically changed and perhaps even choked off. The amounts of accretion onto planets and the solar photosphere (points a and b) were investigated by Fahr (1980). Concerning the third point (c), Ripken and Fahr (1981) used the characteristic equation for a heavily mass-loaded solar wind to study the response of the critical point, i.e. the transition from subsonic to supersonic flow, to interstellar density. They found that, for higher ISM densities, the location of the critical point moves towards the solar corona, but that it never can be moved back to the base of the corona (r_q in Fig. 11.1) unless ISM with densities larger than 10^6 cm^{-3} sweep over the solar system. In other words, the solar wind will be strongly decelerated everywhere as a result of heavy mass loading. However, in no case will it be completely choked off, since even interstellar gas densities of up to 10^6 cm^{-3} will not be able to effectively influence the solar wind region inside the sonic or critical point, or to impede the solar wind from becoming supersonic; i.e., no so-called breeze-solution can be enforced. These arguments, however, do not apply to the essentially different physics governing the supersonic solar wind of the young Sun, which is still within its parental cloud of birth with probable densities of larger than 10^5 cm^{-3}. In this case, the solar corona embedded in a dense and cool interstellar medium may be too strongly cooled by energetic neutral atoms resulting from charge exchange between coronal protons and interstellar atoms.

The question of whether a dense IC would prevent the solar wind (SW) from reaching the Earth, which would result in cloud material directly accreting onto the terrestrial atmosphere, as well as many other aspects of this complex problem, were already discussed in the literature of the past (Fahr, 1968a; Fahr, 1968b; Talbot and Newman, 1977; Holzer, 1977; Fahr, 1980; Ripken and Fahr, 1981; Zank and Frisch, 1999; Scherer, 2000; Scherer et al., 2002; Frisch, 2004 and references therein). Such a scenario is considered as a possible trigger of

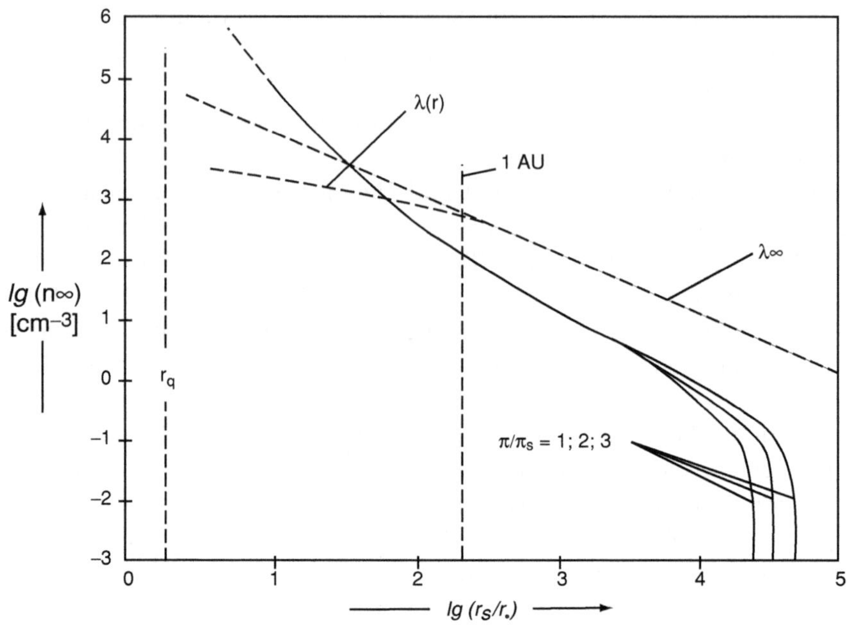

Figure 11.1. Shown is the upwind distance of the solar wind termination shock r_s (Eq. 1.8) as function of the LISM gas density, both in logarithmic units (measured in solar radii, r_q and "1 AU" show the location of the base of the corona and the Earth, respectively). At very low densities the curves split because of different LISM magnetic pressures (π) applied. The value π is measured in units of a standard interstellar pressure π_s (Parker, 1963). Also shown is the associated LISM collisional mean free path λ (Eq. 1.6), given as $10\lambda/r_\odot$ so that it can be given together with r_s on the same abscissa, again in logarithmic units. (Figure is taken from Fahr, 1980).

global glaciations (McCrea, 1975), depositions of a prebiotic material on the primordial Earth (Greenberg, 1981), possible ecological repercussions for the Earth because of an accretion of the cloud's matter, and of course, bio-mass extinction (Yabushita and Allen, 1997).

The mean travel time τ between encounters of the Sun with an IC, a key parameter in this discussion, has been estimated in many works (e.g. Talbot and Newman, 1977; Scoville and Sanders, 1986 and Clube and Napier, 1986), based on available observational data on the distribution of neutral and molecular hydrogen in the Galaxy, and on a refined statistical analysis. Summarizing, the results of these estimations may be represented as follows:

$$\tau = \frac{\delta}{\overline{v}}. \tag{0.1}$$

The δ is a mean free path between encounters and $\overline{v}$ - rms velocity of the Sun relative to the IC, which is the quadratic mean of the cloud-cloud velocity dispersion and the solar peculiar velocity. It is straightforward to estimate δ roughly as:

$$\delta = \frac{1}{\pi D_{IC}^2 N}. \tag{0.2}$$

Here D_{IC} is the radius of the IC and N is the number density of clouds. We estimate the mean travel time between encounters of the Sun with neutral (HI) and molecular (H_2) clouds, respectively, as 27 Myr and 460 Myr. These estimates are based on HI and GMC mean values of 2 and 5 pc for D_{IC}, $15 \cdot 10^{-5}$ and $0.14 \cdot 10^{-5}$ pc^{-3} for N, (Talbot and Newman, 1977), as well as for the $\overline{v} \sim 20$ km/s (Scoville and Sanders, 1986). Scoville and Sanders (1986) have mentioned that the actual time between collisions with GMC will be $\sim$2 times larger ($\sim$ 1 Gyr), while Clube and Napier(1986) have argued for shorter periods, of about 700 Myr.

Thus one may conclude that neutral HI clouds, having a mean density of 10 to 100 cm^{-3}, and a radius of about a few pc, are objects frequently encountered by the Sun during its galactic journey (more than 100 times since its birth $\sim$4.6 Gyr ago), while the more dense GMC must have been encountered just a few times.

Nevertheless Yabushita and Allen (1983, 1989, 1997) have pointed out that just such rare passage through the core of the GMC may have substantially depleted the oxygen of the Earth's atmosphere to cause the bio-mass extinction at the Cretaceous/Tertiary boundary. It should be mentioned also that Wimmer-Schweingruber and Bochsler (2000) have recently interpreted data showing implanted gas sticking in surface layers of lunar soil grains, and Florinski et al. (2003) have proposed a possible explanation for the enhancement of cosmogenic elements in terrestrial ice and sediment samples as a record of an encounter with a dense IC.

The consequences of intensive neutral hydrogen flows, and high H-densities at 1 AU, for the chemistry of the terrestrial atmosphere have been envisioned as considerable (Fahr, 1968a; McKay and Thomas, 1978) and have been repeatedly discussed in the scientific literature (see Scherer, 2000; Scherer et al., 2002; Frisch, 2004 for recent reviews). Concerning the climatic changes, McCrea (1975) has emphasized that just such a triggering mechanism is needed to explain ice epochs that are separated by a few hundred million years. Also, Yabushita and Allen (1989) have pointed out that a passage through the core of a GMC with a density $n \sim 10^5$ cm^{-3} could give rise to an inflow of hydrogen atoms into the terrestrial atmosphere of up to 10^{12} cm^{-2}s^{-1}. As a consequence, the Earth's atmosphere may have been substantially depleted of oxygen by strongly activated H_2O formation, and so may have, for instance, caused a long term bio-mass extinction at the Cretaceous/Tertiary boundary $\sim$65 Myr ago. Also, only such cataclysmic (not catastrophic) events are compatible with the well known high concentration of iridium deposited in clay layers of these paleogeologic boundaries. Moreover, Yabushita and Allen (1997) have stressed that the reduced abundance of oxygen contained in ancient amber may also be taken as a record of such an event of atmospheric oxygen depletion 65 Myr ago.

Our final aim is to show that the mechanism behind the theory of terrestrial atmospheric oxygen depletion (Yabushita and Allen, 1989) has severe limitations, and might not work even in the extremely rare case of an encounter with the core of a GMC. On the other hand, we show that one consequence of this mechanism, first mentioned by McKay and Thomas (1978) concerning the simultaneous mesospheric ozone reduction, may already occur in the much more likely case of an encounter with any part of a gas-dust GMC with a density of at least 1000 cm^{-3}. This is important because, as is mentioned above, cores of GMCs and clumps within GMCs occupy only ~ 0.1 % and $\sim$1% of the volume of the GMC, respectively.

11.1 How does an Interstellar Cloud Touch the Solar System and the Earth?

To investigate the behavior of neutral matter inside the heliopause, which is a separatrix between the solar wind and the ambient interstellar medium, one usually treats the interaction of the solar wind with the countermoving flow as the problem of the penetration of neutral atoms from the local ISM (LISM) into the heliosphere (Baranov and Malama, 1993; Pauls et al., 1995; Zank, 1999; Fahr et al., 2000; see Fahr, 2004 for a recent review).

Results of 5-fluid hydrodynamic model calculations in which, for the first time, all dynamically relevant particle populations were included in a self-consistent manner, namely protons in the solar wind and LISM, neutral hydrogen atoms, pick-up ions due to charge exchange and photoionization processes of the atoms, galactic and anomalous cosmic ray particles (GCR and ACR), are shown in Fahr et al. (2000). As shown on the basis of this model, an increase of the LISM H-atom density by a factor of about 10 (to 1 cm^{-3}) pushes the termination shock inward to a region close to Saturn's orbit (Scherer, 2000). Solar gravity also was ignored in this model, and these authors considered only regions with heliocentric distances larger than 3 AU, i.e. outside of the corresponding accretion radius r_g (for its definition see below in this Section).

But the influence of both gravity and photoionization on the neutral hydrogen flow into the heliosphere seems to be important for quantitative determinations of the flow parameters at regions near the Earth's orbit. Because the kinetic treatment requires very time-consuming numerical codes, we therefore prefer a hydrodynamic modelling here, which is justified for IC densities of, and greater than, a few 100 cm^{-3} (see e.g. Yeghikyan and Fahr, 2003). The prevailing physical conditions in the reduced heliosphere suggest that the idealized hydrodynamic equations, with the inclusion of effects of gravity and ionization, can be used here. Thus, to describe the behavior of neutrals outside of the reduced heliosphere, one may use a simple gasdynamic interaction model, where the system of fluid equations both for neutral interstellar material and solar wind plasma are decoupled, but gravity and photoionization are taken into account. Such an approach was used to study the time–indepedent flow parameters in the heliosphere (Whang, 1996).

Below we use the closed system of fluid equations (Whang, 1996; Fahr, 2000; Yeghikyan and Fahr, 2004a) to calculate a 2D gasdynamic time-dependent simulation of neutral flow parameters, when the heliosphere is drastically reduced in size to a dimension of about 1 AU or less, as a consequence of increased densities of inflowing IC, i.e. 100-1000 times greater than at present.

Not included in the present model are external magnetic field influences on the structure of the compressed heliosphere, because we are interested in the neutral flow outside of the heliosphere. Indeed, strong magnetic fields of up to 8000 μG are found in small dense clumps of GMC, whereas usual GMCs reveal average fields within strengths of 100 μG (Myers and Goodman 1988). HI clouds have weaker fields (Heiles et al., 1991). Thus, in the GMC case the solar heliosphere could, in principle, be subject to comparable dynamic and magnetic pressures caused by the cloud's material (Parker 1963). This might affect the heliospheric structure, but obviously could not influence the outer neutral flow behavior.

It should be noted here that the Sun was probably more active in its early history than it is now (Guinan and Ribas, 2002, and references therein). To be more specific, about 4.6 Gyr ago the Sun had an initial luminosity of $\sim$70% of the present solar luminosity, and was about 200 K cooler, but the UV luminosity in the range 150-250 nm may have been enhanced by a factor of 2 during the first few 10^8 yr (Zahnle and Walker, 1982). These values for the earlier solar wind intensity are more uncertain. They are based on observational studies of stellar winds from young solar-type stars with different ages, which indicate that the winds of the young T-Tauri Sun could possibly be 100-1000 times more intense than at present (Fahr and Scherer, 1995; Fahr et al., 1997; Lammer et al., 2002). Finally, Sackmann and Boothroyd (2003) quite have recently shown that the current accurate helioseismic data are consistent with high-precision solar evolutionary models, which predict a more luminous Sun with a stronger wind than envisaged by the standard model (for the first 1-2 Gyr after the origin of the Sun). Such powerful (2-3 orders of magnitude) solar wind encounters with an HI cloud would enlarge the heliosphere beyond 10-100 AU, and greatly increase the ionization degree of neutral matter at 1 AU. Dense GMC's probably would be able to suppress the powerful solar wind inside of 1 AU because of stronger dynamic counterpressure.

In the following discussion, we will adopt the present parameters of the solar output, which is valid for the most frequent events under consideration (encounters with IC during last 2-3 Gyr), while more specific models should be developed for earlier events.

The encounter of the solar system with a dense interstellar cloud would of course lead to a strongly time-dependent reaction of its extended heliospheric plasma structures as they exist, for instance, at present. When the extended heliosphere, with an upwind extension of about 150 AU, makes first contact with the predominantly neutral gas material of an approaching cloud, then the extended supersonic and subsonic solar wind becomes neutralized by rapid charge exchange with incoming neutral H-atoms. As a consequence of this event, energetic neutral atoms with energies of the order of 1 KeV are produced and travel radially outwards from the heliosphere into the arriving cloud material. By elastic H-H collisions, these energetic atoms strongly heat the cloud material arriving at the heliospause. A strong pressure gradient will thus be produced, which reduces the inflow velocity of the cloud material on the axis, and pushes the cloud gas away from the stagnation point to the heliopause flanks. In addition, as a result of the increasing outer interstellar ram pressure, the heliosphere shrinks to very small dimensions.

Here we do not intend to consistently describe this strongly time-dependent adaptation history, but we shall presume that a nearly stationary counterflow configuration will be eventually established, with the main cloud material deviating around a strongly reduced heliopause (see illustrative representation in Fig. 11.2). We predefine this resulting heliopause configuration by assuming that it is determined by pressure equilibrium between the strongly coupled inner solar wind plasma flow and the outer interstellar flow constituted by ionized cloud material.

To conceive adequate descriptions of the resulting interstellar gas flows around astrospheres, it is first of all important to compare the following relevant scales: the interstellar mean free path λ_{ISM}, the typical dimension of the heliosphere roughly given by the shock distance r_s, and the accretion radius $r_g = GM_\odot(1-\Gamma)/(V_\infty^2 + c_\infty^2)$ of the flow (Edgar, 2004). Here G, $M_\odot$ and Γ denote the gravitational constant, the solar mass, and the degree of gravity compensation by solar radiation pressure; V_∞ and c_∞ are the cloud velocity relative to the Sun and the sound velocity of the cloud gas (for cold clouds it is expected that always $V_\infty \gg c_\infty$). As evident in Figure 11.1, the interstellar mean free path is always larger than the shock distance r_s, unless the interstellar gas densities become larger than 5000 cm^{-3}. Under such undercritical conditions when Knudsen numbers $Kn \simeq \lambda_{ISM}/r_s \geq 1$ prevail, kinetic model approaches on the basis of the Boltzmann-Vlasov theorem are needed. Such models have also been developed very early by Danby and Camm (1957) and Fahr (1968). For supercritical densities, on the other hand, hydrodynamical modelings can be used like those already presented in the past by Parker (1963), Ruderman and Spiegel (1971), Hunt (1971), or Baranov and Krasnobaev (1977).

At very low gas temperatures, when pressure forces are negligible, kinetic solutions and solutions of the Euler set of hydrodynamical equations become identical, yielding a circumsolar gas density distribution given by Fahr (1980), where the following formulae 1.1–1.5 and their derivations can be found (see also Edgar, 2004):

$$\frac{\rho(r,\theta)}{\rho_\infty} = \frac{1}{\sin\theta} \cdot \left[\frac{2(1-\cos\theta) + \overline{r}\sin^2\theta}{2(4\overline{r}(1-\cos\theta) + \overline{r}^2\sin^2\theta)^{1/2}} + \sin\frac{\theta}{2}\right]. \tag{1.1}$$

Here $\overline{r}$ denotes the solar distance measured in accretion radii $r_g = GM_\odot/V_\infty^2$ (Edgar 2004), θ denotes the spacepoint angle measured from the upwind axis, and V_∞ is the velocity of the interstellar gas relative to the Sun. For the upwind direction, $\theta = 0$ the above formula delivers the following result:

$$\frac{\rho(r,\theta=0)}{\rho_\infty} = \frac{1+\overline{r}\left(1+\sqrt{1+\frac{2}{\overline{r}}}\right)}{2\overline{r}\sqrt{1+\frac{2}{\overline{r}}}}. \tag{1.2}$$

For small values of $\overline{r}$, an asymptotic density increase is seen with (Eq. 1.3).

$$\frac{\rho(\overline{r}\ll 1,\theta=0)}{\rho_\infty} = \frac{1}{2\overline{r}^{1/2}}. \tag{1.3}$$

At these distances it strongly deviates from the formula for a radially symmetric accretion flow given by Butler et al. (1978) and Edgar (2004) in the form:

$$\frac{\rho(r,\theta=0)}{\rho_\infty} = \frac{\sqrt{3}}{\overline{r}^2\sqrt{1+\frac{2}{\overline{r}}}}, \tag{1.4}$$

yielding an asymptotic density increase of

$$\frac{\rho(\overline{r}\ll 1,\theta=0)}{\rho_\infty} = \frac{\sqrt{3}}{\overline{r}^{3/2}\sqrt{2}}. \tag{1.5}$$

Equations 1.1 through 1.3 are derived as solutions of the Euler equations, with gravitational attraction included, but with pressure forces excluded. As a result of the strong density increases, the pressure forces can not be ignored for the cone region along the downwind axis. Here the work done against pressure forces reduces the angular momentum of the flow, and thus binds the flow to the solar gravity. From this critical cone region matter shall accrete onto the Sun.

When neutral interstellar matter approaches the Sun, it is evident that it systematically becomes more ionized by photoionization due to the solar EUV radiation. On the other hand, for ISM H-atom densities of the order of $n_H = 10^2$–$10^3\mathrm{cm}^{-3}$, the free passage of a 1 KeV proton (i.e. solar wind proton) is restricted, by charge exchange reactions, to a mean free path of less than

$$\lambda_{ex} = (n_H\sigma_{ex})^{-1} = 2\cdot(10^{-1}\text{–}10^{-2})\mathrm{AU}. \tag{1.6}$$

For such solar distances of less than 1 AU, the interstellar matter is, however, essentially photoionized. Hence at these small distances, in a first-order view, solar wind protons interact with the ionized component of the ISM rather than with the neutral ISM component. Adopting an H-ionization fraction ε of the ISM at the border of the plasma-plasma interaction configuration, one can then estimate a corresponding stand-off distance L of the resulting heliopause by (see Yeghikyan and Fahr, 2003):

$$L^2 = r_0 r_s \sqrt{\frac{\rho_{w0} V_{w0}^2}{4\epsilon \rho_{H\infty} V_\infty^2}}, \tag{1.7}$$

where the density of "hydrodynamically" interacting interstellar protons is $\rho_p = \epsilon \cdot \rho_{H\infty}$. Here r_0 is an inner solar reference distance where the unperturbed solar wind has a density ρ_{w0} and a velocity V_{w0}, respectively. The unperturbed ISM is characterized by a density $\rho_{H\infty}$ and a velocity V_∞ relative to the solar system. The value r_s denotes the termination shock distance from the Sun. The expected termination shock of the solar wind then is located at:

$$r_s = r_0 \sqrt{\frac{3K(M_1, \gamma) \rho_{w0} V_{w0}^2}{4C_{ISM}}}. \tag{1.8}$$

The value C_{ISM} is the Bernoulli constant of the ionized ISM given by:

$$C_{ISM} = \varepsilon \rho_{H\infty} \left[\frac{1}{2} V_\infty^2 + \frac{\gamma - 1}{\gamma} P_{H\infty} / \rho_{H\infty} \right]. \tag{1.9}$$

Here $K(M_1, \gamma) = P_1 / P_0$ is the pressure adaptation function given by (see e.g. Fahr et al., 1981, or Fahr, 2000):

$$K(M_1, \gamma) = \frac{\gamma(\gamma + 1) M_1^2}{2\gamma M_1^2 - (\gamma - 1)} \left[\frac{(\gamma + 1)^2 M_1^2}{4\gamma M_1^2 - 2(\gamma - 1)} \right]^{-\frac{\gamma}{\gamma - 1}}. \tag{1.10}$$

The value $\gamma = 5/3$ is the polytropic constant, M_1 is the preshock sonic Mach number, P_1 is the preshock total pressure, and P_0 is the interstellar total pressure at the heliopause stagnation point.

On the basis of the above heliospheric dimensions and, with the assumption of a Parker-type plasma-plasma interaction configuration, the heliopause geometry is determined by:

$$R(\phi) = R_0 f(\phi), \tag{1.11}$$

where $R_0 = L$, $\phi = \pi - \theta$, and θ is the angle between the streamline and the direction of the Sun's motion. Here and below the value π has its usual meaning ≈3.14 rad. The function $f(\phi)$ can be modeled based on existing analytical representations for the above addressed configuration (Fahr et al., 1988; Ratkiewicz, 1992; Wilkin, 1996). We have used the function given by Wilkin (1996):

$$R(\phi) = R_0 \csc\phi[3(1 - \phi\cot\phi)]. \tag{1.12}$$

We now start our calculations by assuming that the ISM approaching the solar system, but not yet at the heliopause, is subject only to solar photoionization processes. Only after passage through the heliopause will the neutral ISM H-atom gas become strongly subject to charge exchange processes with the solar wind protons. We can argue that the heliopause dimension and geometry are barely influenced by the H-atoms penetrating into the inner heliosphere. This is because of the fact that H-atoms, undergoing charge exchange with a supersonic solar wind in the inner heliospheric region, produce H pick-up ions that not only act upon the solar wind by momentum loading, but also by their pressure gradient. It can then be shown that the effect of H pick-up ions, which remain nearly isothermal with the solar wind expansion, is to cause the cancellation of these two oppositely directed forces, which then guarantees a nearly undecelerated solar wind flow and keeps the ram pressure at a constant level (see Fahr and Rucinski, 2002, Fahr, 2002, Fahr and Chashei, 2002). In the region downstream of the termination shock the charge-exchange processes of H-atoms with solar wind protons have little dynamical and thermodynamical effects, because in this subsonic plasma region H pick-up ions hardly can be disentangled from original solar wind protons. Our conclusion thus is that neutral H-atoms penetrating into the heliosphere, even though not treated quantitatively here, can be expected to hardly change the heliosphere.

11.2 Change of the Ionization Degree and Chemical State in the Circumsolar Flow

If the Sun were at rest with respect to the LISM, a stationary state would require that a local equilibrium prevail up to distances where photo-dissociation (photo-ionization) and recombination rates cancel locally. The degree of ionization of the interstellar gas surrounding the Sun at rest has been determined for modest densities in earlier works of Williams (1965) and Krasnobaev (1971), using the assumptions that there is no hydrodynamic motion of the LISM and effects of the solar wind are negligible. These authors have shown similar results for the density range $1\ \mathrm{cm}^{-3} \leq n \leq 500\ \mathrm{cm}^{-3}$: gradual decrease of the ionization degree with solar distance up to a few 1000 AU, in accordance with a simple estimation of the thickness of the transition region, $d \sim (a \cdot n)^{-1}$, where $a = 6 \cdot 10^{-18}\ \mathrm{cm}^2$ is the photoionization cross section of a neutral hydrogen atom. In the more dense molecular environments existing in the cores of GMCs, and their clumpy substructures with densities of about $10^5\ \mathrm{cm}^{-3}$, an analogous photodissociation region should be developed as a result of photodestruction of the H_2 being produced by photon absorption in UV lines that have 2-3 orders of magnitude larger cross-sections compared to photoionization.

In contrast to the static case, if the Sun is moving fresh molecular (neutral) hydrogen of the IC must be permanently approaching the solar system from the upwind direction. Thus, the extent L_c of the dissociation (ionization) cavity in upwind direction is defined as the upwind distance where the rate of freshly convection-imported hydrogen molecules (atoms) equals the local photodissociation (photoionization) rate. This, evidently, leads to the following expression for L_c (the index c is equal to m, for molecules, and to n, for neutral atoms, respectively, see for analogy Fahr, 2004):

$$L_c \simeq \frac{4\beta_{0,c} r_0^2}{3V_{\mathrm{H}}}. \tag{2.1}$$

Here $\beta_{0,m}$ ($\beta_{0,n}$) is the solar photodissociation (photoionization) rate at some reference distance r_0. The photoionization rate $\beta_{0,n}$ is equal to $8 \cdot 10^{-8}\ \mathrm{s}^{-1}$ at the reference point $r_{\mathrm{E}} = 1$ AU (Fahr et al., 2000). The dissociation of molecular hydrogen is complicated due to different possible branches of solar photon interaction with H_2 molecules: the direct photodissociation rate is $3 \cdot 10^{-7}\ \mathrm{s}^{-1}$ at 1 AU for the moderate solar activity conditions, while the rates for photoionization of H_2 and photodissociation with ionization are by 1 order of magnitude less than the dominant direct photodissociation rate (Gruntman, 1996). Thus one finds $L_m \simeq 2.4$ AU and $L_n \simeq 0.61$ AU for the cases of molecular and atomic flow, respectively. Hence the state of the neutral material when arriving at the Earth's orbit is essentially not changed with respect to the interstellar conditions. A treatment of molecular hydrogen as a new fluid, in addition to atoms and ions, is not considered here. In the following sections we will assume that the flow is initially completely atomic at infinity.

11.3 Model of the Neutral Gas Flow

We approximate the situation as follows: let the Sun be in the origin of a system of cylindrical coordinates (r, ζ, z), and let the axis of symmetry z be antiparallel to the motion of the Sun, so that the countermoving cloud's matter flows from negative z-values to positive ones. Such a picture schematically is illustrated in (Fig. 11.2), where the streamlines of the interstellar neutral and ionized wind passing around the strongly reduced heliosphere are shown.

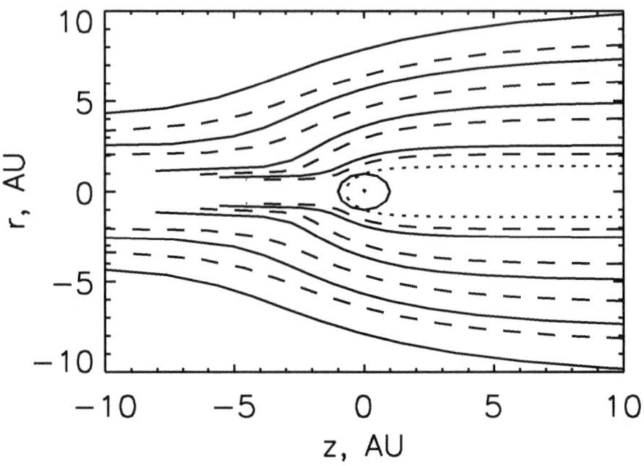

Figure 11.2. Qualitative view of the streamlines of the neutral (solid lines) and ionized (dashed lines) flow of the IC wind in the presence of the solar wind, compressed in the upwind direction to inside of 1 AU. The Sun is in the origin of the coordinate system. The heliopause is shown by the dotted line and the Earth's orbit (shown by solid circle) is outside of it in the upwind direction. (Figure is taken from Yeghikyan and Fahr, 2004a).

It should be emphasized here that we have focused our consideration on a completely neutral inflow material, and on IC densities larger than $1000\ \mathrm{cm}^{-3}$. For such densities, the hydrodynamic approximation is justified for the upwind side of the heliosphere, excluding the region with the heliocentric distances less than a few 0.1 AU. At these densities, as already mentioned, the mean free path of the particles is less than 0.1 AU and Knudsen numbers are of the order of $Kn \sim t_{\mathrm{hydro}}/t_{\mathrm{coll}} \sim 10$, where t_{hydro} and t_{coll} are the characteristic hydrodynamic and collision times, respectively.

To describe the behavior of the flow, one can choose a two-fluid model of hydrogen atoms and protons, which is an extension of the one-fluid model of Whang (1996) and Yeghikyan and Fahr (2003). Using this as a model, a set of conservation laws for mass-momentum-energy-flows can be rewritten, with the right-hand-sides containing corresponding source terms caused by gravitation, photoionization and the mutual charge-exchange, in the form:

$$\frac{\partial \rho_j}{\partial t} + \nabla(\rho_j \mathbf{V}_j) = S_{\rho,j} \qquad (3.1)$$

$$\frac{\partial(\rho_j \mathbf{V}_j)}{\partial t} + \nabla \mathbf{\Pi}_j = \rho_j(1-\mu_j)\nabla\left(\frac{GM_\odot}{R}\right) + S_{V,j} \tag{3.2}$$

$$\frac{\partial E_j}{\partial t} + \nabla[\mathbf{V}_j(E_j + P_j)] \tag{3.3}$$

$$= \rho_j(1-\mu_j)\mathbf{V}_j\nabla\left(\frac{GM_\odot}{R}\right) + S_{E,j}.$$

Here ρ_j, $\mathbf{V}_j$ and P_j are the density, velocity, and scalar pressure of the species $j = n$ (neutral hydrogen) and $j = p$ (protons), respectively. The values $\mathbf{\Pi}_{lk,j} = \rho_j V_l V_k + P_j \delta_{lk}$ and $E_j = \frac{1}{2}\rho_j V_j^2 + \frac{P_j}{\gamma-1}$ are the hydrodynamical-stress tensor and the total energy per unit volume, respectively. The value $R = \sqrt{(r^2+z^2)}$ is the solar distance in a cylindrical coordinate system (r, ζ, z) with symmetry axis z. Other symbols have the following means: γ = 5/3 is the ratio of specific heats for atomic hydrogen gas, μ_j is a factor denoting the ratio of the repulsion force by solar Ly_α radiation to the gravitational attraction force ($\mu_j = 0$ for $j = p$), and $M_\odot$ and G are the solar mass and the gravitational constant, respectively.

The source terms in Eqs. (3.1)-(3.3) result from the interaction of the flow components with the radiation and gravitational fields of the Sun, with an additional contribution from their mutual conversion caused by charge-exchange. Taking all of this together leads to the following source terms (Fahr et al., 2000):

$$S_{\rho,n} = -\beta_{\text{phi}}\rho_n + \rho_p^2\alpha/m_p \tag{3.4}$$

$$S_{\rho,p} = \beta_{\text{phi}}\rho_n - \rho_p^2\alpha/m_p \tag{3.5}$$

$$S_{\mathbf{V},n} = -\beta_{\text{phi}}\rho_n\mathbf{V}_n - \nu_{\text{ce}}^-\rho_n(\mathbf{V_n} - \mathbf{V_p}) + \rho_p^2\alpha/m_p \tag{3.6}$$

$$S_{\mathbf{V},p} = \beta_{\text{phi}}\rho_n\mathbf{V}_n + \nu_{\text{ce}}^-\rho_n(\mathbf{V_n} - \mathbf{V_p}) - \rho_p^2\alpha/m_p \tag{3.7}$$

$$S_{E,n} = -\beta_{\text{phi}}E_n + \nu_{\text{ce}}^-(E_p - E_n) \tag{3.8}$$

$$S_{E,p} = \beta_{\text{phi}}E_n - \nu_{\text{ce}}^+(E_p - E_n) \tag{3.9}$$

where charge-exchange collision rates of hydrogen atoms and protons are given by $\nu_{\text{ce}}^- = \sigma_{\text{ce}} \cdot v_{\text{rel}} n_p$ and $\nu_{\text{ce}}^+ = \sigma_{\text{ce}} \cdot v_{\text{rel}} n_n$, respectively. The value $v_{\text{rel}} = \sqrt{3.76 \cdot 10^8(T_p + T_n) + (\mathbf{V_n} - \mathbf{V_p})^2}$ (in cm s^{-1}). The charge-exchange cross-section is given by $\sigma_{\text{ce}} = [2.1 \cdot 10^{-7} - 9.2 \cdot 10^{-9} ln(v_{\text{rel}})]^2$ (σ_{ce} in cm^2), where n_p and n_n are number density for protons and neutral atoms, respectively. The value $\alpha = 4.0 \cdot 10^{-10}/T_p^{0.8}$ is a photorecombination coefficient in a recombination process of protons with electrons (assumed that $n_p \sim n_{\rm e}$, e.g. Osterbrock, 1989).

The ionization rate of neutral hydrogen atom by photoionization for the optically thin case is approximately defined as

$$\beta_{\rm phi} = \beta_0 (\frac{r_{\rm E}}{R})^2, \tag{3.10}$$

with $\beta_0 = \beta_{0,\rm phi} = 8 \cdot 10^{-8}\ {\rm s}^{-1}$. This is about one order of magnitude less than the ionization rate caused by charge exchange with solar protons (i.e. inside the heliosphere), $\beta_{0,\rm ce}$, which is equal to $7.68 \cdot 10^{-7}\ {\rm s}^{-1}$ at the same reference point $r_{\rm E}$ (Fahr et al., 2000). In the following we will use $\beta_0 = \beta_{0,\rm phi}$, according to the above condition that neutral flow material in the region described in this paper is completely shielded from solar wind protons.

The flow is assumed to be symmetrical with respect to the z-axis, so the results shown here can be rotated by an azimuthal angle ζ with respect to the z-axis.

For an axis-symmetrical case, when all derivatives in ζ vanish, using dimensionless variables and omitting "j indices", we introduce

$$\tilde{\rho} = \rho/\rho_i, \tilde{u} = u/V_i, \tilde{v} = v/V_i, \tilde{P} = P/\rho_i V_i^2,$$

$$\tilde{E} = E/E_i, \tilde{t} = tV_i/\lambda, \lambda = \beta_0 r_{\rm E}^2/V_i, \tilde{z} = z/\lambda, \tilde{r} = r/\lambda,$$

where u and v are the corresponding components of velocity $\mathbf{V}$. This system has a standard vector form of a hyperbolic set of partial differential equations (LeVeque et al., 1998) with relevant source terms on the right sides given in the form:

$$\mathbf{q}_t + \mathbf{A}(\mathbf{q})_z + \mathbf{B}(\mathbf{q})_r = \mathbf{\Psi}(r, z, t, \mathbf{q}). \tag{3.11}$$

The explicit expressions for the $\mathbf{A}$, $\mathbf{B}$ and $\mathbf{\Psi}$, and other details, can be found in (Yeghikyan and Fahr, 2003; Yeghikyan and Fahr, 2004a).

The initial data describe an inner cavity with shape given by Eq. 1.12, and the neutral flow number density from Fahr (1968a). We adopted the boundary conditions of Fahr et al. (2000), and assumed that the inflow of completely neutral material is from the left boundary, and that the bottom boundary (i.e. the axis) is a "reflecting" boundary by symmetry reasons, and that the top and right boundaries are outflow boundaries (zero-order extrapolation). We have limited our calculations here to the upwind side only, up to the reduced heliosphere before the heliopause. The reader should keep in mind that the

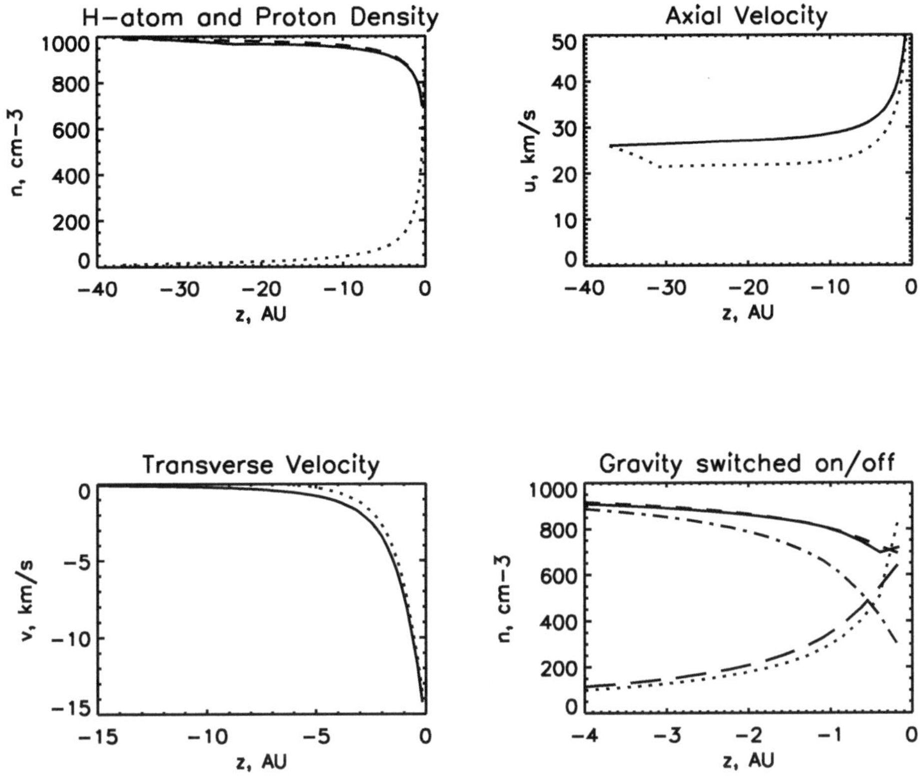

Figure 11.3. Number densities of the IC H-atom and proton flow versus axial distance z, at radial coordinate $r = 0$ (transverse velocities are shown at $r = 1$ AU). Solid and dotted lines are used to distinguish between atoms and protons, respectively. The values $n_{\rm i} = 1000\ {\rm cm}^{-3}$, $V_{\rm i} = 26$ km/s, $T_{\rm i} = 100$ K. Here $\beta_0 = \beta_{\rm phi}$, $\mu = 0.1$, the spatial resolution along the z-axis is 0.2 AU. The density distributions in the cold kinetic model (see text) with the same input parameters are also shown in the top, left and bottom, right, by the dashed lines. (Figure is taken from Yeghikyan and Fahr, 2004a).

condition inside the heliosphere has no influence on the supersonic flow outside of this region.

In Fig. 11.3 we compare the calculated distribution of density for different (gravitating, not-gravitating) cases with that of a cold (collisionless) kinetic model (e.g. see Holzer, 1977, Eq. 21) for the same input parameters and $\beta = \beta_{\rm phi} = 8 \cdot 10^{-8}\ {\rm s}^{-1}$. The density decrease (increase) for the atoms (protons) close to the Sun is evidently caused by the photoionization, while the role of the gravity is visible only inside the region, limited by a so-called accretion radius $r_g = 2GM_\odot/V_i^2$ (Hunt, 1971, cf. the cases $G \neq 0$ and $G = 0$ in Fig. 11.3). A complete coincidence of the density distribution in the cold kinetic and hydrodynamic models justifies our hydrodynamic approach in the calculational domain.

The optical depth in the $Ly - \alpha$ hydrogen line, even at the 1 AU, would always be greater than one for $n_i \simeq 100\ \mathrm{cm}^{-3}$. At the same time, the contribution of diffusely scattered $Ly - \alpha$ photons to the repulsion force would be nearly negligible, mainly because the first scattering of the solar $Ly - \alpha$ photon contributes to the radiation pressure. Thus $\mu = 0.1$ seems to be an appropriate approximation.

It is clear that only when $R \geq r_g$ will the flow be unaffected by gravitation (Talbot and Newman, 1977; Edgar 2004). For $V_i = 26$ km/s one finds $r_g = 2.6$ AU. This means that the case of Bondi's flow (e.g. Edgar 2004) $\rho \simeq \rho_i (r_g/R)^{3/2}$ and $V^2 \simeq 2GM_\odot/R$ is applicable inside the distance r_g (see Fig. 11.3).

A key parameter, $\omega = R_g/\lambda^*$, which determines the net density enhancement of neutral matter, is the ratio of the accretion radius R_g to the typical photoionization scale $\lambda^* = \beta_E r_E^2/V_i$ (Yeghikyan and Fahr, 2003), where $\beta_E = 8 \cdot 10^{-8}\ \mathrm{s}^{-1}$ is the photoionization rate for H-atoms at r_E. The quantity "ω" does not depend on the heliocentric distance; it depends on the relative velocity between the Sun and the cloud, and may vary between 5 and 25 for possible values of V_i ranging from 5 to 30 km/s (Talbot and Newman, 1977). Finally, provided that the most probable values of V_i are 20–26 km/s (see, e.g. Yeghikyan and Fahr, 2003 and references therein), the GMC material may give rise to a neutral hydrogen flux enhancement of up to 6 orders of magnitude as compared to the present-day values.

11.4 Amount of Neutral Gas, Accreted by the Earth

We can now estimate the amount of neutral gas, accreted by the Earth. An interaction between the charged particles and the magnetosphere is more complicated, and we simply assume here that all of them will be deflected. By order of magnitude accuracy one may write for the amount of accreted matter (Butler et al., 1978): $dM/dt \simeq \pi R_E^2 \cdot V_{rel} \cdot n \cdot m_p$ (g/s), where R_E is the Earth's radius. It practically coincides with the Earth's accretion radius because of the relative velocity of the H-atom with respect to the Earth, V_{rel}, of about 40 km/s. It is clear that there is an annual variation in V_{rel} between 10 to 70 km/s because of the orbital velocity of the Earth, equal to about 30 km/s. An average value of 40 km/s, therefore, would be quite reasonable. Because of the many uncertainties involved with evaluating the flow in the downstream region, where the direct influence of the solar wind protons is essential, we have used an upstream density value at 1 AU. For example, for $n_i \simeq 1000\ \mathrm{cm}^{-3}$, using calculated values of n at 1 AU (about $800\ \mathrm{cm}^{-3}$, see Fig. 11.2, 11.3), one finds that during the crossing time of about 1 million years the Earth may accrete about $2 \cdot 10^{17}$ g per one passage.

We have obtained a calculated value n that is one order of magnitude less than estimated in Butler et al. (1978), because lower relative velocities of the encounter (about 2 km/s with corresponding $r_g \sim 400$ AU) were considered. In other words, they have considered an almost radially symmetric accretion, instead of a density distribution formed by a monodirectional flow (see comparison of formulae 1.3 and 1.5). Besides, the case without photoionization of the neutral flow has been considered by Butler et al. (1978).

It should be highlighted here that a sink mechanism for neutral atoms other than photoionization does not exist in the heliosphere (electron impact ionization at 1 AU would be negligible). Newly created fast neutrals (as a result of charge exchange with solar wind protons) also may be accreted by the Earth. The quantitative investigation of such a model will be addressed separately.

A response of the terrestrial atmosphere to such an amount of accreted neutral hydrogen gas may be considerable. Neutral hydrogen fluxes at the Earth would increase by up to $10^{10}\ \mathrm{atoms/cm^2/s}$ ($n_H = 10^3\ \mathrm{cm}^{-3}$), which is greater by 4 orders of magnitude than at present. This highly reactive hydrogen could reach stratospheric heights at 40–50 km by direct inflow, and accumulate at levels that may be relevant for oxygen blocking and ozone depletion, for example, inevitably causing strong influences on the terrestrial atmosphere in general, and on the terrestrial climate in particular (Fahr, 1968a; Fahr, 1968b; Bzowski et al., 1996; Frisch et al., 2002).

Finally, one can show (see below) that in the case of a countermoving GMC with a density of $n_H = 10^3\ \mathrm{cm}^{-3}$, there would be a considerable decrease of the ozone concentration in the mesosphere (Yeghikyan and Fahr, 2004b).

11.5 Atmospheric Effects

This highly reactive hydrogen could penetrate the upper terrestrial atmosphere up to some height where the collisional "optical" thickness becomes 1, and at that altitude could be involved in chemical reactions, regulating the abundances of the important atmospheric species such as oxygen and ozone (McKay and Thomas, 1978). These authors have argued that in the middle atmosphere (between 50 and 150 km above sea-level) the reactions that interchange odd-hydrogen compounds (H, OH, HO_2) are much faster than those that couple odd-hydrogen to even-hydrogen compounds (H_2, H_2O). This makes it possible to balance the incident flux of interstellar (atomic/molecular) hydrogen by the upward escape flux of H atoms and the downward flux of H_2O in the following form (see McKay and Thomas, 1978):

$$\begin{aligned}\Phi &= 0.5 \cdot \Phi_{\rm esc}(\rm H) + \Phi_{\rm down}(\rm H_2O) \\ &= 5.9 \cdot 10^{11} f^{1/2} + 1.9 \cdot 10^{13} f.\end{aligned} \tag{5.1}$$

Here Φ is the flux of incoming interstellar H_2, while f, as explained in detail by McKay and Thomas (1978), is the ratio of H_2O number density to the total atmospheric number density at 100 km above sea level. Atomic hydrogen atoms move upwards and are lost by exospheric escape. Water molecules, which tend to reconstitute the local homospheric density equilibrium distribution, diffuse downward towards the tropospheric levels, form clouds, and finally are lost to the ground through water precipitation. Later Yabushita and Allen (1983, 1989) showed that, on the basis of this scheme, all atmospheric oxygen must be removed by interstellar hydrogen in a ratio 2:1 by number, when the dense core of a GMC with a density of $10^5\ \rm cm^{-3}$ is encountered by the Earth.

It should be noted that, for a complete removal of the present column of $N_{\rm O} = 4.4 \cdot 10^{43}$ atoms of atmospheric oxygen (in the form of O_2), it is necessary to have critical fluxes of interstellar hydrogen at 1 AU, as given by the following simple relation:

$$F_{\rm c} = \frac{2 \cdot N_{\rm O}}{4\pi R_{\oplus}^2 \cdot \Delta T} = \frac{1.5 \cdot 10^{13}}{D(\rm pc)} \frac{\rm atoms}{\rm cm^2 \cdot s}. \tag{5.2}$$

These fluxes are maintained over a certain time period $\Delta T = D/V_{\rm i}$ that is characteristic for the solar system passage through a GMC with a dimension D. The critical flux level $F_{\rm c}$ then amounts to $2.9 \cdot 10^{12}$ and $1.5 \cdot 10^{12}\ \rm cm^{-2}\ s^{-1}$ for passage times of 0.2 and 0.4 Myr of 5 pc - and 10 pc - sized GMCs, respectively (the relative velocity $V_{\rm i}$ between the Sun and the GMC is assumed to be 26 km/s).

Our scenario described by the two-fluid model predicts a neutral hydrogen flux of $3 \cdot 10^9\ \rm cm^{-2}s^{-1}$ for a GMC density of $10^3\ \rm cm^{-3}$ (Fig. 11.3). We assume that, under these conditions, the incoming neutral hydrogen flow plays the same role as considered in the mentioned even-odd reaction scheme of McKay and Thomas (1978) for molecular hydrogen, i.e. Eq. 5.1 is valid. In this case, according to Yabushita and Allen (1989), for a given incoming atomic hydrogen flux Φ Eq. 5.1 results in a value of f, which in turn yields values for $\Phi_{\rm down}(\rm H_2O)$.

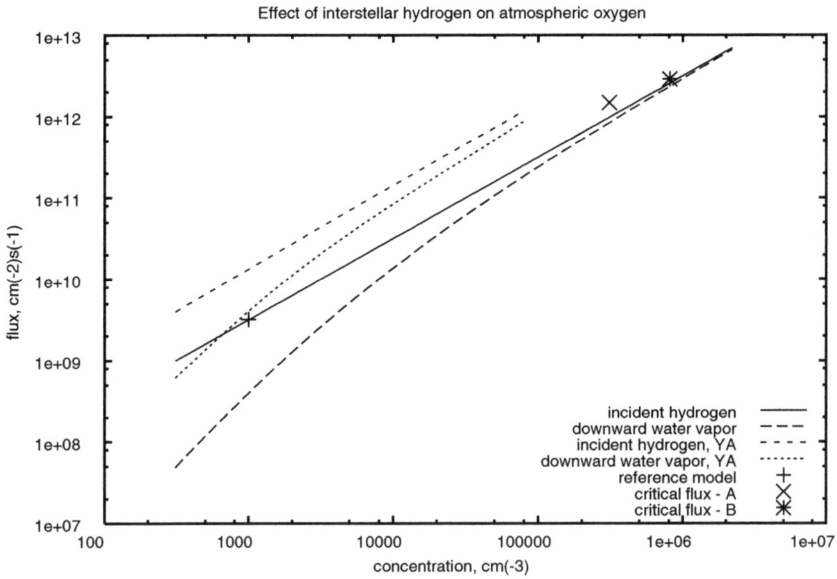

Figure 11.4. Flux of incident hydrogen at 1 AU, Φ, and downward flux captured as water, $\Phi_{\rm down}(\rm H_2O)$ versus the density of a GMC, are shown by curves marked YA (Yabushita and Allen's model) and by curves without marks (our hydrodynamical model), respectively. The critical values of incident hydrogen, corresponding to the 5 pc (B) and 10 pc (A) sized GMCs, respectively, are estimated from Eq. 5.2 and also are shown. The cross is located at the flux level of 3×10^9 cm^{-2}s^{-1} corresponding to the case of the flow mentioned in Fig. 11.3 (i.e. $n_{\rm i} = 1000\,\rm cm^{-3}$). (Figure is taken from Yeghikyan and Fahr, 2004b).

In Fig. 11.4 we show the incident neutral hydrogen flux, Φ, and the downward flux captured as water, $\Phi_{\rm down}(\rm H_2O)$, versus density of GMC. The results from our hydrodynamical calculations are compared with those of Yabushita and Allen (1989). A difference is caused by the more refined hydrodynamical and ionizational model that we have used here. The critical fluxes $F_{\rm c}$ of 5 and 10 pc sized GMC are calculated by Eq. 5.2. Such clouds correspond to the cases B and A mentioned in Fig. 11.4, with concentrations of $9.1 \cdot 10^5$ and $4.7 \cdot 10^5$ cm^{-3}, respectively. The hydrogen flux level, F, corresponding to the model case of Fig. 11.3 is shown by a cross. It can be clearly seen that only gaseous material of GMCs, with an average density of $5 \cdot 10^5$ cm^{-3} or more, is sufficient (i.e. $F > F_{\rm c}$) to completely remove all atmospheric oxygen.

At the same time, such densities correspond to cores and clumps of GMCs with dimensions that are one or two orders of magnitude smaller ($D \sim 0.1$ pc) than those of the whole GMC; i.e. in this case Eq. 5.2 would require a critical flux that is one or two orders greater than those presented in Fig. 11.4, because of the shorter times of passage. Thus we conclude that this mechanism should not work for any real GMC with parameters $n \sim 1000\ \mathrm{cm}^{-3}$ and $D \sim$ 5–10 pc.

The remainder of the chapter will consist of an attempt to estimate one other consequence of increased hydrogen content in the terrestrial upper atmosphere, by focusing on its effect on the global ozone concentration.

11.6 Ozone Concentration in the Mesosphere

In the mesosphere, between 50–100 km, the increased influence of hydrogen becomes significant for ozone destruction, as first mentioned by McKay and Thomas (1978). Note that in general the chemistry of a hydrogen-oxygen-nitrogen atmosphere is governed by very complicated equations. However, when we eliminate the less probable reactions from the reaction system, it is possible to develop a general idea of the role of the principal reactions. In particular, since ozone has a short lifetime in the mesosphere and consequently is very close to photochemical equilibrium, its local concentration can conveniently be written as a simple function of water vapor concentration $n(\mathrm{H_2O})$ (Nicolet, 1971; Chamberlain, 1978; Brasseur and Solomon, 1984):

$$n(\mathrm{O_3}) = \frac{Y}{[n(\mathrm{H_2O})]^{\alpha}}. \tag{6.1}$$

Here Y denotes a known analytical function that is dependent on the reaction rates involved, and on the total and molecular oxygen concentrations at a given altitude. According to Brasseur and Solomon (1984), α is equal to 1, 1/2 and 1/3 at altitudes 80, 70 and 50 km, respectively. Ozone is assumed to be formed by three-body recombination between atomic and molecular oxygen, and is destroyed by bimolecular processes, where hydrogen acts as a catalyst, as well as by photodissociation. Again, destruction of odd oxygen (ozone) by $\mathrm{HO_x}$-particles (H, OH, $\mathrm{HO_2}$) cannot be important below 45 km (e.g. Crutzen, 1971).

Adopting the values of the reaction rates and their distributions with height from the existing atmospheric models (e.g. Hunten and Donahue, 1976, and references therein), the ozone concentration can now be calculated as a function of altitude if the height distribution of water vapor is known. The vertical profile of water vapor is governed, in general, by the competition between regeneration, transport and removal processes. The total downward flux through the mesosphere must be constant because loss and production processes of $\mathrm{H_2O}$ are sufficiently rapid (Nicolet, 1971). Because the horizontal gradients of the concentrations obviously are relatively small (Brasseur and Solomon, 1984), and since in the mesosphere molecular diffusion is negligible (Hunten and Donahue, 1976), the water vapor height distribution is primarily determined by vertical transport in the eddy-diffusion-controlled regime. By requiring that the downward flux of H_2O is constant at a given incoming flux of extraterrestrial hydrogen, one can relate the value of $\Phi_{\text{down}}(\text{H}_2\text{O})$ to the concentration of water vapor $n(\text{H}_2\text{O})$ at the reference altitude of 100 km (McKay and Thomas, 1978). Indeed,

$$\Phi_{\text{down}}(\text{H}_2\text{O}) = \frac{n(\text{H}_2\text{O})_{100} \cdot K_{100}}{2H_{\text{av}}}. \tag{6.2}$$

This in turn gives $n(\text{H}_2\text{O})$ at the given altitude in the mesosphere:

$$n(\text{H}_2\text{O}) = \frac{n(\text{H}_2\text{O})_{100} \cdot K_{100}}{K}. \tag{6.3}$$

In this equation, $n(\text{H}_2\text{O})_{100}$ is the perturbed water vapor concentration at 100 km and $H_{\text{av}} = 6$ km is the average mesospheric scale height in the atmosphere, as defined by the barometric law (Nicolet, 1971; Brasseur and Solomon, 1984).

It must be stressed that the distribution of H_2O in the mesosphere will strongly depend on the choice made for the vertical profiles of the eddy diffusion coefficients in the calculations of the diffuse downward current. Here, the eddy diffusion coefficient at 100 km, K_{100}, is taken to be $2 \cdot 10^6\ \text{cm}^2\text{s}^{-1}$, varying with altitude according to the profile proposed in the U.S. Standard Atmosphere (Brasseur and Offerman, 1986). For comparison we will also use the profile proposed by Lindzen (1971), which varies as the square root of density.

Thus the ratio δ of the ozone concentration $n(\text{O}_3)$ in the mesosphere to its unperturbed value at the same altitude, $n^*(\text{O}_3)$, is determined by

$$\delta = \frac{n(\text{O}_3)}{n(\text{O}_3)^*} = \frac{n(\text{H}_2\text{O})^{\alpha}}{n(\text{H}_2\text{O})^{*\alpha}}, \tag{6.4}$$

where α is equal to 1, 1/2 and 1/3 at altitudes 80, 70 and 50 km, respectively, and $\text{n}(\text{H}_2\text{O})^*$ is the known unperturbed value of $\text{n}(\text{H}_2\text{O})$.

11.7 Results and Discussion

The equilibrium concentrations for perturbed ozone in the mesosphere, between 50 and 100 km, evaluated by means of the adopted unperturbed values from the atmospheric model of Liu & Donahue (see, e.g. Hunten and Donahue,

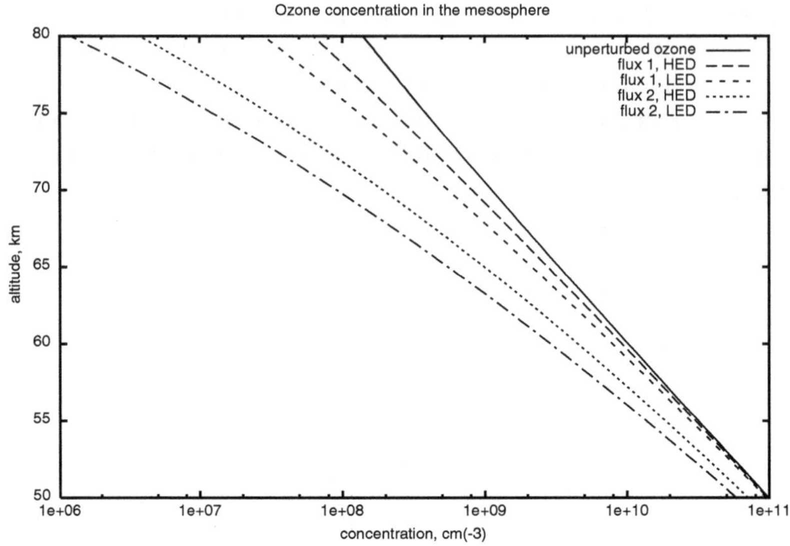

Figure 11.5. Distribution of ozone for the unperturbed atmosphere and for two values of incident interstellar hydrogen flux "1" (5×10^9 cm^{-2}s^{-1}) and "2" (5×10^{10} cm^{-2}s^{-1}). The curves labeled HED correspond to the eddy diffusion profile USSA 76 (Brasseur and Offerman, 1986), while the curves labeled LED refer to the profile from Lindzen (1971) (Figure is taken from Yeghikyan and Fahr, 2004b).

1976), are given in Fig. 11.5 for two values of incoming neutral ISM hydrogen densities, $\sim 2 \cdot 10^3$ cm^{-3} and $\sim 2 \cdot 10^4$ cm^{-3} respectively.

It can be seen that in all cases a significant reduction of the ozone concentration occurs, owing to the influence of the increased hydrogen fluxes. The total O_3 content, as compared with unperturbed values, decreased from 50–80% near 80 km to 2–4% at the stratopause ($\sim$50 km), and from 2 orders of magnitude near $\sim$80 km to 30–80% at the stratopause, for incident fluxes of $5 \cdot 10^9$ cm^{-2}s^{-1} and $5 \cdot 10^{10}$ cm^{-2}s^{-1}, respectively. Such fluxes correspond to GMC densities of 1–$2 \cdot 10^3$ to 1–$2 \cdot 10^4$ cm^{-3}, respectively.

Thus, in the Earth's atmosphere above 45 km, H-atoms are very important in limiting the natural content of ozone. An important consequence of the increased hydrogen fluxes may be that, as argued McKay and Thomas (1978), the creation of saturation conditions over much of the mesosphere should result in a permanent layer of ice clouds of nearly world-wide extent. The earlier analysis (see discussion in McKay and Thomas, 1978, and references therein) revealed that a lowering of the average height of the mesopause, usually located at an altitude of 85–90 km, as well as a lowering of the average mesopause temperature would be significant.

Indeed, radiative heating in the stratosphere and mesosphere is mostly caused by the absorption by ozone of solar radiation (Roble 1995). This reaches a maximum at about 45 km, but falls off in altitude with the decrease in ozone concentration. Above 80 km there is a contribution as a result of the absorption by molecular oxygen. An important heating mechanism in the mesosphere, which dominates over solar heating between 80–90 km, is the exothermic chemical reaction of odd-oxygen constituents (Mlynczak and Solomon 1993). At the same time, cooling in the mesosphere is dominated by 15 μm infrared band emission of CO_2, with a much smaller contribution caused by 9.6 μm emission of ozone. This process gives rise to the maximum in CO_2 cooling at about 100 km, which is responsible for the mesopause. The global mean mesospheric temperature structure, as calculated by the numerical models (e.g. Roble, 2000), clearly shows that just the stratopause corresponds to a maximum in heating due to absorption of solar energy by ozone. Above the stratopause, temperature falls off with the decrease in ozone density and increase in CO_2 radiative cooling to the mesopause. As is well known, CO_2 is produced in the troposphere by plants and anthropogenic sources, principally fossil fuel combustion. CO_2 is un-reactive, resulting in a near constant volume mixing ratio in the middle atmosphere-mesosphere: only in the lower thermosphere does it undergo a small amount of solar dissociation (Brasseur and Solomon 1984).

As argued by McKay and Thomas (1978), in a perturbed state there is very little ozone in the neighbouring mesosphere so that hardly any solar UV radiation is absorbed there. In fact, the solar UV radiation would penetrate more deeply (by approximately one scale height or ~6 km, see McKay and Thomas, 1978). For this reason, and because of CO_2 cooling, which can not be perturbed by hydrogen, the mesopause is the coldest place in the terrestrial atmosphere and the temperature here decreases and reaches the lowest values. In the unperturbed atmosphere, mesospheric (noctilucent) clouds are formed at high latitudes in summer, in the mesopause region. Such a cloud layer under perturbed conditions would be of practically world-wide extent, and would result an equilibrium global surface temperature decrease of 1 K. Because the Earth's climate is very sensitive to small reductions in the surface temperature, a reduction of 1 K sustained over a period of several thousand years, was probably enough to trigger an ice age of substantially long duration (McKay and Thomas, 1978). The reason for this sensitivity is the strong positive feedback between temperature and surface albedo, since an ice-covered surface sends much more solar radiation back into space than an ice-free surface (McCrea, 1975).

For the present-day point of view, the possibility that processes responsible for changing mesospheric ozone may also alter the mesopause temperature should be considered with caution. It is more likely that small-scale gravity waves play a crucial role in the dynamics and thermal structure of the mesosphere (Roble 2000). As the mentioned atmospheric waves are triggered by the absorbed solar radiation in the middle atmosphere, dominated by ozone absorption, it becomes more likely that ozone changes lower down might affect the dynamics, and thus affect the mesopause temperature indirectly. A quantitative description of gravity wave spectra and occurrence frequencies, which will lead to a greater understanding of the energy budget and coupling between regions of the perturbed atmosphere, requires a 3D general circulation model combined with chemistry (e.g. Roble 2000) that is, of course, beyond the scope of this work.

11.8 Summary

We have discussed the problem of solar system passages through dense interstellar gas clouds with non-negligible dust content. We have shown that such clouds increase the inflow density and push the heliopause down to the orbit of the Earth, with the result that counterstreaming IC material directly influences the Earth. In contrast to the static case, when the Sun is moving fresh molecular (neutral) hydrogen material of the IC is permanently approaching the solar system from the upwind direction, and thus the extent of the dissociation (ionization) cavity in the upwind direction will be defined as the distance at which the rate of freshly imported hydrogen molecules (atoms) equals the local photodissociation (photoionization) rate. It is shown that close to the Sun, at 1 AU, the cloud's matter is predominantly neutral and is completely shielded from the direct solar wind influence. At IC H-atom densities of the order of $n_{\mathrm{H}} = 10^3\ \mathrm{cm}^{-3}$ and more, incoming matter that is approaching the Sun, but still outside the reduced heliopause, will be subject only to photoionization processes, which then simplifies the hydrodynamic model of the flow.

We have used a simple two-fluid treatment of the incoming flow, ionized by the solar UV, while taking into account the conversion between hydrogen atoms and protons caused by charge exchange. We have carried out the time-dependent simulation of flow parameters by means of a 2D-gasdynamic numerical code that takes into account both photoionization and gravity of the Sun, as well as the mentioned charge-exchange process. These processes are important to quantitatively estimate the total amount of neutral material, accreted from the cloud by the Earth during the cloud crossing time of about

1 million years. We have estimated this amount to be greater than $2 \cdot 10^{17}$ g for a single passage.

The terrestrial atmospheric response to the inflow of greatly enhanced fluxes of neutral hydrogen atoms may be considerable, and definitely worthwhile to study. It appears that an enhanced flux of interstellar hydrogen at the terrestrial orbit can produce important climatic effects on the Earth through complete ozone destruction in the mesosphere. This mechanism, if true, may explain, in particular, the ice ages that are separated by roughly a few hundred million years (McCrea, 1975); McKay and Thomas, 1978; Yabushita and Allen, 1989).

Other consequences of an enhanced interstellar hydrogen flux for the terrestrial atmosphere, the climate, and for mass-extinctions are discussed in McKay and Thomas (1978) and Yabushita and Allen (1983, 1985, 1989, 1997).

Acknowledgments: It is a pleasure for A.G.Y to acknowledge the receipt of financial support from the Alexander von Humboldt Foundation and to thank the Institute of Astrophysics and Extraterrestrial Research, University of Bonn for hospitality where a large part of this work was developed. Financial support by the Deutsche Forschungsgemeinschaft in the frame of the project "Helio-trigger" (Fa 97/28-1) is also gratefully acknowledged. The authors thank I. Chashey, S.Grzedzielski and K. Scherer for constructive review comments and enlightening discussions. We thank the Editor, P. Frisch, and two anonymous referees for useful comments.

References

Allen's Astrophysical Quantities, A.N.Cox, Ed., Springer, 2000

Baranov, V. B., Krasnobaev, K. V., *Hydrodynamic theory of a cosmic plasma*, Moscow, Izdatel'stvo Nauka, In Russian, 1977

Baranov, V. B. and Malama, Yu. G., Model of the solar wind interaction with the Local Interstellar Medium: numerical solution of self-consistent problem, J. Geophys. Res., 98, 15157-15163, 1993

Baranov, V.B., Krasnobaev, K.V., Kulikovskiy, A model of the interaction of the solar wind with the interstellar medium, Sov. Phys. Dokl., 15, 791-793, 1971

Blum, P.W. and Fahr, H., Interaction between Interstellar Hydrogen and the Solar Wind, Astron. Astrophys. 4, 280-286, 1970

Brasseur, G. and Offermann, D., J. Geophys. Res., 91, 10818-10822, 1986

Brasseur, G. and Solomon, S. 1984, *Aeronomy of the middle atmosphere*, D. Reidel Publ. Comp., 1984

Butler, D.M., Newman, M.J., Talbot, Jr. R.J., Interstellar cloud material: contribution to Planetary Atmospheres, Science, 201, 522-525, 1978

Bzowski, M., Fahr, H.-J., Rucinski, D., Interplanetary neutral particle fluxes influencing the Earth's atmosphere and the terrestrial environment, Icarus, 124, 209-219, 1996

Chamberlain, J.W. *Theory of planetary atmospheres*, Acad. Press, 1978

Clube, S.V.M. and Napier, W.M., Giants comets and the Galaxy: implications of the terrestrial record, In: *The Galaxy and the Solar System*, Eds., R.Smoluchowski, R., Bahcall, J.N., Mattews, M.S.,The Univ. Arizona Press, Tucson, 69-82, 1986

Crutzen, P.J. 1971, J. Geophys. Res., 76, 7311, 1971

Danby, J.M.A., and Camm, G.L., Statistical dynamics and accretion, Mon. Not. Roy. Astron. Soc., 117, 50-95, 1957

Edgar, R., A review of Bondi-Hoyle-Lyttleton accretion, New Astronomy Reviews, 48, 843-859, 2004

Fahr, H.J., On the influence of neutral interstellar matter on the upper atmosphere, Astrophys. Space Sci., 2, 474-495, 1968a

Fahr, H.J., Neutral corpuscular energy flux by charge-transfer collisions in the vicinity of the Sun, Astrophys. Space Sci., 2, 496-503, 1968b

Fahr, H.J., Aspekte zur Wechselwirkung von Sternen mit dichten interstellaren Gaswolken, Mitt. Astron. Gesellschaft, 47, 233-276, 1980

Fahr, H., Formation of the Heliospheric Boundaries and the Induced Dynamics of the Solar System: A Multifluid View, In *The Outer Heliosphere: Beyond the Planets*, Eds., Scherer, K., Fichtner. H., Marsch, E., Copernicus Gesellschaft e.V., 67-89, 2000

Fahr, H.J., Solar wind heating by an embedded quasi-isothermal pick-up ion fluid, Annales Geophysicae, 20, 1509-1518, 2002

Fahr, H.-J., Global structure of the heliosphere and interaction with the local interstellar medium: three decades of growing knowledge, Adv. Space Res., 34 (1), 3-13, 2004

Fahr, H.J. and Scherer, K., On the motion of wind-driving stars relative to the ambient interstellar medium, Astrophys. Space Sci., 225, 21-45, 1995

Fahr, H.J. and Chashei, I., On the thermodynamics of MHD wave-heated solar wind protons, Astron. Astrophys., 395, 991-1000, 2002

Fahr, H.J. and Rucinski, D., Modification of properties and dynamics of distant solar wind due to its interaction with neutral interstellar gas, Nonlinear Proc. in Geophysics, 9, 377-384, 2002

Fahr, H.J., Petelski, E.F., and Ripken, H.W., Weak Shock Termination of the Solar Wind, In: "Solar Wind-IV", Burghausen, Ed., Rosenbauer, H., 543-549, 1981

Fahr, H.J., Grzedzielski, S. and Ratkiewicz, R., Magnetohydrodynamic modeling of the 3-dimensional heliopause using the Newtonian approximation, Annales Geophysicae, 6, 337-354, 1988

Fahr, H.J., Fichtner, H. and Scherer, K., The Influence of the Local Interstellar Medium on the Solar Wind Dynamics in the Inner Heliosphere, Space Sci. Rev., 79, 659-708, 1997

Fahr, H.J., Kausch, T., and Scherer, H., A 5-fluid hydrodynamic approach to model the solar system-interstellar medium interaction, Astron. Astrophys., 357, 268-282, 2000

Florinski, V., Zank, G.P., Axford, W.I., The Solar System in a dense interstellar cloud: Implications for cosmic-ray fluxes at Earth and ^{10}Be records, Geophys. Res. Lett., 30 (23), 2206, 2003

Frisch, P.C., LISM structure - fragmented superbubble shell? Space Sci. Rev., 78, 213-222, 1996

Frisch, P.C., Why study interstellar matter very close to the Sun? Adv. Space Res., 34 (1), 20-26, 2004

Frisch, P.C., Muller, H.R., Zank, G.P. and Lopate C., astro-ph/0208556, 2002

Greenberg, J.M., Chemical evolution of interstellar dust - a source of prebiotic material?, In: *Comets and the Origin of Life*, Ed., Ponnamperuma, C., D.Reidel Publ. Comp., 111-127, 1981

Gruntman, M., The effect of the neutral solar wind component upon the interaction of the Solar System with the interstellar gas stream, Sov. Astron. Lett., 8(1), 24-28, 1982

Gruntman, M., Neutral solar wind properties: Advance warning of major geomagnetic storms, J. Geophys. Res., 99(A10), 19213-19227, 1994

Gruntman, M., H2+ pickup ions in the solar wind: Outgassing of interplanetary dust, J. Geophys. Res., 101(A7), 15555-15568, 1996

Gruntman, M., Leonas, V.B. and Grzedzielski, S., Neutral solar wind experimention, In: *Physics of the outer heliosphere*, Eds., Grzedzielski, S. and Page, D.E., Pergamon Press, NY, 355-358, 1990

Guinan, E.F. and Ribas, I., Our changing Sun: the role of solar nuclear evolution and magnetic activity on Earth's atmospheres and climate, In: *The evolving Sun and its influence on planetary environments*, Eds., Montesinos, B., Gimenez, A., Guinan, E.F., ASP Conf. Series, 269, 85-106, 2002

Heiles, C., Goodman, A.A., McKee, C.F. and Zweibel, E.G., Magnetic fields in dense regions, In: *Fragmentatation of molecular clouds and star formation*, Eds.: Falgarone, E., Boulander, F., G.Duvert, G., IAU Symp. 147, Kluwer AP, 43-60, 1991

Holzer, T.E., Neutral hydrogen in interplanetary space, Rev. Geophys. Space Phys., 15, 467-490, 1977

Hunt, R., A fluid dynamical study of the accretion process, Mon. Not. R. Astron. Soc., 154, 141-165, 1971

Hunten, D.M. and Donahue, T.M., Hydrogen loss from the terrestrial planets, Ann. Rev. Earth. Planet. Sci., 4, 265-292, 1976

Innanen, K.A. The Solar System in the Galactic Environment, Earth, Moon and Planets, 72, 1-6, 1996

Krasnobaev, K.V., Static solar HII zone, Soviet Astronomy, 14, 840-844, 1971

Krimigis, S.M., Decker, R.B., Hill, M.E., Armstrong, T.P., Gloeckler, G., Hamilton, D.C., Lanzerotti, L.J., Roelof, E. C., Voyager 1 exited the solar wind at a distance of 85 AU from the Sun , Nature, 426, Issue 6962, 45-48, 2003

Lammer, H., Stumptner, W., Molina-Cuberos, G.J, Lara, L.M., and Tehrany, M.G., From atmospheric isotope anomalies to a new perspective on early solar activity: consequences for planetary paleoatmospheres, In: *The evolving Sun and its influence on planetary environments*, Eds., Montesinos, B., Gimenez, A., Guinan, E.F., ASP Conf. Series, 269, 249-270, 2002

Leitch, E.M. and Vasisht, G., Mass extinctions and the sun's encounters with spiral arms, New Astronomy, 3, 51-56, 1998

LeVeque, R.J., Nonlinear conservation laws and finite volume methods, In: *Computational methods for astrophysical fluid flow*, Eds., Steiner, O., Gautschy, A., Saas-Fee Advanced Course 27, Springer, 1-160, 1998

Lindzen, R.S. Tides and Gravity Waves in the Upper Atmosphere, In: *Mesospheric models and related experiments*, Ed., Fiocco, G., D. Reidel Publ., 122-130, 1971

McCrea, W.H., Ice ages and the Galaxy, Nature, 255, 607-609, 1975

McKay, C.P. and Thomas, G.E. Consequences of a past encounter of the Earth with an interstellar cloud, Geophys. Res. Letters, 5 (3), 215-218, 1978

McKee, C.F., The multiphase interstellar medium, In: *The Physics of the Interstellar medium and Intergalactic Medium*, Eds., Ferrara, A., McKee, C.F., Heiles, C., Shapiro, P.R., ASP Conf. Ser., 80, 292-316, 1995

Mlynczak, M.G., and Solomon, S. A detailed evaluation of the heating efficiency in the middle atmosphere, J. Geophys. Res., 98, 10,517-10,541, 1993

Myers, P.C. and Goodman, A.A. Evidence for magnetic and virial equilibrium in molecular clouds, Astrophys. J., 326, L27-L30, 1988

Nicolet, M. 1971, Aeronomic Reactions of Hydrogen and Ozone, In: *Mesospheric models and related experiments*, Ed., Fiocco, G., D. Reidel Publ., 1-22, 1971

Osterbrock, D.E. *Astrophysics of Gaseous Nebulae and Active Galactic Nuclei*, University Science Books, Mill Valley, 1989

Parker, E.N., *Interplanetary dynamical processes*, Interscience Publishers, 1963

Pauls H.L., Zank, G.P., and Williams, L.L., Interaction of the solar wind with the local interstellar medium, J. Geophys. Res., 100, 21595-21604, 1995

Ratkiewicz, R., An analytical solution for the heliopause boundary and its comparison with numerical solutions, Astron. Astrophys., 255, 383-387, 1992

Ripken, H.W. and Fahr H.J., Solar wind interactions with neutral hydrogen inside the orbit of the Earth, In: *Solar Wind Four Conf.*, Burghausen, Aug. 28 - Sept. 1, 1978, 528-534, 1981

Roble, R.G., Energetics of the mesosphere and thermosphere, AGU Geophysical Monograph on the Upper Mesosphere and Lower Thermospere, R.M. Johnson and T. L. Killeen eds., 1995

Roble, R.G., On the feasibility of developing a global atmospheric model extending from the ground to the exosphere, In: Atmospheric Science Across the Stratopause, AGU Geophysical Monograph 123, American Geophysical Union, Washington, D.C., 53-68, 2000

Ruderman, M.A. and Spiegel, E.H., Galactic Wakes, Astrophys. J., 165, 1-14, 1971

Sackmann, I. and Boothroyd, A.I., Our Sun. V. A bright young Sun consistent with helioseismology and warm temperatures on ancient Earth and Mars, Astrophys. J., 583, 1024-1039, 2003

Scherer, K., Variation of the heliospheric shield and its influence on Earth, In: *The Outer Heliosphere: Beyond the Planets*, Eds., Scherer, K., Fichtner, H., Marsch, E., Copernicus Gesellschaft e.V., 327-356, 2000

Scherer, K., Fichtner H., and Stawicki O., Shielded by the wind: the influence of the interstellar medium on the environment of Earth, J. Atmosph. Solar-Terr. Phys., 64, 795-804, 2002

Scoville, N.Z., and Sanders, D.B., Observational constraints on the interaction of giant molecular clouds with the Solar System, In: *The Galaxy and the Solar System*, Eds., R.Smoluchowski, R., Bahcall, J.N., Mattews, M.S., The Univ. Arizona Press, Tucson, 69-82, 1986

Shaviv, N.J., The spiral structure of the Milky Way, cosmic rays, and ice age epochs on Earth, New Astronomy, 8, 39-77, 2003

Talbot, R.J., Jr., and Newman, M.J., Encounters between stars and dense interstellar clouds, Astroph. J. Suppl., 34, 295-308, 1977

Whang, Y.C., Ionization of interstellar hydrogen, Astrophys. J., 468, 947-954, 1996

Wilkin, F.P., Exact analytic solutions for stellar wind bow shocks, Astrophys. J., 459, L31-L34, 1996

Williams, R.E., The size of a solar HII region, Astrophys. J., 142, 314-320, 1965

Wimmer-Schweingruber, R.F. and Bochsler, P., Is there a record of interstellar pick-up ions in lunar soils? In: *Acceleration and Transport of Energetic Particles Observed in the Heliosphere*: ACE 2000 Symposium, Eds., Mewaldt, R.A. et al., 270-273, 2000

Yabushita, S. and Allen, A.J., On the effect of interstellar gas on atmospheric oxygen and terrestrial life, Observatory, 103, 249-251, 1983

Yabushita, S. and Allen, A.J., On the effect of interstellar matter on terrestrial climate, Observatory, 105, 198-200, 1985

Yabushita, S. and Allen, A.J., On the effect of accreted interstellar matter on the terrestrial environment, Mon. Not. R. Astron. Soc., 238, 1465-1478, 1989

Yabushita, S. and Allen, A.J., Did an impact alone kill the dinosaurs?, Astronomy and Geophysics, 38 (2), 15-19, 1997

Yeghikyan, A.G. and Fahr, H.J., Consequences of the Solar System passage through dense interstellar clouds, Annales Geophysicae, 21, 1263-1273, 2003

Yeghikyan, A.G. and Fahr, H.J., Effects induced by the passage of the Sun through dense molecular clouds. I. Flow outside of the compressed heliosphere, Astron. Astrophys., 415, 763-770, 2004a

Yeghikyan, A.G. and Fahr, H.J., Terrestrial atmospheric effects induced by counterstreaming dense interstellar cloud material, Astron. Astrophys., 425, 1113-1118, 2004b

Zahnle, K.J. and Walker, J.C.G., The evolution of solar ultraviolet luminosity, Rev. Geophys. Space Phys., 20, 280-292, 1982

Zank, G.P., Interaction of the solar wind with the local interstellar medium: a theoretical perspective, Space Sci. Rev., 89, 413-688, 1999

Zank, G.P. and Frisch, P.C., Consequences of a change in the galactic environment of the Sun, Astrophys. J., 518, 965-973, 1999

Chapter 12

VARIATIONS OF GALACTIC COSMIC RAYS AND THE EARTH'S CLIMATE

Jasper Kirkby
CERN
Geneva 23, Switzerland
jasper.kirkby@cern.ch

Kenneth S. Carslaw
University of Leeds, School of Earth & Environment
Leeds LS2 9JT, United Kingdom
carslaw@env.leeds.ac.uk

Abstract The galactic environment of Earth appears to have varied only subtly over mankind's time scale. Have these variations had any perceptible effect on Earth's climate? Surprisingly, they may; palaeoclimatic evidence suggests that the climate may be influenced by solar/cosmic ray forcing on all time scales from decades to billions of years. However, despite an intense scientific interest spanning two centuries, the mechanism underlying solar-climate variability has remained a mystery. Recent satellite data, however, have provided an intriguing new clue: low cloud cover may be influenced by galactic cosmic rays. Since cosmic rays are modulated by the solar wind, this would link Earth's climate to solar *magnetic* activity, which is known to be highly variable. On longer time scales, it would also imply that the climate responds to changes in the geomagnetic field and the galactic environment of Earth, both of which affect the cosmic ray flux. The least understood aspect of this sequence of processes is the microphysical mechanism by which cosmic rays may affect clouds. Physical mechanisms have been proposed and modelled, but definitive experiments are lacking. However, the presence of ion-induced nucleation of new particles in the atmosphere is supported by recent observations, and new experiments are underway to investigate the nature and significance of cosmic ray-cloud-climate interactions.

P. C. Frisch (ed.), Solar Journey: The Significance of our Galactic Environment for the Heliosphere and Earth, 349–397.

Keywords: Solar-cosmic ray-climate variability, solar irradiance, galactic cosmic rays, palaeoclimate, cosmic ray-cloud-climate mechanisms, ion-induced nucleation, ice nucleation, global electric circuit

1. Introduction

The first observation of solar-climate variability was recorded two centuries ago by the astronomer William Herschel [Herschel, 1801] who noticed that the price of wheat in England was lower when there were many sunspots, and higher when there were few. The most well-known example of a solar-climate effect is known as the Maunder Minimum [Eddy, 1976], the period between 1645 and 1715—which ironically coincides with the reign of Louis XIV, *le Roi Soleil*, 1643–1715—during which there was an almost complete absence of sunspots (Fig. 12.1). This marked the most pronounced of several prolonged cold spells in the period between about 1450 and 1850 which are collectively known as the Little Ice Age. During this period the River Thames in London regularly froze across, and fairs complete with swings, sideshows and food stalls were a standard winter feature. Numerous studies of palaeoclimatic proxies have both confirmed that the Little Ice Age was a global phenomenon and, moreover, that it was but one of around 10 occasions during the Holocene when the Sun entered a grand minimum for a centennial-scale period, and Earth's climate cooled (§4.3.0).

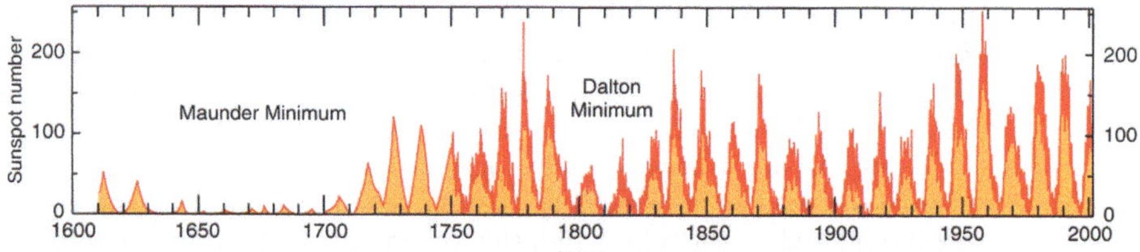

Figure 12.1. Variation of the sunspot number from 1610 to 2001. The record starts 3 years after the invention of the telescope by Lippershey in Holland.

However, despite the widespread evidence, solar variability remains controversial as a source of climate change since no causal mechanism has been established to link the two phenomena. The most obvious mechanism to consider would be the possibility of a variable solar irradiance, but estimates of long-term irradiance changes based on sun-like stars appear to be insufficient to cause the observed climate variations (§2.2.0). The physical mechanism for solar-climate variability therefore remains a mystery.

However, an important new clue has recently been provided by satellite data which reveal a surprising correlation between galactic cosmic ray (GCR) intensity and the fraction of Earth covered by low clouds [Svensmark and Friis-Christensen, 1997, Marsh and Svensmark, 2003]. Since changes of solar magnetic activity affect the solar wind and thereby modulate the GCR intensity, this could provide the long-sought missing link between solar- and climate variability. If more cosmic rays lead to more low clouds, this would exert a strong cooling influence on the radiative energy balance of Earth. It would explain how a tiny energy input from GCRs to the atmosphere—comparable to starlight—could produce such large changes in the climate. Clouds cover a large fraction of Earth's surface (a global annual mean of about 65%) and exert a strong net cooling effect of about 28 Wm^{-2}[Hartmann, 1993], so long-term variations of only a few per cent would be climatically significant.

Moreover, if this mechanism were to be established, it could have significant consequences for our understanding of the solar contribution to the present global warming. The globally-averaged GCR intensity declined by about 15% during the 20th century due to an increase of the solar open magnetic flux by more than a factor of two [Lockwood *et al.*, 1999]. If the cosmic ray-cloud effect is real then the implied reduction of low cloud cover by about 1.3% absolute could have given rise to a radiative forcing of about $+0.8 Wm^{-2}$, which is comparable to the estimated total anthropogenic forcing of about $+1.3$ Wm^{-2} [IPCC, 2001].

A GCR-cloud effect would provide an indirect mechanism to amplify solar forcing, bringing additional uncertainties to climate change projections. The Intergovernmental Panel on Climate Change (IPCC) stated in its Third Assessment Report (2001) that "mechanisms for the amplification of solar forcing are not well established". The solar amplification mechanism–which appears to drive the large centennial-scale climate variations seen, for example, in Fig. 12.8—is absent from present climate models. Since these natural climate variations are comparable to the present global warming, they must be properly accounted for before we can reliably interpret 20th century climate change and confidently predict future warming due to anthropogenic greenhouse gases.

2. Solar Irradiance

2.1 Climate Sensitivity to Radiative Forcing

To interpret how Earth would respond to a radiative forcing, it is useful to consider the climate sensitivity. Climate models indicate an approximately linear relationship between global mean radiative forcing, Q (Wm^{-2}), and the equilibrium global mean surface temperature change, ΔT (K),

$$\Delta T = \lambda Q$$

where λ (K/Wm^{-2}) is the climate sensitivity parameter. This parameter is relatively insensitive to the nature of the forcing—for example, greenhouse gases or solar irradiance—provided the forcing agent is not highly variable spatially (like, for example, aerosols). All climate feedback processes, such as changes in water vapour, clouds or ice sheet albedo, are implicitly included in λ.

The value of λ can be inferred in several ways. Climate models indicate a doubling of the concentration of atmospheric CO_2 from pre-industrial levels (280 ppm) produces +4 Wm^{-2} forcing and a mean temperature rise ranging from 1.5 K to 4.5 K, with a central value of 3 K [IPCC, 2001]. Therefore $\lambda \simeq (3 \pm 1.5)/4 = (0.75 \pm 0.4)\,\mathrm{K/Wm^{-2}}$. This is consistent with the estimate, $\lambda \simeq 5$ K / 7 Wm^{-2} = 0.7 K/Wm^{-2}, obtained from ice core samples for the transition between glacial and interglacial periods. Using this value, the possible contribution of +0.8 Wm^{-2} from a cosmic ray-cloud effect during the 20th century (§1) would translate to a temperature rise of around $0.7 \times 0.8 = 0.6$ K, which would be highly significant.

These figures can be compared with the response of Earth if it were simply to act as a black body. In this case the radiant emittance is $R = \sigma T^4$, where σ is the Stefan-Boltzmann constant. The radiation from a black body varies as $\Delta R/R = 4\Delta T/T$, so that $\Delta T = (T/4R)\,\Delta R$. Since $\Delta R/R = \Delta I/I$, the fractional change in solar irradiance, it follows that $\lambda_0 = T/4I$. The global mean solar irradiance reaching the lower troposphere is $I \simeq 0.7 \times 1366/4 \simeq 240\,\mathrm{Wm^{-2}}$ (the factor 0.7 accounts for shortwave albedo and the factor 4 averages the solar irradiance of 1366 Wm^{-2} over the full surface area of Earth). The effective radiating temperature of Earth is then given by $240 = \sigma T^4$, indicating $T = 255$ K. (This is the effective blackbody temperature of Earth as seen from a satellite, corresponding to radiation from the final scattering layer in the upper troposphere.) We thereby estimate $\lambda_0 = 255/(4 \cdot 240) = 0.26\,\mathrm{K/Wm^{-2}}$ for Earth in the absence of any feedbacks. Therefore Earth's climate feedback factor is about 2.5, which indicates the presence of positive feedbacks that amplify the response to a radiative forcing.

2.2 Solar Irradiance Variability

Solar cycle variability. Solar radiometry measurements made by satellites over the last 25 years have shown that the so-called solar constant varies by about 0.08% over the solar cycle (Fig. 12.2). Counter-intuitively, the Sun radiates more during periods of high solar magnetic activity since, although the sunspots are cooler than the mean temperature of the photosphere (around 5840 K), they are more than compensated by the associated bright *faculae*.

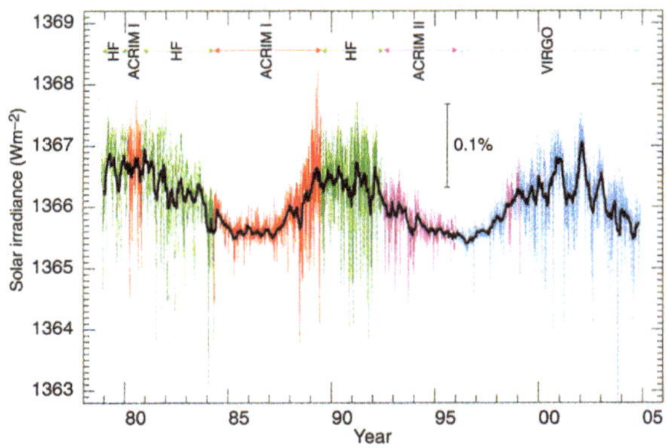

Figure 12.2. Total solar irradiance at the top of Earth's atmosphere over the last 2.5 solar cycles (updated from [Fröhlich, 2000]). The curve is a composite of several satellites, each covering a different time interval, as indicated by the different coloured regions of the curve and the labels shown above it. Peak irradiance corresponds to sunspot maximum (and GCR minimum). The large high-frequency fluctuations are due to sunspots rotating in and out of the field of view.

A variation of solar irradiance by 0.08% ($1.1\,\mathrm{Wm^{-2}}$) over the sunspot cycle corresponds to a global mean variation of $0.7 \times 1.1/4 = 0.2\,\mathrm{Wm^{-2}}$ at Earth's surface. This would be expected to produce an *equilibrium* temperature change $\Delta T = 0.7 \times 0.2 = 0.14$ K. However the large thermal inertia of the well-mixed upper 100 m layer of the oceans will reduce the actual temperature response. This damping effect can be estimated by considering an analogous electromagnetic RC-circuit equivalent for the ocean surface layer: a resistor and capacitor in series, which are set into forced oscillation by a sinusoidal voltage [Kirkby, 2002]. In this analogy, forcing heat fluxes are analogous to current, temperature to voltage, the ocean to a capacitor and the climate sensitivity parameter to a resistor. The estimated attenuation factors are large, e.g. a factor 5 for $\lambda = 0.7$ K/Wm^{-2}, corresponding to an 8.4 yr time constant and phase lag of about 80°. This implies the expected response of Earth's mean surface temperature to irradiance variability over the solar cycle is only about 0.03 K.

Although this is a challenging measurement, sea-surface temperature data over the last century do indeed reveal signals of the 11-yr solar cycle in the Indian, Pacific and Atlantic Oceans [White *et al.*, 1997, White *et al.*, 1998]. The solar cycle peak-to-peak amplitude is (0.06 ± 0.005) K and it lags the changes in solar irradiance by about 0–65°.

Secular Solar Variability. Irradiance variations over the 11-yr sunspot cycle are now well established and, moreover, reasonably well explained in terms of sunspot darkening and faculae brightening. However, the important question for climate is whether or not there are also underlying secular variations on longer time scales. Even a variation of comparable magnitude—0.1%—would be significant since the climate response would not be damped by the thermal inertia of the oceans.

The Lean *et al.* (1995) estimate of the long-term variation of solar irradiance has been widely used in climate models to calculate the solar contribution to climate change in the 20th century [IPCC, 2001] and in the period since the Maunder Minimum. It is based on observations of luminosity variations in Sun-like stars and on a speculated long-term relationship between solar luminosity and solar magnetic activity. Cosmic ray flux, recorded in ^{10}Be concentrations in ice cores (§3.3), are used as a proxy for solar magnetic activity. The estimated secular increase of irradiance since 1700 is about 2.6 Wm^{-2}, with a superimposed variation of up to about 1 Wm^{-2} due to the sunspot cycle [Lean *et al.*, 1995]. This translates to a top-of-the-atmosphere long-term solar radiative forcing of $(0.7/4) \cdot 2.6 = 0.46$ Wm^{-2} (§2.1), corresponding to a temperature change of $0.7 \times 0.46 = 0.32$ K.

This figure may be compared with palaeoclimatic reconstructions, which suggest global temperatures were around 0.7 K cooler during the Maunder Minimum, after subtracting the estimated anthropogenic contributions during the last century (§4.2). The estimated irradiance variations therefore appear to be too small by about a factor two to account for the observed temperature change. Moreover, new results on Sun-like stars and new modelling studies have brought the Lean *et al.* (1995) estimate into question [Foukal *et al.*, 2004]. Indeed, the original authors have now accepted [Lean *et al.*, 2002] that their 1995 calculations of long-term irradiance variations may be over-estimated by about a factor five. This would imply a temperature change of only 0.06 K from solar irradiance variations since the Maunder Minimum, which would account for only a small fraction—about 10%—of the observed change.

The absence of evidence for long-term solar irradiance changes does not, of course, rule them out. It does, however, underscore how little is understood about the mechanism for solar-climate variability. Moreover, it implies that current general circulation model (GCM) simulations of 20th century warming may substantially over-estimate the contribution from solar irradiance variability. The clear solar signal in the 20th century temperature record (§4.1)—and also at earlier times—is therefore unexplained. A solar amplification mechanism could solve this puzzle.

3. Galactic Cosmic Rays

3.1 Solar-climate Mechanisms

There are only two physical paths that could connect variations of the Sun to Earth's climate: 1) solar electromagnetic radiation (either irradiance or else a spectral component such as the ultraviolet, UV), or 2) galactic cosmic rays, whose intensity is modulated by the solar wind. A direct interaction of the solar wind with the troposphere can probably be excluded since the charged particles of the solar wind have very low energy (mainly protons, with energies of a few keV) and are easily shielded by Earth's magnetosphere. In the polar regions, where the magnetic shielding is weak, the particles range out in the thermosphere at an altitude of about 100 km. This is far from the troposphere—the layer of the atmosphere most responsible for Earth's climate—which lies below about 10–15 km altitude.

Since variations of total solar irradiance appear to be too low to account for observed climate variations, attention has focused on changes of the UV component of the solar spectrum [Haigh, 1996, Larkin *et al.*, 2000] which, although it carries only a small fraction of the total energy (about 0.1%), shows a much larger variation of several per cent over the solar cycle. The UV spectrum is absorbed at altitudes above 30 km by oxygen (<240 nm wavelength) and ozone (200–300 nm), and causes measurable heating of the thin atmosphere in the upper stratosphere. A positive feedback mechanism exists since the increased UV creates more ozone and therefore more UV absorption. However, the measured change of ozone concentration from solar minimum to maximum is small (about 1–2%). Modelling reproduces these variations, and some studies (e.g. [Shindell, 1999]) suggest that the resultant changes in stratospheric heating and circulation could dynamically and radiatively couple to the troposphere, which could in turn influence cloudiness.

In contrast with solar UV radiation, GCRs directly penetrate the lower troposphere where the cloud variation is observed, and they have an appreciable intensity variation over the solar cycle (§3.2). At low altitudes over land (<1 km), natural radioactivity contributes about 20% of atmospheric ionisation. Over oceans, the contribution of radioisotopes is negligible and so, averaged over the entire troposphere, GCRs are by far the dominant source of ionisation (more than 99%).

The data presented in §4 show that GCR variations are clearly connected with past climate change. However the interpretation is often ambiguous; this may indicate either a direct effect of GCRs on the climate or else the GCRs may merely be acting as a proxy for solar irradiance/UV changes. In principle, this ambiguity can be resolved since, on longer time scales, the GCR intensity is also affected by the geomagnetic field and by the galactic environment (§3.2), whereas the solar irradiance/UV is not. Indeed, there is suggestive evidence linking the climate both with the geomagnetic field and with the galactic environment (§4). For the remainder of this paper, we will focus on a possible direct influence of cosmic rays on the climate.

3.2 Cosmic Ray Modulation

Cosmic rays are accelerated by supernovae and other energetic sources in our galaxy. On entering the heliosphere, charged cosmic rays are deflected by the magnetic fields of the solar wind. The transport problem of the GCRs through the heliosphere was first solved by Parker (1965). It involves several processes of which the dominant is scattering off the magnetic irregularities, which produces a random walk or diffusion effect. It has been shown theoretically that the heliospheric shielding can be approximated by an equivalent heliocentric retarding (i.e. decelerating) electric potential [Gleeson and Axford, 1967]. This retarding potential varies between about 1000 MV during periods of very high solar activity and zero during grand minima such as the Maunder Minimum. The solar wind therefore partly shields the Earth from the lower energy GCRs and affects the flux at energies below about 10 GeV. The effective retarding potential over the present eleven-year solar cycle averages about 550 MV, ranging from about 450 MV at the minimum to 850 MV at maximum. This leads to a distinct solar modulation of the GCR intensity (Fig. 12.3).

The geomagnetic field also partially shields the Earth from GCRs. The dipole field imposes a minimum vertical momentum of about 13 GeV/c at the equator, 3 GeV/c at mid latitudes, and falling essentially to zero at the geomagnetic poles. In consequence, the GCR intensity is about a factor 3.6 higher at the poles than at the equator, and there is a more marked solar cycle variation at higher latitudes. Over the solar cycle, the variation of GCR intensity at the top of the atmosphere is about 15%, globally averaged, and ranges from ~5% near the geomagnetic equator to ~50% at the poles (Fig. 12.3). At lower altitudes both the GCR intensity and its fractional solar modulation decrease. These are consequences of the absorption of low energy GCRs and their secondary particles by the atmospheric material, which totals about 11 nuclear interaction lengths. Balloon measurements (Fig. 12.4) show solar cycle variations of about 10% at low altitudes around 3 km (for a 2.4 GeV/c rigidity cutoff).

The sources of GCR modulation are summarised in Table 12.1. On the longest timescales, the GCR intensity is modulated by passage through

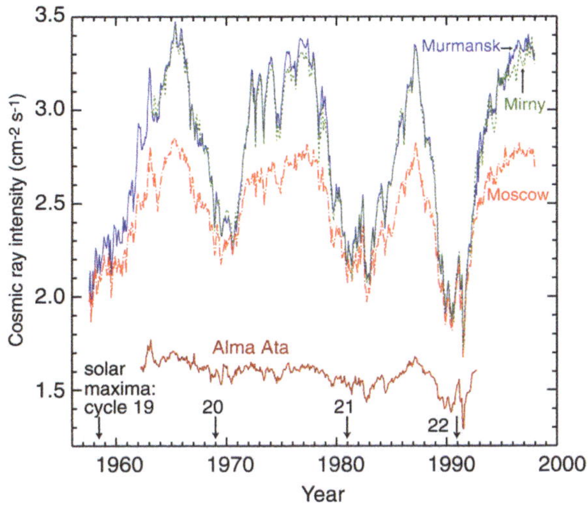

Figure 12.3. Balloon measurements of the cosmic ray intensity at shower maximum (15–20 km altitude) for the period 1957–1998, measured by the Lebedev Physical Institute. The curves correspond to four different locations for the balloon flights: Mirny-Antarctica (0.03 GeV/c rigidity cutoff), Murmansk (0.6 GeV/c), Moscow (2.4 GeV/c) and Alma-Ata (6.7 GeV/c). Due to atmospheric absorption, the data of Murmansk and Mirny practically coincide with each other. The approximate times of the sunspot maxima for the last 4 solar cycles are indicated.

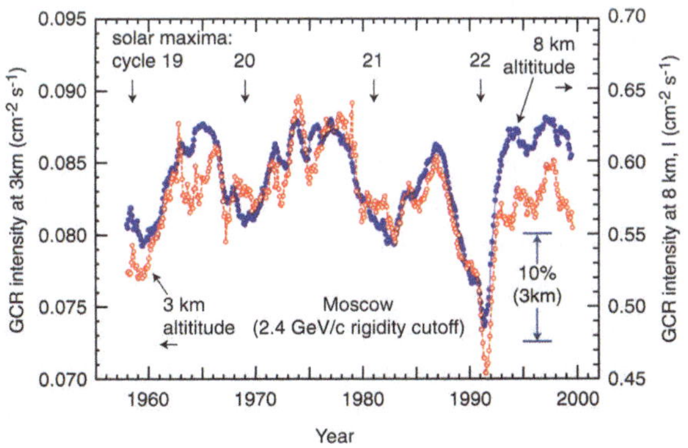

Figure 12.4. Balloon measurements of the GCR intensity at 3 km and 8 km altitudes from 1957 to 2000 (2.4 GeV/c rigidity cutoff), measured by the Lebedev Physical Institute. The approximate dates of the sunspot maxima for the last 4 solar cycles are indicated.

Table 12.1. Sources of GCR Modulation and their Typical Characteristics.

Source	*Variability timescale*	*GCR change*
Solar wind	10 yr – 1 kyr	±10–30%
Geomagnetic field	100 yr – 100 Myr	±20–100%
Galactic variations	>10 Myr	±50–100%

the spiral arms of the Milky Way [Shaviv, 2002]. The local GCR intensity is higher within the spiral arms owing to the relative proximity of the supernovae generators and the subsequent diffusive trapping of the energetic charged particles by the interstellar magnetic fields. Superimposed on this diffuse GCR rate are possible isolated events caused by nearby supernovae. For example, evidence has recently been obtained for an enhancement in the fraction of the ^{60}Fe radio-isotope ($\tau_{1/2}$ = 1.5 Myr) in a deep-sea ferromagnetic crust [Knie *et al.*, 2004]. This suggests a nearby supernova, at a distance of several 10's pc, which increased the GCR intensity by ~15% for a few 100 kyr around 2.8 Myr ago.

3.3 Cosmic Ray Archives

On reaching Earth, cosmic rays interact with nuclei in the atmosphere, creating showers of secondary particles, and dissipating energy by ionisation. Among the products are light radio-isotopes, of which the two most abundant are ^{14}C ($\tau_{1/2}$ = 5730 yr; global mean production rate ~2.0 atoms cm^{-2}s^{-1}) and ^{10}Be (1.5 Myr; ~1.8×10^{-2} atoms cm^{-2}s^{-1}). These settle on the surface of Earth either via the carbon cycle ($^{14}CO_2$) or in rain and snow (^{10}Be). Since this is their only terrestrial source, the resultant archives of light radio-isotopes found in tree rings, ice cores and marine sediments provide direct records of the past GCR flux. For the long-lifetime ^{10}Be, these records at present go back about 1 Myr, and are likely to be extended in future.

Once produced, ^{14}C is rapidly oxidised to $^{14}CO_2$. The turnover time of CO_2 in the atmosphere is quite short—about 4 years—mostly by absorption in the oceans and assimilation in living plants. However, because of recirculation between the oceans and the atmosphere, changes in the ^{14}C fraction on timescales less than a few decades are smoothed out. Plant material originally contains the prevailing atmospheric fraction of ^{14}C and, subsequently, since the material is not recycled into the atmosphere, the fraction decreases with the characteristic half life of ^{14}C. By analysing the ^{14}C content in the rings of long-lived trees such as the California bristlecone pine, and also ancient tree samples, a detailed record of GCR intensity over the last 40 kyr has been assembled.

In the case of ^{10}Be [Beer, 2000], after production it rapidly attaches to aerosols and follows the motion of the surrounding air masses. Since the production of ^{10}Be follows the intensity profile of the cosmic ray hadronic showers, about 2/3 is produced in the stratosphere and 1/3 in the troposphere, globally averaged. Due to the tropopause barrier, aerosols in the stratosphere take about 1–2 years to settle on the Earth's surface, whereas the mean residence time in the troposphere is only about a week. If the sedimentation occurs in the form of snow in a permanently frozen and stable region such as Greenland or Antarctica then the subsequent compacted ice preserves a temporal record in layers according to their depth. If, on the other hand, sedimentation occurs into rivers and oceans, then eventually (after about 1000 yr mean time) the ^{10}Be settles in ocean sediments. Cores retrieved from polar ice or the ocean floor thereby constitute an archive of past GCR intensity. Ice cores have the advantage of higher time resolution, whereas ocean sediments provide longer time records and a globally-averaged measurement.

3.4 Climate Archives

Many ingenious proxies have been developed to reconstruct the climate prior to the instrumental record of the last 150 years. Cultural records over the last millennium are an important source since humans are sensitive to climate change, especially when prolonged drought, cold or flooding is involved. These documents include the dates when the first cherry blossoms appeared each spring in China, the records of grape harvests in Europe, and the waxing and waning of Alpine glaciers. Other records (with their approximate time span before present) are tree rings (10 kyr), mosses (10 kyr), coral terraces (500 kyr), ice cores (700 kyr), speleothems (1 Myr), loess deposits (wind-blown silt; 1 Myr), pollens (1 Myr), ocean sediments ($>$500 Myr) and geomorphology (4 Byr).

Ice cores are an especially valuable record of past climate since they have high resolution and excellent preservation of a wide range of climatically-sensitive material. Trapped air bubbles preserve the atmospheric composition at earlier times, measuring gases such as carbon dioxide (ocean ventilation, volcanic and biomass activity), methane (wetland extent), deuterium (temperature), nitrous oxide (bacterial activity and stratospheric air exchange) and methanesulphonic acid (phytoplankton activity); layer thicknesses measure precipitation rate; dust content measures wind, aridity, land biota and volcanic activity; the radio-isotope ^{10}Be measures solar/GCR activity; non-sea-salt sulphate measures sulphuric acid content of the atmosphere (phytoplankton and volcanic activity); nitrates measure ionising radiation (UV and cosmic

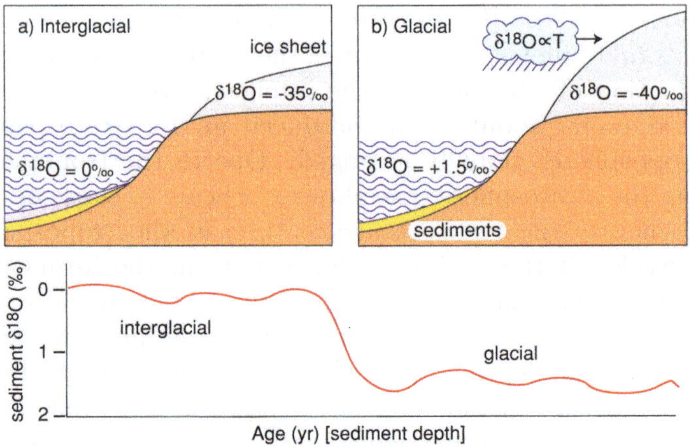

Figure 12.5. Principle of the global ice volume proxy obtained from the ^{18}O fraction in ocean sediments (lower panel) during a) interglacial and b) glacial conditions. The physical basis for proxy temperature measurements from the stable ^{18}O isotope is that the vapour pressure of $H_2{}^{18}O$ is lower than that of $H_2{}^{16}O$. Evaporated water is therefore ^{18}O-depleted and the subsequent rainfall from a cloud progressively reduces the ^{18}O fraction of the remaining water vapour. Ice formed from snow is therefore isotopically light (depleted in ^{18}O)—and even lighter during glacial climates since the lower temperatures further deplete the clouds of ^{18}O. A large global ice volume therefore leaves the oceans relatively enriched in ^{18}O. Since the prevailing isotopic fraction of the water is transferred to organisms living in the oceans, the ^{18}O fraction found in ocean sediments provides a proxy for global ice volume.

ray bursts); and, of particular importance, the stable ^{18}O isotope measures past temperatures, rainfall and global ice volume (Fig. 12.5).

4. Solar/GCR-climate Variability

The GCR and climate archives provide extensive evidence for significant correlations between cosmic ray flux and climate, on both short and long time scales. The pattern is summarised in Table 12.2. We will present a brief review of solar/GCR-climate observations in this section, progressively looking further back in time.

Table 12.2. Observed Correlation of Cosmic Rays and Climate.

Cosmic ray flux	*Climate*
high	cool
low	warm

4.1 Twentieth century

Global cloud cover. Based on satellite data from the ISCCP [Rossow, 1996], Svensmark *et al.* have reported a correlation between GCR intensity and the fraction of Earth covered by low clouds (Fig.12.6b) [Svensmark and Friis-Christensen, 1997, Marsh and Svensmark, 2003]. These correlations have been subjected to intense scrutiny and criticism (e.g. [Kernthaler *et al.*, 1999, and others]). The measurements are difficult, involving the inter-calibration of instruments on board several geostationary satellites and polar-orbiting satellites, over a period of 20 years. The low-cloud data appear to show a poor correlation with GCR intensity after 1994 (or perhaps a GCR modulation superimposed on a linearly-decreasing trend). However, questions have been raised about an apparent break in the cloud measurements at the end of 1994 (Figs. 12.6a and b) which coincides with a period of a few months during which no polar-orbiting satellite was available for inter-calibration [Marsh and Svensmark, 2003].

Although not conclusive, the low cloud data nevertheless suggest a significant solar imprint. The low cloud modulation represents around 1 Wm^{-2} variation in net radiative forcing at Earth's surface over the solar cycle, which is about a factor 5 larger than the variation of solar irradiance (§2.2.0). Furthermore, the two effects are in phase; increased sunspot activity corresponds to increased irradiance, decreased GCR

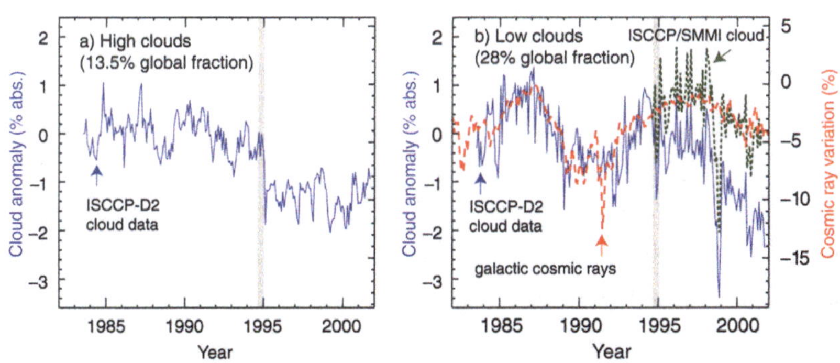

Figure 12.6. Satellite measurements of global cloud anomalies over the last 20 years: a) high clouds (above about 6.5 km), and b) low clouds (below about 3.2 km) [Marsh and Svensmark, 2003]. The vertical grey bar around the end of 1994 corresponds to a period without any polar-orbiting satellite available to inter-calibrate the geostationary satellites used for the ISCCP cloud measurements. In panel b) the dashed red curve shows the variation of GCR intensity, and the dotted green curve shows the ISCCP data after correcting by its difference with SMMI cloud data after 1994.

intensity and decreased low cloud amount. Since low clouds have a net cooling effect, this implies increased radiative forcing and a warmer climate. Finally, Usoskin *et al.* (2004) report that the solar/GCR-cloud correlation shows some evidence for an increased amplitude at higher latitudes. Since the GCR intensity and modulation is more pronounced at higher latitudes, whereas solar/UV intensity variations are the same at all latitudes, this observation favours a GCR mechanism rather than a solar/UV effect.

Sea surface temperatures. An important component of the temperature reconstruction during the current warming is sea-surface temperatures (SSTs), which have been measured on a routine basis by ocean-going ships since the mid 19th century. The SST record is a particularly valuable measure of global climate since it represents over 70% of Earth's surface and is much more spatially and temporally homogeneous than the land surface. It is also free of the uncertainties of temperature increases arising from population growth in 'urban heat islands'. The mean SSTs over the period 1860–1985 for the Atlantic, Pacific and Indian Oceans are shown in Fig. 12.7, together with the global mean SST [Reid, 1987]. All oceans show a temperature rise that levels off in the

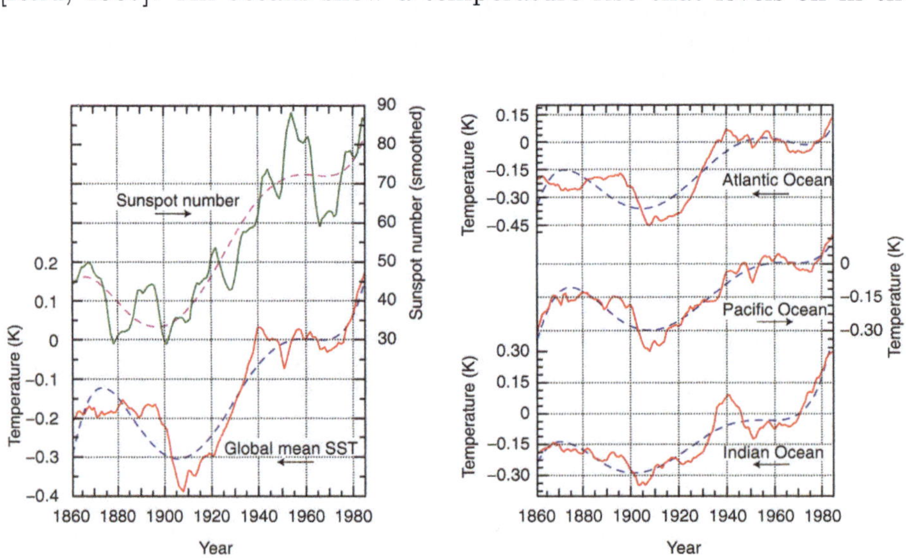

Figure 12.7. Annual mean sea-surface temperatures (SST), 1860–1985, for the Atlantic, Pacific and Indian Oceans (right-hand panel) and the global mean (lower curve in the left-hand panel) [Reid, 1987]. The temperatures are shown relative to their 1951–1980 averages. Also shown is the 11-year running mean of the annual sunspot numbers (upper curve in the left-hand panel). The smooth dashed curves are 7th order polynomial fits to the data.

same period around 1945–1980, as well as a cooling around the beginning of the last century. Both of these features are characteristic of solar activity and GCR intensity, as can be seen in the smoothed sunspot number (Fig. 12.7).

A world-wide simultaneous variation of SST puts severe constraints on a possible forcing mechanism. Since the same characteristic features are seen in all oceans, they are unlikely to be caused by changes such as El Niño events, shifts in wind patterns or changes in the thermohaline circulation, which would lead to differences between the oceans. The mechanism could in principle be increases of anthropogenic greenhouse gases, but the variation in the first half of the 20th century occurred before these were significant. There were insufficient volcanic events to account for the mid-century cooling. In conclusion, these data provide quite strong evidence that solar variability was the primary cause of the warming during at least the first half of the last century. Indeed, this is now accepted by general circulation models. However, the apparent good agreement has recently been thrown into doubt since the climate models use reconstructions of irradiance variations (e.g. [Lean *et al.*, 1995]) that may be over-estimated by a factor of five (§2.2.0).

4.2 Last millennium

Numerous palaeoclimatic reconstructions have shown evidence for a Medieval Warm period between about 1000 and 1270, followed by a prolonged cold period known as the Little Ice Age between about 1450 and 1850. Temperatures during the Medieval Warm period were elevated above normal, allowing the Vikings to colonise Greenland and wine-making to flourish in England. It was followed by a period of about 4 centuries during which—save for a few short interruptions—the glaciers advanced and a cooler, harsher climate predominated.

A recent multi-proxy reconstruction of Northern Hemisphere temperatures estimates that the Little Ice Age was about 0.7 K below the 1961–1990 average, and that climate during the Medieval Warm period was comparable to the present (Fig. 12.8a) [Moberg *et al.*, 2005]. This reconstruction contrasts with a widely-quoted earlier reconstruction [Mann *et al.*, 1998, Mann *et al.*, 1999], known as the *hockey-stick* curve, which shows a rather flat temperature variation prior to 1900 (Fig. 12.8a). However the methodology of the principal-component analysis used to derive the hockey-stick curve has recently been questioned and, when properly analysed, the data show larger long-term temperature variations, characteristic of the Little Ice Age and Medieval Warm period [McIntyre and McKitrick, 2005].

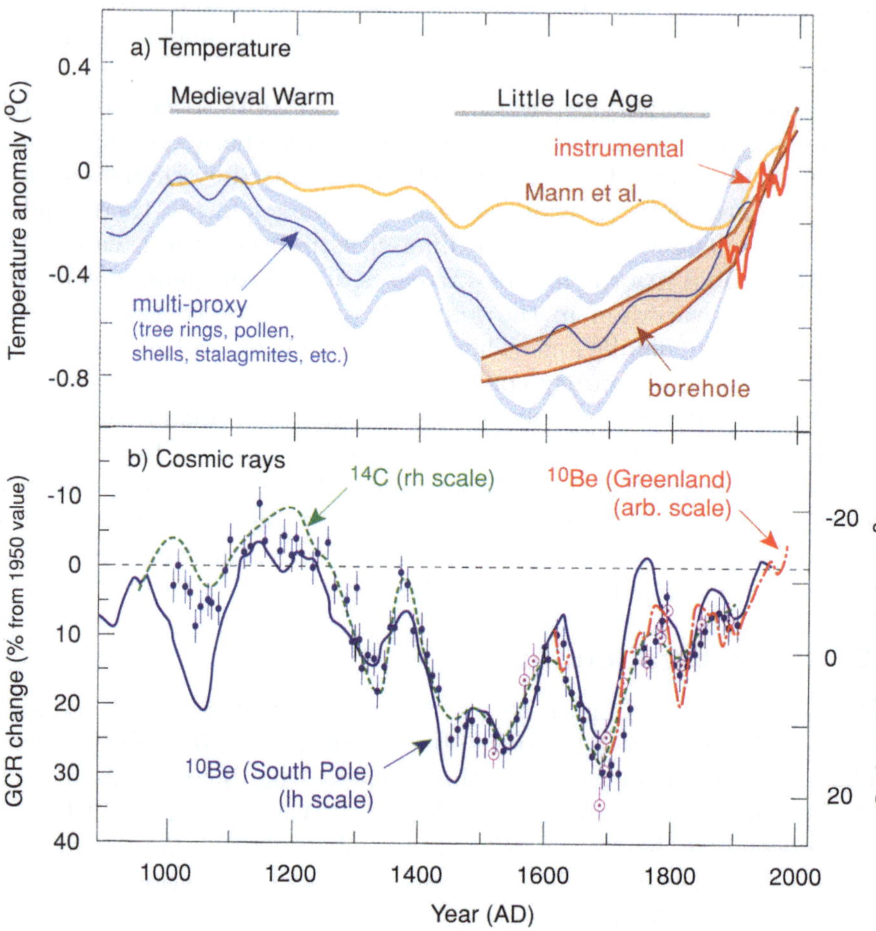

Figure 12.8. Comparison of variations during the last millennium of a) temperature (with respect to the 1961–1990 average) and b) galactic cosmic rays (note the inverted scale; high cosmic ray fluxes are associated with cold temperatures). The temperature curves comprise a recent multi-proxy reconstruction of northern hemisphere temperatures (the full band shows the 95% confidence interval; [Moberg *et al.*, 2005]), the so-called 'hockey-stick' curve [Mann *et al.*, 1998, Mann *et al.*, 1999], borehole temperature measurements [Pollack and Smerdon, 2004], and smoothed instrumental measurements since 1860. The cosmic ray reconstructions are based on several ^{14}C measurements: tree rings (shown by the data points and the dashed green curve) [Klein *et al.*, 1980], and ^{10}Be concentrations in ice cores from the South Pole (solid blue curve) [Raisbeck *et al.*, 1990] and Greenland (dot-dashed red curve) [Usoskin *et al.*, 2002].

The variation of GCR intensity over the last millennium is shown in Fig. 12.8b), as reconstructed from ^{14}C in tree rings, and ^{10}Be in ice cores from the South Pole and Greenland. Close similarities are evident between the temperature and GCR records. Most of the GCR variation over this period is due to solar magnetic variability (see, for example, Fig. 12.1), so these data cannot distinguish between a direct effect of GCRs on climate, or a GCR proxy for solar irradiance/UV.

4.3 Holocene

Indian Ocean monsoon. Neff *et al.* (2001) have measured the $\delta^{18}O$ composition in the layers of a stalagmite from a cave in Oman, which are U-Th dated to cover the period from 9,600 to 6,200 yr BP. The $\delta^{18}O$ is measured in calcium carbonate, which was deposited in isotopic equilibrium with the water that flowed at the time of formation of the stalagmite. The data are shown in Fig. 12.9 together with $\Delta^{14}C$ measured in tree rings such as the California bristlecone pine. The two

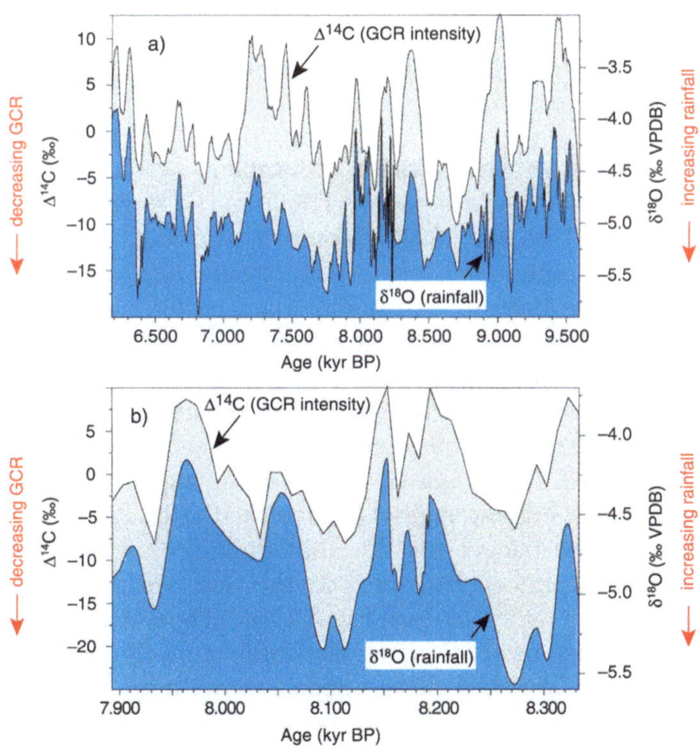

Figure 12.9. A comparison of variations of $\delta^{18}O$ in a U-Th-dated stalagmite from a cave in Oman with variations of $\Delta^{14}C$ from tree rings elsewhere in the world, for a) the 3.4 kyr period from 9,600 to 6,200 yr BP and b) the 430 yr period from 8,330 to 7,900 yr BP [Neff *et al.*, 2001].

timescales have been tuned to match bumps within the known experimental errors (smooth shifts have been applied to the U-Th dates up to a maximum of 190 yr). During a 430-yr period centred around 8.1 kyr BP, the stalagmite grew at a rate of 0.55 mm/yr—an order of magnitude faster than at other times—which allowed a high resolution δ^{18}O measurement to be made (Fig. 12.9b).

Oman today has an arid climate and lies beyond the most northerly excursion of the inter tropical convergence zone (ITCZ), which determines the region of heavy rainfall of the Indian Ocean monsoon system. However there is evidence that the northern migration of the ITCZ reached higher latitudes at earlier times and, in consequence, that Oman had a wetter climate. In this region, the temperature shifts during the Holocene are estimated to account for only 0.25 per mil variation in δ^{18}O [Neff *et al.*, 2001]. However the δ^{18}O values of monsoonal rainfall associated with the ITCZ show an inverse correlation with amount of rainfall (increased rainfall results in a decreased ^{18}O fraction; see Fig. 12.5). For these data, the δ^{18}O variations are therefore ascribed to changes of rainfall. Higher rainfall is associated with reduced GCR intensity (Fig. 12.9).

The striking similarity between the δ^{18}O and Δ^{14}C data indicates that that solar/GCR activity tightly controlled the monsoon rainfall of the Indian Ocean region during this 3,000-year period.

Ice-rafted debris in the North Atlantic. Bond *et al.* have analysed sediments of ice rafted debris (IRD) in the North Atlantic [Bond *et al.*, 1997a, Bond *et al.*, 1997b, Bond *et al.*, 2001]. The latter are found in deep sea cores as layers of tiny stones and micro-fossils that were frozen into the bases of advancing glaciers and then rafted out to sea by glaciers. These reveal abrupt episodes when cool, ice-bearing waters from the North Atlantic advanced as far south as the latitude of southern Britain, coincident with changes in the atmospheric circulation recorded in Greenland. A quasi-cyclic occurrence of IRD events with a periodicity of 1470 ± 530 yr has been found, during which temperatures dropped and glacial calving suddenly increased (Fig. 12.10).

Until recently the trigger for this millennial-scale climate cycle was unknown. Orbital periodicities around the Sun are too long to cause millennial-scale climate cycles (see Fig. 12.14). There are several reasons to rule out ice sheet oscillations as the forcing mechanism. First, the rafting icebergs are launched simultaneously from more than one glacier, and so the driving mechanism cannot be ascribed to a single ice sheet. It requires a common climate forcing mechanism that induces the release of ice over a large region. Second, the events continue with the same periodicity through at least three major climate transitions: the Younger Dryas-Holocene transition, the glacial-interglacial transition, and the

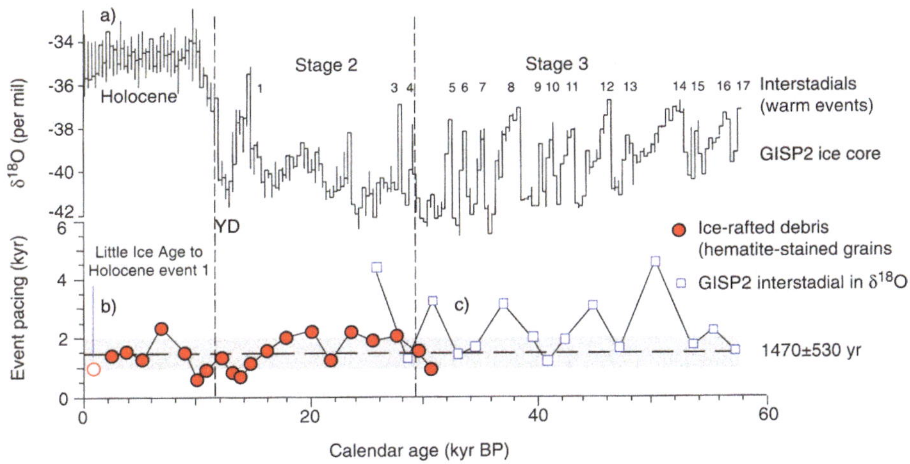

Figure 12.10. Periodicity of ice rafted debris (IRD) events in the North Atlantic [Bond *et al.*, 1997b]. The curves show a) the GISP2 ice core δ^{18}O record of Greenland temperatures over the last 60 kyr (Holocene, late glacial/Stage 2 and mid glacial/Stage 3 periods), b) the periodicity of IRD events from 32 kyr BP to the present, measured from haematite-stained grains (other tracers give similar results), and c) the periodicity of Dansgaard-Oeschger warm events in the GISP2 δ^{18}O data from 58 kyr to 26 kyr BP.

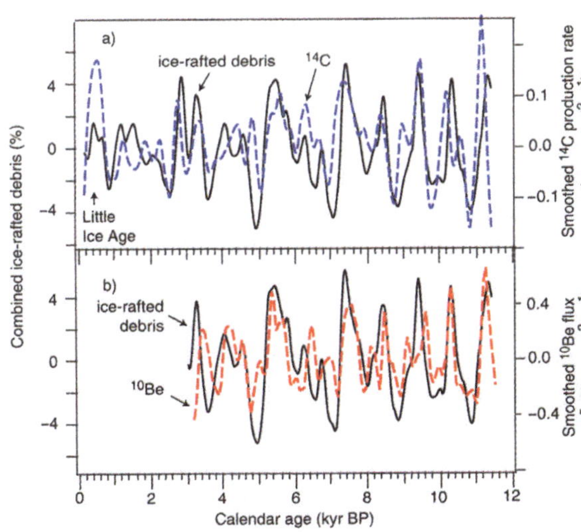

Figure 12.11. Correlation of solar/GCR variability (dashed curves) with ice-rafted debris (solid curves) in the North Atlantic during the Holocene [Bond *et al.*, 2001]: a) the ^{14}C cosmic ray record (correlation coefficient 0.44) and b) the ^{10}Be cosmic ray record (0.56), together with the combined ice-rafted-debris tracers. The Little Ice Age is indicated in the upper panel.

boundary within the glacial age between the marine isotope stages 2 and 3. Even though the ice conditions during these transitions were changing dramatically, the IRD events continued with the same periodicity.

Third, and especially surprising, is the evidence that the IRD cold events have continued through the Holocene (Fig. 12.10), with the same periodicity (but with a lower amount of IRD material). The events were abrupt during both the glacial and Holocene periods, generally switching on and off within one or two centuries. The estimated decreases in North Atlantic Ocean temperatures during the Holocene IRD events are 2 K, or about 15–20% of the full Holocene-to-glacial temperature difference.

A recent study has shown that solar/GCR variability is highly correlated with the ice rafted debris events during the Holocene phase (Fig. 12.11) [Bond *et al.*, 2001]. This correlation between high GCR flux and cold North Atlantic temperatures embraces the Little Ice Age, which is seen not as an isolated event but rather as the most recent of around ten such events during the Holocene. A high GCR flux is associated with a cold climate, and a low flux with a warm climate. On these 100 yr timescales, variations of the cosmic ray flux are thought to reflect changing solar activity. However, recent high-resolution palaeomagnetic studies suggest that short-term geomagnetic variability may in fact control a significant fraction of the GCR modulation within the Holocene, even on 100 yr timescales [St-Onge *et al.*, 2003]. This suggests that GCRs directly influence the climate, rather than merely serve as a proxy for solar variability.

Figure 12.11 provides convincing evidence that solar/GCR forcing is responsible for at least the Holocene phase of this 1500 yr climate cycle. Moreover, the similarity of the pacing of these events with the warm Dansgaard-Oeschger events during the ice age suggests that they too may be initiated by solar/GCR forcing. In this case, changes of North Atlantic Deep Water (NADW) production are a positive feedback in response to the solar/GCR forcing, rather than being the primary driver of climate change. Simulations [Ganopolski and Rahmstorf, 2001] show that a small decrease of freshwater into the North Atlantic is sufficient to increase salinity of the North Atlantic and trigger the 'warm' heat conveyor mode with NADW formation further north, in the Nordic Seas. This would suggest that solar/GCR forcing may have initially modified the hydrological cycle in the North Atlantic region, which then triggered the warm NADW mode.

Biogenic activity in the North Pacific region. Recent evidence shows that that these millennial scale shifts in Holocene climate were also present in the sub-polar North Pacific [Hu *et al.*, 2003]. Variations of biogenic silica were measured in the sediment from Arolik Lake in a tundra region of south-western Alaska. Biogenic silica reflects the sedimentary abundance of diatoms—single-celled algae which dominate lake primary productivity. High-resolution analyses reveal cyclic variations in climate and ecosystems during the Holocene with periodicities similar to those of the North Alantic drift ice and the cosmogenic nuclides ^{14}C and ^{10}Be (Fig. 12.12). High GCR flux is associated with low biogenic activity. Taken together, these results imply that solar/GCR forcing of Holocene climate occurred over a large fraction of the high-latitude northern region.

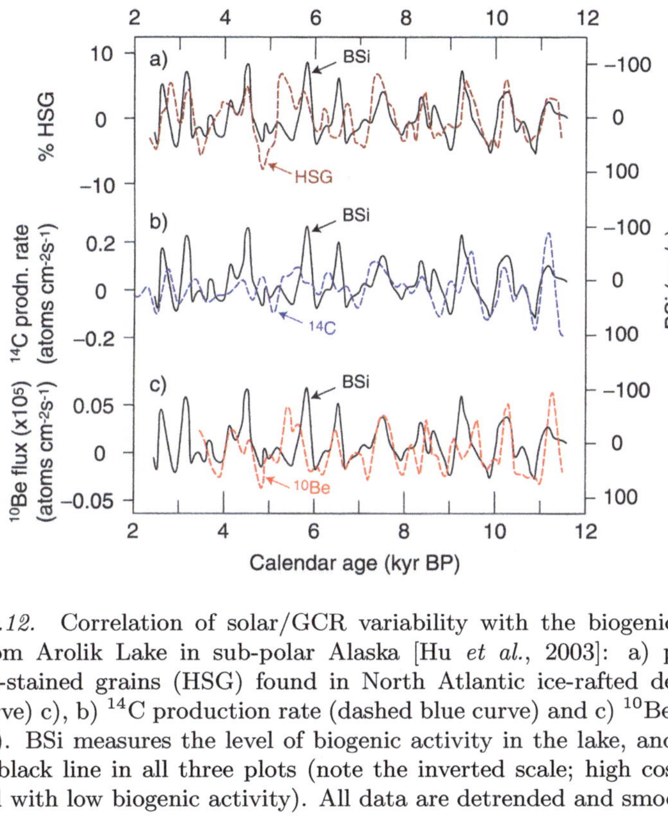

Figure 12.12. Correlation of solar/GCR variability with the biogenic silica (BSi) record from Arolik Lake in sub-polar Alaska [Hu *et al.*, 2003]: a) percentage of haemetite-stained grains (HSG) found in North Atlantic ice-rafted debris (dashed brown curve) c), b) ^{14}C production rate (dashed blue curve) and c) ^{10}Be flux (dashed red curve). BSi measures the level of biogenic activity in the lake, and is shown as the solid black line in all three plots (note the inverted scale; high cosmic rays are associated with low biogenic activity). All data are detrended and smoothed.

4.4 Ice ages

The most important clue for identifying the cause of the glacial cycles is the spectral purity of their periodicity [Muller and McDonald, 2000]. For the past million years, the glacial pattern is dominated by a precise 100 kyr cycle, and for the million years before that, by an equally precise 41 kyr cycle (Fig. 12.13). These match the major frequencies of

Earth's orbital cycles (Fig. 12.14), and so it is generally accepted that the primary driver for the ice ages was astronomical. A linkage between orbit and climate is provided by the Milankovitch model, which states that melting of the northern ice sheets is driven by peaks in Northern Hemisphere summer insolation (solar heating). This has become established as the standard model of the ice ages since it naturally includes spectral components at the orbital modulation frequencies.

However, high precision palaeoclimatic data have revealed serious discrepancies with the Milankovitch model that fundamentally challenge its validity and re-open the question of what causes the glacial cycles [Muller and McDonald, 2000]. It has been proposed that cosmic rays are

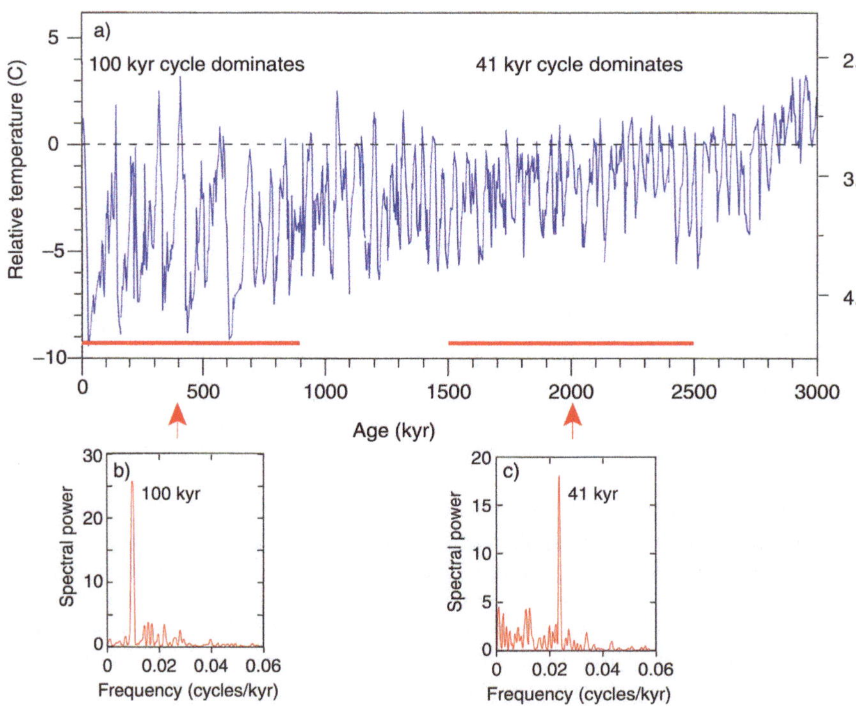

Figure 12.13. a) Climate for the last 3 Myr, as derived from the benthic $\delta^{18}O$ record from DSDP site 607 in the North Atlantic. For the past 1 Myr, climate has followed a precise 100-kyr cycle, and prior to that, an equally-precise 41 kyr cycle, as seen in the two periodograms: b) spectral power for ODP site 659, from 0 to 900 kyr and c) spectral power for DSDP site 607, from 1.5 to 2.5 Myr [Muller and McDonald, 1997]. The time scale for both periodograms assumes a constant sedimentation rate, i.e. the data are untuned. The narrow spectral widths imply that the glacial cycles are driven by an astronomical force, regardless of the detailed mechanism; oscillations purely internal to Earth's climate system could not maintain such precise phase coherency.

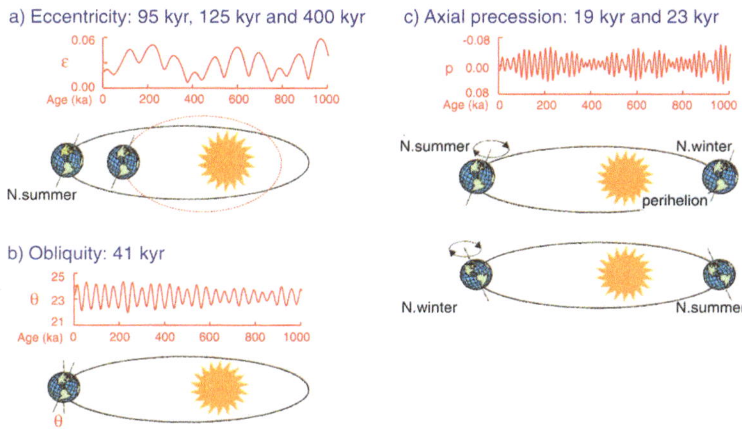

Figure 12.14. Earth's orbital cycles (caused by the gravitational pulls of the planets and by gravitational torques exerted on Earth's equatorial bulge): a) eccentricity (95, 125 and 400 kyr periodicity): the departure of Earth's orbit from a perfect circle, b) obliquity (41 kyr): the angle of tilt of Earth's axis towards the Sun, which affects the seasonal insolation contrast, and c) precession (19 and 23 kyr): the 'wobble' of Earth's axis with respect to the stars, which causes the perihelion (position of closest approach to the Sun) to cycle through the seasons.

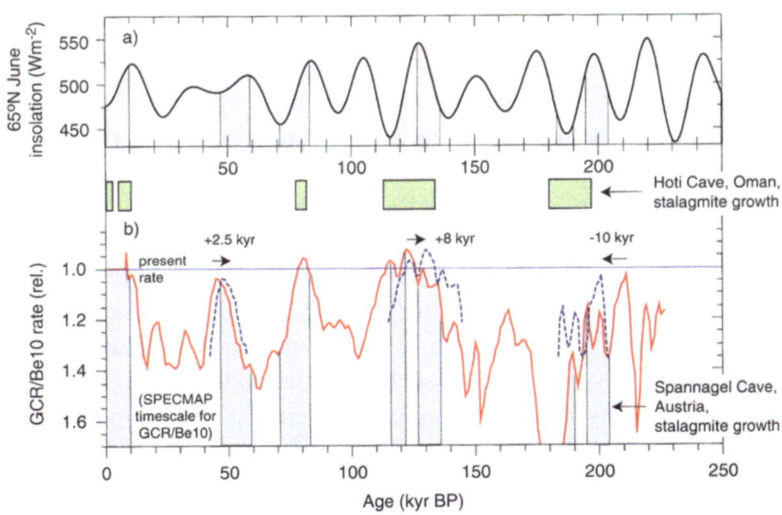

Figure 12.15. Comparison of the growth periods of stalagmites in Austria and Oman with a) 65°N June insolation and b) the relative GCR flux (^{10}Be ocean sediments; note inverted scale) [Christl *et al.*, 2004, Kirkby *et al.*, 2004]. The growth periods are indicated by shaded bands (Spannagel cave) or boxes (Hoti cave). Growth periods at Spannagel cave require warm temperatures, close to the present climate; growth periods at Hoti cave require a moist climate. Periods without stalagmite growth are unshaded. The dashed curves in b) indicate the estimated corrections of systematic errors in the SPECMAP timescale, on which the GCR record is based. The growth periods are associated with intervals of low GCR flux, close to present values.

a primary trigger of glacial-interglacial conditions [Kirkby *et al.*, 2004]. This proposal is based on a wide range of evidence, some of which is discussed in this paper, including the records of speleothem growth in caves in Austria and Oman [Christl *et al.*, 2004], and the record of cosmic ray flux over the past 220 kyr obtained from the ^{10}Be composition of deep-ocean sediments [Christl *et al.*, 2003] (Fig. 12.15).

The measurements of ^{10}Be production show that the GCR flux over the last 220 kyr was generally around 20–40% higher than today, due to a reduced geomagnetic field. However there were several relatively short periods when the GCR flux returned to present levels. During these periods, stalagmite growth was observed in the Austrian and Oman caves. The growth of stalagmites at each of these locations is especially sensitive to climate since it requires temperatures at least as warm as today's (within 1.5±1°C). In contrast, the stalagmite growth periods show no clear pattern of association with 65°N June insolation, contrary to the expectations of the standard Milankovitch model (Fig. 12.15a).

A further problem for the Milankovitch model concerns the timing of Termination II—the penultimate deglaciation. The growth period of stalagmite SPA 52 from Spannagel Cave began at 135±1.2 kyr [Spötl *et al.*, 2002] (Fig. 12.15b). So, by that time, temperatures in central Europe were within 1°C or so of the present day. This corroborates Henderson and Slowey's conclusion, based on sediment cores off the Bahamas, that warming was well underway at 135±2.5 kyr [Henderson and Slowey, 2000]. Furthermore, dating of a Barbados coral terrace shows that the sea level had risen to within 20% of its peak value by 135.8±0.8 kyr [Gallup *et al.*, 2002]. These results confirm the 'early' timing of Termination II, originally discovered at Devil's Hole Cave, Nevada, by Winograd *et al.* (1992). In summary, the warming at the end of the penultimate ice age was underway at the *minimum* of 65°N June insolation, and essentially complete about 8 kyr prior to the insolation maximum (Fig. 12.15a). The Milankovitch model therefore suffers a causality problem at Termination II: the deglaciation precedes its supposed cause. Furthermore, based on an analysis of deep ocean cores, Visser *et al.* (2003) report that warming of the tropical Pacific Ocean at Termination II preceded the northern ice sheet melting by 2–3 kyr. So the northern ice sheets were not the primary driver of the penultimate deglaciation—contrary to the expectations of the Milankovitch model—but rather a response to some other initial forcing.

The ^{10}Be data, on the other hand, show that the GCR flux began to decrease around 150 kyr, and had reached present levels by about 135 kyr (Fig. 12.15b). This is compatible with Termination II being driven in part by a reduction of cosmic ray flux—albeit within large experimental errors. A similar reduction of GCR flux also occurred around 20 kyr BP due to a rise in the geomagnetic field strength towards present values. This was coincident with the first signs of warming at the end of the last glaciation, as recorded in the Antarctic ice at about 18 kyr BP.

At present the GCR archive data do not have the precision to compare to the spectral purity observed in the climate record (Fig. 12.13). Moreover, a mechanism would be needed to explain how the orbital cycles could be imprinted on the GCR intensity. One possibility would be an orbital influence on the geomagnetic field. Indeed, some measurements suggest that long-term records of variations of Earth's magnetic field—in both strength and magnetic inclination—show the presence of orbital frequencies [Channell *et al.*, 1998, Yamazaki and Oda, 2002]. The effects persist when the archives are corrected for climatic influences. Although it is speculative that orbital variations could modulate the Earth's dipole field, such a linkage is plausible.

If cosmic rays are indeed affecting climate on glacial time scales, then there are definite predictions that can be tested by further observations and experiments. For example, there should be a climatic response to geomagnetic reversals or excursions ('failed reversal' events during which the geomagnetic field dips to a low value but returns with the same polarity). A relatively recent excursion was the Laschamp event, when the geomagnetic field briefly ($\lesssim$1 kyr duration) dropped to around 10% of its present strength and the GCR flux approximately doubled [Wagner *et al.*, 2000]. Based on several independent radio-isotope measurements, the Laschamp event has been precisely dated at (40.4 ± 2.0) kyr ago (2σ error) [Guillou *et al.*, 2004].

No evidence of climate change was observed in the GRIP (Greenland) ice core during the Laschamp event [Wagner *et al.*, 2001]. However several climatic effects were recorded elsewhere. A pronounced reduction of the East Asia monsoon was registered in Hulu Cave, China [Wang *et al.*, 2001]. At the same time, a brief wet period was recorded in speleothems found in tropical northeastern Brazil—a region that is presently semi-arid [Wang *et al.*, 2004]. The wet period was precisely U/Th dated to last 700 ± 400 yr from 39.6 to 38.9 kyr ago; this was the only recorded period of stalagmite growth in the 30 kyr interval from 47 to 16 kyr ago. Further evidence has recently been found in deep-sea cores from the South Atlantic, by analysing Nd isotopes as a sensitive proxy of the thermohaline ocean circulation [Piotrowski *et al.*, 2005]. A brief, sharp reduction of North Atlantic Deep Water (NADW) was recorded (i.e. towards colder conditions) coincident with the Laschamp event. In summary, there appears to be evidence for climatic responses at the time of the Laschamp event associated with the hydrological cycle.

4.5 Galactic variability

On much longer timescales, up to 1 Gyr, Shaviv (2002) has reported evidence for the occurrence of ice-age epochs on Earth during crossings of the solar system with the galactic spiral arms. The GCR flux reaching the solar system should periodically increase with each crossing of a galactic spiral arm due to the locality of the GCR sources and the subsequent 'diffusive' trapping of charged particles in the interstellar magnetic fields. This expectation is supported by the GCR exposure age recorded in iron meteorites [Shaviv, 2002].

This observation has been strengthened by Shaviv and Veizer (2003), who find a close correlation between the GCR flux and ocean temperatures reconstructed from $\delta^{18}O$ in calcite and aragonite shells found in sediments from the tropical seas (30°S–30°N), deposited during the Phanerozoic eon (the past 550 Myr, corresponding to the age of multicellular animal life on Earth) [Veizer *et al.*, 2000]. Both the GCR flux and

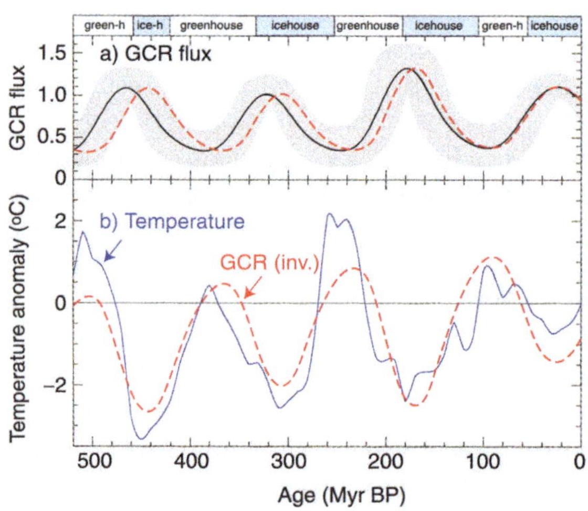

Figure 12.16. Correlation of cosmic rays and climate over the past 500 Myr [Shaviv and Veizer, 2003]: a) GCR mean flux variations as the solar system passes through the spiral arms of the Milky Way, reconstructed from iron meteorite exposure ages [Shaviv, 2002], and b) ocean temperature anomalies reconstructed from $\delta^{18}O$ in calcite shells found in sediments from the tropical seas [Veizer *et al.*, 2000]. Panel a) shows the nominal reconstructed GCR flux (solid black curve) and the error range (grey band). The dashed red curves in panels a) and b) shows the best fit of the GCR flux to the temperature data (which is given by the solid blue curve in panel b), within the allowable error (note inverted GCR scale in panel b). The data are de-trended and smoothed. The shaded blue bars at the top represent cool climate modes for Earth (icehouses) and the white bars are warm modes (greenhouses), as established from sediment analyses elsewhere.

ocean temperatures appear to follow the same periodicity of about 140 Myr and the same phase (Fig. 12.16). Extended periods of high GCR flux are found to coincide with cold epochs on Earth, when glaciation reached to low latitudes ('icehouse' climate). Conversely, during periods of low GCR flux, the climate was warm ('greenhouse'). Alternative explanations, such as increased CO_2 levels, fail to account for a 140 Myr periodicity.

5. GCR-cloud-climate Mechanisms

Mechanisms have been proposed to link the GCR flux in the atmosphere to atmospheric processes that can influence the climate. These mechanisms involve two effects of cosmic rays on the atmosphere, both involving changes in clouds. Firstly, the ionisation rate of air in the troposphere (the lowest 10–15 km of the atmosphere) varies with the cosmic ray flux. One hypothesis is that the ionisation rate influences the production of new aerosol particles in the atmosphere, and that these new particles can ultimately affect the formation of cloud droplets. Secondly, cosmic ray ionisation appears to modulate the entire ionosphere-earth electrical circuit and, in particular, the current flowing between the ionosphere and Earth. Note that, whereas solar UV influences the ionisation of the ionosphere, the current flowing between the ionosphere and Earth is determined by the tropospheric ionisation (impedance), which is unaffected by solar UV. The second hypothesis is that changes in this current influence the properties of clouds through processes involving charge effects on droplet freezing. Both mechanisms involve a number of steps between the initial effect of cosmic rays on the atmosphere and the ultimate effect on clouds and climate.

The effect of GCRs on the atmosphere is one mechanism by which solar variations can affect the climate system. Another mechanism, which has also received much attention, is the influence of variations in ultraviolet radiation on upper atmospheric ozone concentrations, and the subsequent effect of these variations on the general pattern of heating and circulation in the atmosphere (§3.1).

5.1 The importance of clouds in the climate system

Clouds account for a global average 28 Wm^{-2} net cooling of the climate system [Hartmann, 1993] so even small changes in their average coverage or their reflectivity of solar radiation can have significant effects on the climate. They absorb longwave terrestrial radiation, which leads to a heating of Earth's surface at night, but they also reflect incoming solar shortwave radiation, leading to a cooling of the surface during the day. The net effect of clouds has been shown by models and observations to be cooling.

Layered clouds covering large areas—such as marine stratus and strato-cumulus—make the greatest contribution to the cooling of the climate system, by reflection of solar radiation. Climate scientists are interested in how such clouds might respond to changes in air pollution [IPCC, 2001]. Increases in aerosol particle concentrations over the last century has led to an increase in cloud droplet concentrations in polluted regions. The net effect on cloud reflectivity is estimated to have produced a globally-averaged radiative forcing of as much as -2 Wm^{-2} since the start of the Industrial Revolution (the negative sign indicates a reduction, i.e. a cooling effect). The GCR-cloud-climate hypothesis is that long term changes in GCR flux might also have contributed to changes in cloud properties, although the effects are likely to be masked in regions where air pollution can lead to large changes in cloud properties.

Clouds also have important effects on the redistribution of energy in the climate system. In particular, release of latent heat in the mid troposphere supplies energy to the atmospheric general circulation. Processes that affect the release of latent heat have the potential to affect weather patterns and the climate more generally (e.g. [Nober *et al.*, 2003]). Ice formation is especially important in the energy budget of clouds. One hypothesis connecting GCR flux and cloud processes involves effects on ice formation at the tops of clouds.

5.2 GCRs in Earth's atmosphere

Cosmic rays comprise high energy (GeV and above) particles (mostly protons and helium nuclei) which ionise the atmosphere and produce air ions or "small ions". Away from continental sources of radon, cosmic rays dominate the production of air ions all the way to Earth's surface. The combination of two geophysical factors, geomagnetism and solar variability, modulate the ionisation rate (§3.2 and Figs. 12.3, 12.4 and 12.17). These factors, together with the cosmic ray energy spectrum, lead to a temporal and spatial distribution of cosmic rays entering the atmosphere, with about a factor 4 higher flux at the poles than at the equator. Solar magnetic activity controls the heliospheric magnetic field and modulates the cosmic ray flux, primarily below energies of about 10 GeV per nucleon. The amplitude of solar-cycle variation in ionisation rate is approximately 15% peak-to-peak, globally averaged. A long term decrease in cosmic ray flux during the last century (and starting at the

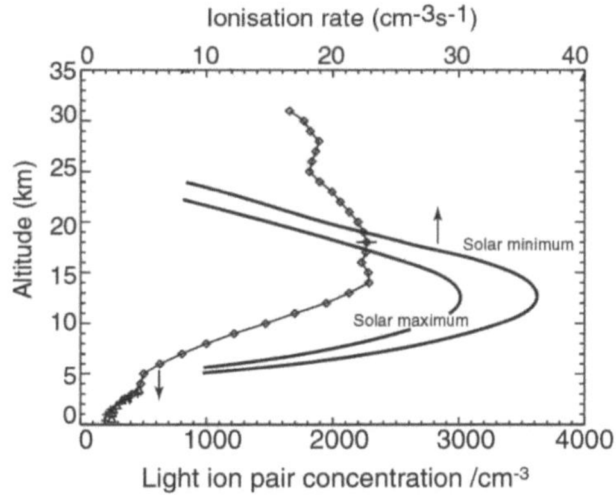

Figure 12.17. Variation with altitude of the ion-pair production rate in the atmosphere, and typical steady state ion-pair concentrations [Harrison and Carslaw, 2003]. The variation of ionisaton rate between the maximum and minimum of the 11 year solar cycle is indicated (see also Fig. 12.3).

end of the Little Ice Age; see Fig. 12.8) has also been detected, based on a number of reconstructions [Lockwood, 2003, Carslaw *et al.*, 2002].

The ion-pair production rate is balanced by processes of recombination and scavenging by aerosol particles and clouds, leading to steady state bipolar ion concentrations in the lowest part of the atmosphere of about 500 cm^{-3}, rising to a maximum of about 3000–4000 cm^{-3} at around 15 km. In the lower atmosphere the scavenging by aerosols dominates removal of ions and reduces steady state ion concentrations to typically less than one-fifth of what would occur in aerosol-free air.

5.3 Ion-induced particle formation

One mechanism for a GCR-climate interaction is that cosmic rays might influence the atmospheric concentration of cloud condensation nuclei. Cloud condensation nuclei are the aerosol particles upon which cloud droplets form, so any change in their abundance will influence the microphysical and even dynamical properties of clouds.

The continuous processes of particle removal in the atmosphere by deposition to the surface, cloud scavenging and self coagulation mean that the rate at which particles are produced influences the particle concentration at any one time. Particles are produced in the atmosphere through primary mechanisms (direct injection of new, mostly large, particles into the atmosphere, such as sea spray) or through gas-to-particle conversion (nucleation, often termed secondary particle formation). It has been proposed that ions in the atmosphere can affect the rate of formation of particles by this latter mechanism. Unfortunately, the mechanisms of particle nucleation even in the neutral atmosphere are poorly understood [Kulmala *et al.*, 2004], making it difficult to quantify the contribution of ion-mediated processes. Despite these uncertainties, however, an increasing number of experimental and modelling studies suggest that ions can play a significant role in particle formation, at least under some conditions.

To understand the influence of ions on particle formation requires a brief introduction to the mechanism involving neutral molecules. Nucleated particles are formed in the atmosphere from condensable vapours. The range of molecules suitable for particle formation is very limited; laboratory and theoretical studies suggest that only sulphuric acid and ammonia are effective, although a range of organic species probably also contribute to growth of the new nm-sized molecular clusters. New particles consist of clusters of molecules as small as two in number. In a nucleation burst, particle concentrations at 3 nm size (the smallest size observable with commercially-available instruments) may be as high as 10^6 cm^{-3}, although concentrations of around 10^4 cm^{-3} are more typical. To be effective as cloud condensation nuclei (CCN) on which water can condense under typical atmospheric conditions requires growth of these new particles to around 40 nm diameter or more, which occurs through further condensation over several hours or days after nucleation. Cloud condensation nucleus concentrations rarely exceed a few hundred per cubic centimetre, less than one-hundredth the concentration of particles produced during nucleation. The survival rate of new particles as they grow to CCN sizes is therefore a critical parameter, and likely to be influenced by ionisation. Particle formation tends to occur readily and frequently in the upper troposphere where the condensable vapours are most supersaturated (due primarily to low ambient temperatures). In the lower atmosphere, nucleation tends to occur during 'aerosol bursts', involving long periods of no new production followed by brief (several hour) periods of rapid production.

Although only a small fraction of nucleated particles grow to CCN sizes, and only a small fraction of these are likely to have been caused by ion processes, it is worth noting that cloud reflectivity, which is important in climate change, is highly sensitive to very small changes in CCN concentration. The change in cloud reflectivity, R, approximately depends on the number of droplets, N, according to

$$\frac{\Delta R}{R} = (1 - A) \cdot \frac{\Delta N}{N}$$

Thin stratocumulus clouds that cover an appreciable fraction of the ocean typically have a reflectivity of around 0.5 and a droplet concentration of 100 cm^{-3}. For such clouds, a change of only 1% in droplet number ($\Delta N = 1$ cm^{-3}) would change the reflectivity by 0.5%.

There are two ways in which ions can increase the effective production rate of particles at readily-observable sizes of about 3 nm diameter (Fig. 12.18). They can *induce* particle formation by stabilising a cluster of condensable molecules such as sulphuric acid and ammonia through Coulomb attraction. They can also *mediate* the formation process by influencing the rate at which polar molecules subsequently condense [Nadykto and Yu, 2003, Nadykto and Yu, 2004, Laakso *et al.*, 2003]. The acceleration of cluster growth increases the fraction of particles that survive removal by coagulation before reaching observable sizes of 3 nm. Beyond about 5 nm the electrostatic effects become negligible in influencing subsequent coagulation and growth rates. However, coagulational loss rates of new particles fall dramatically beyond this size also. So with regard to the survival of particles up to CCN sizes, the critical size range is below 5 nm, and this is exactly the range for which ions have the greatest effect.

The contribution of ionisation to particle formation will be difficult to establish unambiguously because of the great variability of observed

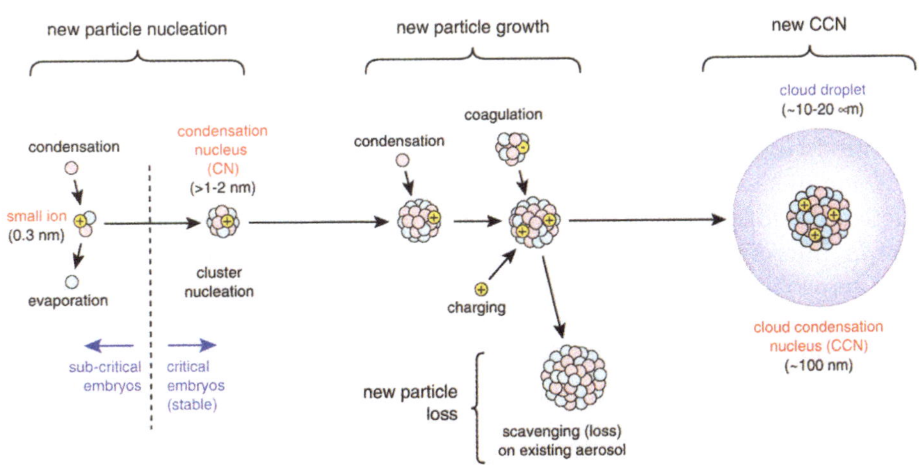

Figure 12.18. Schematic showing the nucleation and growth of new particles in the atmosphere from trace condensable vapours. Under certain conditions, the critical embryonic stage involving the nucleation and growth to sizes around 1–2 nm is enhanced by the presence of ionisation from cosmic rays. Only a small fraction of new particles reach the size of cloud condensation nuclei (CCN); most are removed by coagulation or scavenging on existing aerosol.

formation rates and because several very poorly constrained neutral mechanisms may offer alternative explanations. An upper limit to the ion-induced particle production rate close to Earth's surface (the boundary layer) is approximately 10 cm^{-3} s^{-1}, which is the maximum cosmic ray-induced ionisation rate. Observed particle production rates cover a very wide range, from as low as 0.001 to 10^4 or more cm^{-3} s^{-1} [Kulmala *et al.*, 2004], although the majority of events lie in the range 0.1-10 cm^{-3} s^{-1}. Clearly only some of the observed nucleation events can be explained in terms of ions. A further complication in assessing the potential contribution of ions to particle production rates is that particles can be observed using current instrumentation only down to 3 nm diameter, while cluster formation and growth probably begins at 0.5 nm. A large but uncertain fraction of the new clusters below 1 nm are lost through coagulation with existing particles before they grow to observable sizes [Dal Maso *et al.*, 2002].

Large positive cluster ions with atomic mass numbers up to 2500 have been detected in the upper troposphere using an aircraft-based large-ion mass spectrometer [Eichkorn *et al.*, 2002]. The charged clusters were composed of hydrogen atoms, acetone, sulphuric acid and water. The largest ions are probably very small charged aerosols formed through coagulation of neutral sulphuric acid-water clusters with charged clusters. The observations do not allow the origin of the neutral clusters to be identified, but one possible source is ion-ion recombination. These observations provide strong evidence for the ion-mediated formation and growth of aerosol particles in the upper troposphere. Cluster ions have also been observed near Earth's surface [Horrak *et al.*, 1998] in urban air, and have been shown [Yu and Turco, 2000] to be consistent with expected charged cluster growth rates.

A number of modelling studies have examined how air ions can influence the formation rate of new particles and their subsequent growth rate in different parts of the atmosphere, and have attempted to quantify how aerosol concentrations might respond to changes in ionisation rate.

The model simulations have typically used box models in which particle formation and growth is simulated under realistic atmospheric conditions for a period of several hours to days. The models include the processes of cluster formation, growth, coagulation between the clusters (an effective cluster growth mechanism) and coagulation loss of the clusters to existing, much larger, particles. The model experiments have been conducted using prescribed ionisation rates for a range of atmospheric conditions and for cases with and without charge effects. Our quantitative understanding of the kinetics and thermodynamics of neutral and charged clusters, and their interaction with larger particles and other molecules, is currently limited. As a result, independent model simulations currently disagree regarding the regions of the atmosphere where ion-induced nucleation is likely to be important in determining the production rate of new particles.

The first simulations of ion cluster formation and growth [Yu and Turco, 1997] examined their role in aircraft exhausts. Results suggested that the number of ions produced in the exhaust was an important factor in the determining the number of particles. Later box model simulations [Yu and Turco, 2001] under normal atmospheric conditions examined how the presence of ions could enhance the formation and growth rate of sulphuric acid and water clusters. They treated particle formation as a kinetic process and adjusted the coagulation and condensation rates to take into account electrostatic forces. The model results of Yu and Turco (2001) emphasise the important effect of enhanced condensation of sulphuric acid vapour for the growth to observable sizes.

An important quantity is the fraction of ionisation events resulting in new observable particles. Yu and Turco (2001) estimate that a 25% increase in the ionisation rate would result in a 7–9% increase in the total concentration of particles between 3 and 10 nm. Thus, although every ion serves as a site for cluster formation, only approximately 1 in 3 of the new ion-induced clusters is able to grow to observable sizes of 3–10 nm. Even fewer are likely to survive to typical CCN sizes of 40 nm. Although Yu and Turco (2001) simulated particle growth out to such sizes, the results are difficult to put into context because in the real atmosphere, over several hours or days, many other processes will affect the survival of the particles than are represented in the simple box model.

Yu (2002) has also used the model to identify regions of the atmosphere where ion-mediated processes might be most important. He concluded (again, using a box model) that particle concentrations in the upper troposphere were unlikely to change significantly due to small changes in ionisation rate because particle formation is already fast there. He identified the atmospheric boundary layer as one region where changes in ionisation rate could lead to changes in particle concentration (Fig. 12.19).

A second set of model simulations and comparisons with observations has been undertaken by Laakso and co-workers [Laakso *et al.*, 2002, Laakso *et al.*, 2003, Laakso *et al.*, 2004]. They compared their model calculations of positive and negative ions and neutral clusters with observations made above a Finnish forest canopy. Interestingly, they observed an excess of negative ions in the size range 1.5–3 nm and a greater charge on 3–5 nm particles than expected simply assuming

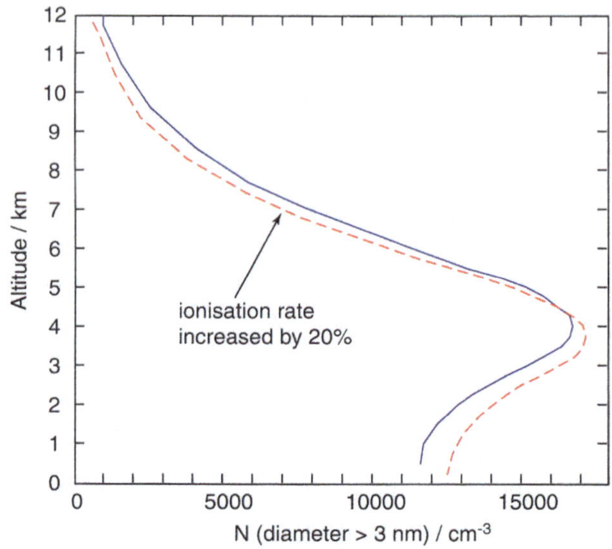

Figure 12.19. Results of a model simulation showing the effect of changes in the ionisation rate on particle concentrations resulting from sulphuric acid and water clusters [Yu, 2002]. The figure shows the concentration of particles larger than 3 nm diameter after a 3 hour simulation period. The results suggest that the greatest sensitivity is close to the Earth's surface. The model simulations assumed a realistic vertical profile of ionisation rate, increasing from 2 $cm^{-3}s^{-1}$ at the surface to 30 $cm^{-3}s^{-1}$ at 12 km. These model results differ from [Lovejoy *et al.*, 2004], who suggest that ion-induced nucleation of H_2SO_4/H_2O clusters is an efficient source of new particles in the middle and upper troposphere, but negligible in the lower troposphere.

diffusion charging equilibrium, which indicates that ions are involved in new particle formation. Overall, they conclude that ion-induced nucleation can account for some nucleation events (up to about 1 cm^{-3} s^{-1}) under conditions with low existing particle surface area, where coagulational loss of the new clusters is small.

Lee *et al.* (2003) have shown that observed ultrafine particles (<9 nm diameter) in the upper troposphere can be explained using an ion-induced mechanism [Lovejoy *et al.*, 2004]. However, high particle formation rates are also predicted assuming classical homogeneous nucleation of sulphuric acid and water under upper tropospheric conditions [Vehkamaki *et al.*, 2002]. At the low temperatures of the upper troposphere (190–230 K) homogeneous nucleation becomes very rapid and particles are formed at close to the molecular collision rate, so the upper troposphere is not a good place to test the importance of ion mechanisms. Ion-induced nucleation is likely to be distinguishable only in atmospheric regions where other mechanisms can be excluded.

Lovejoy *et al.* (2004) have performed similar modelling experiments to Laakso and Yu based on their own laboratory measurements of cluster thermodynamics, which was a major uncertainty in the previous simulations. The laboratory experiments [Froyd and Lovejoy, 2003a, Froyd and Lovejoy, 2003b] measured the thermodynamics for the binding of H_2SO_4 and H_2O in charged clusters of the form $HSO_4^-(H_2SO_4)_x(H_2O)_y$ and $H^+(H_2SO_4)_n(H_2O)_m$, which are believed to occur in the atmosphere (x, y, n and m are variable numbers of molecules attached to the cluster). The model of Lovejoy *et al.* (2004) differs from that of either Laakso *et al.* or Yu *et al.* because it treats cluster formation as a reversible process, with H_2SO_4 condensing and evaporating from the clusters as a kinetic process constrained by the thermodynamics of the clusters. In contrast, in the absence of such thermodynamic measurements, Yu and Turco (2001) treated cluster growth as an irreversible process, but reduced the growth rates to compensate for lack of evaporation by reducing the sticking rate of condensing H_2SO_4 molecules onto the clusters.

When the Lovejoy *et al.* (2004) model is applied to the atmosphere it produces quite a different result from Yu and Turco and Laakso *et al.* They find that ionisation is a significant source of new particles in the mid and upper troposphere but does not explain nucleation events close to the surface. They predict ion-induced nucleation rates for the upper troposphere of >100 cm^{-3}d^{-1} (0.001 cm^{-3}s^{-1}), which are broadly consistent with observations. One major obstacle to confirming the accuracy of the model is the sensitivity of calculated nucleation rates to the input parameters of temperature, relative humidity and H_2SO_4 gas phase concentration, with the latter being the most difficult to measure. In general, though, the Lovejoy *et al.* (2004) model can explain the limited available observed nucleation rates for temperatures below 270 K, but it under-predicts rates for the higher temperatures found closer to the surface. Closer to the surface, it is likely that other factors accelerate cluster growth, such as condensation of organic molecules.

In summary, three independent models of ion-induced particle formation have been developed based on different assumptions. Currently, all of the models suggest that ions play an important role in particle formation somewhere in the atmosphere, but the models disagree on exactly where in the atmosphere ions are likely to be most important. If ion-induced nucleation is influencing the production rate of new particles then it is reasonable to expect that CCN concentrations will respond in some way to changes in the ionisation rate. Further model simulations taking into account a wider range of realistic atmospheric processes will be needed—as well as further laboratory and field measurements—to quantify the expected CCN production rates.

5.4 The global electric circuit and effects on clouds

A second mechanism for a GCR-climate interaction concerns the influence of cosmic rays on the global electrical current flowing between the ionosphere and Earth's surface. Decadal changes in the ionospheric potential and surface potential gradient have been observed [Markson, 1981, Harrison, 2004] and explained in terms of changes in the cosmic ray flux. A number of studies have suggested that the associated change in current flowing in the fair weather part of the atmosphere may influence cloud processes.

Figure 12.20 shows the main features of the global electric circuit [Rycroft *et al.*, 2000]. Thunderstorms generate an electric current that flows to the conducting ionosphere. The ionospheric potential drives a current through the fair weather part of the atmosphere with a global mean current density of approximately 2 pA m^{-2}. A surface potential gradient of about 100 Vm^{-1} is established across a near surface resistance, where ion concentrations are low due to efficient removal of free ions to aerosol particles.

Harrison (2004) has shown that the surface potential gradient at Eskdalemuir, Scotland (55° 19', 3° 12' W) decreased by ~25% between 1920 and 1950 and, ignoring the period 1950–1970 dominated by nuclear

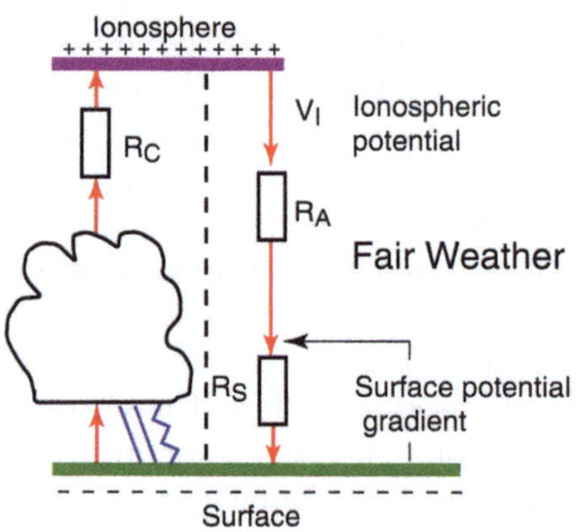

Figure 12.20. Simplified global electric circuit. Thunderstorms generate an electric current that flows through a resistance R_C to the conducting ionosphere. This establishes a positively-charged ionosphere with respect to Earth's surface. The ionospheric potential V_I drives a current through the fair weather part of the atmosphere, represented by the resistance $R_A + R_S$. A surface potential gradient of about 100 V/m is generated across the near-surface resistance, R_S. The fair weather current density is approximately 2 pA m^{-2}.

weapons tests, continued to decrease through the 1970s. Beryllium-10 concentrations in the Greenland ice core decreased by 34% over the same period, and other reconstructions suggest a decrease of around 20% [Lockwood, 2003, Carslaw *et al.*, 2002]. Figure 12.20 helps to explain how changes in cosmic rays could lead to changes in the surface potential gradient. First of all, it is worth noting that a decrease in cosmic ray flux would reduce the local ionisation rate close to the surface, resulting in an increase in electrical resistance and potential gradient there, opposite to what has been observed. So a direct local effect of cosmic rays can be excluded as an explanation for the long term changes in surface potential gradient. The positive correlation between cosmic rays and the surface potential gradient implies a control of the electric circuit in the charging part of the circuit; that is cosmic ray ionisation controls the electrical resistance between the top of thunderstorms (the current generators) and the ionosphere. Reductions in the potential gradient arise from a decrease in the current flowing in the electrical circuit, as a result of the increased resistance R_C.

The ionospheric potential has also been observed to correlate with surface cosmic ray flux over a shorter period from the 1950s to 1980s [Markson, 1981], amounting to a 10–20% variation in potential for a 10% change in surface cosmic ray flux. The response of the ionospheric potential to changes in cosmic ray flux is somewhat less than implied by the study of Harrison.

Williams (2003) has suggested that long-term decreases in aerosol pollution at the Scottish site could be an alternative explanation for the decrease in surface potential gradient. Aerosol particles reduce the conductivity of air by scavenging ions, so a long-term decrease in atmospheric particulate loading at the Scottish site would be needed to explain the decrease in potential gradient. The local evidence suggests that aerosol changes alone are unlikely to explain the long-term decrease in potential gradient observed, and that changes in cosmic rays have influenced the long-term properties of Earth's electric circuit [Harrison, 2004].

What effect might these changes in the electrical circuit have on the atmosphere? Physical mechanisms linking changes in the ionosphere-earth current to changes in the properties of clouds have been developed in a number of papers by B. Tinsley and co-workers [Tinsley, 1991, Tinsley, 1996a, Tinsley, 1996b, Tinsley, 2000, Tinsley and Deen, 1991, Tinsley *et al.*, 1989, Tinsley and Heelis, 1993, Tinsley *et al.*, 2000, Tinsley *et al.*, 2001], and recently reviewed by Carslaw *et al.* (2002) and Harrison and Carslaw (2004). The central hypothesis is that the electrical current flowing into the tops of supercooled clouds leads to relatively highly charged cloud droplets and aerosol particles which can affect the formation of ice crystals.

The tops of many clouds exist in a supercooled state, with water droplets persisting to temperatures far below the freezing point of water. Droplet freezing between 0°C and about −37°C requires ice nuclei—a small subset of atmospheric aerosol particles able to induce ice crystallization. The formation of ice crystals in such clouds causes the release of latent heat and greatly influences the generation of precipitation since ice crystals in the presence of liquid droplets grow rapidly. The effect of cosmic rays on cloud processes is proposed to proceed through the following steps: 1) variations in cosmic rays modulate the magnitude of the ionosphere-earth current and the current flowing into clouds; 2) this increases the charges on cloud droplets and aerosol particles at cloud boundaries, which may then be entrained inside clouds; 3) changes in the charge carried by aerosol particles and cloud droplets influences the ice formation rate in clouds below 0°C; and 4) changes in cloud development, precipitation and release of latent heat result from changes in freezing rates.

Direct observation of changes in cloud properties due to changes in the ionosphere-earth current have not been made, and are unlikely to succeed in the near future due to our very poor understanding of what controls ice formation, even in the absence of weak cloud electrification. In addition, cloud properties are highly variable, being driven mostly by large scale meteorological effects, so any cosmic ray modulation of their properties will be masked. Confirmation of the Tinsley hypothesis therefore relies on observing and explaining statistical relationships between atmospheric variables and cosmic ray flux, and a number of such attempts have been made by Tinsley and co-workers.

The proposed mechanism linking cosmic rays, atmospheric electrical current and cloud droplet freezing is plausible. Both theory and observation support the build-up of space charge at the top and bottom of clouds (regions of net unipolar charge carried by ions, aerosol and droplets; see Fig. 12.21). Space charge is created because cloud particles efficiently scavenge small ions, thereby reducing the local conductivity, and leading to an increased local potential gradient through the cloud to drive a constant current density. Observed space charge densities can be as high as $100e$ cm^{-3} at cloud upper and lower surfaces [Harrison and Carslaw, 2003]. Although not yet confirmed by observations, the magnitude of space charge is likely to be modulated by the ionosphere-earth current density flowing into cloud tops.

The mechanism therefore hinges upon an enhanced freezing probability of supercooled droplets in the elevated space charge environment.

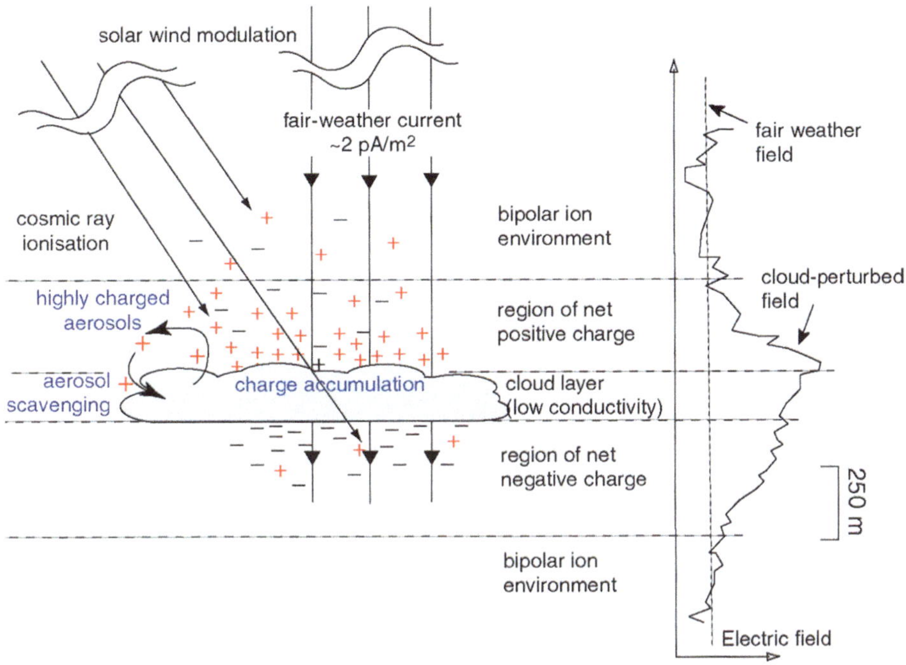

Figure 12.21. Schematic showing the creation of space charge at the top and bottom of clouds. This space charge can become attached to droplets and aerosol particles, and then entrained within clouds.

Charge carried by droplets themselves does not affect the droplet freezing probability. The more likely possibility is that charged aerosol particles (some of which will be ice nuclei) will be scavenged by droplets more rapidly than their uncharged counterparts [Tinsley *et al.*, 2000, Tripathi and Harrison, 2002]. This so-called *contact* ice nucleation, involving the scavenging of an aerosol particle by a supercooled droplet, is one of several mechanisms that can lead to freezing (the others involve ice-forming nuclei already within the droplet). Observations have allowed simple parameterisations of the number of such contact nuclei to be fitted in terms of the supersaturation of the vapour with respect to ice [Meyers *et al.*, 1992], although field observations of ice formation rates and of ice nuclei themselves show great variability. Tripathi and Harrison (2002) estimated, for a droplet of radius 26 μm, that a particle charge of $10e$ would be comparable to a change in ice nucleus concentration expected from a 1 K reduction in temperature. Thus, any effect of changes in space charge on ice formation in cloud tops are likely to be very small for changes in the ionosphere-earth current of a few per cent.

In summary, changes in the global electric circuit over decadal periods have been observed and attributed to changes in the cosmic ray flux, although there is uncertainty regarding alternative explanations. Changes in the surface potential gradient and ionospheric potential imply a change in the fair weather electric current. The current flowing into cloud tops may influence the production of ice crystals, although any effect is likely to be small compared with the very large and poorly understood variability in ice formation rates. It is certainly not contentious to state that changes in ice formation would affect the release of latent heat, precipitation, general atmospheric dynamics and climate. However, the coupling of changes in Earth's electric circuit and atmospheric dynamics involves small estimated forcings that are likely to be difficult to observe in the real atmosphere.

6. Conclusions and Future Prospects

During the last few years, a wide range of new palaeoclimatic evidence has shown that solar/GCR forcing has exerted a major influence on past climate. Although this evidence is largely based on correlations between cosmic ray- and climate archives—which do not necessarily imply a casual connection—the sheer diversity and quality of the correlations preclude mere chance association.

However, although the evidence for solar/GCR forcing of climate appears beyond doubt, the mechanism remains unclear. There are only two candidates for the forcing agent: a) the solar irradiance, or a spectral component such as the UV, or b) galactic cosmic rays, which are modulated by the solar wind. In the former case, GCR variability is considered to be a proxy for changes of solar irradiance or UV, which are assumed to vary hand-in-hand with long-term changes of solar magnetic activity. Although secular changes of solar magnetic activity are well established, its link with changes of irradiance and UV is not; beyond the 11-yr sunspot cycle, there is no direct or indirect evidence (from sun-like stars) that there are secular variations of the Sun's irradiance on timescales shorter than about 10 Myr. This does not rule them out, of course, but it does illustrate how little is understood at present of the solar forcing mechanism.

An influence of GCRs on climate can in principle be unambiguously resolved from a direct solar irradiance/UV effect since the former implies that climate will also be affected by variations of the geomagnetic field or, on a longer time scale, of the galactic environment. Indeed, some evidence has been reported of both such associations. However, at this stage, the data are insufficient either to confirm or to rule out either candidate as the forcing agent.

Satellite cloud observations over the last 20 years may have provided an important clue on the climate mechanism: low cloud amount appears to be modulated by the solar cycle. If confirmed by further observations, this would be climatically significant since clouds exert a strong control on Earth's radiative energy balance. Present low cloud data favour an association with GCR intensity, although they are also consistent with irradiance/UV variations (but with the opposite sign; increased low clouds are associated with increased GCRs but *decreased* irradiance/UV).

Theoretical and modelling studies suggest that cosmic rays may influence cloud microphysics in several ways. These include ion-induced nucleation of new aerosols from trace condensable vapours, and the formation of relatively highly charged aerosols and cloud droplets at cloud boundaries, which may enhance the formation of ice particles, thereby releasing latent heat and affecting precipitation. Recent observations support the presence of ion-induced nucleation in the atmosphere, although its significance in the presence of other sources of variability is not yet established.

The key to further progress on the cosmic ray-cloud-climate question is to understand the physical mechanism. This requires an experimental study of the fundamental microphysical interactions between cosmic rays and clouds. Demonstrating overall cause and effect in the atmosphere beginning with changes in ionisation rate and ending with observations of perturbed clouds will be quite challenging. For this reason an experiment is under construction by a multi-disciplinary team [CLOUD collaboration] to study GCR-cloud microphysics under carefully controlled conditions in the laboratory using a CERN particle beam as an artificial source that closely simulates natural cosmic rays [CLOUD proposal, 2000, 2004, Kirkby, 2002]. The apparatus includes an advanced cloud chamber and reactor chamber where the atmosphere is realistically represented by moist air charged with aerosol and trace gases. The chambers are equipped with a wide range of external instrumentation to monitor and analyse their contents. The thermodynamic conditions anywhere in the troposphere and stratosphere can be re-created within the chambers. In contrast with experiments in the atmosphere, CLOUD can compare processes when the cosmic ray beam is present and when it is not, and all experimental parameters can be controlled.

With the expected advances in experiments, observations and theory in the next few years, there are good prospects that we may finally be able to answer William Herschel's two-centuries-old question of why the price of wheat in England was lower when there were many sunspots.

References

Beer, J., (2000), Long-term indirect indices of solar variability, *Space Sci. Rev.* **94**, 53–66.

Bond, G.C., and R. Lotti (1997a), Iceberg discharges into the North Atlantic on millennial time scales during the last glaciation, *Science* **267**, 1005–1010.

Bond, G.C., W. Showers, M. Cheseby, R. Lotti, P. Almasi, P. deMenocal, P. Priore, H. Cullen, I. Hajdas and G. Bonani (1997b), A pervasive millennial-scale cycle in North Atlantic Holocene and glacial climates, *Science* **278**, 1257–1266.

Bond, G.C., B. Kromer, J. Beer, R. Muscheler, M.N. Evans, W. Showers, S. Hoffmann, R. Lotti-Bond, I. Hajdas and G. Bonani (2001), Persistent solar influence on North Atlantic climate during the Holocene, *Science* **294**, 2130–2136.

Carslaw, K.S., R.G. Harrison and J. Kirkby (2002), Cosmic rays, clouds, and climate, *Science* **298**, 1732–1737.

Channell, J.E.T., D.A. Hodell, J. McManus and B. Lehman (1998), Orbital modulation of the Earth's magnetic field intensity, *Nature* **394**, 464–468.

Christl, M., C. Strobl and A. Mangini (2003), Beryllium-10 in deep-sea sediments: a tracer for the Earth's field intensity during the last 200,000 years, *Quat. Sc. Rev.* **22**, 725–739.

Christl, M., A. Mangini, S. Holzkämper, C. Spötl (2004), Evidence for a link between the flux of galactic cosmic rays and Earth's climate during the past 200,000 years, *J. Atm. Sol. Terr. Phys.* **66**, 313–322.

CLOUD proposal (CLOUD collaboration, 2000–06), A study of the link between cosmic rays and clouds with a cloud chamber at the CERN PS, CERN-SPSC-2000-021, SPSC-P-317; CERN-SPSC-2000-030, SPSC-P-317-Add-1; CERN-SPSC-2000-041, SPSC-P-317-Add-2; SPSC-M-721, CERN-SPSC-2004-023. `http://www.cern.ch/cloud/iaci_workshop/cloud.html`

CLOUD collaboration: *Univ. Aarhus*, Denmark; *Univ. Bergen*, Norway; *California Institute of Technology*, USA; *CERN*, Switzerland; *Danish National Space Center*, Denmark; *Finnish Meteorological Institute*, Finland; *Univ. Helsinki*, Finland; *Univ. Kuopio*, Finland; *Lebedev Physical Institute*, Russia; *Univ. Leeds*, UK; *Univ. Mainz*, Germany; *Max-Planck Institute for Nuclear Physics - Heidelberg*, Germany; *Univ. Missouri-Rolla*, USA; *Paul Scherrer Institute*, Switzerland; *State University of New York at Albany*, USA; *Univ. Reading*, UK; *Rutherford Appleton Laboratory*, UK; *Tampere University of Technology*, Finland; and *Univ. Vienna*, Austria.

Dal Maso, M., M. Kulmala, K.E.J. Lehtinen, J.M. Makela, P. Aalto and C.D. O'Dowd (2002), Condensation and coagulation sinks and formation of nucleation mode particles in coastal and boreal forest boundary layers, *J. Geophys. Res.-Atmos.*, **107**, art. no.-8097.

Eddy, J.A., (1976), The Maunder minimum, *Science* **192**, 1189.

Eichkorn, S., S. Wilhelm, H. Aufmhoff, K.H. Wohlfrom, and F. Arnold (2002), Cosmic ray-induced aerosol formation: First observational evidence from aircraft-based ion mass spectrometer measurements in the upper troposphere, *Geophys. Res. Lett.* **29**, 43.

Foukal, P., G. North, T. Wigley (2004), A stellar view on solar variations and climate, *Science* **306** (2004) 68.

Fröhlich, C. (2000), Observations of irradiance variability, *Space Science Reviews* **94**, 15–24.

Froyd, K.D., and E.R. Lovejoy (2003a), Experimental thermodynamics of cluster ions composed of H_2SO_4 and H_2O. 1. Positive ions, *J. Phys. Chem. A*, **107**, 9800–9811.

Froyd, K.D., and E.R. Lovejoy (2003b), Experimental thermodynamics of cluster ions composed of H_2SO_4 and H_2O. 2. Measurements and *ab initio* structures of negative ions, *J. Phys. Chem. A*, **107**, 9812–9824.

Gallup, C.D., H. Cheng, F.W. Taylor and R.L. Edwards, Direct determination of the timing of sea level change during Termination II, *Science* **295**, 310–313 (2002).

Ganopolski, A. and S. Rahmstorf (2001), Rapid changes of glacial climate simulated in a coupled climate model, *Nature* **409**, 153–158.

Gleeson, L.J., and W.I. Axford (1967), Cosmic rays in the interplanetary medium, *Astrophys. Journal* **149**, 1115–1118.

Guillou, H., B.S. Singer, C. Laj, C. Kissel, S. Scaillet and B.R. Jicha (2004), On the age of the Laschamp geomagnetic excursion, *Earth and Planet. Sci. Lett.* **227**, 331–343.

Haigh, J.D., (1996), The impact of solar variability on climate, *Science* **272**, 981.

Harrison, R.G., and K.S. Carslaw (2003), Ion-aerosol-cloud processes in the lower atmosphere, *Rev. Geophys.*, **41**, art. no.-1012.

Harrison, R.G., (2004), Long-range correlations in measurements of the global atmospheric electric circuit, *J. Atmos. Sol.-Terr. Phys.*, **66**, 1127–1133.

Hartmann, D.L., Radiative effects of clouds on Earth's climate, in *Aerosol-Cloud-Climate Interactions*, International Geophysics Series **54**, ed. P.V. Hobbs, Academic Press Inc., San Diego (1993), 151–173.

Henderson, G.M., and N.C. Slowey (2000), Evidence from U/Th dating against Northern Hemisphere forcing of the penultimate deglaciation, *Nature* **404**, 61–66.

Herschel, W., (1801), *Philosophical Transactions of the Royal Society*, **91**, 265–283.

Horrak, U., A. Mirme, J. Salm, E. Tamm and H. Tammet (1998), Air ion measurements as a source of information about atmospheric aerosols, *Atmos. Res.*, **46**, 233–242.

Hu, F.S., D. Kaufman, S. Yoneji, D. Nelson, A. Shemesh, Y. Huang, J. Tian, G.C. Bond, B. Clegg and T. Brown (2003), Cyclic variation and solar forcing of Holocene climate in the Alaskan sub-Arctic, *Science* **301**, 1890–1893.

Huang, S., H.N. Pollack and P.-Y. Shen (2000), Temperature trends over the past five centuries reconstructed from borehole temperatures, *Nature* **403**, 756-758.

IPCC Third Assessment Report, *Climate Change 2001: The Scientific Basis*, Intergovernmental Panel on Climate Change, eds. J.T. Houghton *et al.*, Cambridge University Press, UK.

Kernthaler, S.C., R. Toumi and J.D. Haigh (1999), *Geophys. Res. Lett.* **26**, 863; T.B. Jorgensen and A.W. Hansen (2000), *J. Atm. Sol. Terr. Phys.* **62**, 73; J.E. Kristjánsson and J. Kristiansen (2000), *J. Geophys. Res.* **105**, 11851; J.E. Kristjánsson, A. Staple and J. Kristiansen (2002), *Geophys. Res. Lett.* **29**, 10.1029/2002GL015646; B. Sun and R.S. Bradley (2002), *J. Geophys. Res.* **107**, D14, 10.1029/2001-JD000560.

Kirkby, J. (2002), CLOUD: a particle beam facility to investigate the influence of cosmic rays on clouds, *CERN-EP-2002-019* and *Proc. of the Workshop on Ion-Aerosol-Cloud Interactions* (2001), ed. J. Kirkby, CERN, Geneva, *CERN 2001-007*, 175–248. http://cloud.web.cern.ch/cloud/iaci_workshop/proceedings.html

Kirkby, J., A. Mangini and R.A. Muller (2004), The glacial cycles and cosmic rays, *CERN-PH-EP-2004-027*. http://arxiv.org/abs/physics/0407005

Klein, J., J.C. Lerman, P.E. Damon and T. Linick (1980), Radiocarbon concentrations in the atmosphere: 8000 year record of variations in tree rings, *Radiocarbon* **22**, 950–961.

Knie, K., G. Korschinek, T. Faestermann, E.A. Dorfi, G. Rugel and A. Wallner (2004), ^{60}Fe anomaly in a deep-sea manganese crust and implications for a nearby supernova source, *Phys. Rev. Lett.* **93**, 171103.

Kulmala, M., H. Vehkamaki, T. Petajda, M. Dal Maso, A. Lauri, V.M. Kerminen, W. Birmili and P.H. McMurry (2004), Formation and growth rates of ultrafine atmospheric particles: a review of observations, *J. Aerosol. Sci.*, **35**, 143–176.

Laakso, L., J.M. Mäkelä, L. Pirjola and M. Kulmala (2002), Model studies on ion-induced nucleation in the atmosphere, *J. Geophys. Res.* D20, 10.1029/2002JD002140.

Laakso, L., M. Kulmala and K.E.J. Lehtinen (2003), Effect of condensation rate enhancement factor on 3-nm (diameter) particle formation in binary ion-induced and homogeneous nucleation, *J. Geophys. Res.-Atmos.*, **108**, art. no.-4574.

Laakso, L., T. Anttila, K.E.J. Lehtinen, P.P. Aalto, M. Kulmala, U. Horrak, J. Paatero, M. Hanke and F. Arnold (2004), Kinetic nucleation and ions in boreal forest particle formation events, *Atmos. Chem. Phys.*, **4**, 2353-2366.

Larkin, A., J.D. Haigh and S. Djavidnia (2000), The effect of solar UV irradiance variations on the Earth's atmosphere, in *Solar Variability and Climate*, eds. E. Friis-Christensen *et al.*, *Space Science Reviews* **94**, Nos. 1–2, 145–152.

Lean, J.L., J. Beer, R. Bradley (1995), Reconstruction of solar irradiance since 1610: Implications for climatic change, *Geophys. Res. Lett.* **22** (1995) 3195-3198.

Lean, J.L., Y.-M. Wang and N.R. Sheeley (2002), The effect of increasing solar activity on the Sun's total and open magnetic flux during multiple cycles: Implications for solar forcing of climate, *Geophys. Res. Lett.* **29**, 2224.

Lee, S.H., J.M. Reeves, J.C. Wilson, D.E. Hunton, A.A. Viggiano, T.M. Miller, J.O. Ballenthin and L.R. Lait (2003), Particle formation by ion nucleation in the upper troposphere and lower stratosphere, *Science*, **301**, 1886–1889.

Lockwood, M., R. Stamper and M.N. Wild (1999), A doubling of the Sun's coronal magnetic field during the past 100 years, *Nature* **399**, 437.

Lockwood, M., (2003), Twenty-three cycles of changing open solar magnetic flux, *J. Geophys. Res-Space Phys.*, **108**, art. no.-1128.

Lovejoy, E.R., J. Curtius and K.D. Froyd (2004), Atmospheric ion-induced nucleation of sulphuric acid and water, *J. Geophys. Res.* **109**, D08204, doi:10.1029/2003JD004460.

Mann, M.E., R.S. Bradley and M.K. Hughes (1998), Global-scale temperature patterns and climate forcing over the past six centuries, *Nature* **392**, 779–787.

Mann, M.E., R.S. Bradley and M.K. Hughes (1999), Northern Hemisphere temperatures during the past millennium: inferences, uncertainties, and limitations, *Geophys. Res. Lett.* **26**, 759–762.

Markson, R., (1981), Modulation of the Earth's electric-field by cosmic-radiation, *Nature*, **291**, 304–308.

Marsh, N., and H. Svensmark (2003), Galactic cosmic ray and El Niño-Southern Oscillation trends in International Satellite Cloud Climatology Project D2 low-cloud properties, *J. Geophys. Res.* **108** D6, 4195.

Meyers, M.P., P.J. Demott and W.R. Cotton (1992), New primary ice-nucleation parameterisations in an explicit cloud model, *J. Appl. Meteorol.*, **31**, 708–721.

McIntyre, S., and R. McKitrick (2005), Hockey sticks, principal components and spurious significance, *Geophys. Res. Lett.*, doi: 2004GL012750.

Moberg, A, D.M. Sonechkin, K. Holmgren, N.M. Datsenko and W. Karlén, (2005), Highly variable Northern Hemisphere temperatures reconstructed from low- and high-resolution proxy data, *Nature* **433**, 613—618.

Muller, R.A., and G.J. MacDonald (1997), Glacial cycles and astronomical forcing, *Science* **277**, 215–218.

Muller, R.A., and G.J. MacDonald (2000), *Ice Ages and Astronomical Causes*, Springer Praxis, Chichester, UK.

Nadykto, A.B., and F.Q. Yu (2003), Uptake of neutral polar vapour molecules by charged clusters/particles: Enhancement due to dipole-charge interaction, *J. Geophys. Res.-Atmos.*, **108**, art. no.-4717.

Nadykto, A.B., and F.Q. Yu (2004), Dipole moment of condensing monomers: A new parameter controlling the ion-induced nucleation, *Phys. Rev. Lett.*, **93**, art. no.-016101.

Neff, U., *et al.* (2001), Strong coincidence between solar variability and the monsoon in Oman between 9 and 6 kyr ago, *Nature* **411**, 290–293.

Nober, F.J., H.F. Graf and D. Rosenfeld (2003), Sensitivity of the global circulation to the suppression of precipitation by anthropogenic aerosols, *Glob. Planet. Change*, **37**, 57–80.

Parker, E.N., (1965), The passage of energetic charged particles through interplanetary space, *Planet. Space Sc.* **13**, 9–49.

Piotrowski, A.M., S.L. Goldstein, S.R. Hemming and R.G. Fairbanks (2005), Temporal relationships of carbon cycling and ocean circulation at glacial boundaries, *Science* **307**, 1933–1938.

Pollack, H.N., and J.E. Smerdon (2004), Borehole climate reconstructions: spatial structure and hemispheric averages, *J. Geophys. Res.* **109**, doi:10.1029/2003JD004163.

Raisbeck, G.M., F. Yiou, J. Jouzel and J.-R. Petit (1990), ^{10}Be and ^{2}H in polar ice cores as a probe of the solar variability's influence on climate, *Phil. Trans. Roy. S. Lond.* **A300**, 463–470.

Reid, G.C. (1987), Influence of solar variability on global sea surface temperatures, *Nature* **329**, 142–143.

Rossow, W.B., A.W. Walker, D.E. Beuschel, and M.D. Roiter (1996), International Satellite Cloud Climatology Project (ISCCP): documentation of new cloud datasets, WMO/TD **737**, World Meteorological Organization, Geneva.

Rycroft, M.J., S. Israelsson and C. Price (2000), The global atmospheric electric circuit, solar activity and climate change, *J. Atm. Sol. Terr. Phys.* **62**, 1563-1576.

Shaviv, N.J., (2002), Cosmic ray diffusion from the galactic spiral arms, iron meteorites, and a possible climatic connection, *Phys. Rev. Lett.* **89**, 051102.

Shaviv, N.J., and J. Veizer (2003), Celestial driver of Phanerozoic climate?, *GSA Today*, Geological Society of America, July 2003, 4–10.

Shindell, D. *et al.* (1999), Solar cycle variability, ozone, and climate, *Science* **284**, 305–308.

Spötl, C., A. Mangini, N. Frank, R.,Eichstädter and S.J. Burns (2002), Start of the last interglacial period at 135 ka: Evidence from a high Alpine speleothem, *Geology* **30**, no. 9, 815–818.

St-Onge, G., J.S. Stoner, C. Hillaire-Marcel (2003), Holocene palaeomagnetic records from the St. Lawrence Estuary, eastern Canada: centennial- to millennial-scale geomagnetic modulation of cosmogenic isotopes, *Earth and Planet. Sci. Lett.* **209**, 113–130.

Svensmark, H., and E. Friis-Christensen (1997), Variation in cosmic ray flux and global cloud coverage—a missing link in solar-climate relationships, *J. Atm. Sol. Terr. Phys.* **59**, 1225.

Tinsley, B.A., G.M. Brown and P.H. Scherrer (1989), Solar variability influences on weather and climate - possible connections through cosmic-ray fluxes and storm intensification, *J. Geophys. Res.-Atmos.*, **94**, 14783–14792.

Tinsley, B.A., (1991), Interpretation of short-term solar variability effects in the troposphere, *J. Geomagn. Geoelectr.*, **43**, 775–783.

Tinsley, B.A., and G.W. Deen (1991), Apparent tropospheric response to MeV-GeV particle-flux variations - a connection via electrofreezing of supercooled water in high-level clouds, *J. Geophys. Res.-Atmos.* **96**, 22283–22296.

Tinsley, B.A., and R.A. Heelis (1993), Correlations of atmospheric dynamics with solar-activity evidence for a connection via the solar-wind, atmospheric electricity, and cloud microphysics, *J. Geophys. Res.-Atmos.*, **98**, 10375–10384.

Tinsley, B.A., (1996a), Correlations of atmospheric dynamics with solar wind-induced changes of air-earth current density into cloud tops, *J. Geophys. Res.-Atmos.*, **101**, 29701–29714.

Tinsley, B.A., (1996b), Solar wind modulation of the global electric circuit and apparent effects on cloud microphysics, latent heat release, and tropospheric dynamics, *J. Geomagn. Geoelectr.*, **48**, 165–175.

Tinsley, B.A., (2000), Influence of solar wind on the global electric circuit, and inferred effects on cloud microphysics, temperature, and dynamics in the troposphere, *Space Sci. Rev.*, **94**, 231–258.

Tinsley, B.A., R.P. Rohrbaugh, M. Hei and K.V. Beard (2000), Effects of image charges on the scavenging of aerosol particles by cloud droplets and on droplet charging and possible ice nucleation processes, *Atmos. Res.* **57**, 2118–2134.

Tinsley, B.A., R.P. Rohrbaugh and M. Hei (2001), Electroscavenging in clouds with broad droplet size distributions and weak electrification, *Atmos. Res.* **59**, 115–135.

Tripathi, S.N., and R.G. Harrison (2002), Enhancement of contact nucleation by scavenging of charged aerosol particles, *Atmos. Res.* **62**, 57–70.

Usoskin, I.G., K. Mursula, S.K. Solanki, M. Schüssler and G.A. Kovaltsov (2002), A physical reconstruction of cosmic ray intensity since 1610, *J. Geophys. Res.* **107**, doi:10.1029/2002JA009343.

Usoskin, I.G., N. Marsh, G.A. Kovaltsov, K. Mursula and O.G. Gladysheva (2004), Latitudinal dependence of low cloud amount on cosmic ray induced ionisation, *Geophys. Res. Lett.*, doi: 10.1029/2004GL019507.

Vehkamaki, H., M. Kulmala, I. Napari, K.E.J. Lehtinen, C. Timmreck, M. Noppel and A. Laaksonen (2002), An improved parameterization for sulfuric acid-water nucleation rates for tropospheric and stratospheric conditions, *J. Geophys. Res.-Atmos.*, **107**, art. no.-4622.

Veizer, J., Y. Godderis and L.M. François (2000), Evidence for decoupling of atmospheric CO_2 and global climate during the Phanerozoic eon, *Nature* **408**, 698–701.

Visser, K., R. Thunell and L. Stott (2003), Magnitude and timing of temperature change in the Indo-Pacific warm pool during deglaciation, *Nature* **421**, 152–155.

Wagner, G., D M. Livingstone, J. Masarik, R. Muscheler and J. Beer (2001), Some results relevant to the discussion of a possible link between cosmic rays and the Earth's climate, *J. Geophys. Res.* **106**, D4, 3381–3387.

Wagner, G., J. Masarik, J. Beer, S. Baumgartner, D. Imboden, P.W. Kubik, H.-A. Synal and M. Suter (2000), Reconstruction of the geomagnetic field between 20 and 60 kyr BP from cosmogenic radionuclides in the GRIP ice core, *Nuc. Inst. Meth. Phys. Res.* **B172**, 597–604.

Wang, Y.J., H. Cheng, R.L. Edwards, Z.S. An, J.Y. Wu, C.-C. Shen, J.A. Doral (2001), A high-resolution absolute-dated late Pleistocene monsoon record from Hulu Cave, China, *Science* **294**, 2345–2348.

Wang, X., A.S. Auler, R.L. Edwards, H. Cheng, P.S. Cristalli, P.L. Smart, D.A. Richards and C.-C. Shen (2004), Wet periods in northeastern Brazil over the past 210 kyr linked to distant climate anomalies, *Nature* **432**, 740–743.

White, W.B., J. Lean, D.R. Cayan and M.D. Dettinger (1997), Response of global upper ocean temperature to changing solar irradiance, *J. Geophys. Res.* **102**, C2, 3255–3266.

White, W.B., D.R. Cayan and J. Lean (1998), Global upper ocean heat storage response to radiative forcing from changing solar irradiance and increasing greenhouse gas/aerosol concentrations, *J. Geophys. Res.* **103**, C10, 21333–21366.

Williams, E.R., (2003), Comment to "Twentieth century secular decrease in the atmospheric potential gradient" by Giles Harrison, *Geophys. Res. Lett.* **30**, art. no.-1803.

Winograd, I.J., T.B. Coplen, J.M. Landwehr, A.C. Riggs, K.R. Ludwig, B.J. Szabo, P.T. Kolesar, and K.M. Revesz (1992), Continuous 500,000-year climate record from vein calcite in Devils Hole, Nevada, *Science* **258**, 255–260.

Yamazaki, T., and H. Oda (2002), Orbital influence on Earth's magnetic field: 100,000-year periodicity in inclination, *Science* **295**, 2435–2438.

Yu, F.Q., and R.P. Turco (1997), The role of ions in the formation and evolution of particles in aircraft plumes, *Geophys. Res. Lett.* **24**, 1927–1930.

Yu, F.Q., and R.P. Turco (2000), Ultrafine aerosol formation via ion-mediated nucleation, *Geophys. Res. Lett.* **27**, 883-886.

Yu, F., and R.P. Turco (2001), From molecular clusters to nanoparticles: The role of ambient ionisation in tropospheric aerosol formation, *J. Geophys. Res.* **106**, 4797–4814.

Yu, F.Q., (2002), Altitude variations of cosmic ray induced production of aerosols: Implications for global cloudiness and climate, *J. Geophys. Res.-Space Phys.*, **107**, art. no.-1118.

Index

A

B

C

D

E

F

G

H

I

T

U

V

W

X

Y

Astrophysics and Space Science Library

Volume 338: ***Solar Journey: The Significance of our Galactic Environment for the Heliosphere and Earth***, edited by P.C. Frisch. Hardbound ISBN 1-4020-4397-X, May 2006

Volume 337: ***Progress in the Study of Astrophysical Disks***, edited by A.M. Fridman, M.Y. Marov, I.G. Kovalenko. Hardbound ISBN 1-4020-4347-3, March 2006

Volume 336: ***Scientific Detectors for Astronomy 2005***, edited by J.E. Beletic, J.W. Beletic, P. Amico. Hardbound ISBN 1-4020-4329-5, December 2005

Volume 335: ***Organizations and Strategies in Astronomy 6***, edited by A. Heck. Hardbound ISBN 1-4020-4055-5, November 2005

Volume 334: ***The New Astronomy: Opening the Electromagnetic Window and Expanding our View of Planet Earth***, edited by W. Orchiston. Hardbound ISBN 1-4020-3723-6, October 2005

Volume 333: ***Planet Mercury***, by P. Clark and S. McKenna-Lawlor. Hardbound ISBN 0-387-26358-6, November 2005

Volume 332: ***White Dwarfs: Cosmological and Galactic Probes***, edited by E.M. Sion, S. Vennes, H.L. Shipman. Hardbound ISBN 1-4020-3693-0, September 2005

Volume 331: ***Ultraviolet Radiation in the Solar System***, by M. Vázquez and A. Hanslmeier. Hardbound ISBN 1-4020-3726-0, November 2005

Volume 330: ***The Multinational History of Strasbourg Astronomical Observatory***, edited by A. Heck. Hardbound ISBN 1-4020-3643-4, June 2005

Volume 329: ***Starbursts – From 30 Doradus to Lyman Break Galaxies***, edited by R. de Grijs, R.M. González Delgado. Hardbound ISBN 1-4020-3538-1, May 2005

Volume 328: ***Comets***, by J.A. Fernández. Hardbound ISBN 1-4020-3490-3, July 2005

Volume 327: ***The Initial Mass Function 50 Years Later***, edited by E. Corbelli, F. Palla, H. Zinnecker. Hardbound ISBN 1-4020-3406-7, June 2005

Volume 325: ***Kristian Birkeland – The First Space Scientist***, by A. Egeland, W.J. Burke. Hardbound ISBN 1-4020-3293-5, April 2005

Volume 324: ***Cores to Clusters – Star Formation with next Generation Telescopes***, edited by M.S. Nanda Kumar, M. Tafalla, P. Caselli. Hardbound ISBN 0-387-26322-5, October 2005

Volume 323: ***Recollections of Tucson Operations***, by M.A. Gordon. Hardbound ISBN 1-4020-3235-8, December 2004

Volume 322: ***Light Pollution Handbook***, by K. Narisada, D. Schreuder Hardbound ISBN 1-4020-2665-X, November 2004

Volume 321: ***Nonequilibrium Phenomena in Plasmas***, edited by A.S. Shrama, P.K. Kaw. Hardbound ISBN 1-4020-3108-4, December 2004

Volume 320: ***Solar Magnetic Phenomena***, edited by A. Hanslmeier, A.Veronig, M. Messerotti. Hardbound ISBN 1-4020-2961-6, December 2004

Volume 319: ***Penetrating Bars through Masks of Cosmic Dust***, edited by D.L. Block, I. Puerari, K.C. Freeman, R. Groess, E.K. Block. Hardbound ISBN 1-4020-2861-X, December 2004

Volume 318: ***Transfer of Polarized light in Planetary Atmospheres***, by J.W. Hovenier, J.W. Domke, C. van der Mee. Hardbound ISBN 1-4020-2855-5. Softcover ISBN 1-4020-2889-X, November 2004

Volume 317: ***The Sun and the Heliosphere as an Integrated System***, edited by G. Poletto, S.T. Suess. Hardbound ISBN 1-4020-2830-X, November 2004

Volume 316: ***Civic Astronomy - Albany's Dudley Observatory, 1852-2002***, by G. Wise
Hardbound ISBN 1-4020-2677-3, October 2004

Volume 315: ***How does the Galaxy Work - A Galactic Tertulia with Don Cox and Ron Reynolds***, edited by E. J. Alfaro, E. Pérez, J. Franco
Hardbound ISBN 1-4020-2619-6, September 2004

Volume 314: ***Solar and Space Weather Radiophysics- Current Status and Future Developments***, edited by D.E. Gary and C.U. Keller
Hardbound ISBN 1-4020-2813-X, August 2004

Volume 313: ***Adventures in Order and Chaos***, by G. Contopoulos. Hardbound ISBN 1-4020-3039-8, January 2005

Volume 312: ***High-Velocity Clouds***, edited by H. van Woerden, U. Schwarz, B. Wakker
Hardbound ISBN 1-4020-2813-X, September 2004

Volume 311: ***The New ROSETTA Targets- Observations, Simulations and Instrument Performances***, edited by L. Colangeli, E. Mazzotta Epifani, P. Palumbo
Hardbound ISBN 1-4020-2572-6, September 2004

Volume 310: ***Organizations and Strategies in Astronomy 5***, edited by A. Heck
Hardbound ISBN 1-4020-2570-X, September 2004

Volume 309: ***Soft X-ray Emission from Clusters of Galaxies and Related Phenomena***, edited by R. Lieu and J. Mittaz
Hardbound ISBN 1-4020-2563-7, September 2004

Volume 308: ***Supermassive Black Holes in the Distant Universe***, edited by A.J. Barger
Hardbound ISBN 1-4020-2470-3, August 2004

Volume 307: ***Polarization in Spectral Lines***, by E. Landi Degl'Innocenti and M. Landolfi
Hardbound ISBN 1-4020-2414-2, August 2004

Volume 306: ***Polytropes – Applications in Astrophysics and Related Fields***, by G.P. Horedt
Hardbound ISBN 1-4020-2350-2, September 2004

Volume 305: ***Astrobiology: Future Perspectives***, edited by P. Ehrenfreund, W.M. Irvine, T. Owen, L. Becker, J. Blank, J.R. Brucato, L. Colangeli, S. Derenne, A. Dutrey, D. Despois, A. Lazcano, F. Robert
Hardbound ISBN 1-4020-2304-9, July 2004
Paperback ISBN 1-4020-2587-4, July 2004

Volume 304: ***Cosmic Gammy-ray Sources***, edited by K.S. Cheng and G.E. Romero
Hardbound ISBN 1-4020-2255-7, September 2004

Volume 303: ***Cosmic rays in the Earth's Atmosphere and Underground***, by L.I. Dorman
Hardbound ISBN 1-4020-2071-6, August 2004

Volume 302: ***Stellar Collapse,*** edited by Chris L. Fryer
Hardbound, ISBN 1-4020-1992-0, April 2004

Volume 301: ***Multiwavelength Cosmology***, edited by Manolis Plionis
Hardbound, ISBN 1-4020-1971-8, March 2004

Volume 300: ***Scientific Detectors for Astronomy,*** edited by Paola Amico, James W. Beletic, Jenna E. Beletic
Hardbound, ISBN 1-4020-1788-X, February 2004

Volume 299: ***Open Issues in Local Star Fomation,*** edited by Jacques Lépine, Jane Gregorio-Hetem
Hardbound, ISBN 1-4020-1755-3, December 2003

Volume 298: ***Stellar Astrophysics - A Tribute to Helmut A. Abt,*** edited by K.S. Cheng, Kam Ching Leung, T.P. Li
Hardbound, ISBN 1-4020-1683-2, November 2003

Volume 297: ***Radiation Hazard in Space,*** by Leonty I. Miroshnichenko
Hardbound, ISBN 1-4020-1538-0, September 2003

Volume 296: ***Organizations and Strategies in Astronomy, volume 4,*** edited by André Heck
Hardbound, ISBN 1-4020-1526-7, October 2003

Volume 295: ***Integrable Problems of Celestial Mechanics in Spaces of Constant Curvature,*** by T.G. Vozmischeva
Hardbound, ISBN 1-4020-1521-6, October 2003

Volume 294: ***An Introduction to Plasma Astrophysics and Magnetohydrodynamics,*** by Marcel Goossens
Hardbound, ISBN 1-4020-1429-5, August 2003
Paperback, ISBN 1-4020-1433-3, August 2003

Volume 293: ***Physics of the Solar System,*** by Bruno Bertotti, Paolo Farinella, David Vokrouhlický
Hardbound, ISBN 1-4020-1428-7, August 2003
Paperback, ISBN 1-4020-1509-7, August 2003

Volume 292: ***Whatever Shines Should Be Observed,*** by Susan M.P. McKenna-Lawlor
Hardbound, ISBN 1-4020-1424-4, September 2003

Volume 291: ***Dynamical Systems and Cosmology***, by Alan Coley
Hardbound, ISBN 1-4020-1403-1, November 2003

Volume 290: ***Astronomy Communication***, edited by André Heck, Claus Madsen
Hardbound, ISBN 1-4020-1345-0, July 2003

Volume 287/8/9: ***The Future of Small Telescopes in the New Millennium***, edited by Terry D. Oswalt
Hardbound Set only of 3 volumes, ISBN 1-4020-0951-8, July 2003

Volume 286: ***Searching the Heavens and the Earth: The History of Jesuit Observatories***, by Agustín Udías
Hardbound, ISBN 1-4020-1189-X, October 2003

Volume 285: ***Information Handling in Astronomy - Historical Vistas***, edited by André Heck
Hardbound, ISBN 1-4020-1178-4, March 2003

Volume 284: ***Light Pollution: The Global View***, edited by Hugo E. Schwarz
Hardbound, ISBN 1-4020-1174-1, April 2003

Volume 283: ***Mass-Losing Pulsating Stars and Their Circumstellar Matter***, edited by Y. Nakada, M. Honma, M. Seki
Hardbound, ISBN 1-4020-1162-8, March 2003

Volume 282: ***Radio Recombination Lines***, by M.A. Gordon, R.L. Sorochenko
Hardbound, ISBN 1-4020-1016-8, November 2002

Volume 281: ***The IGM/Galaxy Connection***, edited by Jessica L. Rosenberg, Mary E. Putman
Hardbound, ISBN 1-4020-1289-6, April 2003

Volume 280: ***Organizations and Strategies in Astronomy III***, edited by André Heck
Hardbound, ISBN 1-4020-0812-0, September 2002

Volume 279: ***Plasma Astrophysics , Second Edition***, by Arnold O. Benz
Hardbound, ISBN 1-4020-0695-0, July 2002

Volume 278: ***Exploring the Secrets of the Aurora***, by Syun-Ichi Akasofu
Hardbound, ISBN 1-4020-0685-3, August 2002

Volume 277: ***The Sun and Space Weather,*** by Arnold Hanslmeier
Hardbound, ISBN 1-4020-0684-5, July 2002

Volume 276: ***Modern Theoretical and Observational Cosmology,*** edited by Manolis Plionis, Spiros Cotsakis
Hardbound, ISBN 1-4020-0808-2, September 2002

Volume 275: ***History of Oriental Astronomy,*** edited by S.M. Razaullah Ansari
Hardbound, ISBN 1-4020-0657-8, December 2002

Volume 274: ***New Quests in Stellar Astrophysics: The Link Between Stars and Cosmology,*** edited by Miguel Chávez, Alessandro Bressan, Alberto Buzzoni,Divakara Mayya
Hardbound, ISBN 1-4020-0644-6, June 2002

Volume 273: ***Lunar Gravimetry,*** by Rune Floberghagen
Hardbound, ISBN 1-4020-0544-X, May 2002

Volume 272: ***Merging Processes in Galaxy Clusters,*** edited by L. Feretti, I.M. Gioia, G. Giovannini
Hardbound, ISBN 1-4020-0531-8, May 2002

Volume 271: ***Astronomy-inspired Atomic and Molecular Physics,*** by A.R.P. Rau
Hardbound, ISBN 1-4020-0467-2, March 2002

Volume 270: ***Dayside and Polar Cap Aurora,*** by Per Even Sandholt, Herbert C. Carlson, Alv Egeland
Hardbound, ISBN 1-4020-0447-8, July 2002

Volume 269: ***Mechanics of Turbulence of Multicomponent Gases,*** by Mikhail Ya. Marov, Aleksander V. Kolesnichenko
Hardbound, ISBN 1-4020-0103-7, December 2001

Volume 268: ***Multielement System Design in Astronomy and Radio Science,*** by Lazarus E. Kopilovich, Leonid G. Sodin
Hardbound, ISBN 1-4020-0069-3, November 2001

Volume 267: ***The Nature of Unidentified Galactic High-Energy Gamma-Ray Sources,*** edited by Alberto Carramiñana, Olaf Reimer, David J. Thompson
Hardbound, ISBN 1-4020-0010-3, October 2001

Volume 266: ***Organizations and Strategies in Astronomy II,*** edited by André Heck
Hardbound, ISBN 0-7923-7172-0, October 2001

Volume 265: ***Post-AGB Objects as a Phase of Stellar Evolution,*** edited by R. Szczerba, S.K. Górny
Hardbound, ISBN 0-7923-7145-3, July 2001

Volume 264: ***The Influence of Binaries on Stellar Population Studies,*** edited by Dany Vanbeveren
Hardbound, ISBN 0-7923-7104-6, July 2001

Volume 262: ***Whistler Phenomena - Short Impulse Propagation,*** by Csaba Ferencz, Orsolya E. Ferencz, Dániel Hamar, János Lichtenberger
Hardbound, ISBN 0-7923-6995-5, June 2001

Volume 261: ***Collisional Processes in the Solar System,*** edited by Mikhail Ya. Marov, Hans Rickman
Hardbound, ISBN 0-7923-6946-7, May 2001

Volume 260: ***Solar Cosmic Rays,*** by Leonty I. Miroshnichenko
Hardbound, ISBN 0-7923-6928-9, May 2001

For further information about this book series we refer you to the following web site:
www.springer.com

To contact the Publishing Editor for new book proposals:
Dr. Harry (J.J.) Blom: harry.blom@springer-sbm.com
Sonja japenga: sonja.japenga@springer-sbm.com

MIX
Papier aus verantwortungsvollen Quellen
Paper from responsible sources
FSC® C105338

If you have any concerns about our products,
you can contact us on
ProductSafety@springernature.com

In case Publisher is established outside the EU,
the EU authorized representative is:
Springer Nature Customer Service Center GmbH
Europaplatz 3, 69115 Heidelberg, Germany

Printed by Libri Plureos GmbH
in Hamburg, Germany